Instructor's Resource Manual

Kevin Bodden
Randy Gallaher

Lewis and Clark Community College

Elementary and Intermediate Algebra

for College Students

■ Third Edition

Allen R.
Angel

PEARSON
Prentice
Hall

Upper Saddle River, NJ 07458

Vice President and Editorial Director, Mathematics: Christine Hoag
Editor-in-Chief, Developmental Mathematics: Paul Murphy
Project Manager, Editorial: Dawn Nuttall
Editorial Assistant: Georgina Brown
Senior Managing Editor: Linda Mihatov Behrens
Associate Managing Editor: Bayani Mendoza de Leon
Project Manager, Production: Kristy Mosch
Supplement Cover Manager: Paul Gourhan
Supplement Cover Designer: Victoria Colotta
Operations Specialist: Ilene Kahn
Senior Operations Supervisor: Diane Peirano

Printed in the United States of America

10 9 8 7 6 5 4 3 2 1

ISBN 13: 978-0-13-614775-6

ISBN 10:   0-13-614775-5

Pearson Education Ltd., *London*
Pearson Education Australia Pty. Ltd., *Sydney*
Pearson Education Singapore, Pte. Ltd.
Pearson Education North Asia Ltd., *Hong Kong*
Pearson Education Canada, Inc., *Toronto*
Pearson Educación de Mexico, S.A. de C.V.
Pearson Education—Japan, *Tokyo*
Pearson Education Malaysia, Pte. Ltd.

Table of Contents

Chapter 1 Pretest Form A

Consider the set of numbers: $\left\{-12, 31, -\sqrt{3}, 0, \sqrt{7}, \frac{3}{4}, -\frac{1}{8}, 2.25\right\}$. List those that are:

1. rational numbers

2. integers

3. irrational numbers

1. _____

2. _____

3. _____

What symbol (<, >, or =) will make each of the following expressions true?

4. $\left|-6\right|$ _____ $\left|-7\right|$

5. -5 _____ -10

4. _____

5. _____

Evaluate.

6. $-16 + 4$

7. $5 - 16 - 11$

8. $\frac{5}{12} + \frac{5}{6}$

9. $\frac{11}{14} - \frac{1}{2}$

10. $(-4)(-6)$

11. $\frac{5}{8} \div \frac{3}{4}$

12. $\frac{0}{5}$

13. 10^3

14. Write $3 \cdot 3 \cdot 3 \cdot 3 \cdot 3 \cdot 5 \cdot 5 \cdot y \cdot y \cdot y \cdot z \cdot z$ in exponential form.

15. Evaluate $(5r + 5s)^2$ for $r = -2$ and $s = 4$.

16. Evaluate $4x^2 - 7$ when $x = -3$.

6. _____

7. _____

8. _____

9. _____

10. _____

11. _____

12. _____

13. _____

14. _____

15. _____

16. _____

Name the property illustrated.

17. $4 \cdot 5 = 5 \cdot 4$

18. $6(5 + 7) = 6(5) + 6(7)$

17. _____

18. _____

Karen spent $69.29 at the store.

19. If sales tax is 6%, how much sales tax did she pay?

20. How much is her total bill, including tax?

19. _____

20. _____

Chapter 1 Pretest Form B

Name:_____

Date:_____

Consider the set of numbers: $\left\{-4, -\sqrt{25}, 0, \sqrt{6}, \dfrac{7}{8}, 9.35, 15\right\}$. List those that are:

1. integers

2. natural numbers

3. irrational numbers

1. _____

2. _____

3. _____

What symbol ($<$, $>$, or $=$) will make each of the following expressions true?

4. $\left|-13\right|$ _____ $\left|-15\right|$

5. -9 _____ -5

4. _____

5. _____

Evaluate.

6. $6 + (-9)$

7. $9 - 15 - 6$

8. $\dfrac{3}{10} + \dfrac{3}{20}$

9. $\dfrac{7}{10} - \dfrac{1}{2}$

10. $8(-5)$

11. $\dfrac{5}{7} \div \dfrac{10}{11}$

12. $\dfrac{14}{0}$

13. 4^3

14. Write $11 \cdot 11 \cdot 11 \cdot m \cdot m \cdot n \cdot n \cdot n \cdot n$ in exponential form.

15. Evaluate $(4a + 6b)^2$ for $a = -5$ and $b = 3$.

16. Evaluate $5x^2 - 12$ when $x = -2$.

6. _____

7. _____

8. _____

9. _____

10. _____

11. _____

12. _____

13. _____

14. _____

15. _____

16. _____

Name the property illustrated.

17. $3 + (-7) = (-7) + 3$

18. $5 \cdot (4 \cdot 3) = (5 \cdot 4) \cdot 3$

17. _____

18. _____

Kathy spent \$24.53 at the store.

19. If sales tax is 7%, how much sales tax did she pay?

20. How much is her total bill, including tax?

19. _____

20. _____

Mini-Lecture 1.1
Study Skills for Success in Mathematics

Learning Objectives:

1. Recognize the goals of this text.
2. Learn proper study skills.
3. Prepare for and take exams.
4. Learn to manage time.
5. Purchase a calculator.

Examples:

1. Goals of textbook
   a) To present traditional algebra topics
   b) To prepare students for advanced mathematics courses
   c) To build student confidence and enjoyment of mathematics
   d) To improve student reasoning and critical thinking skills
   e) To stress importance of mathematics in solving real-life problems
   f) To help students think mathematically and enhance problem-solving skills

2. Proper Study Skills
   a) Maintain a positive attitude
   b) Properly prepare for and attend class
   c) Read the textbook
   d) Complete homework assignments
   e) Find a proper place to study
   f) Form a study group
   g) Seek help right away when needed

3. Preparing for and Taking Exams
   a) Review previous homework, class notes, quizzes, etc.
   b) Study relevant formulas, definitions, and procedures.
   c) Read the Avoiding Common Errors boxes and Helpful Hint boxes.
   d) Complete the Chapter Review, Mid-Chapter Test and Chapter Practice Test.
   f) When taking the exam, read the directions and problems carefully.
   g) Pace yourself and use all available time. Attempt every problem.

4. Managing Time
   a) Make a list of weekly commitments with estimated time allotments. Schedule study time.
   b) Be organized to avoid wasting time.

5. Choose an appropriate calculator

Teaching Notes:

- Many developmental students have math anxiety and hesitate to ask questions.
- Discuss any resources that are available on your campus where students can get help with mathematics (such as a math lab or a tutoring center).
- Point out the student supplements that are available for this textbook.
- Recommending a specific model of calculator to the students will help to insure that students have one that is appropriate.

Mini-Lecture 1.2
Problem Solving

Learning Objectives:

1. Learn the five-step problem solving procedure.
2. Solve problems involving bar, line, and circle graphs.
3. Solve problems involving statistics.
4. Key vocabulary: *algebraic expression, measures of central tendency, mean, median*

Examples:

1. Clint is purchasing a new car with a $19,900 sticker price. He has two financing options from which to pick. Option 1: $0 money down and 60 monthly payments of $465.50. Option 2: $5000 down and 48 monthly payments of $409.75.
   a) How much total money would Clint save if he chooses option 2?
   b) How much would Clint pay in interest if he chooses option 1?
   c) How much would Clint pay in interest if he chooses option 2?

2. On an interstate, Clint's new car gets 33 miles per gallon of gasoline. How far on an interstate can the car be driven on 0.25 gallon of gasoline?

3. In 2005 a person with an adjusted gross income between $29,700 and $71,950 who filed a single return had to pay federal tax in the amount of $4090 plus 25% of income in excess of $29,700. If Jerome's 2005 adjusted gross income was $68,345, what was his federal tax if he filed a single return?

4. A group of 300 movie goers were asked which type of movie they preferred. The circle graph shows the resulting distribution.
   a) How many of the movie goers preferred drama?
   b) How many more of the movie goers preferred action than comedy?

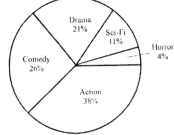

5. Lance Armstrong won the Tour de France seven consecutive times (1999 to 2005) before retiring. The margins of victory (in minutes) by which he won the seven races are: 7.617, 6.033, 6.733, 7.283, 1.017, 6.317, and 4.667
   a) Find the mean of Armstrong's margins of victory.
   b) Find the median of the margins of victory.

6. Kathy's first four exam scores were 72, 83, 88, and 74. She has one exam remaining and would like to earn a B in the course which requires a mean of at least 80. What is the minimum score that Kathy can earn on the last exam in order to get a B in the course?

Teaching Notes:

- Many students find application problems very challenging. Encourage a positive attitude.
- Encourage students to draw diagrams, charts, etc. when possible.
- Encourage students to practice using algebra to solve problems even though they may be able to solve the problem without it.
- Refer students to the *Guidelines for Problem Solving* in section 1.2 of the textbook.

Answers: 1a) $3262; 1b) $8030; 1c) $4768; 2) 8.25 miles; 3) $13,751.25; 4a) 63; 4b) 36; 5a) 5.667 minutes; 5b) 6.317 minutes; 6) 83

Mini-Lecture 1.3
Fractions

Learning Objectives:

1. Learn multiplication symbols and recognize factors.
2. Simplify fractions.
3. Multiply fractions.
4. Divide fractions.
5. Add and subtract fractions.
6. Change mixed numbers to fractions and vice versa.
7. Key vocabulary: *variable, factor, fraction, numerator, denominator, simplified, greatest common factor (GCF), whole numbers, least common denominator (LCD), mixed number*

Examples:

1. Simplify.

   a) $\dfrac{25}{40}$

   b) $\dfrac{78}{192}$

2. Multiply.

   a) $\dfrac{6}{25} \cdot \dfrac{5}{9}$

   b) $\dfrac{21}{32} \cdot \dfrac{12}{35}$

3. A children's book weighs $\dfrac{9}{16}$ pound. How much will a case of 24 books weigh?

4. Divide.

   a) $\dfrac{6}{13} \div \dfrac{9}{14}$

   b) $\dfrac{9}{10} \div \dfrac{15}{8}$

5. Add or Subtract as indicated.

   a) $\dfrac{11}{16} + \dfrac{3}{16}$ b) $\dfrac{8}{9} - \dfrac{2}{9}$ c) $\dfrac{5}{12} + \dfrac{3}{8}$ d) $\dfrac{8}{9} - \dfrac{5}{6}$

6. Change the mixed number to a fraction.

   a) $5\dfrac{3}{4}$

   b) $2\dfrac{5}{8}$

7. Change the fraction to a mixed number.

   a) $\dfrac{29}{6}$

   b) $\dfrac{137}{11}$

8. Add, subtract, multiply, or divide as indicated.

   a) $5\dfrac{2}{3} + 2\dfrac{3}{4}$ b) $5\dfrac{1}{8} - 2\dfrac{1}{6}$ c) $4\dfrac{1}{5} \cdot 2\dfrac{1}{6}$ d) $3\dfrac{4}{5} \div 1\dfrac{3}{10}$

Teaching Notes:

- Some students may need a visual demonstration illustrating equivalent fractions.
- Emphasize that fraction answers should be written in simplified form.
- Comment that addition and subtraction of fractions require that we find a common denominator but multiplication and division do not.

Answers: 1a) $\dfrac{5}{8}$; 1b) $\dfrac{13}{32}$; 2a) $\dfrac{2}{15}$; 2b) $\dfrac{9}{40}$; 3) $13\dfrac{1}{2}$ pounds; 4a) $\dfrac{28}{39}$; 4b) $\dfrac{12}{25}$; 5a) $\dfrac{7}{8}$; 5b) $\dfrac{2}{3}$; 5c) $\dfrac{19}{24}$; 5d) $\dfrac{1}{18}$; 6a) $\dfrac{23}{4}$; 6b) $\dfrac{21}{8}$; 7a) $4\dfrac{5}{6}$; 7b) $12\dfrac{5}{11}$; 8a) $8\dfrac{5}{12}$; 8b) $2\dfrac{23}{24}$; 8c) $9\dfrac{1}{10}$; 8d) $2\dfrac{12}{13}$

Mini-Lecture 1.4

The Real Number System

Learning Objectives:

1. Identify sets of numbers.
2. Know the structure of the real numbers.
3. Key vocabulary: *set, elements, empty set, counting numbers, whole numbers, integers, rational numbers, irrational numbers, real numbers*

Examples:

1. Consider the following set of numbers.

$$\left\{-34, \sqrt{7}, -\frac{9}{16}, -8.5, -6, 4\frac{5}{8}, 0, 6, \pi, \sqrt{25}, 127\right\}$$

List the elements that are
   a) natural numbers.
   b) whole numbers.
   c) integers.
   d) rational numbers.
   e) irrational numbers.
   f) real numbers.

2. Indicate whether each statement is true or false.
   a) Every integer is a whole number.
   b) Every natural number is a rational number.
   c) Every real number is either a rational number or an irrational number.
   d) Every rational number is also a natural number.

Teaching Notes:

- Emphasize to students that for a number to be classified as a counting number, whole number, integer, etc., it only needs to be able to be written in the proper form, but it does not have to be in that form. For example, $\frac{10}{2}$ is a whole number because it can be written as 5.

- Point out to students that if a rational number is written in decimal form, it will either terminate or repeat. If an irrational number is written in decimal form, it will neither terminate nor repeat.

Answers: 1a) $\left\{6, \sqrt{25}, 127\right\}$; 1b) $\left\{0, 6, \sqrt{25}, 127\right\}$; 1c) $\left\{-34, -6, 0, 6, \sqrt{25}, 127\right\}$; 1d) $\left\{-34, -\frac{9}{16}, -8.5,\right.$

$\left. -6, 4\frac{5}{8}, 0, 6, \sqrt{25}, 127\right\}$; 1e) $\left\{\sqrt{7}, \pi\right\}$; 1f) $\left\{-34, \sqrt{7}, -\frac{9}{16}, -8.5, -6, 4\frac{5}{8}, 0, 6, \pi, \sqrt{25}, 127\right\}$; 2a) *False;*

2b) *True;* 2c) *True;* 2d) *False*

Mini-Lecture 1.5
Inequalities

Learning Objectives:

1. Determine which is the greater of two numbers.
2. Find the absolute value of a number.
3. Key vocabulary: *absolute value*

Examples:

1. Insert either $<$ or $>$ between the pair of numbers to make a true statement.

   a) -7____-4
   b) $-\dfrac{5}{2}$____0
   c) $\dfrac{2}{5}$____$\dfrac{2}{7}$
   d) -3.1____-3.01

2. Evaluate each absolute value expression.

   a) $|-17|$
   b) $-|-21|$
   c) $|0|$
   d) $\left|\dfrac{5}{8}\right|$

3. Insert either $<$, $>$, or $=$ between the pair of numbers to make a true statement.

   a) $|-8|$____$|8|$
   b) $-|-3|$____-1
   c) $\left|-\dfrac{1}{2}\right|$____$0$
   d) $-(-10)$____$-|-10|$

Teaching Notes:

- Students often confuse the inequality symbols. Point out that the inequality symbol should always point towards the smaller number.
- Remind students that absolute value can be thought of as the number of units the number is from 0 on the number line. The absolute value cannot be negative because it is a distance.
- We say that the absolute value cannot be negative (rather than saying it is always positive) because it can be 0.

Answers: *1a) $<$; 1b) $<$; 1c) $>$; 1d) $<$; 2a) 17; 2b) -21; 2c) 0; 2d) $\dfrac{5}{8}$; 3a) $=$; 3b) $<$; 3c) $>$; 3d) $>$*

Mini-Lecture 1.6
Addition of Real Numbers

Learning Objectives:

1. Add real numbers using a number line.
2. Add fractions.
3. Identify opposites.
4. Add using absolute values.
5. Add using calculators.
6. Key vocabulary: *operations, opposites, additive inverses, sum*

Examples:

1. Evaluate using a number line.
   a) $5+(-3)$
   b) $-2+4$
   c) $-1+(-4)$
   d) $4+(-4)$

2. Add.
   a) $-\dfrac{1}{2}+\dfrac{2}{3}$
   b) $\dfrac{3}{8}+\left(-\dfrac{5}{6}\right)$
   c) $-\dfrac{2}{5}+\dfrac{3}{4}$
   d) $-\dfrac{2}{9}+\left(-\dfrac{4}{15}\right)$

3. Find the opposite of each number.
   a) -8
   b) 7.2
   c) $-\dfrac{7}{25}$
   d) 0

4. Add using absolute values.
   a) $7+8$
   b) $-5+(-6)$
   c) $-10+2$
   d) $4+(-7)$
   e) $-3.5+8.2$
   f) $-2.34+(-7.15)$
   g) $-\dfrac{1}{5}+\left(-\dfrac{1}{4}\right)$
   h) $-\dfrac{4}{6}+\dfrac{10}{15}$

5. Find the sum using your calculator.
   a) $-832+526$
   b) $-241+(-351)$
   c) $-892+(-167)$
   d) $1248+(-1659)$

Teaching Notes:

- Point out that the negative sign can be thought of as meaning "the opposite of." Show how this meaning is useful in cases such as $-(-6)$ which means the opposite of negative 6, or 6.
- Students may find it helpful to think of negative numbers as debts and positive numbers as savings.

Answers: 1a) 2; 1b) 2; 1c) -5; 1d) 0; 2a) $\dfrac{1}{6}$; 2b) $-\dfrac{11}{24}$; 2c) $\dfrac{7}{20}$; 2d) $-\dfrac{22}{45}$; 3a) 8; 3b) -7.2;

3c) $\dfrac{7}{25}$; 3d) 0; 4a) 15; 4b) -11; 4c) -8; 4d) -3; 4e) 4.7; 4f) -9.49; 4g) $-\dfrac{9}{20}$; 4h) 0; 5a) -306;

5b) -592; 5c) -1059; 5d) -411

Mini-Lecture 1.7
Subtraction of Real Numbers

Learning Objectives:

1. Subtract numbers.
2. Subtract numbers mentally.
3. Evaluate expressions containing more than two numbers.
4. Key vocabulary: *opposite, additive inverse, difference*

Examples:

1. Evaluate.

   a) $10 - (+4)$ b) $12 - 8$ c) $4 - 10$ d) $-3 - 9$

   e) $4.6 - 9.4$ f) $-\dfrac{1}{6} - \dfrac{2}{3}$ g) $4 - (-5)$ h) $-8 - (-5)$

   i) $-6 - (-10)$ j) $2.1 - (-5.7)$ k) $\dfrac{7}{10} - \left(-\dfrac{2}{15}\right)$ l) $-\dfrac{3}{4} - \left(-\dfrac{7}{12}\right)$

   m) Subtract 12 from 9.   n) Subtract -10 from -2.

   o) The lowest point in the state of California occurs in Death Valley and is 280 feet below sea level. The highest point in California is Mt. Whitney, which is 14,495 feet above sea level. What is the difference in elevation between these two locations in California?

2. Evaluate mentally.

   a) $-9 - 5$ b) $-6 - 7$ c) $2 - 5$ d) $3 - 13$

3. Evaluate.

   a) $-4 - 16 + 12$ b) $-7 + 6 - 8$ c) $-5 - (-7) - 4$ d) $4 - 9 + (-6) - (-2)$

4. Find the difference using your calculator.

   a) $-1907 - 837$ b) $-298 - (-527)$ c) $2341 - 5120$ d) $-374 - (-213)$

Teaching Notes:

- Remind students to always change subtraction to addition by "adding the opposite."
- Emphasize that when an expression contains both addition and subtraction, it should be evaluated from left to right.

Answers: 1a) 6; 1b) 4; 1c) -6; 1d) -12; 1e) -4.8; 1f) $-\dfrac{5}{6}$; 1g) 9; 1h) -3; 1i) 4; 1j) 7.8; 1k) $\dfrac{5}{6}$;
1l) $-\dfrac{1}{6}$; 1m) -3; 1n) 8; 1o) 14,775 feet; 2a) -14; 2b) -13; 2c) -3; 2d) -10; 3a) -8; 3b) -9;
3c) -2; 3d) -9; 4a) -2744; 4b) 229; 4c) -2779; 4d) -161

Mini-Lecture 1.8
Multiplication and Division of Real Numbers

Learning Objectives:

1. Multiply numbers.
2. Divide numbers.
3. Remove negative signs from denominators.
4. Evaluate divisions involving 0.
5. Key vocabulary: *like signs, unlike signs, product, quotient*

Examples:

1. Evaluate.
   a) $6(-8)$ b) $(-3)(-7)$ c) $(-4)(10)$ d) $0(-7)$

   e) $\left(-\dfrac{5}{8}\right)\left(-\dfrac{14}{15}\right)$ f) $(0.8)(-0.9)$ g) $(-3)(-4)(-6)$ h) $(-2)(-1)(-8)(-3)(4)$

2. Evaluate.
   a) $\dfrac{20}{-4}$ b) $\dfrac{-32}{-8}$ c) $\dfrac{-55}{5}$ d) $29.44 \div (-3.2)$

3. Evaluate.
   $\left(\dfrac{5}{6}\right) \div \left(\dfrac{-10}{9}\right)$

4. Indicate whether the quotient is 0 or undefined.
   a) $\dfrac{0}{7}$ b) $\dfrac{0}{-10}$ c) $\dfrac{12}{0}$ d) $\dfrac{-5}{0}$

5. Find the product or quotient using your calculator.
   a) $\dfrac{2184}{-12}$ b) $-268(-125)$ c) $\dfrac{-182,592}{-288}$ d) $(-85)(76)$

Teaching Notes:

- Students often are confused by division involving 0. Particularly, many students mistakenly think that an expression such $5 \div 0$ will simplify to 0. Be sure to warn students about this error in thinking.
- To help explain why an expression such as $5 \div 0$ is undefined, try dividing 5 by numbers that approach 0. For example, have students use calculators to evaluate each of the following: $5 \div 0.1$, $5 \div 0.01$, $5 \div 0.001$, etc. to see that the quotient approaches infinity as the divisor approaches 0.

Answers: 1a) -48; 1b) 21; 1c) -40; 1d) 0; 1e) $\dfrac{7}{12}$; 1f) -0.72; 1g) -72; 1h) 192; 2a) -5; 2b) 4;

2c) -11; 2d) -9.2; 3) $-\dfrac{3}{4}$; 4a) 0; 4b) 0; 4c) undefined; 4d) undefined; 5a) -182; 5b) 33,500;

5c) 634; 5d) -6460

Mini-Lecture 1.9
Exponents, Parentheses, and the Order of Operations

Learning Objectives:

1. Learn the meaning of exponents.
2. Evaluate expressions containing exponents.
3. Learn the difference between $-x^2$ and $(-x)^2$.
4. Learn the order of operations.
5. Learn the use of parentheses.
6. Evaluate expressions containing variables.
7. Key vocabulary: *base, exponent, order of operations, grouping symbols*

Examples:

1. Evaluate.

   a) $(-8)^2$　　　b) -8^2　　　c) $(-5)^4$　　　d) -5^4

   e) $(-10)^3$　　　f) -10^3　　　g) $\left(\dfrac{3}{4}\right)^2$　　　h) $\left(-\dfrac{1}{2}\right)^3$

2. Evaluate.

   a) $6+3\cdot2^3-12$

   b) $-9+3\left[-5+\left(36\div2^2\right)\right]$

   c) $(20\div5)+4(5-2)^2$

   d) $-12-54\div2\cdot3^2+8$

   e) $-4^2+20\div4$

   f) $(-4)^2+20\div4$

   g) $\dfrac{5}{6}-\dfrac{7}{8}\cdot\dfrac{2}{5}$

   h) $\left\{7-3\left[5-(8-2)\right]^2\right\}^2$

3. Write the following statements as mathematical expressions using parentheses and brackets and then evaluate.
   Multiply 8 by 3. To this product add 16. Divide this sum by 4. Multiply this quotient by 7.

4. Evaluate a) x^2, b) $-x^2$, and c) $(-x)^2$ for $x=5$.

5. Evaluate a) x^2, b) $-x^2$, and c) $(-x)^2$ for $x=-6$.

6. Evaluate each expression for the given value of the variable or variables.

   a) $3x^2-4x+8$ when $x=\dfrac{1}{3}$　　　b) $x^2-5xy+6y^2$ when $x=-2$ and $y=-3$

Teaching Notes:

- The acronym PEMDAS may mislead some students to believe that multiplication must always be completed before division and that addition must always completed before subtraction. Emphasize that this is incorrect.

Answers: *1a) 64; 1b) -64; 1c) 625; 1d) -625; 1e) -1000; 1f) -1000; 1g) $\dfrac{9}{16}$; 1h) $-\dfrac{1}{8}$;*

2a) 18; 2b) 3; 2c) 40; 2d) -247; 2e) -11; 2f) 21; 2g) $\dfrac{29}{60}$; 2h) 16; 3) $\{[(8\cdot3)+16]/4\}\cdot7=70$;

4a) 25; 4b) -25; 4c) 25; 5a) 36; 5b) -36; 5c) 36; 6a) 7; 6b) 28

11

Mini-Lecture 1.10
Properties of the Real Number System

Learning Objectives:

1. Learn the commutative property.
2. Learn the associative property.
3. Learn the distributive property.
4. Learn the identity properties.
5. Learn the inverse properties.

Examples:

1. Name the property illustrated.
   a) $(x+8)+5 = x+(8+5)$
   b) $6 \cdot 7 = 7 \cdot 6$
   c) $3 \cdot (10 \cdot 5) = (3 \cdot 10) \cdot 5$

   d) $3(x+8) = 3x + 3 \cdot 8$
   e) $-4m + 4m = 0$
   f) $-9 + 0 = -9$

   g) $6 + 10 = 10 + 6$
   h) $\dfrac{7}{10} \cdot 1 = \dfrac{7}{10}$
   i) $\left(-\dfrac{7}{9}\right) \cdot \left(-\dfrac{9}{7}\right) = 1$

   j) $6 + (7+8) = 6 + (8+7)$
   k) $\left(\dfrac{1}{2} \cdot 2\right) \cdot 9 = 1 \cdot 9$
   l) $\dfrac{2}{3} \cdot \left(\dfrac{1}{2}x - \dfrac{1}{4}\right) = \dfrac{2}{3} \cdot \left(\dfrac{1}{2}x\right) - \dfrac{2}{3} \cdot \left(\dfrac{1}{4}\right)$

2. Determine a) the additive inverse, and b) the multiplicative inverse for $-\dfrac{3}{8}$.

3. The name of a property is given followed by part of an equation. Complete the equation, to the right of the equals sign, to illustrate the given property.
   a) Associative property of addition
      $(x+3)+9 =$
   b) Distributive property
      $3(2x-8) =$

   c) Inverse property of multiplication
      $[2(0.5)]x =$
   d) Commutative property of multiplication
      $x \cdot 2 - 9 =$

Teaching Notes:

- Discuss the similarities between the distributive property and multiplying a multi-digit number by a single digit number. For example compare $2(x+4)$ and $2(14)$.

Answers: 1a) associative property of addition; 1b) commutative property of multiplication; 1c) associative property of multiplication; 1d) distributive property; 1e) inverse property of addition; 1f) identity property of addition; 1g) commutative property of addition; 1h) identity property of multiplication; 1i) inverse property of multiplication; 1j) commutative property of addition; 1k) inverse property of multiplication; 1l) distributive property; 2a) $\dfrac{3}{8}$; 2b) $-\dfrac{8}{3}$; 3a) $x+(3+9)$; 3b) $3(2x)-3(8)$ or $6x-24$; 3c) $1x$ or x; 3d) $2x-9$

Additional Exercises 1.1

Instructor Information:

Name: _____

Office location: _____

Office hours: _____

_____

_____

Phone number: _____

Email: _____

Classmate Information:

Obtain the names of at least two classmates whom you can contact for information or study questions.

1. Name: _____

    Phone number: _____

    Email address: _____

2. Name: _____

    Phone number: _____

    Email address: _____

Math Lab:

Location: _____

Hours: _____

Phone number: _____

Tutoring Services:

Location: _____

Hours: _____

Phone number: _____

Recommended Supplements:

_____

_____

_____

_____

Additional Exercises 1.2

1. The selling prices of six homes shown by a real estate agent are: $156,000; $120,000; $140,000; $110,000; $150,000; and $130,000. Find the median price of these six homes.

2. A city's high temperature readings over 7 days were 67°, 75°, 68°, 69°, 72°, 70°, and 66°. Find the median high temperatures over the 7 days.

3. On five exams, Audrey's scores were 76, 92, 84, 88 and 90. Find the mean score of Audrey's exams.

4. A taxi driver's tips over a 4-day period were $69, $57, $62, and $59. Find the driver's mean daily tip over the 4-day period.

5. A doctor makes a down payment of 25% of the $145,000 purchase price of an office building. How much is the down payment?

6. A real estate broker receives a commission of 3% of the selling price of a house. Find the commission received by the broker for selling a house for $90,000.

7. Marilyn had a checking account balance of $1780.42 before making a deposit of $510.00. She then wrote checks for $20.00 and $405.12. Find her current checkbook balance.

8. Katrin wants to buy a recliner that sells for $500. She can pay the total amount at the time of purchase or $45 a month for 12 months. How much can she save by paying the total amount at the time of purchase?

9. When the odometer in Dillon's car reads 18,522.8 miles, he fills his gas tank. The next time he fills his tank, it takes 12.2 gallons, and his odometer reads 18,815.6 miles. Find the number of miles per gallon his car gets.

10. A portion of the federal income tax rate schedule for a single return in 2005 is shown in the table

Adjusted Gross Income	Taxes
$0–$7,300	10% of income
$7,300–$29,700	$730.00 + 15% in excess of $7,300
$29,700–$71,950	$4,090.00 + 25% in excess of $29,700
$71,950–$150,150	$14,652.50 + 28% in excess of $71,950

If Harold's adjusted gross income in 2005 was $84,600, determine his tax.

11. A sales executive earns a salary of $26,000 per year plus a 4% commission on sales. During the year, the executive's sales were $280,000. Find the executive's total income for the year.

12. A little theater group sells adult tickets for $6.50 each. A patron pays $40 for 4 adult tickets plus voting privileges. A regular membership is $25 for voting privileges only. If Rudi plans to attend 4 plays and he wants voting privileges, how much could he save by becoming a patron?

1. _____

2. _____

3. _____

4. _____

5. _____

6. _____

7. _____

8. _____

9. _____

10. _____

11. _____

12. _____

Additional Exercises 1.2 *(cont.)*   Name:_____

The table shows how a family's monthly income of $4200 is budgeted:

Category	Percentage
Food	22%
Housing	35%
Savings	9%
Transportation	18%
Miscellaneous	16%

13. How much of the family's income is budgeted for food each month?   13. _____

14. How much does the family budget for transportation each month?   14. _____

15. How much more is budgeted for housing than food each month?   15. _____

The circle graph shows the distribution of deductions and take-home pay of an employee's monthly income of $2800.

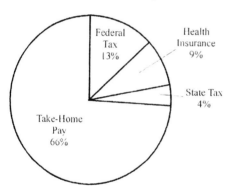

16. Find the employee's monthly cost for health insurance.   16. _____

17. Find the employee's monthly take-home pay.   17. _____

18. A mean of 90 is needed to make an A in a course. On his first 4 exams, Jeremy made 92, 94, 88, and 98. What is the minimum grade Jeremy can receive on the fifth exam to get an A in the course?   18. _____

19. Sarah's average cost for lunch one week was $4.75. Find the amount she spends on lunches during a five-day workweek.   19. _____

20. On a shopping trip, Anna made purchases totaling $35.53. If the sales tax rate is 6.5%, how much change should she receive from a $50 bill?   20. _____

Additional Exercises 1.3

Simplify each fraction.

1. $\dfrac{16}{40}$

2. $\dfrac{24}{36}$

3. $\dfrac{18}{45}$

4. Write the fraction $\dfrac{25}{6}$ as a mixed number.

5. Write the mixed number $8\dfrac{5}{9}$ as a fraction.

Find the product or quotient. Simplify each answer.

6. $\dfrac{1}{8} \cdot \dfrac{4}{5}$

7. $\dfrac{6}{35} \cdot \dfrac{5}{18}$

8. $\dfrac{15}{22} \cdot \dfrac{2}{65}$

9. $\dfrac{2}{9} \div \dfrac{18}{7}$

10. $\dfrac{9}{2} \div \dfrac{2}{3}$

11. $\dfrac{16}{5} \div \dfrac{2}{3}$

12. $\left(3\dfrac{3}{7}\right)\left(1\dfrac{1}{4}\right)$

13. $\left(5\dfrac{5}{8}\right) \div \left(2\dfrac{11}{12}\right)$

Add or subtract. Simplify each answer.

14. $\dfrac{5}{12} + \dfrac{5}{8}$

15. $5\dfrac{1}{4} + 3\dfrac{1}{8}$

1. _____

2. _____

3. _____

4. _____

5. _____

6. _____

7. _____

8. _____

9. _____

10. _____

11. _____

12. _____

13. _____

14. _____

15. _____

Additional Exercises 1.3 *(cont.)*

16. $\dfrac{1}{10} + \dfrac{5}{6}$

16. _____

17. $\dfrac{5}{6} - \dfrac{1}{2}$

17. _____

18. $\dfrac{11}{14} - \dfrac{1}{2}$

18. _____

19. $6\dfrac{1}{3} - 1\dfrac{4}{5}$

19. _____

20. At his annual physical checkup, 8-year-old Benjamin measured $51\dfrac{1}{2}$ inches tall. At his checkup the previous year, Benjamin measured $48\dfrac{7}{8}$ inches tall. How much did Benjamin grow between checkups?

20. _____

Additional Exercises 1.4

Consider the set of numbers: $\left\{0.5, 13, 0, -\sqrt{13}, -5, \sqrt{7}, \frac{1}{4}\right\}$.

Choose the elements of that set that are:

1. Integers

2. Rational numbers

3. Irrational numbers

1. _____

2. _____

3. _____

Consider the set of numbers: $\left\{-16, 40, -\sqrt{7}, 0, \sqrt{3}, \pi, \frac{1}{5}, -\frac{1}{5}, 3.25\right\}$

List those numbers that are:

4. Whole numbers

5. Positive integers

6. Irrational

7. Real

8. The number $\sqrt{3}$ belongs to which of these sets?
   Natural numbers, whole numbers, integers, rational numbers,
   irrational numbers, and real numbers.
   Name all that apply.

4. _____

5. _____

6. _____

7. _____

8. _____

State whether each statement is true or false.

9. 5 is a rational number.

10. −6 is a whole number.

11. 0.5 is a positive integer.

12. Every natural number is an integer.

13. $\sqrt{36}$ is an irrational number.

14. Some integers are not whole numbers.

9. _____

10. _____

11. _____

12. _____

13. _____

14. _____

Consider the set of numbers: $\left\{-7, -\sqrt{5}, \pi, 0, 4, \frac{5}{6}, -0.72\right\}$

List the numbers from the set that satisfy the conditions.

15. An integer, but not a natural number.

16. A rational number but not an integer.

17. A rational number but not a negative number.

18. A whole number but not a positive integer.

19. An irrational number and a positive number.

20. A rational number but not a whole number.

15. _____

16. _____

17. _____

18. _____

19. _____

20. _____

Additional Exercises 1.5

Evaluate.

1. $|-22|$

2. $-|22|$

3. $-|-14|$

4. $|-6|$

Insert <, or > to make a true statement.

5. 12 __ 17

6. 5 __ -8

7. -3 __ 0

8. -7 __ -9

9. $\dfrac{3}{10}$ __ $-\dfrac{2}{5}$

10. 8.05 __ 8.005

11. $-\dfrac{5}{8}$ __ $-\dfrac{1}{8}$

12. $-(-5)$ __ $|-9|$

Insert >, <, or = to make a true statement.

13. $|-7|$ __ $|-3|$

14. $|-0.6|$ __ $\left|\dfrac{3}{5}\right|$

15. -2.6 __ $-\dfrac{13}{5}$

16. $\left|-\dfrac{5}{9}\right|$ __ $\left|-\dfrac{9}{5}\right|$

17. $\dfrac{3}{5}+\dfrac{3}{5}+\dfrac{3}{5}+\dfrac{3}{5}$ __ $4 \cdot \dfrac{3}{5}$

18. $6 \div \dfrac{3}{5}$ __ $\dfrac{3}{5} \div 6$

19. $\dfrac{1}{2} \cdot \dfrac{1}{4}$ __ $\dfrac{1}{2} - \dfrac{1}{4}$

20. $\dfrac{3}{4} \cdot \dfrac{3}{4}$ __ $\dfrac{3}{4} \div \dfrac{3}{4}$

1. _____

2. _____

3. _____

4. _____

5. _____

6. _____

7. _____

8. _____

9. _____

10. _____

11. _____

12. _____

13. _____

14. _____

15. _____

16. _____

17. _____

18. _____

19. _____

20. _____

Additional Exercises 1.6

Name:_____

Date:_____

Write the opposite of each number.

1. −9

2. $\dfrac{22}{3}$

3. $-2\dfrac{3}{8}$

4. −3.4

5. $\dfrac{3}{8}$

1. _____

2. _____

3. _____

4. _____

5. _____

Add.

6. $86 + (-33)$

7. $-10 + 2$

8. $8 + (-14)$

9. $-17 + 19$

10. $25 + (-47)$

11. $-7 + (-12)$

12. $-78 + (-129)$

13. $-593 + 281$

14. $-\dfrac{5}{6} + \dfrac{3}{4}$

15. $-\dfrac{1}{8} + \left(-\dfrac{5}{12}\right)$

16. $\dfrac{3}{10} + \left(-\dfrac{11}{15}\right)$

17. $-2.5 + 7.41$

6. _____

7. _____

8. _____

9. _____

10. _____

11. _____

12. _____

13. _____

14. _____

15. _____

16. _____

17. _____

Write an expression that can be used to solve each problem and then solve.

18. At 6:00 a.m., the temperature was 15° F. By 10:00 a.m., it had dropped another 23°. What was the temperature at 10:00 a.m.?

19. A submarine descended 218 feet from the surface and then descended another 54 feet. What was its position at that point?

20. A football team lost 12 yards on one play and gained 5 yards the next play. What was its total yardage on the two plays?

18. _____

19. _____

20. _____

Additional Exercises 1.7

Evaluate.

1. $-16 - 10$

2. $-21 - (-10)$

3. $12 - (-8)$

4. $-80 - 69$

5. $-77 - 96$

6. $2 - 11$

7. $-5 - (-16)$

8. $\dfrac{1}{2} - \dfrac{2}{3}$

9. $\dfrac{1}{4} - \left(-\dfrac{1}{6}\right)$

10. $-\dfrac{4}{9} - \left(-\dfrac{5}{12}\right)$

11. $19 - 29 - 10$

12. $-5 + (-6) - 8$

13. $-7 - (-2) - (-10)$

14. $2.9 - 5.2 - (-1.5)$

15. $27 - 18 + (-4) - 6$

16. $-6 + (-8) + (+9) - (-3)$

17. $-2 - (-12) + (-5) + 6$

18. How much is a drop in temperature from 12°F to –16°F?

19. One submarine is 346 feet below sea level. Another submarine is directly overhead at 176 feet below sea level. How far are the submarines apart?

20. One city is 12 feet below sea level. A nearby city is 109 feet above sea level. What is the difference in the elevations of the two cities?

1. _____

2. _____

3. _____

4. _____

5. _____

6. _____

7. _____

8. _____

9. _____

9. _____

11. _____

12. _____

13. _____

14. _____

15. _____

16. _____

17. _____

18. _____

19. _____

20. _____

Additional Exercises 1.8

Name:_____

Date:_____

Find the product.

1. $(-7)(-5)$

2. $-3(8)$

3. $(-5.8)(2.3)$

4. $(-1)(4)(-6)$

5. $(-2)(-5)(-3)$

6. $\left(-\dfrac{8}{15}\right)\left(-\dfrac{10}{9}\right)$

7. $\left(-\dfrac{2}{11}\right)\left(\dfrac{9}{12}\right)$

8. $\left(\dfrac{-9}{11}\right)\left(\dfrac{12}{-2}\right)$

Find the quotient.

9. $\dfrac{-48}{6}$

10. $(18) \div (-3)$

11. $-3.2 \div 6.4$

12. $\dfrac{-138}{-2}$

13. $\dfrac{-444}{-6}$

14. $-\dfrac{3}{7} \div \left(-\dfrac{10}{21}\right)$

15. $\dfrac{-8}{9} \div \dfrac{71}{72}$

Evaluate.

16. $(-48) \div (-12)$

17. $(5)(-1)(-2)(-6)$

18. Divide -180 by -30

Indicate whether each of the following is zero or undefined.

19. $\dfrac{-7}{0}$

20. $\dfrac{0}{25}$

1. _____

2. _____

3. _____

4. _____

5. _____

6. _____

7. _____

8. _____

9. _____

10. _____

11. _____

12. _____

13. _____

14. _____

15. _____

16. _____

17. _____

18. _____

19. _____

20. _____

Additional Exercises 1.9

Evaluate.

1. 2^5

2. 3^4

3. -10^2

4. -5^4

5. $(-3)^2$

6. $(-7)^3$

7. $\left(\dfrac{2}{3}\right)^3$

8. $\left(-\dfrac{5}{6}\right)^2$

9. $3 \cdot 4^2$

10. $\left(-180 \cdot \dfrac{1}{4}\right) \div 5$

11. $8 \cdot 8^2 - 4 \cdot 5^2$

12. $25 \div 5 \cdot 5 + 8 - 5$

13. $4 \div 2 \cdot 2 + 6 - 3$

14. $-[12 - (-4 - 3)]^2$

15. $834 - 5(19 + 16)$

16. $964 - 6(20 + 16)$

17. Evaluate $(2m - 4n)^2$ for $m = 5$ and $n = -3$.

18. Evaluate $(3c - d)^2$ for $c = -1$ and $d = 4$.

19. Evaluate $\dfrac{gh}{g + h}$ when $g = 9$ and $h = 14$.

20. Evaluate $\dfrac{fg}{f + g}$ when $f = 10$ and $g = 11$.

1. _____

2. _____

3. _____

4. _____

5. _____

6. _____

7. _____

8. _____

9. _____

10. _____

11. _____

12. _____

13. _____

14. _____

15. _____

16. _____

17. _____

18. _____

19. _____

20. _____

Additional Exercises 1.10

Name:_____

Date:_____

Determine (a) the additive inverse and (b) the multiplicative inverse.

1. -13

2. $\dfrac{7}{9}$

3. 2.5

4. $-\dfrac{5}{16}$

1. (a) _____ (b) _____

2. (a) _____ (b) _____

3. (a) _____ (b) _____

4. (a) _____ (b) _____

Name the property illustrated.

5. $14 + 7 = 7 + 14$

6. $(8 + x) + 6 = 8 + (x + 6)$

7. $8 \cdot (6 \cdot 2) = (8 \cdot 6) \cdot 2$

8. $5(4 + x) = 20 + 5x$

5. _____

6. _____

7. _____

8. _____

Complete using the property given.

9. $-3 \cdot 5$
   Commutative property of multiplication

10. $(3 + (-7)) + 5$
    Associative property of addition

11. $8(x + y)$
    Distributive property

12. $10 + (-8)$
    Commutative property of addition

13. $-5(2 \cdot 8)$
    Associative property of multiplication

14. $4(y + x + 3)$
    Distributive property

15. $(3 + 5) \cdot 8$
    Commutative property of addition

16. $(-5x)y$
    Associative property of multiplication

17. $7(x + 6)$
    Commutative property of multiplication

18. $(2x + 7) + 6$
    Associative property of addition

19. $6(x + 8)$
    Commutative property of addition

20. $10(x + y + 7)$
    Distributive property

9. _____

10. _____

11. _____

12. _____

13. _____

14. _____

15. _____

16. _____

17. _____

18. _____

19. _____

20. _____

Chapter 1 Test Form A

1. Abigail bought a gallon of milk for $3.09 and a package of cookies for $2.59. If there is a 7% sales tax, what is her final bill?

1. _____

2. Dawn's first four test grades were 65, 75, 90, and 86. Find the mean of her grades.

2. _____

The circle graph shows the distribution of sizes for 150 shirts purchased from a shop during a particular week.

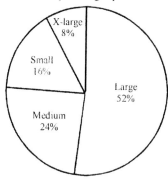

3. How many large-sized shirts were purchased at the shop?

3. _____

4. How many more medium-sized shirts were purchased than small?

4. _____

5. Consider the set of numbers: $\left\{-4, 0, \sqrt{2}, \dfrac{1}{2}, -\dfrac{1}{3}, -0.75\right\}$

   List those that are integers.

5. _____

Insert either $<$, $>$, or $=$ to make a true statement.

6. -1.5 _____ -1.05

6. _____

7. $\left|-4.5\right|$ _____ $-\left(-\dfrac{9}{2}\right)$

7. _____

Evaluate.

8. $-3 - 8$

8. _____

9. $-\dfrac{3}{4} - \left(-\dfrac{1}{10}\right)$

9. _____

10. $-8 + 5 - 2$

10. _____

Chapter 1 Test Form A *(cont.)*

11. $(-6)(-2)(-3)$

11. _____

12. $\left(-\dfrac{3}{2}\right)\left(-\dfrac{5}{8}\right)$

12. _____

13. $-48 \div 8$

13. _____

14. $-6 \div \dfrac{2}{3}$

14. _____

15. $-7 \div 0$

15. _____

16. $-2(-4 + 3)$

16. _____

17. $[-3(2) + 4] - 6$

17. _____

18. $(-3)^4 - 1$

18. _____

19. $(2 - 5)^2 + 4$

19. _____

20. $38 - 24 \div 2 \cdot 3$

20. _____

21. $24 \div 3 \cdot 2^2$

21. _____

Evaluate for the values given.

22. $3x + 4, x = -5$

22. _____

23. $x^2 + 2x + 1, x = -1$

23. _____

24. $x^2 + 3xy + y^2, x = -1, y = 2$

24. _____

25. Name the property illustrated. $(x + 5)3 = 3(x + 5)$

25. _____

Chapter 1 Test Form B

Name:_____

Date:_____

1. The number of pocket cellular phones sold by a store for the first five months of the year were 34, 26, 31, 39, and 45. What was the mean number of cellular phones sold per month?

1. _____

2. Paul bought a loaf of bread for $2.59 and a jar of jelly for $1.59. If the sales tax rate is 6%, how much change should she receive from a $10 bill?

2. _____

The circle graph shows the distribution of fragrances for 120 candles purchased from a candle shop during a particular weekend.

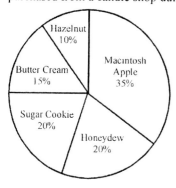

3. How many Macintosh Apple candles were sold?

3. _____

4. How many more Sugar Cookie candles were sold than Butter Cream?

4. _____

5. Consider the set of numbers: $\{-3, -4\frac{2}{3}, \sqrt{13}, -0.3, 0\}$

   List those that are integers.

5. _____

Insert either <, >, or = to make a true statement.

6. -11 _____ -4

6. _____

7. $|-3|$ _____ 2

7. _____

Evaluate.

8. $-5 + 1$

8. _____

9. $3 - (-10)$

9. _____

10. $-7 + 4 - 3$

10. _____

11. $(-5)(4)(-6)$

11. _____

Name:_____

12. $\left(-\dfrac{2}{3}\right)\left(\dfrac{5}{9}\right)$

12. _____

13. $-\dfrac{8}{3} \div \dfrac{6}{5}$

13. _____

14. -4^2

14. _____

15. $0 \div (-4)$

15. _____

16. $-4(5-7)$

16. _____

17. $-1 - (-6 + 9)$

17. _____

18. $-\dfrac{5}{6} + \dfrac{3}{8}$

18. _____

19. $2[6 - (4^2 + 3)]$

19. _____

20. $9 - 36 \div 4 \cdot 2$

20. _____

21. $2^3 - 7 - (3^2 - 5)$

21. _____

Evaluate for the values given.

22. $2x - 10,\ x = -1$

22. _____

23. $x^2 - 3x - 2,\ x = 4$

23. _____

24. $x^2 - xy + y^2,\ x = 2,\ y = 3$

24. _____

25. Name the property illustrated. $2a + 2b = 2(a + b)$

25. _____

Chapter 1 Test Form C

1. A laser printer that regularly sells for $425 has been placed on a clearance rack for 30% off. What is the clearance price?

1. _____

2. The vertical rise, in feet, for eight ski lifts at a resort are 1100, 2050, 980, 900, 1050, 1540, 2010, and 1360. What is the median vertical rise for these ski lifts?

2. _____

A group of 250 movie goers were asked which type of movie they preferred. The circle graph shows the resulting distribution.

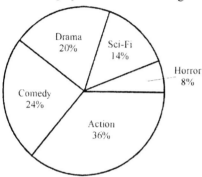

3. How many of the movie goers preferred comedy?

3. _____

4. How many more of the movie goers preferred drama than horror?

4. _____

5. Consider the set of numbers: $\left\{ -\dfrac{3}{8}, \sqrt{7}, 0, \pi, 3.7, 1 \right\}$
   List those that are rational numbers.

5. _____

Insert either <, >, or = to make a true statement.

6. -4 _____ -1

6. _____

7. $-|-5|$ _____ $-(-1)$

7. _____

Evaluate.

8. $-7 - 11$

8. _____

9. $-10 - (-5)$

9. _____

10. $-\dfrac{5}{8} + \dfrac{7}{12}$

10. _____

Chapter 1 Test Form C *(cont.)*

Name:_____

11. $(-3)(-5)(-2)$

11. _____

12. $\left(-\dfrac{3}{5}\right)\left(-\dfrac{2}{11}\right)$

12. _____

13. $-63 \div 7$

13. _____

14. $-18 \div \dfrac{2}{9}$

14. _____

15. $\dfrac{-14}{0}$

15. _____

16. $-5(8 - 10)$

16. _____

17. $11 - [-7 - 1]$

17. _____

18. $\left(-\dfrac{1}{4}\right)^{3}$

18. _____

19. $(-3^{2} - 2^{2}) - (2^{3} - 6)$

19. _____

20. $-36 \div 3 + 9$

20. _____

21. $5(-3) \div (-2 - 3)$

21. _____

Evaluate for the values given.

22. $4x - 9, x = 2$

22. _____

23. $x^{2} - 2x - 4, x = -3$

23. _____

24. $x^{2} - 3xy + 2y^{2}, x = 1, y = -3$

24. _____

25. Name the property illustrated. $(x + 3) + 2 = x + (3 + 2)$

25. _____

Chapter 1 Test Form D

Name:_____

Date:_____

1. Bill and Beverly are buying a new car for $26,500. If they pay 30% down and finance the rest, what amount will they finance?

1. _____

2. Sierra bowled three games. Her scores were 125, 153, and 127. What was Sierra's mean score?

2. _____

A group of 400 children were asked about the type of cookie they preferred. The circle graph shows the resulting distribution.

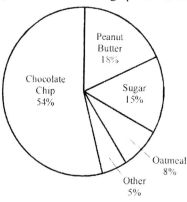

3. How many of the children preferred peanut butter cookies?

3. _____

4. How many more of the children preferred chocolate chip cookies than preferred sugar cookies?

4. _____

5. Consider the set of numbers: $\left\{ \sqrt{5}, -7, 25, -\sqrt{3}, \sqrt{4} \right\}$
   List those that are irrational numbers.

5. _____

Insert either $<$, $>$, or $=$ to make a true statement.

6. $-\dfrac{5}{4}$ _____ $-\dfrac{7}{5}$

6. _____

7. $|-6|$ _____ $|-18|$

7. _____

Evaluate.

8. $10 + (-18)$

8. _____

9. $-\dfrac{7}{10} + \dfrac{8}{15}$

9. _____

31

Chapter 1 Test Form D *(cont.)*

10. $-3 + (-5) - 8 - (-4)$

10. _____

11. $(-11)(-5)(-6)$

11. _____

12. $\left(-\dfrac{3}{10}\right)\left(\dfrac{6}{5}\right)$

12. _____

13. $24 \div (-3)$

13. _____

14. $3\dfrac{1}{7} \div (-11)$

14. _____

15. $0 \div 9$

15. _____

16. $-4 + 5(-3 - 7)$

16. _____

17. $-5[1 - (2 - 6)]$

17. _____

18. $-4^2 + (-2)^2$

18. _____

19. $6[2^3 - 4(2 - 5)^2]$

19. _____

20. $-10 + 16 \div 8 \cdot 2$

20. _____

21. $3 + 4\left[-3 - (-6 + 1)^2\right]$

21. _____

Evaluate for the values given.

22. $-2x - 3,\ x = -4$

22. _____

23. $3x^2 + 4x + 4,\ x = -2$

23. _____

24. $-m^2 + 3mn - 7,\ m = -1, n = -2$

24. _____

25. Name the property illustrated: $7 \cdot (3 \cdot 4) = (7 \cdot 3) \cdot 4$

25. _____

Chapter 1 Test Form E

Name:_____

Date:_____

1. Maggie uses 12% of her monthly salary on health insurance. If she earns $2200 per month, how much does he pay for her insurance?

    1. _____

2. Baker Industries sold 4200 computers in January, 7650 computers in February, 6235 computers in March and 5125 in April. Determine the mean number of computer sold per month during this time period.

    2. _____

3. Consider the set of numbers: $\left\{-5, 0, \frac{1}{8}, \sqrt{4}, 8\right\}$

    List those that are natural numbers.

    3. _____

Insert either $<$, $>$, or $=$ to make a true statement.

4. -8 _____ -4

    4. _____

5. $-|-10|$ _____ $-|10|$

    5. _____

Evaluate.

6. $42 - 61$

    6. _____

7. $17 - (-17)$

    7. _____

8. $9 - (-7) + (-3) - 12$

    8. _____

9. $(7)(-2)(-3)$

    9. _____

10. $\left(\frac{5}{11}\right)\left(-\frac{22}{35}\right)$

    10. _____

11. $-66 \div (-3)$

    11. _____

12. $-30 \div \frac{3}{5}$

    12. _____

13. $-6 \div 0$

    13. _____

14. $-4\left(2 - 3^2\right)$

    14. _____

Chapter 1 Test Form E *(cont.)*

15. $7 - [-4 - (-3)]$

15. _____

16. $\left(-\dfrac{2}{3}\right)^4$

16. _____

17. $6[16 \div (-2)(3) + 5]$

17. _____

18. $10 - (-4)^2 \div 2 - 4(-3 + 5)$

18. _____

19. $16 - 5[8 - (1 - 2 \cdot 6)]$

19. _____

20. -3^4

20. _____

21. $4\dfrac{3}{8} - 1\dfrac{5}{6}$

21. _____

Evaluate for the values given.

22. $-8p - 13$, $p = -0.75$

22. _____

23. $-x^2 - 6x + 4$, $x = -2$

23. _____

24. $4x^2 + 3xy + 2y^2$, $x = 2, y = -3$

24. _____

25. Name the property illustrated. $(x + 6) + z = (6 + x) + z$

25. _____

Chapter 1 Test Form F

Name:_____

Date:_____

1. The number of miles driven during each of five days of a business trip was 110, 72, 66, 145, and 92. Find the mean number of miles driven each day.

   1. _____

2. A salesperson receives $500 per month plus 2% commission on sales over $25,000. During the month, the salesperson's sales were $84,000. Find the salesperson's total earnings for the month.

   2. _____

3. Consider the set of numbers: $\left\{\sqrt{23}, 3, -\dfrac{7}{3}, \sqrt{6}, -0.5\right\}$

   List those that are rational numbers.

   3. _____

Insert either <, >, or = to make a true statement.

4. -15 _____ -13

   4. _____

5. $|-7|$ _____ $-|7|$

   5. _____

Evaluate.

6. $-8 + 3$

   6. _____

7. $10 - (-5)$

   7. _____

8. $-8 - 3 + 5$

   8. _____

9. $(4)(-1)(-11)$

   9. _____

10. $\left(-\dfrac{3}{5}\right)\left(-\dfrac{2}{7}\right)$

    10. _____

11. $-12 \div (-4)$

    11. _____

12. $-36 \div \dfrac{9}{4}$

    12. _____

13. $-7 \div 0$

    13. _____

14. $-3(-2 - 4)$

    14. _____

15. $5 - (2 - 3)$

    15. _____

Chapter 1 Test Form F *(cont.)*

16. $\left(-\dfrac{3}{8}\right)^2$

16. _____

17. $3[7-(2^2+5)]$

17. _____

18. $5-24\div 8-2$

18. _____

19. $3^2-4-(4^2-6)$

19. _____

20. -12^2

20. _____

21. $5\dfrac{1}{4}-2\dfrac{7}{10}$

21. _____

Evaluate for the values given.

22. $6q+13$, $q=-\dfrac{2}{3}$

22. _____

23. $x^2-9x-10$, $x=-6$

23. _____

24. $m^2+2mn-3n^2$, $m=2, n=-4$

24. _____

25. Name the property illustrated. $-4(x+5)=-4x-20$

25. _____

Chapter 1 Test Form G

1. A house is purchased for $110,000. The lender requires a down payment of 20%. The remainder of the purchase price will be financed. Find the amount financed.
   (a) $22,000   (b) $107,800   (c) $98,000   (d) $88,000

2. The number of books lent by a local library during a 6-day period was 718, 644, 319, 547, 634, and 814. Find the median number of books lent.
   (a) 644   (b) 639   (c) 612   (d) 635

A group of 400 children were asked about the type of cookie they preferred. The circle graph shows the resulting distribution.

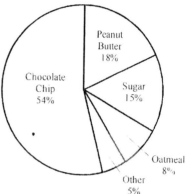

3. How many of the children preferred sugar cookies?
   (a) 60   (b) 72
   (c) 32   (d) 20

4. How many more of the children preferred peanut butter cookies than preferred oatmeal cookies?
   (a) 10   (b) 28
   (c) 12   (d) 40

5. Consider the set of numbers: $\left\{0.5, \sqrt{2}, -\sqrt{3}, -\frac{3}{4}\right\}$. List those that are irrational.

   (a) $\left\{\sqrt{2}, -\sqrt{3}\right\}$ (b) $\left\{-\sqrt{3}, -\frac{3}{4}\right\}$ (c) $\left\{0.5, -\frac{3}{4}\right\}$ (d) $\left\{0.5, \sqrt{2}, -\sqrt{3}, -\frac{3}{4}\right\}$

6. Which statement is true?
   (a) $11 < -8$ (b) $-8 < -11$ (c) $-8 > 0$ (d) $-8 > -11$

7. Which statement is true?
   (a) $|-3| > |-4|$ (b) $-|-3| > -3$ (c) $|-4| > |-3|$ (d) $-|-4| > |-3|$

Evaluate.

8. $14 - 25$
   (a) 11   (b) -11 (c) 39   (d) -39

9. $\dfrac{5}{6} - \dfrac{7}{9}$
   (a) $-\dfrac{2}{3}$ (b) $\dfrac{1}{18}$ (c) $-\dfrac{29}{18}$ (d) $\dfrac{29}{18}$

10. $-8 + 6 - 3$
    (a) -5 (b) 5   (c) 1   (d) -1

11. $(-5)(-2)(-4)$
    (a) -11 (b) 11   (c) 40   (d) -40

12. $\left(-\dfrac{2}{7}\right)\left(\dfrac{3}{7}\right)$
    (a) $\dfrac{1}{7}$ (b) $\dfrac{1}{14}$ (c) $-\dfrac{6}{49}$ (d) $\dfrac{6}{49}$

Chapter 1 Test Form G *(cont.)*

13. $-12 \div \dfrac{4}{3}$

    (a) 1   **(b)** -16 **(c)** 9   **(d)** -9

14. $-6 \div 0$

    (a) 0   **(b)** -6 **(c)** 6   **(d)** undefined

15. $-7(3-4)$

    (a) 49   **(b)** -49 **(c)** 7   **(d)** -7

16. $-3 - (-2 + 5)$

    (a) 0   **(b)** -6 **(c)** -10 **(d)** -4

17. $\left(-\dfrac{2}{3}\right)^3$

    (a) $\dfrac{6}{9}$ **(b)** $-\dfrac{6}{9}$ **(c)** $\dfrac{8}{27}$ **(d)** $-\dfrac{8}{27}$

18. $4[7 - (2^2 + 5)]$

    (a) -8 **(b)** -6 **(c)** 2   **(d)** -64

19. $4 - 24 \div 6 - 3$

    (a) -3 **(b)** -4 **(c)** $-\dfrac{20}{3}$ **(d)** -11

20. $3^2 - 1 - (2^3 - 4)$

    (a) 3   **(b)** -5 **(c)** -4 **(d)** 4

21. Write the expression $mmmnnnnn$ in exponential form.

    (a) $3m \cdot 5n$ **(b)** mn^7 **(c)** $m^3 n^5$ **(d)** $3mn^5$

Evaluate for the values given.

22. $7x - 3,\ x = -2$

    (a) -11 **(b)** -17 **(c)** 2   **(d)** -35

23. $x^2 - 5x + 4,\ x = -2$

    (a) -2 **(b)** 10   **(c)** -10 **(d)** 18

24. $x^2 - 3xy + y^2,\ x = 1,\ y = -3$

    (a) 1   **(b)** -1 **(c)** 19   **(d)** -17

25. Name the property illustrated: $(6 + x) + 7 = (x + 6) + 7$

    (a) commutative, addition   **(b)** associative, addition   **(c)** associative, multiplication   **(d)** distributive

Chapter 1 Test Form H

Name:_____

Date:_____

1. Dillon measures $52\frac{1}{2}$ inches tall. His little sister Holly measures $46\frac{7}{8}$ inches tall. How much taller is Dillon?

   (a) $6\frac{3}{8}$ inches
   (b) $6\frac{5}{8}$ inches
   (c) $5\frac{5}{8}$ inches
   (d) 6 inches

2. The commissions received by a sales representative for the last five months were $2500, $3000, $1700, $3300, and $3000. What is the median commission earned for these five moths?

   (a) $2700
   (b) $2750
   (c) $3000
   (d) $3300

A beer company distributes a $750,000 advertising budget as shown in the circle graph.

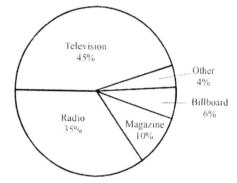

3. How much did the company spend on billboard ads?

   (a) $12,500
   (b) $125,000
   (c) $450,000
   (d) $45,000

4. How much more did the company spend on television ads than on radio ads?

   (a) $75,000
   (b) $337,500
   (c) $10,000
   (d) $262,500

5. Consider the set of numbers: $\left\{-\frac{1}{2}, \sqrt{4}, 0, -3, \sqrt{20}\right\}$

   List those that are irrational numbers.

   (a) $\left\{\sqrt{20}\right\}$
   (b) $\left\{\sqrt{4}, \sqrt{20}\right\}$
   (c) $\left\{0, \sqrt{20}\right\}$
   (d) $\{\ \}$

6. Which statement is true?

   (a) $-7 > 0$
   (b) $-7 < -6$
   (c) $-7 > 6$
   (d) $-7 = 7$

7. Which statement is true?

   (a) $|-5| > |-6|$
   (b) $|-6| < |-5|$
   (c) $|-5| < |5|$
   (d) $|-5| = |5|$

Evaluate.

8. $15 - 32$

   (a) -47
   (b) 47
   (c) -17
   (d) 17

9. $-\frac{1}{9} + \left(-\frac{1}{6}\right)$

   (a) $-\frac{2}{15}$
   (b) $\frac{2}{15}$
   (c) $\frac{5}{18}$
   (d) $-\frac{5}{18}$

10. $-4 + 3 - 1 + 5$

   (a) -3
   (b) 3
   (c) -11
   (d) 13

11. $(-4)(-7)(-5)$

   (a) 16
   (b) -16
   (c) 140
   (d) -140

12. $\left(-\frac{3}{11}\right)\left(-\frac{2}{5}\right)$

   (a) $\frac{6}{55}$
   (b) $-\frac{6}{55}$
   (c) $-\frac{5}{16}$
   (d) $\frac{5}{16}$

Name:_____

13. $-28 \div \dfrac{7}{4}$
   (a) 14   (b) −14   (c) −16   (d) 49

14. $0 \div (-15)$
   (a) 0   (b) 15   (c) −15   (d) undefined

15. $-4(7 - 9)$
   (a) −20   (b) 64   (c) −8   (d) 8

16. $[-2(5) - 4] - 6$
   (a) −17   (b) −20   (c) −5   (d) −7

17. $(-2)^3 - 3^2$
   (a) 0   (b) −12   (c) −17   (d) −1

18. $(3 - 7)^2 + 5$
   (a) 13   (b) 21   (c) −9   (d) −11

19. $16 - 12 \div 6 - 3$
   (a) 11   (b) 12   (c) 10   (d) −1

20. $3^2 - 2^3$
   (a) 0   (b) 1   (c) −2   (d) 3

21. Write $5 \cdot xxyzzz$ in exponential form.
   (a) $5 \cdot 2xy3z$ (b) $5xyz^3$ (c) $5x^2 yz^3$ (d) $5xy^2 z^3$

Evaluate for the values given.

22. $-x + 8,\ x = -2$
   (a) −6   (b) −10   (c) 6   (d) 10

23. $x^2 + x - 13,\ x = 3$
   (a) −4   (b) −1   (c) 4   (d) 1

24. $2x^2 - 3xy + y^2,\ x = 1, y = -1$
   (a) 3   (b) 4   (c) 6   (d) −3

25. Name the property illustrated. $2(x + 3) = 2x + 6$
   (a) distributive   (b) commutative, addition   (c) commutative, multiplication   (d) associative, addition

Chapter 2 Pretest Form A

Name:_____

Date:_____

For questions 1 – 2, use the distributive property to simplify.

1. $8(7x - 4y)$

2. $-(3x + 5y - 9)$

For questions 3 – 7, simplify.

3. $4x - 9x + 3$

4. $7x - 9y - 2x + y$

5. $-6 - (7 + x) - 9x$

6. $\frac{2}{3}m - 2 + m + n - \left(-\frac{3}{5}n\right) + \frac{1}{2}$

7. $-3(2x - 5y) + 2(4x + y)$

For questions 8 – 15, solve.

8. $x + 3 = 18$

9. $\frac{2}{3}x = 36$

10. $6x - 1 = 47$

11. $2x - 5 + 3x = 30$

12. $9x - 3 = x - 35$

13. $3 = 4(x - 1) + 1 - 3x$

14. $0.25(4x - 5) = 1.5x - 0.1(5x + 3)$

15. $\frac{6}{m} = \frac{3}{5}$

16. Solve $z = \frac{x - m}{s}$ for x.

17. The triangles are similar. Find the length of x.

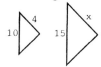

18. A history class consists of 14 males and 11 females. Find the ratio of males to the entire class.

19. Convert 60 ounces to pounds.

20. A worker in an assembly line takes 9 hours to produce 24 parts. At that rate how many parts can she produce in 27 hours?

1. _____

2. _____

3. _____

4. _____

5. _____

6. _____

7. _____

8. _____

9 _____

10. _____

11. _____

12. _____

13. _____

14. _____

15. _____

16. _____

17. _____

18. _____

19. _____

20. _____

Chapter 2 Pretest Form B

Name:_____

Date:_____

For questions 1 – 2, use the distributive property to simplify.

1. $9(6 - 4x)$

2. $-2(3a - 2b + 7)$

1. _____

2. _____

For questions 3 – 7, simplify.

3. $2x - 9x + 10$

4. $\dfrac{3}{4}x + 5 - \dfrac{2}{3}x - \dfrac{3}{7}$

5. $8 - (5 + x) + 3x$

6. $9x + 7y + 7x + y - 8$

7. $-4(3x - 2y) + 5(2x + y)$

3. _____

4. _____

5. _____

6. _____

7. _____

For questions 8 – 15, solve.

8. $20 = m - 3$

9. $\dfrac{4}{5}x = 100$

10. $5y - 4 = 6$

11. $9x + 6 - 3x = 30$

12. $3x - 1 = x + 9$

13. $0.7(2k - 3) = 1.3 + 0.4k$

14. $6(x + 6) = -1 + 6x$

15. $\dfrac{15}{x} = \dfrac{25}{-40}$

8. _____

9. _____

10. _____

11. _____

12. _____

13. _____

14. _____

15. _____

16. Solve $2x + 3y = 12$ for y. Write the answer in $y = mx + b$ form.

16. _____

17. The triangles are similar. Find the length of x.

17. _____

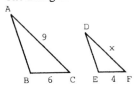

18. Find the ratio of 10 dimes to 3 quarters.

18. _____

19. Convert 20 inches to feet.

19. _____

20. If 8 gallons of insecticide can treat 2 acres of land, how many gallons of insecticide are needed to treat 28 acres of land?

20. _____

Mini-Lecture 2.1
Combining Like Terms

Learning Objectives:

1. Identify terms.
2. Identify like terms.
3. Combine like terms.
4. Use the distributive property.
5. Remove parentheses when they are preceded by a plus or minus sign.
6. Simplify an expression.
7. Key Vocabulary: *variable, algebraic expression, term, numerical coefficient, constant, combine like terms, simplifying an expression*

Examples:

1. Identify the terms.
   a) $3x + 2 - 7y$
   b) $-2m + 5n + 4$
   c) $3a + b - 8$

2. Identify any like terms.
   a) $2a - 5 + 9 - 7a$
   b) $3x^2 + 2x + 4x^2$
   c) $3 + n - 5m + 2$

3. Combine like terms.
   a) $\frac{2}{3}w + \frac{3}{4}w$
   b) $-3x^2 + 5y - 4x^2 - 12 - 3y + 10$

4. Use the distributive property to remove parentheses.
   a) $3(m + 7)$
   b) $-5(s - 4)$

5. Remove the parentheses.
   a) $(4x + 2y - 8)$
   b) $-(3a - 2b + 7)$

6. Simplify.
   a) $-\left(\frac{3}{2}x - \frac{1}{4}\right) + 5x$
   b) $4(2x - 3) - (5y + 4) + 8y$

Teaching Notes:

- When listing terms, students may need to be reminded to include the minus sign for terms with a negative coefficient.
- When using the distributive property, some students may forget to distribute minus signs across all terms inside the parentheses.
- Some students may not understand that terms can involve expressions (e.g. $3(x + 1)$ is one term).

Mini-Lecture 2.2
The Addition Property of Equality

Learning Objectives:

1. Identify linear equations.
2. Check solutions to equations.
3. Identify equivalent equations.
4. Use the addition property to solve equations.
5. Solve equations by doing some steps mentally.
6. Key vocabulary: *equation, linear equation, solution to an equation, solve an equation, equivalent equations, addition property of equality*

Examples:

1. Identify whether each equation is linear.

   a) $3x + 2 = 5x - 1$ b) $2x + \dfrac{1}{x} = 7$ c) $3\sqrt{x} + 4 = 9$

2. Determine whether the given value is a solution to the equation $-3(1-x) - 2x = 7$.

   a) 5 b) 10 c) 15

3. Identify whether the pairs of equations are equivalent.

   a) $2x + 5 = -3$; $4x + 5 = -6$ b) $x + 2 = 1$; $3x + 6 = 3$ c) $4x - 8 = 2$; $4x = 10$

4. Solve the equations:

   a) $n - 3 = -15$ b) $w + 5 = 9$ c) $h - \dfrac{2}{3} = \dfrac{1}{6}$

5. Solve the equations mentally:

   a) $-2 = w - 9$ b) $x + 15 = 27$

Teaching Notes:

- Some students may need a short review on combining like terms.
- Emphasize that checking if a value is a solution is not the same as solving an equation.
- Remind students to check their solutions in the original equation.
- When obtaining equivalent equations, remind students that "*what you do to one side, you must do to the other*".
- Explain to students that solving an equation for a variable just means to isolate the variable on one side of the equation.

Answers: *1a) linear; 1b) not linear; 1c) not linear; 2a) no; 2b) yes; 2c) no; 3a) not equivalent; 3b) equivalent; 3c) equivalent; 4a) $n = -12$; 4b) $w = 4$; 4c) $h = \dfrac{5}{6}$; 5a) $w = 7$; 5b) $x = 12$*

44

Mini-Lecture 2.3
The Multiplication Property of Equality

Learning Objectives:

1. Identify reciprocals.
2. Use the multiplication property to solve equations.
3. Solve equations of the form $-x = a$.
4. Do some steps mentally when solving equations.
5. Key vocabulary: *reciprocal, multiplication property of equality*

Examples:

1. Identify which pairs of numbers are reciprocals.

   a) $\dfrac{2}{3}, -\dfrac{2}{3}$ b) $\dfrac{4}{5}, \dfrac{5}{4}$ c) $-\dfrac{4}{5}, -\dfrac{5}{4}$ d) $3, \dfrac{1}{3}$

2. Solve the equations.

   a) $3y = 27$ b) $\dfrac{t}{6} = 4$ c) $\dfrac{2}{7} w = 6$

   d) $2p = 16$ e) $0.15x = 6$ f) $-7x = -63$

3. Solve the equations.

   a) $-r = 17$ b) $-q = -10$

4. Solve the equations doing at least one step mentally.

   a) $\dfrac{1}{7} h = 12$ b) $-5a = -80$

Teaching Notes:

- Remind students that the product of a number and it's reciprocal is 1, and that 0 has no reciprocal.
- Point out that when the multiplication property is applied, like the addition property, an equivalent equation results.
- Remind students that multiplying both sides of an equation by the reciprocal of a is the same as dividing both sides of the equation by a.
- Emphasize that students should always check their solutions and the check should be done with the original equation.

Answers: 1a) no; 1b) yes; 1c) yes; 1d) yes; 2a) $y = 9$; 2b) $t = 24$; 2c) $w = 21$, 2d) $p = 8$; 2e) $x = 40$; 2f) $x = 9$; 3a) $r = -17$; 3b) $q = 10$; 4a) $h = 84$; 4b) $a = 16$

Mini-Lecture 2.4
Solving Linear Equations with a Variable on Only One Side of the Equation

Learning Objectives:

1. Solve linear equations with a variable on only one side of the equal sign.
2. Solve equations containing decimal numbers or fractions.
3. Key vocabulary: *isolating a variable, addition property, multiplication property*

Examples:

1. Solve the equations.

   a) $4n - 7 = 5$

   b) $2b + 9 = 5$

   c) $3x + 4 = 9$

   d) $-6k + 2 = 7$

   e) $2x + 2 - 3 = 5$

   f) $3x - 5 + 2x = 12$

   g) $-2(x + 3) + 4x = 8$

   h) $10x - (5x + 7) = 1$

2. Solve the equations.

   a) $1.3x + 4.2 - 0.5x = 2.6$

   b) $4.7c - 3.1 - 2.1c = 3.4$

   c) $0.4w + 1.4(w - 3) = -5.1$

   d) $\frac{2}{3}(x - 2) = 5$

   e) $\frac{v}{4} - 5v = 11$

   f) $\frac{5}{4}p + \frac{3}{10}p = \frac{7}{5}$

Teaching Notes:

- Encourage students to review the addition property of equality and the multiplication property of equality.
- Common Error: When multiplying or dividing both sides of an equation by a number, some students do not distribute the number over each term on a side.
- When solving multi-step problems, remind students that they may first need to isolate the variable *terms* using the addition property, then isolate the variable using the multiplication property.
- Remind students that checks should always be made in the original equation.
- Point out that fractions can initially be cleared from an equation by multiplying both sides of the equation by the LCD.

Answers: 1a) $n = 3$; 1b) $b = -2$; 1c) $x = \frac{5}{3}$; 1d) $k = -\frac{5}{6}$; 1e) $x = 3$; 1f) $x = \frac{17}{5}$; 1g) $x = 7$; 1h) $x = \frac{8}{5}$;

2a) $x = -2$; 2b) $c = 2.5$; 2c) $w = -0.5$; 2d) $x = \frac{19}{2}$; 2e) $v = -\frac{44}{19}$; 2f) $p = \frac{28}{31}$

Mini-Lecture 2.5
Solving Linear Equations with the Variable on Both Sides of the Equation

Learning Objectives:

1. Solve equations with the variable on both sides of the equation.
2. Solve equations containing decimal numbers or fractions.
3. Identify identities and contradictions.
4. Key vocabulary: *conditional equations, identity, contradiction*

Examples:

1. Solve the equations.
   a) $7x + 5 = 2x - 10$

   b) $4x - 2x + 9 = 3x + 5x - 15$

   c) $3 - 10a + 5 = 2a + 6 - 4a$

   d) $2(w + 1) = 3w + 7$

   e) $16 - 3(2d + 5) = 4d - 29$

   f) $3r - (6 - r) = 4 + 5(r - 2)$

2. Solve the equations.
   a) $3.75 + 2.50p = 5.50p + 1.50$

   b) $\frac{5}{9}k = \frac{4}{3}k + \frac{7}{6}$

   c) $\frac{1}{5}x + \frac{2}{3} = \frac{3}{2}x - \frac{5}{6}$

   d) $\frac{2}{5}(10n + 2) = \frac{1}{3}(9n + 5) - 1$

3. Solve the equations.
   a) $3x - 8 + 2x = 7x + 5 - 2x$

   b) $3(x + 1) - 4 = -2(2 - x) + 3 + x$

   c) $0.25x - 1.2(5x + 22) = 1.75x - 2.50(3x + 1)$

   d) $\frac{2}{3}x + \frac{1}{9}(3x + 27) = 4x - 3(x - 1)$

Teaching Notes:

- Encourage students to simplify both sides of the equation before using the addition property or multiplication property.
- Remind students that addition and subtraction are inverse operations, as are multiplication and division.
- Remind students to always check their solutions in the original equation. If the equation is an identity, have students check at least two different solutions.
- When isolating the variable, let students know that it does matter on which side the variable is isolated.

Answers: 1a) $x = -3$; 1b) $x = 4$; 1c) $a = \frac{1}{4}$; 1d) $w = -5$; 1e) $d = 3$; 1f) $r = 0$; 2a) $p = 0.75$; 2b) $k = -\frac{3}{2}$;
2c) $x = \frac{15}{13}$; 2d) $n = -\frac{2}{15}$; 3a) *No solution;* 3b) *All real numbers;* 3c) *No solution;* 3d) *All real numbers*

Mini-Lecture 2.6
Formulas

Learning Objectives:

1. Use the simple interest formula and the distance formula.
2. Use geometric formulas.
3. Solve for a variable in a formula.
4. Key vocabulary: *formula, evaluate, simple interest, perimeter, area, circumference, diameter, radius, volume, distance formula*

Examples:

1. Solve.

   a) How much simple interest will be owed on a $25,000 5-year loan at 7.5% simple interest?

   b) What is the principal on a 4-year 6% loan if the simple interest was $2040?

   c) Find the distance traveled if a van travels 4.5 hours at 65 mph.

   d) How fast is a car traveling if it travels 84 miles in 1.5 hours?

2. Solve.

   a) A rectangular tablecloth is 3 yards long and 2 yards wide. Find the perimeter and area.

   b) A rectangular pillowcase has an area of 600 square inches. If the length is 30 inches, what is the width?

   c) A swimming pool in the shape of a right circular cylinder is 4 feet deep and has a diameter of 12 feet. What is the volume of the pool?

   d) Find the circumference and area of a circular sign that has a radius of 70 centimeters.

3. Solve for the indicated variable.

   a) Solve $3x - 2y = 12$ for y.

   b) Solve $t = \dfrac{(2a + nd - d)n}{2}$ for a.

Teaching Notes:

- Remind students to pay attention to units. Sometimes a unit conversion will be necessary before using a formula. Encourage students to examine the variables in a formula to help determine what type of units will be in the final answer.
- When working with formulas, have students begin with the general form and then substitute in known quantities.
- Encourage students to draw a picture, when appropriate, labeling all known and unknown quantities.

Answers: 1a) $9375; 1b) $8500; 1c) 292.5 miles; 1d) 56 miles per hour; 2a) 10 yards, 6 square yards; 2b) 20 inches; 2c) $144\pi \approx 452.39$ cubic feet; 2d) 140π cm, $4900\pi = 15,393.804$ square centimeters;

3a) $y = \dfrac{3}{2}x - 6$; 3b) $a = \dfrac{2t - dn(n-1)}{2n}$

Mini-Lecture 2.7
Ratios and Proportions

Learning Objectives:

1. Understand ratios.
2. Solve proportions using cross-multiplication.
3. Solve applications.
4. Use proportions to change units.
5. Use proportions to solve problems involving similar figures.
6. Key vocabulary: *ratio, terms of the ratio, proportion, extremes, means, cross-multiplication*

Examples:

1. In a survey of 25 algebra students, 11 said they were more visual learners, 8 said they were more auditory learners, and 6 were uncertain. Find
   a) the ratio of the number who were more visual learners to the number who were more auditory learners.

   b) the ratio of the number who were more auditory learners to the total number surveyed.

2. Solve the following by cross multiplying.
   a) $\dfrac{x}{3} = \dfrac{18}{27}$

   b) $\dfrac{-4}{17} = \dfrac{28}{m}$

3. a) Shawn decides to mow lawns to earn some extra money. If it takes him 25 minutes to mow 3000 square feet, how long will it take him to mow 4200 square feet?

   b) A nutritionist recommends her patients consume 25 grams of dietary fiber each day. If 1 ounce of a bran cereal contains 8 grams of dietary fiber, how many ounces of the cereal must be eaten to meet the recommendation?

4. a) There are 2.54 centimeters in an inch. How many inches are in 635 centimeters?

   b) At a downtown bank, Payton is told that she can exchange 79 Euros for $100 U.S. How much will she receive in Euros if she converts $600 U.S.?

5. The given figures are similar. Find the length of the side indicated by the *x*.

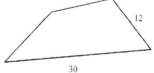

Teaching Notes:

- Remind students that the two ratios in a proportion must have the same units.
- Some students have trouble distinguishing ratios and proportions. Emphasize that a ratio is a comparison of two numbers while a proportion is an equation with a ratio on each side.
- Point out that a *rate* is a ratio that tells how long it takes to perform a task.
- Some students may need to review section 1.2 on problem solving.

Answers: 1a) 11:8 ; 1b) 8:25 ; 2a) x = 2 ; 2b) m = −119 ; 3a) 35 minutes; 3b) 3.125 ounces; 4a) 250 inches; 4b) 474 Euros; 5) x = 5.6

Additional Exercises 2.1

1. List the terms in the expression $4x^3 - 8x^2 - 3x + 1$.

2. List the terms in the expression $-2w^5 + 3w^4 - w^3 - 1$.

3. Determine if x and $-6x^2$ are like terms. Answer yes or no.

Combine like terms when possible.

4. $7x + 2y - 3x + 4y$

5. $3x - 7y + x - 4y - 4$

6. $22p - 17 - 30p + 9$

7. $-8y + (-4) + 5x + (-2y) - 6x$

8. $-14.37 + 2.9x + 5.1x + 12.3$

9. $3p^2 + 4p - 8p^2 + p - 7$

Use the distributive property to remove parentheses.

10. $5(2x + 3y)$

11. $0.4(6 - 1.2k)$

12. $-(3x + 3y - 2)$

13. $3\left(\frac{1}{6}x - 2y + 4\right)$

14. $2.5(4x - 1.2y - 3.8)$

Simplify when possible.

15. $-2(3x + 4y - 1)$

16. $-5(4x - y - 2)$

17. $-4 - (4 + x) - 5x$

18. $5(2a - b) + 4b - 2(3a + 2b)$

19. $-0.4(8x - 3) + 5x$

20. $\frac{1}{3}(x - 5) - \frac{1}{2}(x - 6)$

1. _____

2. _____

3. _____

4. _____

5. _____

6. _____

7. _____

8. _____

9. _____

9. _____

10. _____

11. _____

13. _____

14. _____

15. _____

16. _____

17. _____

18. _____

19. _____

20. _____

Additional Exercises 2.2

1. Is $3x - 5 = -7$ a linear equation? Answer yes or no.

2. Is $8x^2 + 9x + 4 = 13$ a linear equation? Answer yes or no.

3. Is $-\dfrac{1}{6}$ a solution of the equation $6x + 6 = 5$? Y

4. Is 3 a solution of the equation $3x + 6 = -3$? N

5. Is -5 a solution of the equation $8 = 3 - x$? Y

6. Does the number -8 satisfy the equation $1 = 9 - x$? N

Solve each equation.

7. $x + 6 = 8$

8. $28 = m - 7$

9. $58 = t + 23$

10. $x + 5 = 2$

11. $x - (-9) = -33$

12. $x - 22 = -15$

13. $-5 = m - 6$

14. $-25 = 16 + c$

15. $x - 9 = -9$

16. $-42.8 + x = 23.65$

17. $18 + x = -15$

18. $3.14 + x = 2.5$

19. $x - 6.75 = -4.25$

20. $x - 243 = 165$

1. _____

2. _____

3. _____

4. _____

5. _____

6. _____

7. _____

8. _____

9. _____

10. _____

11. _____

12. _____

13. _____

14. _____

15. _____

16. _____

17. _____

18. _____

19. _____

20. _____

Additional Exercises 2.3

Solve each equation.

1. $-\dfrac{x}{4} = 12$

2. $-x = -12$

3. $-24 = -4m$

4. $2x = -14$

5. $\dfrac{1}{7}x = -8$

6. $\dfrac{1}{7}x = 2$

7. $-\dfrac{x}{2} = 4$

8. $3x = \dfrac{1}{3}$

9. $3x = 15$

10. $4x = -32$

11. $\dfrac{1}{3}x = 5$

12. $\dfrac{1}{9}x = 5$

13. $-\dfrac{5}{3}x = -15$

14. $-\dfrac{6}{5}x = -30$

15. $\dfrac{4}{3}p = 28$

16. $\dfrac{1}{3}x = 4$

17. $\dfrac{7}{6}x = 168$

18. $-2.1q = 30.03$

19. $-\dfrac{9}{5}x = -45$

20. $-\dfrac{4}{9}x = -36$

1. _____

2. _____

3. _____

4. _____

5. _____

6. _____

7. _____

8. _____

9. _____

10. _____

11. _____

12. _____

13. _____

14. _____

15. _____

16. _____

17. _____

18. _____

19. _____

20. _____

Additional Exercises 2.4

Solve each equation.

1. $4x - 4 = -40$

2. $9x - 7 = -34$

3. $\dfrac{7}{8}y - 6 = 8$

4. $10 - x = 6$

5. $-2x - 4x = 1$

6. $-9x + 12x = -9$

7. $\dfrac{x}{3} - \dfrac{x}{5} = 2$

8. $\dfrac{x}{7} - \dfrac{x}{9} = 2$

9. $\dfrac{3+z}{4} = -5$

10. $\dfrac{1}{4} = \dfrac{d-9}{2}$

11. $\dfrac{5}{9}y - 4 = 6$

12. $x + 0.4x = 3.5$

13. $5(x - 3) = 45$

14. $-3(x + 7) = 9$

15. $-0.4(0.5x - 8) = 12$

16. $8 = 2(x - 5) + 6x$

17. $2 = 7(x + 4) + 9x$

18. $1 = 3(x - 2) + 3 - 2x$

19. $\dfrac{2}{3}n + \dfrac{1}{4} = \dfrac{2}{5}$

20. $3.65 - 7.4x + 1.12 = 21.76$

1. _____

2. _____

3. _____

4. _____

5. _____

6. _____

7. _____

8. _____

9. _____

10. _____

11. _____

12. _____

13. _____

14. _____

15. _____

16. _____

17. _____

18. _____

19. _____

20. _____

Additional Exercises 2.5

Solve each equation.

1. $5x = 3x - 12$

2. $-8x = 2x + 10$

3. $3x + 1 = x - 4$

4. $-2c - 4 = c + 11$

5. $5x - 3 = x - 4$

6. $4x - 5 = x - 17$

7. $x + 9 = 2(5x - 4)$

8. $2(x - 4) = -6 + 2x$

9. $\dfrac{5}{4} = \dfrac{3m + 1}{8}$

10. $4(x + 8) = 32 + 4x$

11. $x + 3 = 3(4x - 3)$

12. $2(x + 3) = -5 + 2x$

13. $3(x - 6) = -10 + 3x$

14. $\dfrac{m}{3} - 4 = 3m$

15. $2x + 3 - 9x = 3x + 3 - 10x$

16. $3x + 3 + 8x = 5x + 4 + 6x$

17. $-8x + 3 - 2x = -6x + 3 - 4x$

18. $10x + 3 + 10x = 13x - 3 + 7x$

19. $6 + 3x = 5(x - 1) - 3(x - 2)$

20. $\dfrac{1}{5}(30 - 10x) = \dfrac{2}{3}(4x + 2)$

1. _____

2. _____

3. _____

4. _____

5. _____

6. _____

7. _____

8. _____

9. _____

10. _____

11. _____

12. _____

13. _____

14. _____

15. _____

16. _____

17. _____

18. _____

19. _____

20. _____

Additional Exercises 2.6

For questions 1–6, use the formula to find the value of the variable indicated. Round all answers to hundredths.

1. $C = 2\pi r$; (circumference of a circle), find C when $\pi = 3.14$ and $r = 9$ inches.

   1. _____

2. $A = \pi r^2$; (area of a circle), find A when $\pi = 3.14$ and $r = 2.5$ feet.

   2. _____

3. $P = 2l + 2w$; (perimeter of a rectangle), find P when $l = 5.9$ and $w = 4.1$.

   3. _____

4. $P = 2l + 2w$; (perimeter of a rectangle), find P when $l = 4.1$ and $w = 1.7$.

   4. _____

5. $A = \frac{1}{2}bh$; (area of a triangle), find h when $A = 20$ and $b = 5$.

   5. _____

6. $A = \frac{1}{2}bh$; (area of a triangle), find h when $A = 24$ and $b = 12$.

   6. _____

7. Solve for A in $B = \frac{2}{5}(A - 11)$.

   7. _____

8. Solve $5x - 2y = 4$, for y. Write the answer in $y = mx + b$ form.

   8. _____

9. Solve for F in $C = \frac{5}{9}(F - 32)$.

   9. _____

10. Solve $y - 4 = 2(x + 3)$, for y. Write the answer in $y = mx + b$ form.

    10. _____

11. Solve for t in $I = Prt$.

    11. _____

12. Solve for t in the equation $A = 6s^2 t$.

    12. _____

13. Solve $C = 2\pi r$ for r.

    13. _____

14. A package in the shape of a rectangular solid is to be wrapped for shipping. If the length is 8 inches, the height is 5 inches and the width is 4 inches, how much paper is needed to wrap the package?

    14. _____

15. A can of soup in the shape of a right circular cylinder is 5 inches tall and has a radius of 1.5 inches. How much soup will the can hold?

    15. _____

16. A circular swimming pool with a 16-foot diameter is surrounded by a concrete path that is 4 feet wide. What is the area of the path?

    16. _____

17. First State Bank charges 8% simple interest on personal loans. What is the total interest on a $2300 loan over three years?

    17. _____

18. Find the interest earned on $900 invested at 5.5% for 5 years.

    18. _____

19. The formula for finding the simple interest (I) on the loan is $I = PRT$. How much interest will Bill pay on his car loan if he finances $17,000 ($P$) at a 13% simple interest rate (R) for 4 years.

    19. _____

20. Find the interest earned on $2000 invested at 8.5% for 3 months.

    20. _____

Additional Exercises 2.7

Name:_____

Date:_____

For problems 1–6, find the ratio. Write each ratio in lowest terms.

1. A history class consists of 17 males and 15 females. Find the ratio of males to the entire class.

2. Find the ratio of 3 quarters to 10 nickels.

3. Find the ratio of 5 days to one week.

4. A science class consists of 14 males and 13 females. Find the ratio of females to males.

5. Find the ratio of 4 hours to 40 minutes.

6. Find the ratio of 6 days to one week.

1. _____

2. _____

3. _____

4. _____

5. _____

6. _____

Solve for the variable by cross-multiplying.

7. $\dfrac{15}{j} = \dfrac{5}{4}$

8. $\dfrac{10}{x} = \dfrac{10}{-24}$

9. $\dfrac{12}{j} = \dfrac{2}{5}$

10. $\dfrac{-3}{5} = \dfrac{x}{15}$

7. _____

8. _____

9. _____

10. _____

The figures below are similar. For each pair, find the length of the side indicated with an x.

11.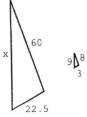

12.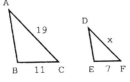

13.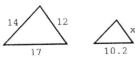

14.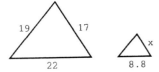

11. _____

12. _____

13. _____

14. _____

56

Additional Exercises 2.7 *(cont.)*

15. A worker in an assembly line takes 6 hours to produce 26 parts. At that rate how many parts can she produce in 30 hours?

15. _____

16. If 20 gallons of insecticide can treat 5 acres of land, how many gallons of insecticide are needed to treat 24 acres of land?

16. _____

17. In 1967 the world population was 3.5 billion and in 2006 it had climbed to 6.5 billion. What is the ratio of the world population in 1967 to the world population in 2006?

17. _____

18. The protein RDA for males is 64 grams per day. Three ounces of a certain product provide 4 grams of protein. How many ounces of the product are needed to provide 64 grams of protein?

18. _____

19. Convert 9 inches to feet.

19. _____

20. Convert 244 ounces to pounds.

20. _____

Chapter 2 Test Form A

Name:_____

Date:_____

For questions 1–2, use the distributive property to simplify.

1. $-3(4-8x)$

2. $-(-2x-5y+4)$

For questions 3–7, simplify.

3. $-x+3-2x+3$

4. $4+2x-7+x$

5. $2(x-2)-4y+4$

6. $\dfrac{2}{5}x+\dfrac{1}{3}-\left(x+\dfrac{3}{7}\right)$

7. $5(x-4)+3(2x+1)$

For questions 8–13, solve.

8. $10x-6=14$

9. $3+4x-x=12$

10. $-2x-4=2(x-8)$

11. $5x+4(3-2x)=3(1-x)$

12. $\dfrac{2}{14}=\dfrac{1}{7x}$

13. $-7(-x+3)-2x=5(x-4)-1$

14. Solve $P=2l+2w$ for w.

15. Solve $-12x+2y=-21$ for y. Write the answer in $y=mx+b$ form.

16. $V=\dfrac{1}{3}\pi r^2 h$; Find V for $r=3$ inches and $h=10$ inches. Assume $\pi=3.14$.

17. Determine the ratio of 20 minutes to 2 hours. Write the ratio in lowest terms.

1. _____

2. _____

3. _____

4. _____

5. _____

6. _____

7. _____

8. _____

9. _____

10. _____

11. _____

12. _____

13. _____

14. _____

15. _____

16. _____

17. _____

Mini-Lecture 2.2
The Addition Property of Equality

Learning Objectives:

1. Identify linear equations.
2. Check solutions to equations.
3. Identify equivalent equations.
4. Use the addition property to solve equations.
5. Solve equations by doing some steps mentally.
6. Key vocabulary: *equation, linear equation, solution to an equation, solve an equation, equivalent equations, addition property of equality*

Examples:

1. Identify whether each equation is linear.

   a) $3x + 2 = 5x - 1$ b) $2x + \dfrac{1}{x} = 7$ c) $3\sqrt{x} + 4 = 9$

2. Determine whether the given value is a solution to the equation $-3(1 - x) - 2x = 7$.

   a) 5 b) 10 c) 15

3. Identify whether the pairs of equations are equivalent.

   a) $2x + 5 = -3$; $4x + 5 = -6$ b) $x + 2 = 1$; $3x + 6 = 3$ c) $4x - 8 = 2$; $4x = 10$

4. Solve the equations:

   a) $n - 3 = -15$ b) $w + 5 = 9$ c) $h - \dfrac{2}{3} = \dfrac{1}{6}$

5. Solve the equations mentally:

   a) $-2 = w - 9$ b) $x + 15 = 27$

Teaching Notes:

- Some students may need a short review on combining like terms.
- Emphasize that checking if a value is a solution is not the same as solving an equation.
- Remind students to check their solutions in the original equation.
- When obtaining equivalent equations, remind students that "*what you do to one side, you must do to the other*".
- Explain to students that solving an equation for a variable just means to isolate the variable on one side of the equation.

Answers: *1a) linear; 1b) not linear; 1c) not linear; 2a) no; 2b) yes; 2c) no; 3a) not equivalent; 3b) equivalent; 3c) equivalent; 4a) $n = -12$; 4b) $w = 4$; 4c) $h = \dfrac{5}{6}$; 5a) $w = 7$; 5b) $x = 12$*

Mini-Lecture 2.3
The Multiplication Property of Equality

Learning Objectives:

1. Identify reciprocals.
2. Use the multiplication property to solve equations.
3. Solve equations of the form $-x = a$.
4. Do some steps mentally when solving equations.
5. Key vocabulary: *reciprocal, multiplication property of equality*

Examples:

1. Identify which pairs of numbers are reciprocals.

   a) $\dfrac{2}{3}, -\dfrac{2}{3}$
   b) $\dfrac{4}{5}, \dfrac{5}{4}$
   c) $-\dfrac{4}{5}, -\dfrac{5}{4}$
   d) $3, \dfrac{1}{3}$

2. Solve the equations.

   a) $3y = 27$
   b) $\dfrac{t}{6} = 4$
   c) $\dfrac{2}{7}w = 6$

   d) $2p = 16$
   e) $0.15x = 6$
   f) $-7x = -63$

3. Solve the equations.

   a) $-r = 17$
   b) $-q = -10$

4. Solve the equations doing at least one step mentally.

   a) $\dfrac{1}{7}h = 12$
   b) $-5a = -80$

Teaching Notes:

- Remind students that the product of a number and it's reciprocal is 1, and that 0 has no reciprocal.
- Point out that when the multiplication property is applied, like the addition property, an equivalent equation results.
- Remind students that multiplying both sides of an equation by the reciprocal of a is the same as dividing both sides of the equation by a.
- Emphasize that students should always check their solutions and the check should be done with the original equation.

Answers: 1a) no; 1b) yes; 1c) yes; 1d) yes; 2a) $y = 9$; 2b) $t = 24$; 2c) $w = 21$; 2d) $p = 8$; 2e) $x = 40$; 2f) $x = 9$; 3a) $r = -17$; 3b) $q = 10$; 4a) $h = 84$; 4b) $a = 16$

Name:_____

18. The following figures are similar figures. Find the length of side *x*.

18. _____

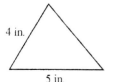

4 in.   *x*

5 in.   3 in.

19. If a batch of 60 cookies requires 2 eggs, how many eggs are required to make 330 cookies?

19. _____

20. If 10 pounds of seed can cover 3000 square feet, how much seed will a 5700 square foot lawn require?

20. _____

Chapter 2 Test Form B

Name:_____

Date:_____

For questions 1–2, use the distributive property to simplify.

1. $-5(-3x+16)$

2. $-2(2x-5+3y)$

For questions 3–7, simplify.

3. $-(x-2y)+4y-3$

4. $4-(y-x)+3x$

5. $-8x-1+x+6$

6. $7x+5-2x-2$

7. $3(3x-7)-4(2x-5)$

For questions 8–13, solve.

8. $4+2x=8$

9. $-2(x+4)+5x=3x-8$

10. $-3x-6=3(x+2)$

11. $-5+8x=3(x-1)$

12. $\dfrac{1}{x}=\dfrac{5}{45}$

13. $7x-8-4(2x-3)=5-x$

14. Solve $V=\dfrac{1}{3}\pi r^2 h$ for h.

15. Solve $5x+4y=22$ for y. Write your answer in $y=mx+b$ form.

16. $P=2l+2w$; Find P for $l=2.5$ yards and $w=3.7$ yards.

17. Determine the ratio of 9 inches to 3 feet. Write the ratio in lowest terms.

1. _____

2. _____

3. _____

4. _____

5. _____

6. _____

7. _____

8. _____

9. _____

10. _____

11. _____

12. _____

13. _____

14. _____

15. _____

16. _____

17. _____

Name:_____

18. The following figures are similar figures. Find the length of side *x*.

    18. _____

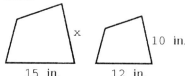

19. If 12 pounds of seed can cover 3500 square feet, how much seed will a 8750 square foot lawn require?

    19. _____

20. If 2 gallons of paint covers 850 square feet, how many gallons of paint will Logan need to paint 3400 square feet of wall surface?

    20. _____

Chapter 2 Test Form C

Name:_____

Date:_____

For questions 1–2, use the distributive property to simplify.

1. $-4(10-3x)$

2. $-3(4+3x-3y)$

For questions 3–7, simplify.

3. $5+8x-2x+1$

4. $-1+3(x-y)+7x$

5. $7(2x-y)+3x-2y$

6. $\dfrac{1}{2}x+\dfrac{2}{3}-\left(\dfrac{1}{4}x-\dfrac{1}{6}\right)$

7. $8(2x-3)-5(3x+2)$

For questions 8–13, solve.

8. $3+3x=-6$

9. $3+2x-x=5$

10. $2(4x-1)=5x+1+3(x-1)$

11. $-(1+2x)=4x-2(1-x)$

12. $-6+8x=3(x-1)+5x$

13. $\dfrac{7}{9}=\dfrac{x}{27}$

14. Solve $x=zs+m$ for z.

15. Solve $4x+16y=11$ for y. Write the answer in $y=mx+b$ form.

16. $I=prt$; Find r when $I=\$234$, $p=\$1300$, and $t=4$ years.

17. Determine the ratio of 12 ounces to 3 pounds. Write the ratio in lowest terms.

1. _____

2. _____

3. _____

4. _____

5. _____

6. _____

7. _____

8. _____

9. _____

10. _____

11. _____

12. _____

13. _____

14. _____

15. _____

16. _____

17. _____

Name:_____

18. The following figures are similar figures. Find the length of side *x*.

18. _____

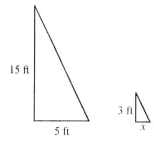

15 ft

5 ft

3 ft

x

19. If 3 gallons of paint covers 950 square feet, how many gallons of paint will Jennifer need to paint 3800 square feet of wall surface?

19._____

20. If Shawn's car uses 4 gallons of gasoline to travel 90 miles, how many miles can he travel with 26 gallons of gasoline?

20. _____

.

Chapter 2 Test Form D

For questions 1–2, use the distributive property to simplify.

1. $-(-2x-3)$

2. $6(10x-7+2y)$

For questions 3–7, simplify.

3. $-10x+2-x-13$

4. $4-6x+2-3x$

5. $-(2+y)-3x+4x$

6. $\dfrac{2}{3}x+\dfrac{1}{2}-\left(\dfrac{1}{6}x-\dfrac{1}{4}\right)$

7. $3(4x+7)-4(2x-5)$

For questions 8–13, solve.

8. $2+x=-9$

9. $4(x+4)=2x+17-(1-2x)$

10. $x-1+3x=7$

11. $-(3-2x)=4x-2(1+x)$

12. $2(1-3x)=3x-16$

13. $\dfrac{x}{3}=\dfrac{4}{6}$

14. Solve $A=2\pi r^2+2\pi rh$ for h.

15. Solve $1.5x-3y=4.2$ for y. Write the answer in $y=mx+b$ form.

16. $A=\dfrac{1}{2}bh$; find b when $A=16.8$ cm^2 and $h=3$ cm .

17. Determine the ratio of 45 minutes to 2 hours. Write the ratio in lowest terms.

1. _____

2. _____

3. _____

4. _____

5. _____

6. _____

7. _____

8. _____

9. _____

10. _____

11. _____

12. _____

13. _____

14. _____

15. _____

16. _____

17. _____

Name:_____

18. The following figures are similar figures. Find the length of side *x*.

18. _____

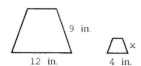

9 in.
12 in.
x
4 in.

19. If a car uses 4 gallons of gasoline to travel 106 miles, how many gallons of gasoline is needed to travel 400 miles?

19. _____

20. If 2 cups of hot chocolate uses 5 tablespoons of chocolate mix, how many cups of hot chocolate can you make with 15 tablespoons of chocolate mix?

20. _____

Chapter 2 Test Form E

Name:_____

Date:_____

For questions 1–2, use the distributive property to simplify.

1. $-2(3x+5)$

2. $-4(x-7-3y)$

For questions 3–7, simplify.

3. $-3+7x-x+5$

4. $3+2(x-2y)+x$

5. $4+2(x-y)-5y$

6. $\dfrac{3}{4}x+\dfrac{5}{6}-\left(\dfrac{1}{8}x+\dfrac{5}{8}\right)$

7. $4(2x-5)-3(x+4)$

For questions 8–13, solve.

8. $1-3x=7$

9. $1+4x-2x=9$

10. $2(x-3)=x-3(x-1)$

11. $4x-2=3(x+3)$

12. $2(x-3)=2+2(x-4)$

13. $\dfrac{2}{9}=\dfrac{10}{x}$

14. Solve $C=2\pi r$ for r.

15. Solve $-5x-10y=28$ for y. Write the answer in $y=mx+b$ form.

16. $P=2l+2w$; Find w if $P=20$ feet and $l=7$ feet.

17. Determine the ratio of 8 inches to 2 feet. Write the ratio in lowest terms.

1. _____

2. _____

3. _____

4. _____

5. _____

6. _____

7. _____

8. _____

9. _____

10. _____

11. _____

12. _____

13. _____

14. _____

15. _____

16. _____

17. _____

Chapter 2 Test Form E *(cont.)*

18. The following figures are similar figures. Find the length of side x

18._____

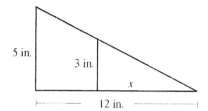

5 in.

3 in.

x

12 in.

19. If a car uses 4 gallons of gasoline to travel 86 miles, how many gallons of gasoline are needed to travel 430 miles?

19. _____

20. If 15 pounds of seed can cover 4300 square feet, how many square feet of lawn can 21 pounds of seed cover?

20. _____

67

Chapter 2 Test Form F

Name:_____

Date:_____

For questions 1–2, use the distributive property to simplify.

1. $-(-3-2x)$

2. $-(-5y-2x+4)$

For questions 3–7, simplify.

3. $28-15x-16+3x$

4. $-3+7x-x+5$

5. $-4(2x-3)-8+7x$

6. $\dfrac{5}{8}x+\dfrac{2}{9}-\left(\dfrac{5}{6}x+\dfrac{1}{6}\right)$

7. $7(2x+8)-5(x+7)$

For questions 8–13, solve.

8. $10x-14=6$

9. $5(x-4)-1=-7(-x+3)-2x$

10. $-2(x+2)=2x-16$

11. $\dfrac{7x}{14}=\dfrac{1}{2}$

12. $4x-2(1+x)=2x-3$

13. $\dfrac{1}{5}x-\dfrac{2}{3}=\dfrac{1}{3}(2-x)$

14. Solve $A=P(1+rt)$ for t.

15. Solve $y-3=2(x+1)$ for y. Write the answer in $y=mx+b$ form.

16. $d=rt$; Find t if $r=25$ mph and $d=160$ miles.

17. Determine the ratio of 9 inches to 2 feet. Write the ratio in lowest terms.

1. _____

2. _____

3. _____

4. _____

5. _____

6. _____

7. _____

8. _____

9. _____

10. _____

11. _____

12. _____

13. _____

14. _____

15. _____

16. _____

17. _____

Name:_____

18. The following figures are similar figures. Find the length of side *x*

18._____

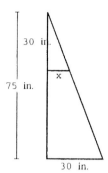

30 in.

75 in.

x

30 in.

19. If Chef Luigi can make 4 large pizzas in 18 minutes, how long will it take him to make 10 pizzas?

19._____

20. If a steam machine uses 5 ounces of detergent to clean a 24 square inch rug, how much detergent will it use for 552 square feet of office space?

20._____

Chapter 2 Test Form G

Name:_____

Date:_____

1. Use the distributive property to simplify.
   $-3(4 - 8x)$

   (a) $-12 - 8x$ (b) $8x - 24$ (c) $-12x + 24$ (d) $24x - 12$

2. Use the distributive property to simplify.
   $-(2x - 5 + 3y)$

   (a) $-2x - 5y + 3$ (b) $2x + 5 + 3y$ (c) $-2x - 3y + 5$ (d) $2y + 3x - 5$

For questions 3–7, simplify.

3. $5 + 8x - 2x + 1$

   (a) $6x + 6$ (b) $4 + 10x$ (c) $4 - 10x$ (d) $6 - 6x$

4. $4 - 6x + 2 - 3x$

   (a) $-9x + 6$ (b) $-9x + 4$ (c) $9x - 6$ (d) $-9x + 2$

5. $2(1 - y) + x + 5x$

   (a) $-y + 6x + 1$ (b) $6x - 2y + 2$ (c) $6x - y - 1$ (d) $6x - 2y - 2$

6. $\dfrac{5}{6}x - \dfrac{7}{8} - \left(\dfrac{1}{3}x - \dfrac{1}{2}\right)$

   (a) $\dfrac{1}{2}x + \dfrac{11}{8}$ (b) $\dfrac{1}{2}x - \dfrac{3}{8}$ (c) $\dfrac{7}{6}x + \dfrac{11}{8}$ (d) $\dfrac{4}{3}x - 1$

7. $4(6x + 3) - 6(x - 3)$

   (a) $18x + 30$ (b) $30x - 6$ (c) $18x - 6$ (d) $30x + 30$

For questions 8–13, solve.

8. $4x + 1 - 2x = 9$

   (a) $x = 4$ (b) $x = -2$ (c) $x = 5$ (d) all real numbers

9. $-6 + 8x = 5x + 3(x - 1)$

   (a) $x = -1$ (b) all real numbers   (c) no solution   (d) $x = 3$

10. $4x - 2 = 3(x + 3)$

    (a) no solution   (b) $x = 5$ (c) $x = 11$ (d) all real numbers

11. $2(4x - 1) = 5x + 1 + 3(x - 1)$

    (a) $x = 0$ (b) all real numbers   (c) $x = -\dfrac{1}{4}$ (d) no solution

12. $\dfrac{10}{x} = \dfrac{2}{9}$

    (a) no solution   **(b)** $x = 18$ **(c)** $x = 27$ **(d)** $x = 45$

13. $4x - 2(1 - x) = -(1 + 2x)$

    (a) $x = -\dfrac{1}{8}$ **(b)** $x = \dfrac{1}{8}$ **(c)** no solution   **(d)** $x = \dfrac{1}{6}$

14. Solve $A = 2lh + 2wh$ for w.

    (a) $w = \dfrac{A}{2lh + 2h}$ **(b)** $w = \dfrac{A - 2h}{2lh}$ **(c)** $w = \dfrac{A - 2lh}{2h}$ **(d)** $w = A - 2lh - 2h$

15. Solve $y + 3 = -2(x + 5)$ for y. Write the answer in $y = mx + b$ form.

    (a) $y = -2x - 7$ **(b)** $y = -2x - 13$ **(c)** $y = -2x + 8$ **(d)** $y = -2x + 2$

16. $A = \dfrac{1}{2}h(b + d)$; Find h if $A = 60$, $b = 5$, and $d = 7$.

    (a) $h = 3$ **(b)** $h = 21$ **(c)** $h = 25$ **(d)** $h = 10$

17. Write the ratio of 12 ounces to 2 pounds in lowest terms.

    (a) $3 : 8$ **(b)** $6 : 1$ **(c)** $1 : 2$ **(d)** $12 : 2$

18. The following figures are similar figures. Find the length of side x.

    (a) 6 ft   **(b)** 8.6 ft   **(c)** 16.6 ft   **(d)** 22 ft

19. If Payton can type 6 pages in 32 minutes, how long will it take her to type a 15-page report?

    (a) 47 min   **(b)** 1 hr   **(c)** 64 min   **(d)** 80 min

20. If a compactor reduces 30 cubic feet of recycling material to 4 cubic feet, how big will 140 cubic feet of material be after going through the same machine?

    (a) 11.67 cubic ft   **(b)** 105 cubic ft   **(c)** 186.67 cubic ft   **(d)** 18.67 cubic ft

Chapter 2 Test Form H

Name:_____

Date:_____

1. Use the distributive property to simplify.
   $-3(-2x+3)$

   (a) $-3-2x+3$ (b) $6x-9$ (c) $3x+6$ (d) $9-6x$

2. Use the distributive property to simplify.
   $-(2x-5y+4)$

   (a) $-2x-5y-4$ (b) $-2x+4$ (c) $-2x+5y-4$ (d) $-5y+2x+8$

For questions 3–7, simplify.

3. $-(x-2y)+4y-3$

   (a) $-x+6y-3$ (b) $-x+2y-3$ (c) $x-8y-3$ (d) $-x+6y+3$

4. $4+2x-7+x$

   (a) $x-11$ (b) $2x-4$ (c) $3x-3$ (d) $3x-4$

5. $7x-8-4(2x-3)$

   (a) $15x-11$ (b) $x-4$ (c) $5x+4$ (d) $-x+4$

6. $\frac{5}{6}x+\frac{3}{4}-\left(\frac{1}{2}x+\frac{1}{6}\right)$

   (a) $\frac{1}{3}x+\frac{7}{12}$ (b) $\frac{3}{4}x-1$ (c) $\frac{4}{3}x+\frac{11}{12}$ (d) $\frac{1}{3}x+\frac{11}{12}$

7. $5(2x+7)-4(x-4)$

   (a) $6x+19$ (b) $14x+51$ (c) $6x+51$ (d) $14x+19$

For questions 8–13, solve.

8. $3+4x-x=12$

   (a) $x=-3$ (b) $x=\frac{1}{3}$ (c) no solution   (d) $x=3$

9. $2x-3+x=-9$

   (a) $x=2$ (b) all real numbers   (c) $x=-2$ (d) $x=-\frac{1}{2}$

10. $-7(-x+3)-2x=5(x-4)-1$

    (a) all real numbers   (b) $x=\frac{1}{3}$ (c) $x=3$ (d) $x=-3$

11. $\dfrac{1}{x} = \dfrac{5}{45}$

   (a) $x = 15$ **(b)** $x = 9$ **(c)** $x = 3$ **(d)** no solution

12. $3x - (6 - 5x) = -\dfrac{8}{3}\left(\dfrac{3}{2} - 3x\right)$

   (a) $x = 2$ **(b)** $x = \dfrac{1}{2}$ **(c)** $x = -2$ **(d)** no solution

13. $-(1 + 2x) = 4x - 2(1 - x)$

   (a) $x = 8$ **(b)** $x = -8$ **(c)** $x = \dfrac{1}{8}$ **(d)** all real numbers

14. Solve $L = 2\pi r h$ for h.

   (a) $h = \dfrac{L}{2\pi r}$ **(b)** $h = L - 2\pi r$ **(c)** $h = 2\pi r L$ **(d)** $h = \dfrac{L - r}{2\pi}$

15. Solve $-3x + 15y = -20$ for y. Write the answer in $y = mx + b$ form.

   (a) $y = 5x + \dfrac{20}{3}$ **(b)** $y = \dfrac{1}{5}x - \dfrac{4}{3}$ **(c)** $y = -5x + \dfrac{5}{3}$ **(d)** $y = -\dfrac{1}{3}x + 20$

16. $V = \dfrac{4}{3}\pi r^3$; Find V when $r = \dfrac{3}{2}$.

   (a) 14.18   **(b)** 42.41   **(c)** 1.57 •   **(d)** 56.55

17. Write the ratio of 10 minutes to 2 hours in lowest terms.

   (a) $5 : 1$ **(b)** $10 : 2$ **(c)** $1 : 6$ **(d)** $1 : 12$

18. The following figures are similar figures. Find the length of side x.

   (a) 12 in.   **(b)** 3 in.   **(c)** 9 in.   **(d)** 15 in.

19. If Chef Luigi can make 3 large pizzas in 14 minutes, how many pizzas can he make in 42 minutes?

   (a) 7   **(b)** $10\dfrac{1}{3}$ **(c)** 12   **(d)** 9

20. If a compactor reduces 45 cubic feet of recycling material to 6 cubic feet, how big will 720 cubic feet of material be after going through the same machine?

   (a) 112 cubic ft   **(b)** 96 cubic ft   **(c)** 270 cubic ft   **(d)** 16 cubic ft

Chapters 1–2 Cumulative Test Form A

1. Multiply: $\dfrac{6}{25} \cdot \dfrac{5}{27}$

   1. _____

2. Divide: $\dfrac{35}{6} \div \dfrac{25}{18}$

   2. _____

3. What symbol (<, >, or =) will make the following expression true?

   $|-7|$ _____ $|-6|$

   3. _____

4. Subtract: $(-4) - (-5)$

   4. _____

5. Evaluate: $13 - 29 - 16$

   5. _____

6. Evaluate: $54 \div 9 \cdot 3 + 12 - 4$

   6. _____

7. Evaluate: $2\left[8 - \left(10 - 4^2\right)\right] - 20$

   7. _____

8. Evaluate: $(3m - 2n)^2$ for $m = 6$ and $n = -5$

   8. _____

9. Which law is illustrated by the following statement?

   $(2 \cdot 8) \cdot 3 = 2 \cdot (8 \cdot 3)$

   9. _____

10. Simplify: $8 + 5 + (-6x) + (-5) - (-3x)$

    10. _____

11. Combine like terms: $9x - 8y - 4x + 9y + 9$

    11. _____

12. Solve: $5x + 7 = 2$

    12. _____

13. Solve: $\dfrac{9}{2}x = -18$

    13. _____

14. Solve: $5x - 9x = 3$

    14. _____

15. Solve: $7 = 8(x - 1) - x$

    15. _____

16. Solve the proportion: $\dfrac{-81}{5} = \dfrac{18}{x}$

    16. _____

17. Solve $2x - 5y = 15$ for y.

    17. _____

18. Determine the ratio of 1 hour to 90 minutes. Write the ratio in lowest terms.

    18. _____

19. The effective property tax rate for 2006 in San Antonio was $5.3873 per $1000 of assessed value. If the Chavez's house is assessed at $172,000, how much property tax will they owe?

    19. _____

20. The protein RDA for females is 57 grams per day. Five ounces of a certain product provide 3 grams of protein. How many ounces of the product are needed to provide 57 grams of protein?

    20. _____

Chapters 1–2 Cumulative Test Form B

1. Find the product and completely reduce the answer.

   $$\frac{96}{44} \cdot \frac{11}{36}$$

   (a) $\dfrac{1}{24}$
   (b) $\dfrac{2}{3}$
   (c) $\dfrac{107}{80}$
   (d) $\dfrac{24}{144}$

2. Divide: $\dfrac{27}{5} \div \dfrac{9}{8}$

   (a) $4\dfrac{4}{5}$
   (b) $1\dfrac{3}{5}$
   (c) $2\dfrac{10}{13}$
   (d) $1\dfrac{13}{27}$

3. Which of the following is a true statement?

   (a) $\left|-8\right| < \left|8\right|$
   (b) $\left|-8\right| < \left|-4\right|$
   (c) $\left|-8\right| > \left|4\right|$
   (d) $0 > \left|-4\right|$

4. Subtract: $-12 - 1$

   (a) -13
   (b) 13
   (c) -11
   (d) 11

5. Evaluate: $58 - 90 + 16 - 3 + 47$

   (a) 34
   (b) -4
   (c) -98
   (d) 28

6. Evaluate: $16 \div 4 \cdot 4 + 9 - 5$

   (a) 5
   (b) 4
   (c) 20
   (d) 8

7. Evaluate: $-[11 - (-4 - 3)]^2$

   (a) -60
   (b) 324
   (c) -324
   (d) -18

8. Evaluate: $(2u + 5x)^2$ for $u = 5$ and $x = -1$

   (a) 125
   (b) 225
   (c) 5
   (d) 25

9. Identify the property of operation that is illustrated: $5(2 \times 9) = (5 \times 2)9$

   (a) Associative property of multiplication
   (b) Commutative property of multiplication
   (c) Commutative property of addition
   (d) Associative property of addition

10. Simplify: $8x + 4 + 6x - 1$

    (a) $12x + 5$
    (b) $3x$
    (c) $-14x + 3$
    (d) $14x + 3$

11. Combine like terms: $7x - 4y - 5x - 5y$

    (a) $2x - 9y$
    (b) $12x - 9y$
    (c) $3x - 10y$
    (d) $2x - 5y$

Chapters 1–2 Cumulative Test
Form B *(cont.)*

12. Solve: $\dfrac{6}{5}y - 8 = 4$

(a) $\dfrac{72}{5}$ (b) $-\dfrac{10}{3}$ (c) $-\dfrac{24}{5}$ (d) 10

13. Solve: $-\dfrac{6}{5}x = -30$

(a) 36 (b) -25 (c) -36 (d) 25

14. Solve: $6x - 4x = -9$

(a) $-4\dfrac{1}{2}$ (b) -11 (c) $4\dfrac{1}{2}$ (d) $-2\dfrac{1}{4}$

15. Solve: $-7 = 3(x + 2) + 5 - 2x$

(a) -18 (b) 4 (c) -14 (d) 0

16. Solve the proportion: $\dfrac{18}{c} = \dfrac{6}{5}$

(a) 108 (b) 90 (c) $\dfrac{108}{5}$ (d) 15

17. Solve $4x - 5y = 30$ for y.

(a) $y = \dfrac{5}{4}x + \dfrac{15}{2}$ (b) $y = \dfrac{4}{5}x - 6$ (c) $y = -\dfrac{4}{5}x + 6$ (d) $y = -\dfrac{4}{5}x - 6$

18. Write the ratio of 8 inches to 4 feet in lowest terms.

(a) $8:4$ (b) $2:1$ (c) $1:6$ (d) $1:5$

19. A worker in an assembly line takes 5 hours to product 23 parts. At that rate how many parts can she produce in 10 hours?

(a) 6 parts (b) 92 parts (c) 230 parts (d) 46 parts

20. A survey indicated that 27 out of 50 U.S. Internet users had downloaded or watched a video clip from the Internet. If 3500 U.S. Internet users were surveyed, how many will have downloaded or watched a video clip from the Internet?

(a) 1890 (b) 700 (c) 2140 (d) 1350

Chapter 3 Pretest Form A

Name:_____

Date:_____

1. Imelda's weekly salary is $25 more than Julian's weekly salary. Let x represent Julian's weekly salary. Write an algebraic expression for Imelda's salary.

2. For her salary this year, a teacher received a 4% raise over her salary last year. Let x represent the salary last year. Write an algebraic expression for her salary this year.

3. Write an algebraic expression that represents the number of inches in f feet.

4. The base of a triangle is 3 inches less than half the height. Let h represent the height. Write an algebraic expression for the base.

5. The sum of a number and 8 is 21. Select a variable and write an equation to represent this situation.

6. Find the number represented in problem 5.

7. One number is seven more than three times the other. The sum of the two numbers is 59. Select a variable and write an equation to represent this situation.

8. Find the two numbers represented in problem 7.

9. Clay is three years younger than his brother Caleb. The sum of their ages is 17. Select a variable and write an equation to represent this situation.

10. Find the ages of Clay and Caleb in problem 9.

11. A number increased by 15% is 230. Select a variable and write an equation to represent this situation.

12. Find the number represented in problem 11.

1. _____

2. _____

3. _____

4. _____

5. _____

6. _____

7. _____

8. _____

9. _____

10. _____

11. _____

12. _____

For problems 13–20, set up an equation and solve each problem.

13. The daily rental cost for a car is $45 plus $0.15 per mile. How many miles can be driven for a cost of $75?

13. _____

14. The sum of three consecutive even integers is 114. What are the three integers?

14. _____

15. Samuel invested $19,000 for one year, part at 11% and part at 13%. If he earned a total interest of $2290, how much was invested at each rate?

15. _____

16. A country's rivers contained 850 billion trout. Assuming no new trout enter the country's rivers, and 17 billion are caught each month, how long will it take for the country's rivers to be empty of all trout?

16. _____

17. Laura is a salesperson in a retail store and earns $80 per week plus 16% of her weekly sales. If Laura made $688 one week, what were her sales that week?

17. _____

18. The length of a rectangle is 5 cm less than 4 times the width. The perimeter of the rectangle is 40 cm. Find the dimensions of the rectangle.

18. _____

19. How many liters of a 24% salt solution must be added to 76 liters of a 64% salt solution to get a 43% salt solution?

19. _____

20. Train A leaves a station traveling at 50 miles per hour. Four hours later, train B leaves the same station traveling in the same direction at 60 miles per hour. How long does it take for train B to catch up to train A?

20. _____

Chapter 3 Pretest Form B

Name:_____

Date:_____

1. Becky's car gets 5 more miles per gallon than Sue's car. Let x represent Sue's gas mileage. Write an algebraic expression for Becky's gas mileage.

  1. _____

2. During a sale, a department store advertised 40% of the regular price of all merchandise. Let x represent the regular price of a particular item. Write an algebraic expression that represents the sale price of that item.

  2. _____

3. Write an algebraic expression that represents the number of minutes in h hours.

  3. _____

4. The width of a rectangle is two more than one-third the length. Let x represent the length. Write an algebraic expression for the width.

  4. _____

5. The difference between a number and 9 is 31. Select a variable and write an equation to represent this situation.

  5. _____

6. Find the number represented in problem 5.

  6. _____

7. One number is 13 less than twice the other. The sum of the two numbers is 50. Select a variable and write an equation to represent this situation.

  7. _____

8. Find the two numbers represented in problem 7.

  8. _____

9. Josh is four years older than his sister Katie. The sum of their ages is 14. Select a variable and write an equation to represent this situation.

  9. _____

10. Find the ages of Josh and Katie in problem 9.

  10. _____

11. A woman's salary was increased by 30% to $33,800. Select a variable and write an equation to represent this situation.

  11. _____

12. Find the woman's original salary from problem 11.

  12. _____

Chapter 3 Pretest Form B *(cont.)*

For problems 13–20, set up an equation and solve each problem.

13. The daily rental cost for a car is $45 plus $0.15 per mile. How many miles can be driven for a cost of $82.50?

13. _____

14. The sum of three consecutive even integers is 144. What are the three integers?

14. _____

15. Roy invested $22,000 for one year, part at 11% and part at 13%. If he earned a total interest of $2670, how much was invested at each rate?

15. _____

16. Jumbo warehouse contains 600 million ball bearings. If no new items are manufactured and 25 million are sold each day, how long will it take to empty the warehouse?

16. _____

17. Terry keeps nickels, dimes and quarters in the console of her car in order to pay road tolls. While cleaning her car, she found that the value of the coins was $3.25. She had two fewer dimes than nickels and one more quarter than nickels. How many of each coin type did Terry have?

17. _____

18. The width of a rectangle is one less than half the length. The perimeter of the rectangle is 34 inches. Find the dimensions of the rectangle.

18. _____

19. How many liters of a 30% alcohol solution must be added to 21 liters of an 80% alcohol solution to get a 60% salt solution?

19. _____

20. Two cars leave a city at the same time and head in opposite directions. The first car travels at 40 mph, and the second car travels at 60 miles per hour. How long does it take for the two cars to be 350 miles apart?

20. _____

Mini-Lecture 3.1
Changing Application Problems into Equations

Learning Objectives:

1. Translate phrases into mathematical expressions.
2. Express the relationship between two related quantities.
3. Write expressions involving multiplication.
4. Translate applications into equations.
5. Key vocabulary: *sum, difference, product, quotient, percent*

Examples:

1. Express each statement as an algebraic expression.
   a) The height, *h*, decreased by ten feet
   b) Ten feet decreased by the height, *h*
   c) Eight inches more than three times the width, *w*

2. For each relationship, select a variable to represent one quantity and state what the variable represents. Then express the second quantity in terms of the variable selected.
   a) Ethan weighs twice as much as his younger brother Ben.
   b) Debbie's weekly pay is $27.50 less than Stacy's weekly pay.
   c) Charlie has $8500 in a bank, part in a checking account and part in a savings account.

3. Write each statement as an algebraic expression.
   a) The cost of *n* T-shirts at $9.25 each
   b) The number of quarts in *g* gallons
   c) A 3% commission earned on the sale of a house for *x* dollars
   d) The price *x* of a new sofa, plus a 6.5% sales tax

4. Select a variable and then write an equation to represent each situation.
   a) One number is five more than three times the other. The sum of the two numbers is 87.
   b) The sum of two consecutive even integers is 138.
   c) Jeff weighs twelve pounds more than twice Amanda weight. The sum of their weights is 387 pounds.
   d) This year the enrollment at a college increased by 4% over last year's enrollment. The enrollment this year is 13,390 students.

Teaching Notes:

- Point out the difference between statements such as "a number decreased by 7" and "7 decreased by a number."
- Point out the significance of comma placement is a statement. For example, "four times a number, increased by 8" is written $4x + 8$ while "four times a number increased by 8" is written $4(x + 8)$.

Answers: 1a) $h - 10$; 1b) $10 - h$; 1c) $3w + 8$; 2a) Let w = Ben's weight, Ethan's weight is 2w; 2b) Let p = Stacy's pay, Debbie's pay is $p - 27.50$; 2c) Let x = the amount in checking, the amount in savings is $8500 - x$; 3a) 9.25n; 3b) 4g; 3c) 0.03x; 3d) $x + 0.065x$; 4a) Let x = the first number, $x + (3x + 5) = 87$; 4b) Let x = the smaller integer, $x + (x + 2) = 138$; 4c) Let x = Amanda's weight, $x + (2x + 12) = 387$; 4d) Let x = last year's enrollment, $x + 0.04x = 13,390$

Mini-Lecture 3.2
Solving Application Problems

Learning Objectives:

1. Use the problem-solving procedure.
2. Set up and solve number application problems.
3. Set up and solve application problems involving money.
4. Set up and solve applications concerning percent.

Examples:

1. Nine less than five times a number is 76. Find the number.

2. The difference between two numbers is 16. Find the two numbers if the smaller number is 4 less than two-thirds of the larger number.

3. The room numbers of two adjacent hotel rooms are two consecutive even numbers. If their sum is 2474, find the room numbers.

4. A serving of Cocoa Puffs contains 10 more calories than a serving of Cheerios (according to the nutrition facts on the side of the cereal boxes). If two servings of Cheerios and four serving of Cocoa Puffs contains a total of 700 calories, find the number of calories in a single serving of each.

5. Josephine ordered some candles from an online company. The company charged $10.25 per candle, plus $15 for shipping and handling. How many candles did Josephine order if her bill was $179?

6. Tim's cell phone company charges $29.99 per month, plus $0.07 for each minute of use over 450 minutes. One month his bill was $45.74. How many minutes did Tim use that month?

7. A bowling alley offers two payment options for its league members. Option 1 charges a monthly fee of $24, plus $2 per game. Option 2 charges no monthly fee but charges $5 per game. How many games would you have to bowl per month for the two options to cost the same?

8. The cost of a new rocking chair, including 6% sales tax, was $284.61. What was the price of the rocking chair before tax.

9. To attract new customers, a health club decreased its annual membership fee by 15%. If the cost of membership is now $119, what was it before the decrease?

Teaching Notes:

- Students sometimes stop too soon when working word problems. Remind students to be careful to answer the question that is being asked in the problem.

Answers: 1) 17; 2) 20 and 36; 3) 1236 and 1238; 4) Cheerios = 110 calories, Cocoa Puffs = 120 calories; 5) 16 candles; 6) 675 minutes; 7) 8 games; 8) $268.50; 9) $140

Mini-Lecture 3.3
Geometric Problems

Learning Objectives:

1. Solve geometric problems.
2. Key vocabulary: *isosceles triangle, equilateral triangle, complementary angles, supplementary angles, vertical angles, parallelogram, trapezoid, rhombus*

Examples:

1. The length of a rectangle is two feet longer than three times the width. The perimeter is 76 feet. Find its dimensions.

2. Each of the two equal angles in an isosceles triangle is 24° less than the third angle. Find all three angles.

3. Each of the two larger angles of a parallelogram measures 9° more than twice the measure of each of the two smaller angles. Find the measures of the four angles.

4. A shoe rack is to have three shelves including the top and bottom. The width of the shoe rack is to be 12 inches more than the height. Find the width and height of the shoe rack if only 96 inches of lumber is available.

Teaching Notes:

- Students may need refreshing on some properties of triangles such as the fact that the angles add up to 180 degrees.
- Refer students back to Section 2.6 in order to brush up on geometric formulas.

Answers: 1) 9 ft by 29 ft; 2) 76°, 52°, 52°; 3) 57°, 57°, 123°, 123°; 4) height = 12 in, width = 24 in

Mini-Lecture 3.4
Motion, Money, and Mixture Problems

Learning Objectives:

1. Solve motion problems involving two rates.
2. Solve money problems.
3. Solve mixture problems.

Examples:

1. A Ford truck and a Chevy truck enter an expressway at the same time and place and head the same direction. The Ford travels at 67 miles per hour and the Chevy travels at 63 miles per hour. In how many hours will they be 10 miles apart?

2. A freight train and a passenger train meet each other on parallel tracks heading in opposite directions. The passenger train travels 15 miles per hour faster than the freight train. After 2 hours, they are 278 miles apart. At what speeds are the two trains traveling?

3. Aaron invested $16,000 for one year, part at 6% and part at 9%. If he earned a total of $1320, how much was invested at each rate?

4. Elizabeth has a total of 40 nickels and dimes in her piggy bank worth $2.60. How many of each type coin does she have?

5. A store owner has 50 pounds of peanuts that sell for $2.80 per pound that he plans to mix with 20 pounds of cashews that sell for $4.20 per pound. How much should he charge per pound for the mix?

6. How many milliliters of a 30% salt solution must be added to 7500 milliliters of a 5% salt solution to yield a 10% salt solution?

7. How many liters each of a 6% acid solution and a 30% acid solution should be mixed in order to obtain 10 liters of a 15% acid solution?

8. How much pure antifreeze should be mixed with 3 quarts of 50% antifreeze in order to get and 80% antifreeze solution?

Teaching Notes:

- Many students find these types of applications very difficult. Emphasize to students the importance of understanding the problems rather than trying to memorize procedures.

Answers: 1) 2.5 hours; 2) freight = 62 mph, passenger = 77 mph; 3) $4000 at 6%, $12,000 at 9%; 4) 28 nickels, 12 dimes; 5) $3.20 per pound; 6) 1875 milliliters; 7) 6.25 liters of 6%, 3.75 liters of 30%; 8) 4.5 quarts

Additional Exercises 3.1

Express the statement as an algebraic expression.

1. Fifteen divided by a number x.

    1. _____

2. The difference between a number x and 50

    2. _____

3. The cost C decreased by 14%

    3. _____

4. The profit P increased by 12%

    4. _____

5. Thirty-five more than half the weight w

    5. _____

6. One-fourth less than the product of a number x and 5

    6. _____

Write the indicated expression.

7. Write an algebraic expression representing the number of cents in n nickels and d dimes.

    7. _____

8. Write an algebraic expression representing the number of seconds in m minutes and s seconds.

    8. _____

9. The number of VCR tapes in a video rental store is 120 more than twice the number of DVDs, d. Write an expression for the number of VCR tapes.

    9. _____

10. The length of a rectangle is 8 inches less than 3 times the width, w. Write an expression for the length.

    10. _____

Write an equation to represent the problem. You will need to select a variable.

11. Four times a number decreased by 13 is 11.

    11. _____

12. Four times a number increased by 14 is 42.

    12. _____

13. A number decreased by 12% is 200.

    13. _____

14. A number decreased by 15% is 180.

    14. _____

15. One number is nine greater than six times the other. Their product is 81.

    15. _____

16. One number is four times another. The sum of the two numbers is 24.

    16. _____

17. Maria is seven years older than her sister. The sum of their ages is 13.

    17. _____

18. The sum of two consecutive integers is 51.

    18. _____

19. Lisa is eight years older than her sister. The sum of their ages is 15.

    19. _____

20. One number is three times another. The sum of the two numbers is 24.

    20. _____

Additional Exercises 3.2

Name:_____

Date:_____

Set up an equation that can be used to solve the problem. Solve the equation and answer the question asked.

1. If the sum of a number and 4 is decreased by 20, the result is 32. Find the number.

2. If the sum of a number and 5 is increased by 40, the result is 20. Find the number.

3. If the product of a number and 9 is decreased by 27, the result is 81. Find the number.

4. If the product of a number and 4 is increased by 12, the result is 8. Find the number.

5. The sum of three consecutive integers is 252. What is the largest of of the three integers?

6. The sum of two integers is 66. Find the two integers if the larger is 6 less than three times the smaller.

7. The sum of three consecutive integers is 111. What is the largest of the three integers?

8. The sum of two integers is 88. Find the two integers if the larger is 7 less than four times the smaller.

9. A country's rivers contained 645 billion sturgeon. Assuming no new sturgeon enter the country's rivers, and 15 billion are caught each month, how long will it take for the country's rivers to be empty of all sturgeon?

10. Jumbo Warehouse contains 700 billion nuts, bolts, and nails. If no new items are manufactured, and 25 billion are sold each month, how long will it take to empty the warehouse?

11. Garrett has $150 in ten- and twenty-dollar bills. The number of ten-dollar bills is one less than twice the number of twenty-dollar bills. How many of each type bill does Garrett have?

12. Emily has nickels, dimes, and quarters in her piggy bank. She has twice as many dimes as nickels and 12 fewer quarters than dimes. The total value of the coins is $9.00. How many of each coin type does Emily have?

13. A rental company rents garden tillers. The charge for a tiller is $40 for the first 5 hours plus $7.50 for each additional hour. If Richard's total rental fee was $92.50, for how long did he have the tiller?

14. The author of a textbook receives 12% of sales as royalties. Find the sales needed if the author wants to earn $75,000 on the book.

15. A country singer was hired to perform a concert at a county rodeo. She was paid $2000 plus 2.5% of admission fees collected at the gate. If the singer received a total of $2590, how much was collected at the gate?

16. Jeff is a salesperson in a retail store and earns $80 per week plus 10% of his weekly sales. If Jeff made $680 one week, what were his sales that week?

1. _____

2. _____

3. _____

4. _____

5. _____

6. _____

7. _____

8. _____

9. _____

10. _____

11. _____

12. _____

13. _____

14. _____

15. _____

16. _____

Additional Exercises 3.2 *(cont.)*

Name:_____

17. Gail's salary was increased by 30% to $45,500. What was her salary before the increase?

17. _____

18. In 2005 the federal income tax rate for a person filing a single return with an adjusted gross income between $29,700 and $71,950 was $4,090 plus 25% of the amount in excess of $29,700. If Herman had to pay $11,255 in federal taxes in 2005, what was his adjusted gross income that year?

18. _____

19. From 2000 to 2006, the number of students enrolled at a state university increased by 40%. If the number of students enrolled at the university in 2006 was 10,325, what was the enrollment in 2000?

19. _____

20. Between 1996 and 2006, the population of a city decreased by 15%. If the population in 2006 was 361,250, what was the population in 1996.

20. _____

Additional Exercises 3.3

Name:_____

Date:_____

Solve the following geometric problems.

1. Two angles are complementary if the sum of their measures is 90°. If angle A and B are complementary angles, and angle A is 9° less than twice angle B, find the measures of angles A and B.

    1. _____

2. Two angles are complementary if the sum of their measures is 90°. If angle A and B are complementary angles, and angle A is 6° more than three times angle B, find the measures of angles A and B.

    2. _____

3. Two angles are supplementary if the sum of their measures is 180°. If angle A and angle B are supplementary angles, and angle B is 12° more than three times angle A, find the measures of angles A and B.

    3. _____

4. Two angles are supplementary if the sum of their measures is 180°. If angle A and angle B are supplementary angles, and angle B is 15° less than twice angle A, find the measures of angles A and B.

    4. _____

5. The sum of the angles of a triangle is 180°. If one angle of a triangle is ten degrees larger than the smallest angle and the third angle is 10° less than three times the smallest angle, find the measures of the three angles.

    5. _____

6. In a parallelogram the opposite angles have the same measures and the sum of the angles is 360°. If each of the two larger angles in a parallelogram is 30° more than each smaller angle, find the measure of each angle.

    6. _____

7. The perimeter of a rectangle is 240 feet. The length is 20 feet more than the width. What are the dimensions?

    7. _____

8. A triangle has a perimeter of 74 inches. Find the three sides if one side is 24 inches larger than the smallest and the third side is three times the smallest.

    8. _____

9. The length of the base of an isosceles triangle is one third the length of one of its legs. If the perimeter of the triangle is 56 inches, what is the length of the base?

    9. _____

10. The perimeter of a rectangle is 200 feet. The width is two-thirds as long as the length. What are the dimensions of the rectangle?

    10. _____

11. The length of an official NBA basketball court is 6 feet less than twice the width. The perimeter is 288 feet. Find the dimensions of the court.

    11. _____

12. The width of an official volleyball court is half its length. The perimeter is 54 meters. Find the dimensions of the court.

    12. _____

13. A triangle has a perimeter of 71 inches. Find the three sides if one side is 31 inches larger than the smallest and the third side is three times the smallest.

    13. _____

14. The length of the base of an isosceles triangle is one fourth the length of one of its legs. If the perimeter of the triangle is 90 cm, what is the length of the base?

    14. _____

15. A book case is to have four shelves including the top and bottom. The height of the book case is to be 2.5 feet more than the width. Find the width and height of the book case if only 26 feet of lumber is available.

15. _____

16. A book case is to have four shelves including the top. The height of the book case is to be 4.25 feet more than the width. Find the width and height of the book case if only 31 feet of lumber is available.

16. _____

17. An official Major League Baseball "diamond" is actually a square with a perimeter of 360 feet. Find the dimensions of the diamond.

17. _____

18. The outer perimeter of the Pentagon building in Washington, D.C. (which houses the U.S. Department of Defense) is 1400 meters. The building is a regular pentagon, which means its five sides are equal in length. Find the length of each outer wall.

18. _____

19. Steve and Mimi are placing a border around a rectangular flower bed. One long side will be next to the house and will not have a border. The longer side is to be 3 feet more than the shorter side, and they have 24 feet of border material. Find the length and width of the flower bed.

19. _____

20. A rectangular room has a 25 yard strip of wallpaper bordering it near the ceiling. If the room is 8 yards long, what is its width?

20. _____

Additional Exercises 3.4

1. A bus travels 150 miles on 10 gallons of gas. How many gallons will it need to travel 225 miles?

    1. _____

2. Marsha can check 13 parts per minute on an assembly line. How many parts will she be able to check in 3 hours?

    2. _____

3. On the first day of their vacation trip the Lopez family traveled 407 miles in 11 hours. What was their average speed?

    3. _____

4. On the fifth day of their vacation trip the Lopez family traveled 216 miles in 6 hours. What was their average speed?

    4. _____

5. A van travels 200 miles on 10 gallons of gas. How many gallons will it need to travel 340 miles?

    5. _____

6. An SUV gets 15 miles per gallon of gasoline. If its tank will hold 22 gallons, how far can the SUV travel on a tank of gas?

    6. _____

7. One day, John worked for 6.5 hours and was paid $92.95. How long did John work the following day if he was paid $114.40?

    7. _____

8. A train traveling at 40 miles per hour leaves for a certain town. One hour later, a bus traveling at 50 miles per hour leaves for the same town and arrives at the same time as the train. If both the train and the bus traveled in a straight line, how far is the town from where they started?

    8. _____

9. Train A leaves a station traveling at 40 mph. Eight hours later, train B leaves the same station traveling in the same direction at 60 mph. How long does it take for train B to catch up to train A?

    9. _____

10. Two trains leave the same station precisely at noon. One train heads due east at 45 miles per hour while the other train heads due west at at 55 miles per hour. At what time will the two trains be 40 miles apart?

    10. _____

11. Two trains, an express and a commuter, are 450 miles apart. Both start at the same time and travel toward each other. They meet 6 hours later. The speed of the express is 25 miles per hour faster than the commuter. Find the speed of each train.

    11. _____

12. George invested $19,000 for one year, part at 11% and part at 12%. If he earned a total interest of $2200, how much was invested at each rate?

    12. _____

13. Cindy invested $24,000 for one year, part at 6% and part at 9%. If she earned a total interest of $1740, how much was invested at each rate?

    13. _____

14. How many liters of 18% salt solution must be added to 92 liters of 61% salt solution to get a 41% salt solution?

    14. _____

15. How many pounds of coffee beans selling for $2.80 per pound should be mixed with 2 pounds of coffee beans selling for $1.60 a pound to obtain a mixture selling for $2.56 per pound?

    15. _____

Additional Exercises 3.4 *(cont.)*

16. How many pounds of salted nuts selling for $3.00 per pound should be mixed with 7 pounds of salted nuts selling for $1.20 a pound to obtain a mixture selling for $1.74 per pound?

16. _____

17. A chemist needs 600 milliliters of a 31% alcohol solution. To obtain this solution, how much of a 10% alcohol solution should be mixed with a 40% alcohol solution?

17. _____

18. Susan charges $25 for a 30-minute private violin lesson. She charges $15 per student for a 60-minute group violin lesson. One day Susan earned $245 from 13 students. How many students did she see for private lessons and how many did she see in the group lesson?

18. _____

19. How many pounds of gourmet candy selling for $2.40 per pound should be mixed with 3 pounds of gourmet candy selling for $1.20 a pound to obtain a mixture selling for $2.04 per pound?

19. _____

20. How many pounds of gourmet candy selling for $2.00 per pound should be mixed with 2 pounds of gourmet candy selling for $3.00 per pound to obtain a mixture selling for $2.20 per pound?

20. _____

Chapter 3 Test Form A

Name:_____

Date:_____

1. The price of a calculus textbook is $35 more than the price of a philosophy textbook. Let x represent the price of the philosophy book. Write an algebraic expression for the price of the calculus textbook.

    1. _____

2. At a football game, the stands were filled with 1200 spectators. Let m represent the number of male fans. Write an algebraic expression for the number of female fans at the game.

    2. _____

3. The larger of two numbers is two more than three times the smaller number. Let n represent the smaller number. Write an expression for the sum of the two numbers.

    3. _____

4. The sales tax in a particular state is 6.5%. Let x represent the price of an item to be purchased in this state. Write an algebraic expression for the complete purchase total.

    4. _____

5. The two equal legs of an isosceles triangle are each 8 cm longer than the base of the triangle. Let b represent the base. Write an expression for the perimeter of the triangle.

    5. _____

6. The sum of two integers is 121. Find the two integers if the larger one is 7 more than twice the smaller.

    6. _____

7. The sum of two consecutive integers is 69. Find the two integers.

    7. _____

8. One number is two less than two-thirds the other. Their sum is 103. Find the two numbers.

    8. _____

9. Ronnie purchased a lawn mower. The cost of the mower, including an 8% sales tax, was $648. Find the cost of the mower before tax.

    9. _____

10. Karen purchased a blouse from a clearance rack. A sign on the rack stated, "The price shown is 70% off the original price." If the clearance price on the blouse was $12.60, what was the original price?

    10. _____

11. From 2005 to 2007, enrollment at a university decreased by 2%. If enrollment in 2007 was 8771 students, what was the enrollment in 2005.

11. _____

12. The Cheap Car Rental Company charges $30 per day, plus $0.25 per mile to rent a car, while Easy Car Rental charges $45 per day, plus $0.10 per mile. If you only need the car for one day, how many miles would you have to drive for the cost to be the same?

12. _____

13. Brenda wants to hire a lawn service to mow her yard for the summer. G&B Lawn Service charges $440 plus $40 for each cutting in excess of 10. How many cuttings can Brenda get for $1000?

13. _____

14. If the largest angle in a triangle is 5 times the measure of the smallest angle, and the middle-sized angle is 20° greater than twice the measure of the smallest angle. Find the three angles.

14. _____

15. A rectangle is 8 cm longer than twice the width. If the perimeter is 88 cm, find the dimensions of the rectangle.

15. _____

16. Each of the two larger angles of a parallelogram measures 30° more than the each of the two smaller angles. Find the measures of the four angles.

16. _____

17. Vince invested $24,000 for one year, part at 10% and part at 8%. If he earned a total of $2280, how much was invested at each rate?

17. _____

18. Two trains leave the same station at the same time but heading in opposite directions. The first train travels 10 miles per hour faster than second train. After 6 hours, they are 720 miles apart. At what speeds are the two trains traveling?

18. _____

19. How many gallons of a 30% salt solution must be added to 5 gallons of a 10% salt solution to yield a solution that is 17.5% salt?

19. _____

20. How many liters each of a 12% acid solution and a 20% acid solution should be mixed in order to obtain 4 liters of a 15% acid solution?

20. _____

Chapter 3 Test Form B

Name:_____

Date:_____

1. The price of a 4-pound bag of C&H sugar is $0.55 more than the price of a 4-pound bag of generic sugar. Let x represent the price of the generic sugar. Write an algebraic expression for the price of the C&H sugar.

2. A cheesecake needs to bake 10 minutes less than twice the time it takes for a pan of brownies to finish baking. Let t represent the time it takes for the brownies to finish baking. Write an algebraic expression for the time the cheesecake needs to bake.

3. Lewis works one-third of the hours that Michelle works each week. Let h represent the number of hours Michelle works. Write an algebraic expression for the number of hours Lewis works.

4. During a basketball game, a total of 155 points were scored. Let p represent the number of points scored by the home team. Write an algebraic expression for the number of points scored by the visitors.

5. The smaller of two numbers is 15 less than one-fourth of the larger number. Let n represent the larger number. Write an algebraic expression for the sum of the two numbers.

6. One number is six more than one-fifth the other. Their sum is 102. Find the two numbers.

7. The sum of two consecutive even integers is 86. Find the two integers.

8. A larger number is 17 less than twice a smaller number. If the smaller number is subtracted from the larger number, the result is 31. Find the two numbers.

9. Kramer purchased a trampoline. The cost of the trampoline, including a 5% sales tax, was $314.79. Find the cost of the trampoline before tax.

10. Norm purchased a pair of shoes that were on sale for 30% off regular price. If the sale price was $49.98, what was the regular price for the shoes?

1. _____

2. _____

3. _____

4. _____

5. _____

6. _____

7. _____

8. _____

9. _____

10. _____

Name:_____

11. In 2006, the number of crimes reported in a particular city dropped by 6% below the previous year. If the number of crimes reported in 2006 was 1081, how many crimes were reported in 2005?

11. _____

12. The Haul-It-All Truck Rental Company charges $65 for a two-day rental, plus $0.15 for each mile over 100 miles. If Burt's total charge for a two-day rental was $83.75, how many miles did he drive the truck?

12. _____

13. Betty has a big yard and wants to hire a lawn service to mow it for the summer. Angel Lawn Service offers two payment options. Option 1 requires a $300 security fee plus $30 per cutting. Option 2 does not require a security fee but costs $50 per cutting. How many cuttings will it take for the two options to cost the same?

13. _____

14. A triangle has a perimeter of 105 cm. Find the three sides if one side is 3 times the smallest side, and the third side is 30 cm larger than the smallest side.

14. _____

15. The width of a rectangle is 1 foot shorter than half the length. If the perimeter is 64 feet, find the dimensions of the rectangle.

15. _____

16. Each of the two smaller angles of a parallelogram measures 18° less than each of the two larger angles. Find the measures of the four angles.

16. _____

17. Tracy invested $38,000 for one year, part at 11% and part at 9%. If he earned a total of $4000, how much was invested at each rate?

17. _____

18. Car A and Car B enter an expressway at the same time and place and head the same direction. Car A travels at 70 miles per hour. If after 2.5 hours Car B is 50 miles behind Car A, what is the speed of Car B?

18. _____

19. How many liters of a 15% salt solution must be added to 2 liters of a 30% salt solution to yield a solution that is 18% salt?

19. _____

20. How many liters each of a 12% hydrochloric acid solution and a 24% hydrochloric acid solution should be mixed in order to obtain 16 liters of a 21% hydrochloric acid solution?

20. _____

Chapter 3 Test Form C

Name:_____

Date:_____

1. A recipe calls for three times as much flour as brown sugar. Let *x* represent the amount of brown sugar. Write an algebraic expression for the amount of flour.

    1. _____

2. Terry's age is 7 years less than twice Abdul's age. Let *a* represent Abdul's age. Write an algebraic expression for Terry's age.

    2. _____

3. During a baseball game, a total of 8 runs were scored. Let *r* represent the number of runs scored by the home team. Write an expression for the number of runs scored by the visiting team.

    3. _____

4. The price for a gallon of super unleaded gasoline is $0.12 more than the price for a gallon of regular unleaded gasoline. Let *g* represent the price for regular unleaded gasoline. Write an algebraic expression for the price of super unleaded gasoline.

    4. _____

5. The larger of two numbers is one more than five times the smaller number. Let *n* represent the smaller number. Write an expression for the sum of the two numbers.

    5. _____

6. The sum of two integers is 131. Find the two integers if the larger one is 8 more than twice the smaller.

    6. _____

7. The sum of three consecutive integers is 36. Find the three integers.

    7. _____

8. One number is five less than three-fourths the other. Their sum is 79. Find the two numbers.

    8. _____

9. Celine purchased a fedora. The cost of the fedora, including a 15% tax, was $316.25. Find the cost of the fedora before the tax.

    9. _____

10. Brandon purchased a pair of shoe that were on sale for 30% off the original price. If the sale price was $104.72, what was the original price?

    10. _____

11. From 2006 to 2007, enrollment at a university increased by 3.6%. If enrollment in 2007 was 12,173 students, what was the enrollment in 2006.

    11. _____

12. Adrian must to choose between two long distance plans. Plan A charges a monthly fee of $15.00, plus $0.07 per minute on every long-distance call. Plan B charges a monthly fee of $50.00 for unlimited long-distance calls (that is, no cost per minute). How many minutes of long-distance calls would it take for the cost of the two plans be the same?

12. _____

13. Rodney is landscaping his yard. He is having 600 paving bricks delivered by the local home improvement store. The store charges $25 for delivery. If the total cost for the bricks and delivery is $673, what is the price for one brick?

13. _____

14. The measure of the smallest angle in a triangle is 7° more than half the measure of the largest angle. The measure of the middle-sized angle is 2° smaller than the measure of the largest angle. Find the three angles.

14. _____

15. The length of a rectangle is four inches longer than twice the width. If the perimeter is 80 inches, find the dimensions of the rectangle.

15. _____

16. Each of the two larger angles of a parallelogram measures 5° more than six times the measure of each of the two smaller angles. Find the measures of the four angles.

16. _____

17. Thelma invested $31,000 for one year, part at 9% and part at 12%. If she earned a total of $3195, how much was invested at each rate?

17. _____

18. Two trains leave the same station at the same time but heading in opposite directions. The first train travels 15 miles per hour faster than second train. After 4 hours, they are 420 miles apart. At what speeds are the two trains traveling?

18. _____

19. How many liters of a 20% salt solution must be added to 6 liters of a 40% salt solution to yield a solution that is 35% salt?

19. _____

20. How many gallons each of a 12% chlorine solution and a 18% chlorine solution should be mixed in order to obtain 9 gallons of a 14% chlorine solution?

20. _____

Chapter 3 Test Form D

Name:_____

Date:_____

1. The total cost for a computer and printer is $1869. Let p represent the cost of the printer. Write an algebraic expression for the cost of the computer.

2. Susan's height (in inches) is 19 more than three times her age (in years). Let a represent Susan's age. Write an algebraic expression for her height.

3. Ethan's age is one year more than three times Anna's age. Let a represent Anna's age. Write an algebraic expression for Ethan's age.

4. Let x represent the price for a 6-pack of soda. Write an algebraic expression for the cost of one soda.

5. The smaller of two numbers is four more than two-thirds of the larger number. Let n represent the larger number. Write an expression for the sum of the two numbers.

6. The sum of two integers is 90. Find the two integers if the larger one is 6 more than twice the smaller.

7. The sum of two consecutive odd integers is 72. Find the two integers.

8. One number is eight less than half the other. Their sum is 52. Find the two numbers.

9. Antonio purchased a new Italian suit. The cost of the suit, including a 25% tax, was $1000. Find the cost of the suit before the tax.

10. After dieting and exercising for several months, Jared now weighs 20% less than he did originally. If Jared now weighs 180 pounds, how much did he weigh originally?

11. Between 2000 and 2006, the population of a town increased by 4%. If the population of the town in 2006 was 50,206, what was the it population in 2000?

1. _____

2. _____

3. _____

4. _____

5. _____

6. _____

7. _____

8. _____

9. _____

10. _____

11. _____

98

Name:_____

12. You are given $45 to buy refreshments for a party. If you spend $20 on a cake, $12 on paper plates, cups and silverware, how many sodas can you buy if they cost $.65 each?

12. _____

13. John needs to rent a moving van for a one-day move. AAA Rentals charges $50, plus $0.20 per mile. Haul-It-All Rentals charges $60, plus $15 per mile. How many miles must be driven for the rental cost to be the same for both companies?

13. _____

14. The measure of the smallest angle in a triangle is one-third the measure of the middle-sized angle. The measure of the largest angle is twice the measure of the middle-sized angle. Find the three angles.

14. _____

15. A triangle has a perimeter of 84 inches. Find the three sides if one side is 24 inches larger than the smallest side, and the third side is triple the smallest side.

15. _____

16. The length of a rectangle is three inches more than six times the width. If the perimeter is 76 inches, find the dimensions of the rectangle.

16. _____

17. Lloyd invested $18,000 for one year, part at 11% and part at 14%. If he earned a total of $2385, how much was invested at each rate?

17. _____

18. Two trains leave the same station at the same time but heading in opposite directions. The first train travels 5 miles per hour faster than second train. After 3 hours, they are 345 miles apart. At what speeds are the two trains traveling?

18. _____

19. How many liters of a 20% salt solution must be added to 8 liters of a 5% salt solution to yield a 10% salt solution?

19. _____

20. How many liters each of a 4% acid solution and a 36% acid solution should be mixed in order to obtain 8 liters of a 15% acid solution?

20. _____

Chapter 3 Test Form E

Name:_____

Date:_____

1. The price of an algebra textbook is $23 less than the price of a history textbook. Let x represent the price of the history book. Write an algebraic expression for the price of the algebra textbook.

    1. _____

2. At a basketball game, the bleachers were filled with 8,270 spectators. Let f represent the number of female spectators. Write an algebraic expression for the number of male spectators at the game.

    2. _____

3. An auto worker gets paid time and a half for overtime work. Let r represent the worker's regular pay rate. Write an expression for the overtime rate.

    3. _____

4. The sales tax in a particular state is 7.5%. Let x represent the price of an item to be purchased in this state. Write an algebraic expression for the complete purchase total.

    4. _____

5. Let x represent the price for a dozen long stem roses. Write an algebraic expression for the cost of one long stem rose.

    5. _____

6. The sum of two integers is 117. Find the two integers if the larger one is 5 more than three times the smaller.

    6. _____

7. The sum of two consecutive integers is 67. Find the two integers.

    7. _____

8. One number is eight less than two-thirds the other. Their sum is 62. Find the two numbers.

    8. _____

9. Blake purchased a new television. The cost of the television, including a 7.5% sales tax, was $322.50. Find the cost of the television before tax.

    9. _____

10. Luke purchased a leaf blower that was on sale for 25% off the original price. If the sale price $24.36, what was the original price?

    10. _____

11. From April 1 to May 1, the price of a particular stock increased by 15%. If the price per share on May 1 was $29.21, what was the price per share on April 1?

    11. _____

Chapter 3 Test Form E *(cont.)*

12. Easy Car Rental Company charges $85 per day, plus $0.16 for each mile in excess of 200 miles. If you only need the car for one day, how many miles can you drive for a total cost of $93?

12. _____

13. Cheyenne requested estimates from two stores on replacing part of the fence around her yard. The first store will charge $50 for a new gate and $3.50 for each foot of fence. The second store charges $75 for a new gate, and $2.25 for each foot of fence. For how many feet of fence will the price from the two stores be the same?

13. _____

14. The measure of the largest angle in a triangle is twice the measure of the smallest angle. The third angle is 45° smaller than the largest angle. Find the three angles.

14. _____

15. A rectangle is 12 cm longer than half the width. If the perimeter is 54 cm, find the dimensions of the rectangle.

15. _____

16. Each of the two larger angles of a parallelogram measures 26° more than the each of the two smaller angles. Find the measures of the four angles.

16. _____

17. Judith invested $13,000 for one year, part at 10% and part at 8%. If he earned a total of $1150, how much was invested at each rate?

17. _____

18. Car A and Car B enter an expressway at the same time and place but travel in opposite directions. Car A's speed is 12 miles per hour faster than Car B's speed. After 2 hours the cars are 248 miles apart. What are the speeds of the two cars?

18. _____

19. How many liters of a 40% salt solution must be added to 2 liters a 20% salt solution to yield a solution that is 30% salt?

19. _____

20. How many liters each of a 12% acid solution and a 20% acid solution should be mixed in order to obtain 8 liters of a 15% acid solution?

20. _____

Chapter 3 Test Form F

1. The price of a 2-liter bottle of A&W root beer is $0.90 more than the price of a 2-liter bottle of generic root beer soda. Let x represent the price of the generic root beer. Write an algebraic expression for the price of A&W.

1. _____

2. Rick's age is nine years less than three times Sam's age. Let a represent Sam's age. Write an algebraic expression for Rick's age.

2. _____

3. Clark works one-fourth of the hours that Lois works each week. Let h represent the number of hours Lois works. Write an algebraic expression for the number of hours Clark works.

3. _____

4. During a football game, a total of 72 points were scored. Let p represent the number of points scored by the home team. Write an algebraic expression for the number of points scored by the visitors.

4. _____

5. The smaller of two numbers is 8 less than one-third of the larger number. Let n represent the larger number. Write an algebraic expression for the sum of the two numbers.

5. _____

6. One number is seven less than four times the other. Their sum is 53. Find the two numbers.

6. _____

7. The sum of two consecutive odd integers is 176. Find the two integers.

7. _____

8. A larger number is 15 less than six times a smaller number. If the smaller number is subtracted from the larger number, the result is 60. Find the two numbers.

8. _____

9. Lamar purchased a textbook. The cost of the book, including a 7% sales tax, was $104.86. Find the cost of the book before tax.

9. _____

10. Cliff purchased a pair of shoes that were on sale for 30% off the regular price. If the sale price was $64.96, what was the regular price for the shoes?

10. _____

Chapter 3 Test Form F *(cont.)*

11. In 2006, the number of crimes reported in a particular city dropped by 5% below the previous year. If the number of crimes reported in 2006 was 969, how many crimes were reported in 2005?

11. _____

12. The Move-It Truck Rental Company charges $85 for a two-day rental, plus $0.18 for each mile over 200 miles. If Karl's total charge for a two-day rental was $106.60, how many miles did he drive the truck?

12. _____

13. Sarah wants to hire a lawn service to mow her yard for the summer. Baker Lawn Service offers two payment options. Option 1 requires a $400 security fee plus $25 per cutting. Option 2 does not require a security fee but costs $50 per cutting. How many cuttings will it take for the two options to cost the same?

13. _____

14. The measure of the smallest angle in a triangle is half the measure of the middle-size angle. The measure of the largest angle is 5° more than the middle-size angle. Find the three angles.

14. _____

15. The length of a rectangle is 3 millimeters shorter than twice the width. If the perimeter is 198 millimeters, find the dimensions of the rectangle.

15. _____

16. The measure of each of the larger angles of a parallelogram is four times the measure of each of the two smaller angles. Find the measures of the four angles.

16. _____

17. Stacy invested $11,000 for one year, part at 7% and part at 9%. If she earned a total of $930, how much was invested at each rate?

17. _____

18. Car A and Car B enter an expressway at the same time and place and head the same direction. Car A travels at 76 miles per hour. If after 3 hours Car B is 42 miles behind Car A, what is the speed of Car B?

18. _____

19. How many liters of a 18% salt solution must be added to 3 liters of a 40% salt solution to yield a solution that is 30% salt?

19. _____

20. How many milliliters each of an 8% acid solution and a 24% acid solution should be mixed in order to obtain 1200 milliliters of a 22% acid solution?

20. _____

Chapter 3 Test Form G

Name:_____

Date:_____

For questions 1 – 3, express the requested quantity in terms of the variable provided.

1. The floor number on which Jim lives in an apartment building is one more than twice the floor number on which Elaine lives. Let x represent Elaine's floor number. Write an expression for Jim's floor number.

   (a) $x + 2 + 1$ (b) $2x + 1$ (c) $2(x + 1)$ (d) $2x - 1$

2. The width of a rectangle is one less than one-third of the length. Let x represent the length. Write an expression for the perimeter of the rectangle.

   (a) $1 - \frac{1}{3}x$ (b) $2x + 2\left(1 - \frac{1}{3}x\right)$ (c) $\frac{1}{3}x - 1$ (d) $2x + 2\left(\frac{1}{3}x - 1\right)$

3. Cierra sold 4 less than twice as many donuts to Rich as she sold to Rachel. Let x represent the number of donuts sold to Rachel. Write an expression for the number of donuts sold to Rich.

   (a) $2x - 4$ (b) $4x - 2$ (c) $2(x - 4)$ (d) $4 - 2x$

For questions 4 – 6, express the verbal statement as an equation. Let x represent "a number."

4. The sum of a number and triple the number decreased by 7 is 19.

   (a) $x + (3x - 7) = 19$ (b) $x + (3x + 7) = 19$ (c) $x + \left(\frac{1}{3}x - 7\right) = 19$ (d) $x + \left(\frac{1}{3}x + 7\right) = 19$

5. The sum of twice a number and the number decreased by 4 is 11.

   (a) $2[x + (x - 4)] = 11$ (b) $2x + (x - 4) = 11$ (c) $2[x + (4x - 1)] = 11$ (d) $2x - (x + 4) = 11$

6. The sum of a number and the number decreased by 3 is 17.

   (a) $x + 3 = 17$ (b) $x + (3 - x) = 17$ (c) $x + (x - 3) = 17$ (d) $x + (3x - 3) = 17$

For questions 7 – 20, solve.

7. The sum of two consecutive even integers is 138. Find the two integers.

   (a) 64, 74   (b) 66, 72   (c) 68, 70   (d) 69, 69

8. The smaller of two numbers is seven more than half the larger number. If the smaller number is subtracted from the larger number, the result is 22. Find the smaller number.

   (a) 10   (b) 58   (c) 12   (d) 36

9. Karen has 1704 books in her library in which the number of fiction books is three times the number of non-fiction books. How many fiction books are in Karen's library?

   (a) 568   (b) 852   (c) 426   (d) 1278

10. Kirsten purchased a new treadmill for her home. The cost of the treadmill, including a 7% tax, was $1284. Find the cost of the treadmill before tax.

    (a) $1194.12   (b) $1200   (c) $1277   (d) $1373.88

11. From 2000 to 2006, the population of a town decreased by 12%. If the town's population in 2006 was 22,682 people, what was its population in 2000?

    (a) 25,775 people   (b) 25,404 people   (c) 22,694 people   (d) 189,017 people

Chapter 3 Test Form G *(cont.)*

12. Trevor is shopping for a plumber to help install a new kitchen sink. AAA Plumbing will charge $175 for parts and $35 per hour for labor. Handy Plumbing will charge $150 for parts and $40 per hour for labor. For how many hours of labor would the cost of the two plumbers be the same?

    (a) 4 hours   **(b)** 5 hours   **(c)** 6 hours   **(d)** 7 hours

13. Acme Car Rental Company charges $75 per day, plus $0.12 for each mile in excess of 100 miles. If you only need the car for one day, how many miles can you drive for a total cost of $90?

    (a) 165 miles   **(b)** 125 miles   **(c)** 750 miles   **(d)** 225 miles

14. A triangle has a perimeter of 68 cm. Find the longest side if one side is three times the smallest side, and the third side is 18 cm larger than the smallest side.

    (a) 10 cm   **(b)** 30 cm   **(c)** 17.2 cm   **(d)** 51.6 cm

15. If the measure of the largest angle in a triangle is nine times the measure of the smallest angle, and the measure of the remaining angle is double the measure of the smallest angle, find the measures of the three angles.

    (a) $10°, 20°, 90°$ **(b)** $15°, 30°, 135°$ **(c)** $20°, 40°, 120°$ **(d)** $12°, 24°, 108°$

16. Each of the two larger angles in a parallelogram measures $100°$ more than the measure of the smaller angles. Find the measures of the four angles.

    (a) $50°, 50°, 130°, 130°$ **(b)** $80°, 80°, 100°, 100°$ **(c)** $40°, 40°, 140°, 140°$ **(d)** $35°, 35°, 135°, 135°$

17. Rhonda invested $25,000 for one year, part at 9% and part at 12%. If she earned a total of $2760, how much was invested at each rate?

    (a) $10,000 at 9%; $15,000 at 12%   **(b)** $8000 at 9%; $17,000 at 12%
    (c) $15,000 at 9%; $10,000 at 12%   **(d)** $17,000 at 9%; $8000 at 12%

18. Car A and Car B enter an expressway at the same time and place but travel in opposite directions. Car A's speed is 8 miles per hour faster than Car B's speed. After 2.5 hours the cars are 330 miles apart. What are the speeds of the two cars?

    (a) 60 mph; 72 mph   **(b)** 55 mph; 63 mph   **(c)** 62 mph; 70 mph   **(d)** 58 mph; 66 mph

19. How many liters of a 30% salt solution must be added to 2 liters of a 10% salt solution to yield a solution that is 20% salt?

    (a) 1 liters   **(b)** 2 liters   **(c)** 2.5 liters   **(d)** 3 liters

20. How many milliliters each of a 10% acid solution and a 25% acid solution should be mixed in order to obtain 1500 milliliters of a 20% acid solution?

    (a) 400 ml of 10%; 1100 ml of 25%   **(b)** 600 ml of 10%; 900 ml of 25%
    (c) 800 ml of 10%; 700 ml of 25%   **(d)** 500 ml of 10%; 1000 ml of 25%

Chapter 3 Test Form H

Date:_____

For questions 1 – 3, express the requested quantity in terms of the variable provided.

1. Twelve thousand dollars is being divided between Lewis and Sonia. Let x represent the amount that goes to Lewis. Write an expression for the amount that goes to Sonia.

   (a) $x - 12,000$ **(b)** $12,000 - x$ **(c)** $\dfrac{12,000}{x}$ **(d)** $x + 12,000$

2. Grace is constructing a fence around her rectangular garden. The length of the garden is 15 meters shorter than triple its width. Let x represent the width. Write an expression for the perimeter of the garden.

   (a) $15 - 3x$ **(b)** $2x + 2(15 - 3x)$ **(c)** $3x - 15$ **(d)** $2x + 2(3x - 15)$

3. Jim weighs 50 pounds less than twice George's weight. Let x represent George's weight.

   (a) $2(50 - x)$ **(b)** $2(x - 50)$ **(c)** $2x - 50$ **(d)** $50 - 2x$

For questions 4 – 6, express the verbal statement as an equation.

4. The sum of a number and twice the number decreased by 5 is 19.

   (a) $x + (5 - 2x) = 19$ **(b)** $x + (2x - 5) = 19$ **(c)** $x + (5x - 2) = 19$ **(d)** $x + (2 - 5x) = 19$

5. Three times a number decreased by half of the number is 12.

   (a) $3x - \dfrac{1}{2}x = 12$ **(b)** $\dfrac{3}{2}x = 12$ **(c)** $3x + \dfrac{1}{2}x = 12$ **(d)** $\dfrac{1}{2}x - 3x = 12$

6. The sum of a number and four times the number decreased by 3 is 12.

   (a) $x + (3x - 4) = 12$ **(b)** $x + (4 - 3x) = 12$ **(c)** $x + (3 - 4x) = 12$ **(d)** $x + (4x - 3) = 12$

For questions 7 – 20, solve.

7. The sum of two consecutive odd integers is 56. Find the two integers.

   (a) 25, 31   **(b)** 27, 29   **(c)** 28, 28   **(d)** 26, 30

8. The smaller of two numbers is five more than one-third of the larger number. If the smaller number is subtracted from the larger number, the result is 33. Find the smaller number.

   (a) 12   **(b)** 57   **(c)** 21   **(d)** 24

9. Regan has 2140 books in her library in which the number of fiction books is four times the number of non-fiction books. How many fiction books are in Regan's library?

   (a) 1070   **(b)** 535   **(c)** 428   **(d)** 1712

10. Philip purchased a new stereo system. The cost of the system, including a 6.5% tax, was $2343. Find the cost of the stereo system before tax.

    (a) $2150.00   **(b)** $2190.71   **(c)** $2200.00   **(d)** $2495.30

11. From 2000 to 2006, the population of a town decreased by 9%. If the town's population in 2006 was 21,294 people, what was its population in 2000?

    (a) 23,400   **(b)** 23,210   **(c)** 236,600   **(d)** 19,378

12. Nova Car Rental Company charges $70 per day, plus $0.14 for each mile in excess of 100 miles. If you only need the car for one day, how many miles can you drive for a total cost of $91?

    (a) 650 miles   **(b)** 250 miles   **(c)** 150 miles   **(d)** 161 miles

Chapter 3 Test Form H *(cont.)*

13. Stephanie is shopping for a plumber to help install a new kitchen sink. AAA Plumbing will charge $220 for parts and $45 per hour for labor. The Main Drain Plumbing Company will charge $175 for parts and $60 per hour for labor. For how many hours of labor would the cost of the two plumbers be the same?

    (a) 2 hours   **(b)** 3 hours   **(c)** 4 hours   **(d)** 5 hours

14. A triangle has a perimeter of 72 cm. Find the smallest side if one side is four times the smallest side, and the third side is 24 cm larger than the smallest side.

    (a) 32 cm   **(b)** 16 cm   **(c)** 8 cm   **(d)** 24 cm

15. If the largest angle in a triangle is 100° larger than the measure of the smallest angle, and the remaining angle is 10° smaller than four times the smallest angle. Find the 3 angles.

    (a) 15°, 50°, 115°   **(b)** 20°, 70°, 120°   **(c)** 20°, 70°, 90°   **(d)** 25°, 90°, 125°

16. Each of the two larger angles in a parallelogram measures 110° more than the measure of the smaller angles. Find the measures of the four angles.

    (a) 45°, 45°, 135°, 135°   **(b)** 80°, 80°, 100°, 100°   **(c)** 35°, 35°, 145°, 145°   **(d)** 30°, 30°, 140°, 140°

17. Chris invested $15,000 for one year, part at 9% and part at 12%. If she earned a total of $1620, how much was invested at each rate?

    (a) $6000 at 9%; $9000 at 12%   **(b)** $10,000 at 9%; $5000 at 12%

    (c) $9000 at 9%; $6000 at 12%   **(d)** $5000 at 9%; $10,000 at 12%

18. Car A and Car B enter an expressway at the same time and place but travel in opposite directions. Car A's speed is 10 miles per hour faster than Car B's speed. After 2 hours the cars are 268 miles apart. What are the speeds of the two cars?

    (a) 61 mph; 73 mph   **(b)** 58 mph; 68 mph   **(c)** 60 mph; 74 mph   **(d)** 62 mph; 72 mph

19. How many gallons of a 50% salt solution must be added to 3 gallons of a 20% solution to yield a solution that is 40% salt?

    (a) 3 gallons   **(b)** 5 gallons   **(c)** 6 gallons   **(d)** 8 gallons

20. How many milliliters each of a 12% acid solution and a 30% acid solution should be mixed in order to obtain 1500 milliliters of a 18% acid solution?

    (a) 1000 ml of 12%; 500 ml of 30%   **(b)** 900 ml of 12%; 600 ml of 30%

    (c) 700 ml of 12%; 800 ml of 30%   **(d)** 400 ml of 12%; 1100 ml of 30%

Chapter 4 Pretest Form A

In which quadrant does each point lie?

1. $(-4, 3)$

1. _____

2. $(-2, -8)$

2. _____

3. Determine if the points given points are collinear:
   $(-1, -1), (8, 5), (-4, -3)$

3. _____

4. Determine whether or not $(15, -2)$ satisfies the equation $2x + 5y = 40$.

4. _____

For questions 5 – 6, find the slope of the line through the given points.

5. $(-1, -2)$ and $(7, -9)$

5. _____

6. $(6, 0)$ and $(-9, -10)$

6. _____

7. Find the slope and y-intercept of $6x - 3y = 6$.

7. _____

For questions 8 – 10, graph the equation using the method of your choice.

8. $y = 3x - 5$

8.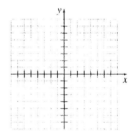

9. $3x + 6y = 18$

9.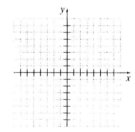

10. $x = -5$

10.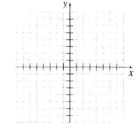

Chapter 4 Pretest Form A *(cont.)*

11. Write the equation of the line.

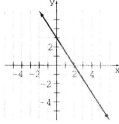

11. _____

For questions 12 – 13, determine if the two given lines are parallel, perpendicular, or neither.

12. $6x + 7y = -35$

$$y = -\frac{6}{7}x - 4$$

12. _____

13. $3y = x + 5$

$$y = 3x + 5$$

13. _____

For questions 14 – 18, find the equation of the line with the given properties.

14. Slope 8, y-intercept –3.

14. _____

15. Slope 5, passing through the point $(1, 6)$

15. _____

16. Passing through the points $(3, -4)$ and $(-6, 11)$

16. _____

17. Horizontal, passing through the point $(-4, -3)$

17. _____

18. Passing through $(1, 1)$, perpendicular to $2y = x + 3$

18. _____

19. Write the equation $y = -\frac{3}{4}x - 10$ in standard form.

19. _____

20. What is the slope of a vertical line?

20. _____

Chapter 4 Pretest Form B

Name:_____

Date:_____

In which quadrant does each point lie?

1. $(-7, -10)$

1. _____

2. $(9, -14)$

2. _____

3. Determine if the points given points are collinear:
   $(0, 5), (-6, 7), (9, 2)$

3. _____

4. Determine whether $(-2, -5)$ satisfies the equation $5y = 9x - 6$.

4. _____

For questions 5 – 6, find the slope of the line through the given points.

5. $(5, 9)$ and $(9, 0)$

5. _____

6. $(6, 9)$ and $(-3, -12)$

6. _____

7. Find the slope and y-intercept of $4x - 2y = -16$.

7. _____

For questions 8 – 10, graph the equation using the method of your choice.

8. $y = -2x + 5$

8.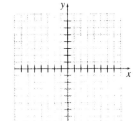

9. $6x - 3y = 12$

9.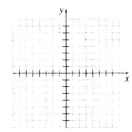

10. $y = -4$

10.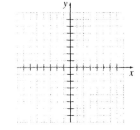

11. Write the equation of the line.   11. _____

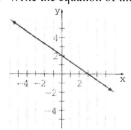

For questions 12 – 13, determine if the two given lines are parallel, perpendicular, or neither.

12. $7x + 8y = 16$ 12. _____
   $y = -7x - 5$

13. $3x + 6y = 1$ 13. _____
   $2x - y = 5$

For questions 14 – 18, find the equation of the line with the given properties.

14. Slope 3, y-intercept -4 14. _____

15. Slope -4, passing through the point $(0, -5)$ 15. _____

16. Passing through the points $(1, 5)$ and $(3, -3)$ 16. _____

17. Vertical, passing through the point $(5, -2)$ 17. _____

18. Passing through $(5, 2)$, parallel to $x + 2y = 3$ 13. _____

19. Write the equation $y = \dfrac{2}{5}x + 3$ in standard form.   19. _____

20. What is the slope of a horizontal line?   20. _____

Mini-Lecture 4.1
The Cartesian Coordinate System and Linear Equations in Two Variables

Learning Objectives:

1. Plot points in the Cartesian coordinate system.
2. Determine whether an ordered pair is a solution to a linear equation.
3. Key Vocabulary: *graph, Cartesian coordinate system, rectangular coordinate system, quadrants, x-axis, y-axis, origin, coordinates, ordered pair, linear equation in two variables, collinear*

Examples:

1. Plot each point on the same axes

   a) $(4,2)$ b) $(-3,-5)$ c) $(2,0)$

   d) $(6,-3)$ e) $(0,5)$ f) $(-4,3)$

2. List the ordered pairs for each point.

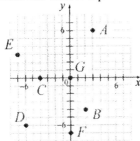

3. Determine whether the three points appear to be collinear.

   a) $(3,5); (6,7); (10,8)$

   b) $(-5,8); (-3,4); (5,-12)$

4. Determine whether each of the following ordered pairs satisfy the equation $3x + y = 12$.

   a) $(0,0)$ b) $(-2,18)$

   c) $(4,0)$ d) $(6,6)$

Teaching Notes:

- Remind students that in an ordered pair the order of the numbers matters.
- Point out how we can graph solutions to a linear equation in one variable on a number line to give a visual representation of the solution set. The same is true for linear equations in two variables but we plot ordered pairs in the Cartesian plane instead of single values on a number line.
- Remind students that two distinct points completely determine a line.

Answers: 1a) – 1f) see graphing answer pages; 2) A(3,6), B(2,−4), C(−4,0), D(−6,−6), E(−7,3), F(0,−7), G(0,0); 3a) no; 3b) yes; 4a) no; 4b) yes; 4c) yes; 4d) no

Mini-Lecture 4.2
Graphing Linear Equations

Learning Objectives:

1. Graph linear equations by plotting points.
2. Graph linear equations of the form $ax + by = 0$.
3. Graph linear equations using the x- and y-intercepts.
4. Graph horizontal and vertical lines.
5. Study applications of graphs.
6. Key Vocabulary: *linear equation in two variables, graph of an equation, x-intercept, y-intercept, horizontal line, vertical line*

Examples:

1. Graph the equations by plotting points. Plot at least three points for each graph.

   a) $y = \dfrac{3}{2}x - 2$

   b) $2x - 4y = 8$

2. Graph the equations by plotting points. Plot at least three points for each graph.

   a) $2x - 3y = 0$

   b) $4x + 2y = 0$

3. Graph using the x- and y-intercepts.

   a) $2x - 2y = 6$

   b) $3y = 4x + 8$

4. Graph the equations.

   a) $x = -2$

   b) $y = 3$

5. Jake Gardners' cell phone plan is for emergencies only. He is charged a monthly fee of $5 plus $0.50 per minute for calls that are made.

   a) Write an equation for the total monthly cost, C, when n minutes are used.

   b) Graph the equation for up to and including 20 minutes used.

   c) Estimate the total monthly cost if 13 minutes are used.

   d) If the total monthly bill is $14, estimate the number of minutes used.

Teaching Notes:

- Students may need to solve for y first when finding ordered pairs to plot. Remind them that this is not necessary but can help identify values for x that will lead to integer values for y.
- Remind students to pick values that are easy to work with, such as 0 or 1, whenever possible.
- Students sometimes get confused when working with equations for vertical and horizontal lines since only one variable is typically written. It might help to show both variables with one of the variables having a 0 coefficient.
- Remind students that every point on a graph represents an ordered pair solution to the equation of the graph.
- Rather than just telling students that the graph of every linear equation of the form $ax + by = 0$ passes through the origin, try using a discovery activity to have them make this observation.

Answers: *1) – 4) graphing answer pages; 5a) $C = 0.50n + 5$; 5b) graphing answer pages; 5c) $11.50; 5d) 18 minutes*

Mini-Lecture 4.3
Slope of a Line

Learning Objectives:
1. Find the slope of a line.
2. Recognize positive and negative slopes.
3. Examine the slopes of horizontal and vertical lines.
4. Examine the slopes of parallel and perpendicular lines.
5. Key vocabulary: *slope, rise, run, positive slope, negative slope, parallel, perpendicular, negative reciprocals*

Examples:
1. Using the slope formula, find the slope of the line through the given points.

   a) $(5,2)$ and $(3,8)$ b) $(-4,7)$ and $(2,10)$

   c) $(9,5)$ and $(9,12)$ d) $(-3,4)$ and $(-3,-4)$

2. By observing the vertical and horizontal change between the two points indicated, determine if the slope of the line is positive or negative, then use the slope formula to compute the slope.

   a) $(5,1)$ and $(3,9)$ b) $(-7,-5)$ and $(-2,3)$

   c) 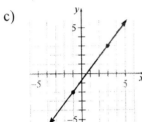   d)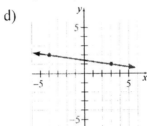

3. Graph the line with the given slope that goes through the given point.

   a) Through $(2,4)$ with $m=0$.   b) Through $(-2,-1)$ with slope undefined.

   c) Through $(-1,2)$ with $m=-\dfrac{3}{4}$

4. Let m_1 represent the slope of line 1, and m_2 represent the slope of line 2. Indicate whether line 1 and line 2 are parallel, perpendicular, or neither.

   a) $m_1=\dfrac{3}{5}$, $m_2=-\dfrac{5}{3}$ b) $m_1=-\dfrac{2}{3}$, $m_2=\dfrac{2}{5}$ c) $m_1=\dfrac{2}{7}$, $m_2=\dfrac{2}{7}$

Teaching Notes:
- Some students confuse the slope formula and put the change of x in the numerator.
- Remind students to be careful when subtracting negatives.
- Students often say a horizontal line has 'no slope'. Remind them that having a slope of 0 is not the same as having no slope.
- Some students confuse the term 'negative reciprocal' and think this means the slope is negative. In such cases, it may be helpful to have them think 'opposite reciprocal' instead.

Answers: *1a)* $m=-3$; *1b)* $m=\frac{1}{2}$; *1c) undefined;* *1d)* $m=0$; *2a) negative,* $m=-4$; *2b) positive,* $m=\frac{8}{5}$; *2c) positive,* $m=\frac{5}{4}$; *2d) negative,* $m=-\frac{1}{7}$; *3) graphing answer pages;* *4a) perpendicular;* *4b) neither;* *4c) parallel*

Mini-Lecture 4.4
Slope-Intercept and Point-Slope Forms of a Linear Equation

Learning Objectives:

1. Write a linear equation in slope-intercept form.
2. Graph a linear equation using the slope and y-intercept.
3. Use the slope-intercept form to determine the equation of a line.
4. Use the point-slope form to determine the equation of a line.
5. Compare the three methods of graphing linear equations.
6. Key vocabulary: *slope, y-intercept, slope-intercept form, point-slope form*

Examples:

1. Determine the slope and y-intercept of the line represented by each equation.
   a) $y = 3x + 7$
   b) $y = 4x - 5$
   c) $3x + 2y = 1$
   d) $5x + 9y = 0$
   e) $-2x + 8y = 16$
   f) $y = -2$

2. Determine whether each pair of lines are parallel, perpendicular, or neither.

   a) $y = 4x - 3$
   $y = -2x + 5$

   b) $5x + 10y = 25$
   $x - 0.5y = 8$

   c) $y = \dfrac{5}{3}x - 7$
   $10x - 6y = 18$

3. Write the equation of each line, with the given properties, in slope-intercept form.
   a) Slope $= -\dfrac{4}{5}$, y-intercept is $\left(0, \dfrac{3}{10}\right)$
   b) Slope $= -2$, through $(-2, 1)$.
   c) Through $(3, 5)$ and $(0, 2)$.
   d) Through $(-4, 3)$ and $(4, 7)$.

Teaching Notes:

- Students often get overwhelmed with having to learn lots of equations or formulas. Showing them how the point-slope form is a variation of the slope formula, or how they can use the slope-intercept form with a point and the slope may be helpful.
- Remind students that if the linear equation does not have a constant term, the y-intercept is the origin, $(0,0)$.
- Review how the slope indicates if two lines are parallel, perpendicular, or neither.

Answers: 1a) slope: 3, y-intercept: $(0,7)$; 1b) slope: 4, y-intercept: $(0,-5)$; 1c) slope: $-\frac{3}{2}$, y-intercept: $\left(0,\frac{1}{2}\right)$; 1d) slope: $-\frac{5}{9}$, y-intercept: $(0,0)$; 1e) slope: $\frac{1}{4}$, y-intercept: $(0,2)$; 1f) slope: 0, y-intercept: $(0,-2)$; 2a) neither; 2b) perpendicular; 2c) parallel; 3a) $y = -\frac{4}{5}x + \frac{3}{10}$; 3b) $y = -2x - 3$; 3c) $y = x + 2$; 3d) $y = \frac{1}{2}x + 5$

Additional Exercises 4.1

1. (−8, 10)

2. (25, −36)

3. (94, 71)

4. (−281, −165)

1. _____

2. _____

3. _____

4. _____

List the ordered pairs corresponding to each point.

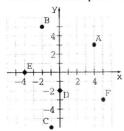

5. *A*

6. *B*

7. *C*

8. *D*

9. *E*

10. *F*

5. _____

6. _____

7. _____

8. _____

9. _____

10. _____

Plot each point.

11. *A* (2, −5)

12. *B* (0, 2)

13. *C* (5, 0)

14. *D* (−2, 4)

11.–14.

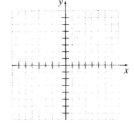

Determine if each set of points is collinear.

15. {(1, −5), (−4, −1), (6, −9)}

16. {(−3, −6), (3, −4), (14, 0)}

17. {(2, 5) , (2, 7) , (2, −3)}

18. Find *u* such that the points (−2, 3), (4, *u*), and (0, −1) are collinear.

19. Determine if the ordered pair (5, −5) is a solution of $4x − 3y = −35$.

20. Determine if the ordered pair (−1, 3) is a solution of $4x − 3y = 13$.

15. _____

16. _____

17. _____

18. _____

19. _____

20. _____

Additional Exercises 4.2

Name:_____

Date:_____

Find the missing coordinate in the solutions for $2x - 4y = 4$.

  1. $(?, 1)$

  2. $(2, ?)$

  3. $(-4, ?)$

Find the missing coordinate in the solutions for $3x - y = -7$.

  4. $(?, 7)$

  5. $(2, ?)$

  6. $(-1, ?)$

Graph by plotting points. Plot at least three points for each graph.

  7. $4x + y = 5$

  8. $y = 2x - 3$

  9. $2y = x - 8$

1. _____

2. _____

3. _____

4. _____

5. _____

6. _____

7.

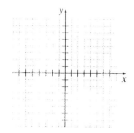

8.

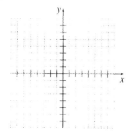

9.

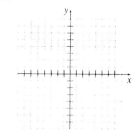

Additional Exercises 4.2 *(cont.)*

Graph each equation.

10. $y = -4$

10.

11. $x = -3$

11.

Graph using the *x* and *y* intercepts.

12. $2x + 3y = 6$

12.

13. $x - 2y = -6$

13.

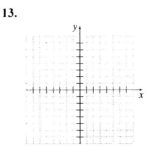

14. $3x - 4y = 0$

14.

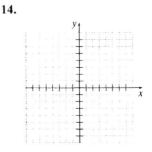

Additional Exercises 4.2 *(cont.)*

Name:_____

Write the equation represented by the given graph.

15.

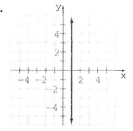

15. _____

16.

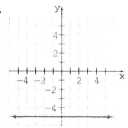

16. _____

Use the following information for questions 17–20.

The cost of renting a U-Drive truck is an initial charge of $120 plus $0.50 per mile.

17. Write an equation for cost, c, in terms of miles driven, m.

17. _____

18. Graph the equation for 0 to 500 miles driven.

18.

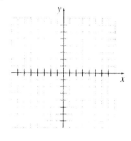

19. Estimate the cost of driving 200 miles.

19. _____

20. If the total cost is $260, estimate the miles driven.

20. _____

Additional Exercises 4.3

Find the slope of the line through the given points.

1. (0, 3) and (–9, 1)

2. (–1, 1) and (8, 4)

3. (8, 5) and (3, –5)

4. (8, –7) and (4, 1)

5. (7, 2) and (7, –4)

6. (4, 7) and (9, 8)

7. (–3, 4) and (7, –4)

8. (3, –2) and (–1, –2)

9. (0, 0) and (2, 10)

10. (–2, –6) and (3, –2)

11. (1, 0) and (0, –7)

1. _____

2. _____

3. _____

4. _____

5. _____

6. _____

7. _____

8. _____

9. _____

10. _____

11. _____

By observing the vertical and horizontal change of the line between the two points indicated, determine the slope of the line.

12.

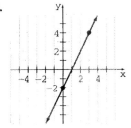

12. _____

13.

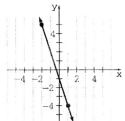

13. _____

14.

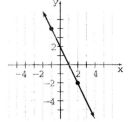

14. _____

Additional Exercises 4.3 *(cont.)*

15.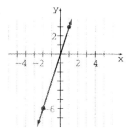

15. _____

16.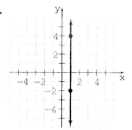

16. _____

17.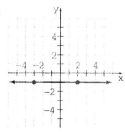

17. _____

18. Find u so that the line through $(1, 4)$ and $(5, 7)$, is parallel to the line through $(-2, -3)$ and $(u, 3)$.

18. _____

19. Find u so that the line through $(7, 3)$ and $(2, 4)$, is perpendicular to the line through $(6, u)$ and $(7, 8)$.

19. _____

20. Find u so that the line through $(1, 2)$ and $(-3, 2)$, is perpendicular to the line through $(3, -2)$ and $(u, 5)$.

20. _____

Additional Exercises 4.4

Find the slope and the *y*-intercept of the line represented by the given equation.

  1. $y = -x + 12$

  2. $y = 7x$

  3. $9x - 3y = -81$

  4. $5x - 4y = 4$

1. _____

2. _____

3. _____

4. _____

Graph using the slope and *y*-intercept.

  5. $y = x + 2$

5.

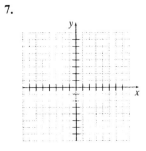

  6. $y = -3x - 3$

6.

  7. $6x + 3y = 9$

7.

Determine the equation of each line.

  8.

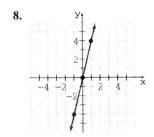

8. _____

Additional Exercises 4.4 *(cont.)*

9.

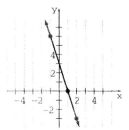

9. _____

10.

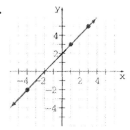

10. _____

Determine if the two given lines are parallel, perpendicular, or neither.

11. $4x + 9y = -27$

$y = -\dfrac{4}{9}x - 9$

11. _____

12. $y = -3x + 6$
$-3y + x = -2$

12. _____

13. $28x + 14y = -3$
$-14y + 28x = -3$

13. _____

14. $2y = 3x - 2$
$8x + 12y = -2$

14. _____

Write the equation of the line with the given properties in slope-intercept form.

15. Having slope 7 and y-intercept -2.

15. _____

16. Having slope 5 and y-intercept -5.

16. _____

17. With slope 1 passing through the point $(2, 3)$.

17. _____

18. With slope 4 passing through the point $(3, 4)$.

18. _____

19. Through $(1, 6)$ and $(5, -2)$.

19. _____

20. Through $(1, -7)$ and $(4, 2)$.

20. _____

Chapter 4 Test Form A

Name:_____

Date:_____

1. In which quadrant does the point $(-2, 3)$ lie?

1. _____

2. In which quadrant does the point $(12, -18)$ lie?

2. _____

3. Determine if the points given are collinear: $(0, -1), (3, 5), (-1, -3)$

3. _____

4. Find the missing coordinate in the following solution for $2x + 3y = 9$: $(-6, ?)$

4. _____

For questions 5 – 8, graph the equation using the method of your choice.

5. $3x - y = 3$

5.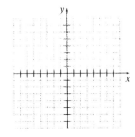

6. $2x + 3y = 12$

6.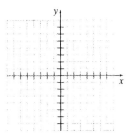

7. $x = -2$

7.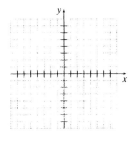

8. $y = -x$

8.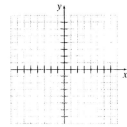

For questions 9 – 10, find the slope of the line through the given points.

9. $(2, 5)$ and $(-3, -5)$

9. _____

10. $(-2, -3)$ and $(4, -9)$

10. _____

Chapter 4 Test Form A *(cont.)*

11. Determine the slope and y-intercept of the graph of the equation
    $4x - 3y = 5$

    11. _____

12. Two distinct lines have slopes $m_1 = \dfrac{3}{4}$ and $m_2 = -\dfrac{3}{4}$. Are the lines
    parallel, perpendicular, or neither?

    12. _____

13. Write the equation of the line.

    13. _____

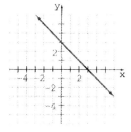

14. Determine if the two lines are parallel, perpendicular, or neither:
    $2x - y = 3$
    $y - 2x = 3$

    14. _____

For questions 15 – 18, find the equation of the line with the given properties.

15. Slope $= -1$, through $(4, 3)$

    15. _____

16. Through $(4, 3)$ and $(-2, -8)$

    16. _____

17. Slope is undefined, through $(5, -1)$

    17. _____

18. Through $(3, 1)$, perpendicular to the line $3x + y = 5$

    • 18. _____

Use the following information to answer questions 19 – 20.

A real estate agent has a weekly income, i, that can be determined by the formula $i = 190 + 0.06s$, where s represents her average weekly sales.

19. Draw a graph of her weekly income for average weekly sales
    ranging from \$0 to \$10,000.

    19.

    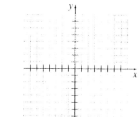

20. Estimate her weekly income if her average sales for the week are
    \$4,000.

    20. _____

Chapter 4 Test Form B

1. In which quadrant does the point $(2, -5)$ lie?

1. _____

2. In which quadrant does the point $(-13, -9)$ $(12, -18)$ lie?

2. _____

3. Determine if the points given are collinear: $(-3, -5), (1, 1), (4, 7)$

3. _____

4. Find the missing coordinate in the following solution for $5x - 2y = 10$: $(?, 5)$

4. _____

For questions 5 – 8, graph the equation using the method of your choice.

5. $3x + 2y = 6$

5.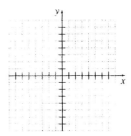

6. $5x + 3y = -15$

6.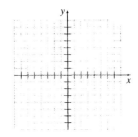

7. $y = 3$

7.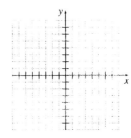

8. $y = -2x + 1$

8.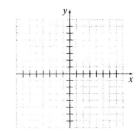

For questions 9 – 10, find the slope of the line through the given points.

9. $(-3, 7)$ and $(4, 2)$

9. _____

10. $(8, -3)$ and $(5, -3)$

10. _____

11. Determine the slope and y-intercept of the graph of the equation.
$-7x + 3y = 63$

11. _____

12. Two distinct lines have slopes $m_1 = \dfrac{5}{8}$ and $m_2 = -\dfrac{5}{8}$. Are the lines parallel, perpendicular, or neither?

12. _____

13. Write the equation of the line.

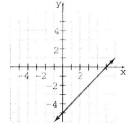

13. _____

14. Determine if the two lines are parallel, perpendicular or neither.
$x + 3y = 2$
$-3x + y = 2$

14. _____

For questions 15 – 18, find the equation of the line with the given properties.

15. Slope = –4, through $(5, -3)$

15. _____

16. Through $(-6, 10)$ and $(9, -10)$

16. _____

17. Slope is undefined, through $(-4, -2)$

17. _____

18. Through $(4, -2)$, parallel to $3x - 2y = 5$

18. _____

Use the following information to answer questions 19 – 20.

A stockbroker charges $30 plus 5 cents per share of stock bought or sold. The broker's commission, c in dollars, is a function of the number of shares n, bought or sold. $c = 30 + 0.05n$.

19. Draw a graph illustrating the broker's commission for up to and including 10,000 shares of stock.

19.

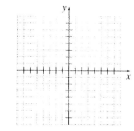

20. If the broker earns a commission of $39. on one sale, how many shares did she sell?

20. _____

Chapter 4 Test Form C

1. In which quadrant does the point $(-1, -5)$ lie?

1. _____

2. In which quadrant does the point $(12, 24)$ lie?

2. _____

3. Determine if the points given are collinear: $(-2, -5), (0, -5), (2, -5)$

3. _____

4. Find the missing coordinate in the following solution for $3x - 4y = 8$: $(?, 4)$

4. _____

For questions 5 – 8, graph the equation using the method of your choice.

5. $3x - 2y = -6$

5.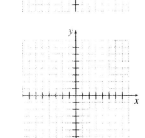

6. $5x - 3y = -15$

6.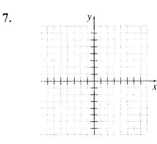

7. $x = 3$

7.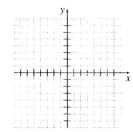

8. $y = -4x$

8.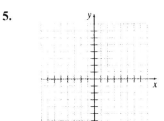

For questions 9 – 10, find the slope of the line through the given points.

9. $(-2, 9)$ and $(5, 2)$

9. _____

10. $(4, -6)$ and $(4, -10)$

10. _____

11. Determine the slope and y-intercept of the graph of the equation.
$6x - 3y = 15$

11. _____

12. Determine the slope of a line that is parallel to the graph of
$5x - 3y = 21$

12. _____

13. Write the equation of the line.

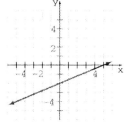

13. _____

14. Determine if the two lines are parallel, perpendicular or neither:
$12x - 3y = 4$

$x + 4y = 1$

14. _____

For questions 15 – 18, find the equation of the line with the given properties.

15. Slope $= -\dfrac{1}{2}$, through (6, 5)

15. _____

16. Through $(2, -3)$ and $(6, -17)$

16. _____

17. Horizontal, through $(-3, -2)$

17. _____

18. Through (1, 0), parallel to the line $5x - y = 3$

18. _____

Use the following information to answer questions 19 – 20.

The monthly profit, p, of a particular store can be estimated by the function $p = 3x - 1800$, where x represents the number of items sold.

19. Draw a graph of the function for up to and including 1000 items sold.

19.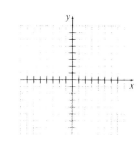

20. Estimate the profit if 600 items are sold.

20. _____

Chapter 4 Test Form D

1. In which quadrant does the point $(-5, 2)$ lie?

   1. _____

2. In which quadrant does the point $(14, -10)$ lie?

   2. _____

3. Determine if the points given are collinear: $(-4, 0), (0, -2), (8, -6)$

   3. _____

4. Find the missing coordinate in the following solution for $5x - 2y = 10$: $(-2, ?)$

   4. _____

For questions 5 – 8, graph the equation using the method of your choice.

5. $4x + y = 4$

   5.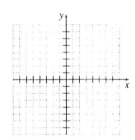

6. $4x + 3y = -12$

   6.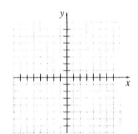

7. $y = -2$

   7.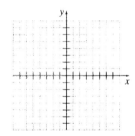

8. $y = 3x - 4$

   8.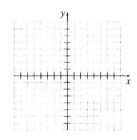

For questions 9 – 10, find the slope of the line through the given points.

9. $(6, -1)$ and $(3, 1)$

   9. _____

10. $(2, 7)$ and $(8, -5)$

   10. _____

11. Determine the slope and *y*-intercept of the graph of the equation.   **11.** _____
$7x - 4y = 16$

12. Two distinct lines have slopes $m_1 = -4$ and $m_2 = \dfrac{1}{4}$. Are the lines   **12.** _____
parallel, perpendicular, or neither?

13. Write the equation of the line.   **13.** _____

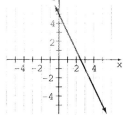

14. Determine if the two lines are parallel, perpendicular or neither: **14.** _____
$9y - 4x = 6$
$4y - 9x = 6$

For questions 15 – 18, find the equation of the line with the given properties.

15. Slope $= -\dfrac{1}{3}$, through $(-6, 2)$ **15.** _____

16. Through $(8, -3)$ and $(-4, -6)$ **16.** _____

17. Vertical, through $(2, 1)$ **17.** _____

18. Through $(10, 3)$, parallel to the line $2x - 5y = 1$ **18.** _____

Use the following information to answer questions 19 – 20.

A salesperson has a weekly income, *m*, that can be defined by the formula $m = 500 + 0.05s$, where *s* is the salesperson's weekly sales.

19. Draw a graph of the weekly income for sales from \$0 to \$10,000.   **19.**

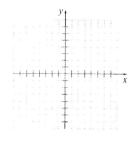

20. Estimate her sales if her weekly income is \$725.   **20.** _____

Chapter 4 Test Form E

1. In which quadrant does the point (4, –1) lie?

    1. _____

2. In which quadrant does the point (–22, –18) lie?

    2. _____

3. Determine if the points given are collinear: (2, 2), (0, –4), (–1, –1)

    3. _____

4. Find the missing coordinate in the following solution for $3x - y = 4$: (?, 5)

    4. _____

For questions 5 – 8, graph the equation using the method of your choice.

5. $y = 3x - 2$

    5.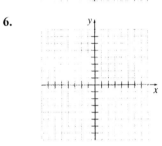

6. $2x - 4y = -8$

    6.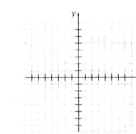

7. $x = -4$

    7.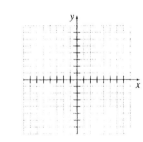

8. $y = -2x + 4$

    8.

For questions 9 – 10, find the slope of the line through the given points.

9. (5, 3) and (7, 1)

    9. _____

10. (–3, –5) and (3, –1)

    10. _____

Chapter 4 Test Form E *(cont.)*

11. Determine the slope and y-intercept of the graph of the equation.
    $-2x + 5y = 10$

    11. _____

12. Two distinct lines have slopes $m_1 = \dfrac{5}{3}$ and $m_2 = \dfrac{3}{5}$. Are the lines parallel, perpendicular, or neither?

    12. _____

13. Write the equation of the line.

    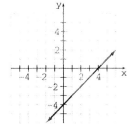

    13. _____

14. Determine if the two lines are parallel, perpendicular or neither.
    $4x = 6y + 3$
    $-2x = -3y + 10$

    14. _____

For questions 15 – 18, find the equation of the line with the given properties.

15. Slope = 5, through $(-1, 1)$

    15. _____

16. Through $(-4, -5)$ and $(6, 0)$

    16. _____

17. Horizontal, through $(1, -8)$

    17. _____

18. Through $(12, -2)$, perpendicular to the line $3y = 4x + 1$

    18. _____

Use the following information to answer questions 19 – 20.

The monthly profit, p, of a small business can be estimated by the function $p = 50x - 24{,}000$, where x represents the number of items sold.

19. Estimate the number of items sold if the monthly profit is \$16,000.

    19. _____

20. Estimate the profit if 600 items are sold.

    20. _____

Chapter 4 Test Form F

1. In which quadrant does the point $(-3, 2)$ lie?

1. _____

2. In which quadrant does the point $(52, 105)$ lie?

2. _____

3. Determine whether the points $(-2, 3)$, $(-1, 5)$, $(0, 7)$ are collinear.

3. _____

4. Find the missing coordinate in the following solution for $-2x + 6y = 8$:
   $(-1, ?)$

4. _____

For questions 5 – 8, graph the equation using the method of your choice.

5. $4x - 5y = 20$

5.

6. $y = -4x + 6$

6.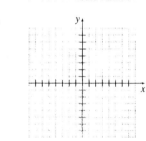

7. $x = 4$

7.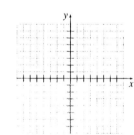

8. $y = 2x$

8.

For questions 9 – 10, find the slope of the line through the given points.

9. $(-5, 7)$ and $(3, -7)$

9. _____

10. $(2, 0)$ and $(0, -3)$

10. _____

Name:_____

11. Determine the slope and *y*-intercept of the graph of the equation $2x + 3y = 18$

11. _____

12. Find the *x*- and *y*-intercepts of $2x + 3y = 9$

12. _____

13. Determine if the following lines are parallel, perpendicular or neither

$$y = -\frac{1}{3}x - 4$$
$$-2y = -6x + 5$$

13. _____

14. Write the equation of the line.

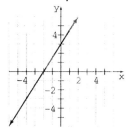

14. _____

For questions 15 – 18, find the equation of the line with the given properties.

15. Slope = –2, through (4, 0)

15. _____

16. Horizontal, through (–3, 5)

16. _____

17. Through (6, 0) and (–3, 3)

17. _____

18. Through (–3, 8), parallel to the line $x - 3y = 3$

18. _____

Use the following information to answer questions 19 and 20.

A real estate agent has a yearly income of $10,000 plus a 2% commission on sales.

19. Write a formula to determine income (*i*) based on sales(*s*).

19. _____

20. Estimate Jai's income if she sells $1,500,000 worth of property in a year.

20. _____

Chapter 4 Test Form G

1. In which quadrant does the point (–4, –1) lie?

   (a) I   (b) II   (c) III   (d) IV

2. In which quadrant does the point (15, –2.5) lie?

   (a) I   (b) II   (c) III   (d) IV

3. Determine which set of points is collinear:

   (a) {(1, 4), (–2, 0), (3, 6)}   (b) {(1, 4), (–2, 0), (3, 8)}
   (c) {(–1, 2), (1, 4), (3, 6)}   (d) {(–1, 2), (–2, 0), (3, 6)}

4. Find the missing coordinate in the following solution for $3x - 5y = 30$: (5, ?)

   (a) –5   (b) –3   (c) –10   (d) 3

5. Graph $3x - 4y = 12$.

   (a) 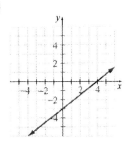   (b) 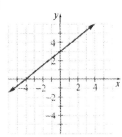   (c) 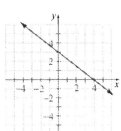   (d)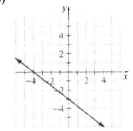

6. Graph $x = 2y$.

   (a) 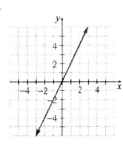   (b) 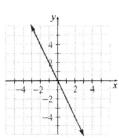   (c) 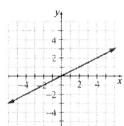   (d)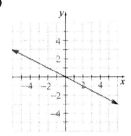

For questions 7 – 9, determine which equation represents the given graph.

7.
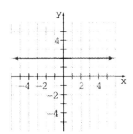

   (a) $y = 2$ (b) $x = 2$ (c) $x + y = 2$ (d) $x - y = 2$

Chapter 4 Test Form G *(cont.)*

8.

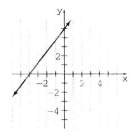

    (a) $4x - 5y = 20$ **(b)** $4x - 5y = -20$ **(c)** $5x - 4y = 20$ **(d)** $5x - 4y = -20$

9.

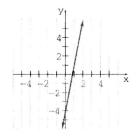

    (a) $y = 5x - 4$ **(b)** $y = 5 - 4x$ **(c)** $y = 4x - 5$ **(d)** $y = 4 - 5x$

For questions 10 – 11, find the slope of the line through the given points.

10. $(2, -5)$ and $(-4, -8)$

    (a) $\dfrac{1}{2}$ **(b)** $\dfrac{2}{3}$ **(c)** $\dfrac{-1}{2}$ **(d)** 2

11. $(3, 0)$ and $(0, 8)$

    (a) $-\dfrac{8}{3}$ **(b)** $\dfrac{8}{3}$ **(c)** $\dfrac{3}{8}$ **(d)** $-\dfrac{3}{8}$

12. Determine the slope and y-intercept of the equation $-6x + 14y = 21$.

    (a) $m = \dfrac{3}{7}; b = \dfrac{3}{2}$ **(b)** $m = \dfrac{-3}{7}; b = \dfrac{3}{2}$ **(c)** $m = \dfrac{7}{3}; b = \dfrac{7}{2}$ **(d)** $m = -\dfrac{7}{3}; b = \dfrac{7}{2}$

13. Find the x- and y-intercepts of $2y = 3x + 2$.

    (a) x-intercept: 2   **(b)** x-intercept: 3   **(c)** x-intercept: 1   **(d)** x-intercept: $-\dfrac{2}{3}$

        y-intercept: 3 y-intercept: 2 y-intercept: $-\dfrac{2}{3}$ y-intercept: 1

14. The lines $y = 5x - 2$ and $-2y = -10x + 7$ are

    (a) perpendicular   **(b)** neither parallel nor perpendicular

    (c) the same line   **(d)** parallel

Chapter 4 Test Form G *(cont.)*   Name:_____

For questions 15 – 18,, find the equation of the line with the given properties.

15. Slope $-\dfrac{1}{2}$, contains (6, 1)

(a) $2x+y=13$ (b) $2x-y=11$ (c) $x-2y=4$ (d) $x+2y=8$

16. Contains (4, 5) and (4, –9).

(a) $x=4$ (b) $y=5$ (c) $y=-\dfrac{7}{4}x+12$ (d) $y=\dfrac{-4}{7}x+\dfrac{105}{13}$

17. Vertical and through the point (–2, 4)

(a) $x=4$ (b) $x=-2$ (c) $y=4$ (d) $y=-2$

18. Through the point (–6, 3) and perpendicular to the line $x+6y=4$

(a) $y=6x+39$ (b) $y=\dfrac{1}{6}x-\dfrac{13}{2}$ (c) $y=6x-24$ (d) $y=\dfrac{1}{6}x+4$

Use the following information to answer questions 19 – 20.

A company earns $16 profit on each unit sold. It also has $48,000 in fixed expenses each month. The graph shows the company's monthly profit up to 6000 units sold.

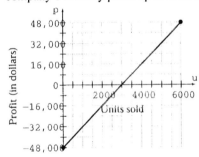

19. Write a formula to determine profit (*p*) based on units sold (*u*).

(a) $p=16(48,000-u)$ (b) $p=16(u-48,000)$ (c) $p=48,000-16u$ (d) $p=16u-48,000$

20. Estimate the company's profit in a month where 4,500 units are sold.

(a) $24,000   (b) $32,000   (c) $16,000   (d) $20,000

Chapter 4 Test Form H

Name:_____

Date:_____

1. In which quadrant does the point $(2, -3)$ lie?

   (a) I   **(b)** II   **(c)** III   **(d)** IV

2. In which quadrant does the point $(-58, -100)$ lie?

   (a) I   **(b)** II   **(c)** III   **(d)** IV

3. Determine which set of points is collinear.

   (a) $\{(-3, -5), (-1, -1), (0, 2)\}$ **(b)** $\{(-3, -5), (-1, -1), (1, 3)\}$
   (c) $\{(-2, -4), (-4, -10), (0, 2)\}$ **(d)** $\{(-2\ -4), (-1, -1), (1, 3)\}$

4. Find the missing coordinate in the following solution for $4x + 3y = 18$:
   $(?, 2)$

   (a) -6 **(b)** -3 **(c)** 3 **(d)** 6

5. Graph $3x + 4y = -12$.

   (a)   **(b)**   **(c)**   **(d)**

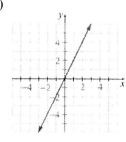 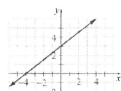 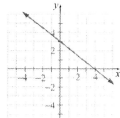 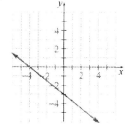

6. Graph $y = -2x$.

   (a)   **(b)**   **(c)**   **(d)**

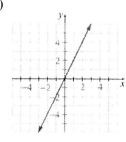 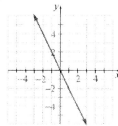 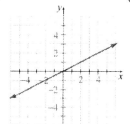 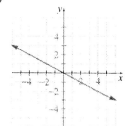

For questions 7 – 9, determine which equation represents the given graph.

7.

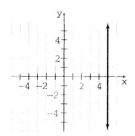

   (a) $x + y = -5$ **(b)** $x - y = -5$ **(c)** $x = 5$ **(d)** $y = 5$

8.

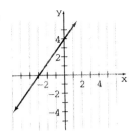

   (a) $4x - 3y = -12$ **(b)** $4x - 3y = 12$ **(c)** $3x - 4y = -12$ **(d)** $3x - 4y = 12$

9.

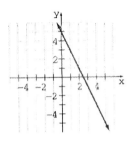

   (a) $y = 2x - 5$ **(b)** $y = 2 - 5x$ **(c)** $y = 5x - 2$ **(d)** $y = 5 - 2x$

For questions 10 – 11, find the slope of the line through the given points.

10. $(-7, 4)$ and $(-1, 3)$

   (a) -6 **(b)** $\dfrac{-1}{6}$ **(c)** $\dfrac{1}{6}$ **(d)** 6

11. $(-3, 0)$ and $(0, 9)$

   (a) 3 **(b)** $\dfrac{1}{3}$ **(c)** $-\dfrac{1}{3}$ **(d)** -3

12. Determine the slope and y-intercept of the equation. $y = -2x + 8$

   (a) $m = -8,\ b = \dfrac{1}{2}$ **(b)** $m = 2,\ b = -8$ **(c)** $m = -2,\ b = 8$ **(d)** $m = -8,\ b = -\dfrac{1}{2}$

13. Find the x- and y-intercepts of $y = -2x + 3$.

   (a) x-intercept: $\dfrac{3}{2}$ **(b)** x-intercept: 3 **(c)** x-intercept: -2 **(d)** x-intercept: 3

       y-intercept: 3 y-intercept: -2 y-intercept: 3 y-intercept: $\dfrac{3}{2}$

14. Which of the following lines is *not* parallel to $y = 4x - 4$?

   (a) $4x - y = 6$ **(b)** $4x + y = 6$ **(c)** $8x - 2y = 6$ **(d)** $y - 4x = -2$

For questions 15 – 18, find the equation of the line with the given properties.

15. Slope $\dfrac{3}{2}$, contains (4, 2)

    (a) $3x + 2y = 16$ (b) $3x - 2y = 8$ (c) $2x - 3y = 2$ (d) $2x + 3y = 14$

16. Through the points $(0, -4)$ and $(-1, 3)$

    (a) $y = -7x - 4$ (b) $y = \dfrac{7}{4}x + 1$ (c) $-7x - 4y = -11$ (d) $y = -4x - 7$

17. Horizontal and through the point $(-2, 4)$

    (a) $y = -2$ (b) $y = 4$ (c) $x = -2$ (d) $x = 4$

18. Through the point $(-1, 7)$ and perpendicular to the line $x + 5y = 2$.

    (a) $y = 5x - 36$ (b) $y = \dfrac{1}{5}x - \dfrac{12}{5}$ (c) $y = 5x + 12$ (d) $y = -\dfrac{1}{5}x + \dfrac{36}{5}$

Use the following information to answer questions 19 – 20.

Mark Marx earns $500 a week plus 4% commission on his sales. The graph shows Mark's weekly income for sales up to $12,000.

Sales (in thousands of dollars)

19. Write a formula for Mark's income (i) based on sales (s).

    (a) $i = 500 + 0.04s$ (b) $i = 500 - 0.04s$ (c) $i = 0.04(s + 500)$ (d) $i = 0.04(s - 500)$

20. Estimate Mark's sales for a week when he has $700. in income.

    (a) $8000   (b) $7000   (c) $4000   (d) $5000

Chapters 1–4 Cumulative Test Form A

1. Evaluate $\dfrac{x-m}{s}$ when $x = 8$, $m = 15$, and $s = 14$.

   1. _____

2. Solve: $1 = 8(x - 8) - 3x$

   2. _____

3. Solve $F = \dfrac{9}{5}C + 32$ for C.

   3. _____

4. Consider the set of numbers:

$$\left\{ -\sqrt{2}, 5.6, \frac{2}{3}, \pi, -\frac{5}{9}, 40 \right\}$$

Which of the numbers are rational numbers?

   4. _____

5. Evaluate: $-2\left[9 - 3\left(4^2 \div 2 \right) + 15 \right]$

   5. _____

6. Simplify: $5a^2 - 4ab + 7ab + b^2$

   6. _____

7. Solve for x: $2(x - 7) + 4x = 3 - (2 - x)$

   7. _____

8. Name the property illustrated:
$6 + x = x + 6$

   8. _____

9. Put $-3x - 7y = 5$ in slope-intercept form.

   9. _____

10. Graph $2x + 5y = 10$ using the intercepts.

   10.

Chapters 1–4 Cumulative Test
Form A *(cont.)*

11. Find the slope and y-intercept of $4x - 6y = 4$.

    11. _____

12. Find the slope of the line through the points $(4, 7)$ and $(-2, 1)$.

    12. _____

13. Find the equation of a line with slope $\dfrac{2}{3}$ that passes through the point $(-6, -5)$. Write the equation in standard form.

    13. _____

14. Solve for x: $\dfrac{2x}{3} - \dfrac{x}{4} = \dfrac{3}{6}$

    14. _____

15. Patrick can type 600 words in 12 minutes. How long will it take him to type 1100 words?

    15. _____

16. If the product of a number and 12 is decreased by 7, the result is -31. Find the number.

    16. _____

17. Jake has $5000 to invest and wants to divide the money between a certificate of deposit (CD) paying 5% interest and some corporate bonds paying 8% interest. After 1 year, he earns $140 more in interest from the corporate bonds than from the CD. How much did he put into each investment?

    17. _____

18. A triangle has a perimeter of 54 inches. Find the length of the three sides if one side is 10 inches longer than the smallest and the third side is twice the smallest side.

    18. _____

19. How many liters of 20% salt solution must be added to 72 liters of 59% salt solution to get a 44% salt solution?

    19. _____

20. Train A leaves a station traveling at 20 mph. Eight hours later, train B leaves the same station traveling in the same direction at 40 mph. How long does it take for train B to catch up to train A?

    20. _____

Chapters 1–4 Cumulative Test Form B

Name:_____

Date:_____

1. Evaluate: $\dfrac{x-m}{s}$ when $x = 50$, $m = 45$, and $s = 15$.

   (a) $-\dfrac{1}{3}$
   (b) $\dfrac{1}{3}$
   (c) $\dfrac{1}{2}$
   (d) $-\dfrac{1}{2}$

2. Solve: $x + 9 = 5(2x - 1)$

   (a) $\dfrac{4}{9}$
   (b) $\dfrac{8}{9}$
   (c) $\dfrac{14}{9}$
   (d) $\dfrac{10}{9}$

3. Solve for t in $I = Prt$.

   (a) $t = \dfrac{I}{Pr}$
   (b) $t = I - Pr$
   (c) $t = \dfrac{Pr}{I}$
   (d) $t = IPr$

4. Simplify: $6x^3 + 4x^2 + 5 - 7x^3 + 9x^2 - 4x + 5$

   (a) $-x^3 + 13x^2 - 4x + 10$
   (b) $-x^3 + 13x^2 - 4x$
   (c) $-x^3 - 13x^2 + 4x + 10$
   (d) $x^3 - 13x^2 + 4x$

5. Evaluate: $-3\left[10 + 4\left(2^3 - 5\right) - 7\right]$

   (a) -87
   (b) -25
   (c) -105
   (d) -45

6. Solve for x: $-4(x + 6) - 3x = 10 + (2 - x)$

   (a) -8
   (b) -6
   (c) 4
   (d) 12

7. Consider $\left\{-\sqrt{3}, 2.4, \dfrac{3}{8}, \pi, -\dfrac{7}{5}, 21\right\}$. Which of the numbers are rational numbers?

   (a) $2.4, \dfrac{3}{8}, -\dfrac{7}{5}, 21$
   (b) $\dfrac{3}{8}, -\dfrac{7}{5}$
   (c) $2.4, \dfrac{3}{8}, -\dfrac{7}{5}$
   (d) $2.4, \dfrac{3}{8}, 21$

8. Name the property illustrated: $4(3 \cdot 7) = (4 \cdot 3)7$

   (a) Associative property of addition
   (b) Commutative property of addition
   (c) Associative property of multiplication
   (d) Commutative property of multiplication

9. Graph $3x + 4y = 12$.

   (a)
   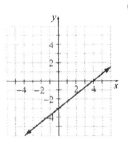
   (b)
   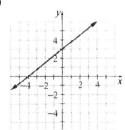
   (c)
   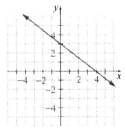
   (d)
   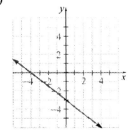

Chapters 1–4 Cumulative Test
Form B *(cont.)*

10. Find the x- and y-intercepts of $y = -5x - 6$.

(a) x-intercept: $(-6, 0)$; y-intercept: $\left(-\dfrac{6}{5}, 0\right)$

(b) x-intercept: $(-5, 0)$; y-intercept: $(0, -6)$

(c) x-intercept: $(-6, 0)$; y-intercept: $(0, -5)$

(d) x-intercept: $\left(-\dfrac{6}{5}, 0\right)$; y-intercept: $(0, -6)$

11. Put $3x + 2y = -3$ in slope-intercept form.

(a) $2y = -3x - 3$　　(b) $y = -3x - 5$　　(c) $x = -\dfrac{2}{3}y - 1$　　(d) $y = -\dfrac{3}{2}x - \dfrac{3}{2}$

12. Find the slope and the y-intercept of the line $9x + 3y = 81$.

(a) $m = -3, (0, 27)$　　(b) $m = -27, \left(0, -\dfrac{1}{3}\right)$　　(c) $m = 3, (0, -27)$　　(d) $m = -27, \left(0, \dfrac{1}{3}\right)$

13. Find the slope of the line through the points $(2, 8)$ and $(7, 3)$.

(a) $m = 1$　　(b) $m = -1$　　(c) $m = 5$　　(d) $m = -5$

14. Solve for x: $\dfrac{3}{5}x + \dfrac{7}{10} = \dfrac{7}{4}$

(a) 0　　(b) $\dfrac{5}{2}$　　(c) $\dfrac{7}{4}$　　(d) $-\dfrac{5}{18}$

15. Find the equation of a line that has slope $-\dfrac{1}{2}$ and passes through the point $(6, 4)$.

(a) $x - 2y = -2$　　(b) $x + 2y = 14$　　(c) $x + 2y = 16$　　(d) $2x - y = 8$

16. On Friday, Shawn drove to the lake in 45 minutes. On Sunday he returned up in 9 less minutes and averaged 12 more miles per hour. How far is the lake from his house?

(a) 48 miles　　(b) 30 miles　　(c) 60 miles　　(d) 42.5 miles

17. The waiting room at Southwest Pediatrics is 10 feet longer than it is wide. During a recent remodeling, the waiting room was wallpapered. A total of 84 feet of wallpaper border was used in the project. What are the dimensions of the waiting room?

(a) 21 feet by 31 feet　　(b) 16 feet by 26 feet　　(c) 18 feet by 24 feet　　(d) 21 feet by 21 feet

18. Town Bank's 7% rate means Sue must spend $6.66 per $1000 of her mortgage each month, and Big Bank's 6.5% rate means she will spend $6.32 per $1000 each month. Big Bank also charges 4 points while Town Bank charges none. Sue needs an $80,000 mortgage. How long will it take for both mortgages to be equal?

(a) about 141 months　　(b) about 3 months　　(c) about 118 months　　(d) about 6 months

19. How many pounds of gourmet candy selling for $2.80 per pound should be mixed with 2 pounds of gourmet candy selling for $1.80 a pound to obtain a mixture selling for $2.60 per pound?

    (a) 10 lb   **(b)** 9 lb   **(c)** 6 lb   **(d)** 8 lb

20. A train traveling at 40 miles per hour leaves for a certain town. One hour later, a bus traveling at 50 miles per hour leaves for the same town and arrives at the same time as the train. If both the train and the bus traveled in a straight line, how far is the town from where they started?

    (a) 20 miles   **(b)** 4 miles   **(c)** 160 miles   **(d)** 200 miles

Chapter 5 Pretest Form A

Name:_____

Date:_____

Simplify each expression.

1. $2x^3 \cdot 3x^2$

2. $\left(2x^2 y\right)^4$

3. $\dfrac{18x^4}{3x}$

4. x^{-9}

5. $9x^0$

6. Express in scientific notation: 2,117,000

7. Express without exponents: 2.40×10^3

8. Multiply: $(5.8 \times 10^{13})(4.3 \times 10^{-16})$; answer in scientific notation.

9. Classify as a monomial, binomial, or trinomial: $4x^2 + 5$

10. Give the degree of the polynomial: $4x^2 + 5$

11. Write the polynomial $5x^2 - x^3 + 2x^4 - 8$ in descending order.

Perform the operations indicated.

12. $(4x - 4x^5 + 9) + (-5x^5 + 5 + 3x)$

13. $(3p^3 + 8) + (8p^3 + 5p + 1)$

14. $(10x + 3) - (7x + 1)$

15. $(7x^2 - 4x + 2) - (3x^2 - 2x - 4)$

16. $3xy^2(8x^2 - 4y)$

17. $(x + 7)(x + 1)$

18. $(3p + 1)(3p - 1)(p + 2)$

19. $\dfrac{15x^2 - 10x + 20}{5x}$

20. $\dfrac{4x^2 + 8x + 1}{2x + 1}$

1. _____

2. _____

3. _____

4. _____

5. _____

6. _____

7. _____

8. _____

9. _____

10. _____

11. _____

12. _____

13. _____

14. _____

15. _____

16. _____

17. _____

18. _____

19. _____

20. _____

Chapter 5 Pretest Form B

Name:_____

Date:_____

Simplify each expression.

1. $4x^2 \cdot 5x^4$

   1. _____

2. $(10ab^3)^4$

   2. _____

3. $\dfrac{12x^7}{4x^4}$

   3. _____

4. x^{-5}

   4. _____

5. $(12x)^0$

   5. _____

6. Express in scientific notation: 2,932,000

   6. _____

7. Express without exponents: 3.91×10^8

   7. _____

8. Multiply: $(5.1 \times 10^{10})(2.8 \times 10^{-12})$; answer in scientific notation

   8. _____

9. Classify as a monomial, binomial, or trinomial: $4x^5 - x^3 + 2x$

   9. _____

10. Give the degree of the polynomial: $4x^5 - x^3 + 2x$

   10. _____

11. Write the polynomial $4x^3 - 7x^2 + x^5 - 19$ in descending order.

   11. _____

Perform the operations indicated.

12. $(4c^5 - 5) + (8c^5 + c - 3)$

   12. _____

13. $(7x^2 - 5x + 2) + (4x - 3x^2 - 9)$

   13. _____

14. $(9x + 4) - (7x + 2)$

   14. _____

15. $(8x^2 - 5x + 3) - (x^2 - 2x - 5)$

   15. _____

16. $-5x^3 y(3x - 2y)$

   16. _____

17. $(x + 6)(x - 5)$

   17. _____

18. $(x - 5)^2 (x + 1)^2$

   18. _____

19. $\dfrac{12x^2 - 28x + 36}{6x + 1}$

   19. _____

20. $\dfrac{3x^2 - 6x + 2}{3x}$

   20. _____

Mini-Lecture 5.1
Exponents

<u>Learning Objectives</u>:

1. Review exponents.
2. Learn the rules of exponents.
3. Simplify an expression before using the expanded power rule.
4. Key Vocabulary: *base, exponent, power, same base, product rule for exponents, quotient rule for exponents, zero exponent rule, power rule, expanded power rule*

<u>Examples</u>:

1. Write the following using exponents.
   a) *aaaaabbbb*　　　　　b) *xxyyyyzz*　　　　　c) *ppqqqqqqrrrr*

2. Simplify each expression.

   a) $7 \cdot 7^3$　　　　　b) $x^6 x^3$　　　　　c) $\dfrac{3^4}{3^2}$

   d) $\dfrac{n^{14}}{n^9}$　　　　　e) $\dfrac{b^4}{b^4}$　　　　　f) $\dfrac{5^3}{5^6}$

   g) $\dfrac{k^3}{k^8}$　　　　　h) $\left(x^2\right)^3$　　　　　i) $\left(-x^3\right)^4$

   j) $\left(4x^2 y\right)^3$　　　　　k) $\left(\dfrac{3c^2}{d^3}\right)^4$　　　　　l) $\left(\dfrac{-2p^4}{3q^2}\right)^3$

   m) $\left(3x^2 y^4\right)^0$　　　　　n) $\dfrac{\left(4ab^2\right)^2}{\left(2a^3 b\right)^4}$

3. Simplify.

   a) $\left(\dfrac{16x^4 y^3}{4x^2 y}\right)^2$　　　　　b) $\left(\dfrac{-30m^3 n^4}{15m^5 n^3}\right)^4$

<u>Teaching Notes</u>:

- Students should be encouraged to do lots of practice to commit the rules to memory.
- Remind students that if all the factors are cancelled out of the numerator (denominator) then the numerator (denominator) is 1, not 0.
- Point out that $x^0 = 1$ only if $x \neq 0$. The zero exponent rule is based on the quotient rule which has the restriction that $x \neq 0$.

Answers: 1a) $a^5 b^4$; 1b) $x^2 y^4 z^2$; 1c) $p^2 q^6 r^4$; 2a) $7^4 = 2401$; 2b) x^9; 2c) $3^2 = 9$; 2d) n^5, 2e) 1; 2f) $\dfrac{1}{5^2} = \dfrac{1}{25}$;

2g) $\dfrac{1}{k^5}$; 2h) x^6; 2i) x^{12}; 2j) $64x^6 y^3$; 2k) $\dfrac{81c^8}{d^{12}}$; 2l) $\dfrac{-8p^{12}}{27q^6}$; 2m) 1; 2n) $\dfrac{1}{a^{10}}$ 3a) $16x^4 y^4$; 3b) $\dfrac{16n^4}{m^8}$

Mini-Lecture 5.2
Negative Exponents

Learning Objectives:

1. Understand the negative exponent rule.
2. Simplify expressions containing negative exponents.
3. Key vocabulary: *negative exponent rule, raised to*

Examples:

1. a) Simplify $\dfrac{y^5}{y^8}$ two ways: using the quotient rule and by dividing out common factors.

   b) Write 4^{-2} using positive exponents.

   c) Write m^{-10} using positive exponents.

2. Simplify. Write your answers using positive exponents.

   a) m^{-5}

   b) $\dfrac{1}{y^{-7}}$

   c) $w^4 \cdot w^{-9}$

   d) $\dfrac{z^8}{z^{-4}}$

   e) $3h^2\left(-2h^{-6}\right)$

   f) $\dfrac{22a^3b^5}{11a^7b^2}$

   g) $\left(\dfrac{3}{r^3}\right)^{-2}$

   h) $\left(\dfrac{2x^{-2}y^4z^{-1}}{4y^9}\right)^3$

Teaching Notes:

- Remind students that the sign of the exponent does not affect the sign of the coefficient.
- Remind students that answers should not contain any negative exponents.
- In problems like $3x^{-2}$ some students incorrectly move the coefficient along with the variable and obtain $\dfrac{1}{3x^2}$.

Answers: 1a) y^{-3}, $\dfrac{1}{y^3}$; 1b) $\dfrac{1}{4^2} = \dfrac{1}{16}$; 1c) $\dfrac{1}{m^{10}}$; 2a) $\dfrac{1}{m^5}$; 2b) m^7; 2c) $\dfrac{1}{w^5}$; 2d) z^{12}; 2e) $-\dfrac{6}{h^4}$;

2f) $\dfrac{2b^3}{a^4}$; 2g) $\dfrac{r^6}{9}$; 2h) $\dfrac{1}{8x^6y^{15}z^3}$

Mini-Lecture 5.3
Scientific Notation

Learning Objectives:

1. Convert numbers to and from scientific notation.
2. Recognize numbers in scientific notation with a coefficient of 1.
3. Do calculations using scientific notation.
4. Key vocabulary: *decimal form, scientific notation, base, exponent*

Examples:

1. Express each number in scientific notation.

   a) 0.000000675

   b) 1,340,000

2. Express each number in decimal form (without exponents).

   a) 5.004×10^{-7}

   b) 3.201×10^{10}

3. Write the quantity without metric prefixes.

   a) 472.1 kilometers

   b) 12.5 nanometers

4. Perform each indicated operation and express each number in decimal form (without exponents)

   a) $\left(2.3 \times 10^3\right)\left(5.0 \times 10^5\right)$

   b) $\left(5.1 \times 10^{-4}\right)\left(6.5 \times 10^3\right)$

   c) $\dfrac{2.8 \times 10^6}{7.0 \times 10^{-2}}$

   d) $\dfrac{1.5 \times 10^{-3}}{5 \times 10^2}$

   e) In the third quarter of 2006, ExxonMobil reported a profit of $\$1.05 \times 10^{10}$. If approximately 6×10^9 shares of stock were outstanding, what was the approximate earnings per share?

5. Perform the indicated operation by first converting to scientific notation. Write the answer in scientific notation.

   a) $\left(320,000\right)\left(2,500,000\right)$

   b) $\dfrac{0.000009}{360,000}$

Teaching Notes:

- Remind students that the numerical value preceding the base must be greater than or equal to 1 but less than 10.
- Point out that the sign of the exponent on the base indicates which direction to move the decimal point.

Answers: *1a)* 6.75×10^{-7}; *1b)* 1.34×10^6; *2a)* 0.0000005004; *2b)* $32,010,000,000$; *3a)* $472,100$ meters; *3b)* 0.0000000125 meters; *4a)* $1,150,000,000$; *4b)* 3.315; *4c)* $40,000,000$; *4d)* 0.000003; *4e)* $\$1.75$; *5a)* 8.0×10^{11}; *5b)* 2.5×10^{-11}

Mini-Lecture 5.4
Addition and Subtraction of Polynomials

Learning Objectives:

   1. Identify polynomials.
   2. Add polynomials.
   3. Subtract polynomials.
   4. Subtract polynomials in columns.
   5. Key vocabulary: *polynomial, term, descending order, monomial, binomial, trinomial, degree of a term, degree of a polynomial, combining like terms, distributive property*

Examples:

   1. Identify which of the following are polynomials. If so, state the degree. If not, state why not.

      a) $-5x^3 + x^2y^2 - xy^4$ b) $2x^5 - 3x^2 + x^{-1}$ c) 12   d) $7x^{3/2} + x - 4$

   2. Add.

      a) $(2x-7)+(4x+9)$ b) $(x^2 - 5x + 10) + (9x^3 - 2x^2 + 4x - 15)$

   3. Subtract.

      a) $(8m+3)-(2m+4)$ b) $(3p^2 - 4p) - (-2p^2 + 14p - 1)$

   4. Subtract using columns.

      a) Subtract $(4k^2 + 3)$ from $(8k^3 - 3k^2 + k + 5)$.

      b) Subtract $(-2c^3 + 5c^2 + 3c + 9)$ from $(6c^2 + 4c + 19)$

Teaching Notes:

- When determining the degree of a term, some students forget to include the implied exponent of 1 in terms such as x^3y^2z.
- It may be helpful to do a quick review of the distributive property and combining like terms.
- Remind students that they can *evaluate* and *simplify* polynomials since they are expressions, but they cannot *solve* a polynomial since it is not an equation.
- Remind students to be careful about distributing the minus sign when subtracting polynomials.

Answers: *1a) polynomial, 5;*   *1b) not a polynomial, negative exponent;*   *1c) polynomial, 0;*   *1d) not a polynomial, fractional exponent;*   *2a)* $6x + 2$;   *2b)* $9x^3 - x^2 - x - 5$;   *3a)* $6m - 1$;   *3b)* $5p^2 - 18p + 1$;   *4a)* $8k^3 - 7k^2 + k + 2$;   *4b)* $2c^3 + c^2 + c + 10$

Mini-Lecture 5.5
Multiplication of Polynomials

Learning Objectives:

1. Multiply a monomial by a monomial.
2. Multiply a polynomial by a monomial.
3. Multiply binomials using the distributive property.
4. Multiply binomials using the FOIL method.
5. Multiply binomials using formulas for special products.
6. Multiply any two polynomials.
7. Key vocabulary: *monomial, binomial, trinomial, polynomial, FOIL, difference of two squares, square of a binomial*

Examples:

1. Multiply.

   a) $\left(4x^2\right)\left(3x^5\right)$ b) $\left(6a^3\right)\left(-5a^4\right)$ c) $(-3m)\left(-7m^5\right)$ d) $\left(8x^2y^3z\right)\left(2x^4yz^3\right)$

2. Multiply.

   a) $3x^4\left(9x^2 - x + 7\right)$ b) $-4y^3\left(2y^3 + 13y^2 - y\right)$ c) $3a^4b\left(a^2b - 3ab + 2b^2\right)$

3. Multiply using the distributive property.

   a) $(x+2)(x-5)$ b) $(4y-3)(2y+3)$ c) $(5a-2b)(a-4b)$

4. Multiply using the FOIL method.

   a) $(2c+5)(c-3)$ b) $(8-4x)(3-2x)$ c) $(3x+2)(3x-2)$

5. Multiply using special products.

   a) $(6x+5)(6x-5)$ b) $(5d-9)(5d-9)$ c) $(3p-2q)^2$

6. Multiply.

   a) $(4w+3)\left(2w^2 - w - 6\right)$ b) $\left(x^3 + 5x^2 - 4x - 7\right)\left(2x^2 - 11\right)$

Teaching Notes:

- Point out that the distributive property extends to the distribution of polynomials as well.
- Point out that a good understanding of the FOIL method can be helpful later when factoring.
- Remind students that exponents do not distribute across a sum or difference. For example, $(x+2)^2 \neq x^2 + 2^2$.
- Encourage students to memorize special products instead of just using the distributive property or the FOIL method.

Answers: 1a) $12x^7$; 1b) $-30a^7$; 1c) $21m^6$; 1d) $16x^6y^4z^4$; 2a) $27x^6 - 3x^5 + 21x^4$; 2b) $-8y^6 - 52y^5 + 4y^4$; 2c) $3a^6b^2 - 9a^5b^2 + 6a^4b^3$; 3a) $x^2 - 3x - 10$; 3b) $8y^2 + 6y - 9$; 3c) $5a^2 - 22ab + 8b^2$; 4a) $2c^2 - c - 15$; 4b) $8x^2 - 28x + 24$; 4c) $9x^2 - 4$; 5a) $36x^2 - 25$; 5b) $25d^2 - 90d + 81$; 5c) $9p^2 - 12pq + 4q^2$; 6a) $8w^3 + 2w^2 - 27w - 18$; 6b) $2x^5 + 10x^4 - 19x^3 - 69x^2 + 44x + 77$

Mini-Lecture 5.6
Division of Polynomials

Learning Objectives:

1. Divide a polynomial by a monomial.
2. Divide a polynomial by a binomial.
3. Check division of polynomial problems.
4. Write polynomials in descending order when dividing.
5. Key vocabulary: *monomial, binomial, trinomial, polynomial, dividend, quotient, divisor, remainder*

Examples:

1. Divide.

   a) $\dfrac{12s + 32}{4}$

   b) $\dfrac{3h^4 - 6h^3 + 10h - 5}{3h^2}$

2. Divide.

   a) $\dfrac{6x^2 - 5x - 21}{2x + 3}$

   b) $\dfrac{5x^2 + 21x + 9}{x + 4}$

3. Divide and check.

   a) $\dfrac{11 + 11x - 12x^2}{-3x - 1}$

   b) $\dfrac{12x^3 + 11x^2 - 17x - 12}{4x + 5}$

4. Divide.

   a) $\dfrac{27 - x - 14x^3 + 4x^4}{x - 3}$

   b) $\dfrac{5m^2 - 15 + 9m^4}{3m - 2}$

Teaching Notes:

- Remind students that the denominator distributes to each term in the numerator. Encourage them to do the distribution before simplifying.

- It is often helpful to work a numeric long division problem (which students are more comfortable with) as you work a polynomial long division problem.

- Remind students to include missing powers of the variable by using coefficients of 0. This helps keep terms lined up when doing long division.

Answers: 1a) $3s + 8$; 1b) $h^2 - 2h + \dfrac{10}{3h} - \dfrac{5}{3h^2}$; 2a) $3x - 7$; 2b) $5x + 1 + \dfrac{5}{x + 4}$; 3a) $4x - 5 + \dfrac{6}{-3x - 1}$;

3b) $3x^2 - x - 3 + \dfrac{3}{4x + 5}$; 4a) $4x^3 - 2x^2 - 6x - 19 - \dfrac{30}{x - 3}$; 4b) $3m^3 + 2m^2 + 3m + 2 - \dfrac{11}{3m - 2}$

Additional Exercises 5.1

Simplify.

1. $3 \cdot 4^3$

    1. _____

2. $4x^3 \cdot 2x^3$

    2. _____

3. $x^5 \cdot x^3$

    3. _____

4. $2x^3 \cdot 2x^2$

    4. _____

5. $\dfrac{6^5}{6^3}$

    5. _____

6. $\dfrac{x^4}{x^7}$

    6. _____

7. $2 \cdot 8^0 \cdot 3^4$

    7. _____

8. $-(9x)^0$

    8. _____

9. $(y^4)^3$

    9. _____

10. $(x^2 y)^4$

    10. _____

11. $\dfrac{6x^7}{2x^4}$

    11. _____

12. $\dfrac{8x^5}{4x^5}$

    12. _____

13. $(2cd^4)^2 (cd)^5$

    13. _____

Additional Exercises 5.1 *(cont.)*

14. $(2fg^4)^4(fg)^6$

14. _____

15. $\dfrac{x^5y^6}{xy^2}$

15. _____

16. $\dfrac{x^2y^5}{xy^4}$

16. _____

17. $\left(\dfrac{4x^5y}{16xy^4}\right)^3$

17. _____

18. $\dfrac{(5x^4y^3)^2(2x^3y^2)^3}{(2x^2y)^2(5x^5y^2)^2}$

18. _____

Write an expression for the total area of the figure or figures shown.

19.

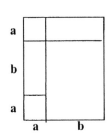

19. _____

20.

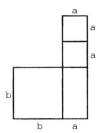

20. _____

Additional Exercises 5.2

Name:_____

Date:_____

Simplify.

1. y^{-7}

2. 7^{-2}

3. $\dfrac{1}{x^{-5}}$

4. $\left(\dfrac{1}{2^4}\right)^{-1}$

5. x^{-6}

6. $x^9 \cdot x^{-7}$

7. $(j^{-13})(j^4)(j^6)$

8. $\dfrac{x^{-1}}{x^{-8}}$

9. $\dfrac{52x^6}{13x^{-7}}$

10. $f^{-3}(f^2)(f^{-3})$

11. $\dfrac{x^{-3}}{x^7}$

12. $\dfrac{24x^6}{12x^{-8}}$

13. $\dfrac{3x^2 y^{-3}}{12x^6 y^3}$

14. $(2x^3 y^{-3})^{-2}$

15. $\dfrac{2x^4 y^{-4}}{8x^7 y^3}$

16. $(4x^4 y^{-4})^3$

17. $5x^2 y(2x^4 y^{-3})$

18. $\left(\dfrac{-7a^2 b^3 c^0}{3a^3 b^4 c^3}\right)^{-4}$

19. $\dfrac{\left(-2a^3 b^2 c^0\right)^{-4}}{\left(3a^2 b^3 c^7\right)^{-2}}$

20. $2^{-1} \cdot 4 - 4^{-1} \cdot 3$

1. _____

2. _____

3. _____

4. _____

5. _____

6. _____

7. _____

8. _____

9. _____

10. _____

11. _____

12. _____

13. _____

14. _____

15. _____

16. _____

17. _____

18. _____

19. _____

20. _____

Additional Exercises 5.3

Express each number in scientific notation.

1. 0.000053

2. 900,000,000

3. 5,500,000

4. 0.0017

5. 3,472,856,921

6. 0.00000189

1. _____

2. _____

3. _____

4. _____

5. _____

6. _____

Express each number without exponents.

7. 4.67×10^7

8. 5.69×10^8

9. 3.13×10^4

10. 6.24×10^0

11. 4.78×10^{-7}

12. 2.01×10^{-5}

7. _____

8. _____

9. _____

10. _____

11. _____

12. _____

Perform the indicated operation and express each number without exponents.

13. $(4.1 \times 10^{14})(3.3 \times 10^{-11})$

14. $(8.5 \times 10^{11})(4.1 \times 10^{-6})$

15. $(2.8 \times 10^{19})(7.6 \times 10^{-19})$

16. $(5.8 \times 10^3)(1.4 \times 10^{-8})$

17. $(34.02 \times 10^3) \div (8.1 \times 10^4)$

18. $(12.1 \times 10^{-2}) \div (2.2 \times 10^5)$

13. _____

14. _____

15. _____

16. _____

17. _____

18. _____

Divide by first converting each number to scientific notation. Write the answer in scientific notation.

19. $\dfrac{0.2233}{1100}$

19. _____

20. Mexico City has an area of approximately 580 sq. miles and a population of approximately 8.7×10^6. What is the approximate number of people per square mile? Write your answer without exponents.

20. _____

Additional Exercises 5.4

Name:_____

Date:_____

Determine if the following is a polynomial. If it has a specific name, give that name.

1. $x^{-2} + 8$

1. _____

2. $4x^4 + 3x$

2. _____

3. $2x^5 + x^3 - 5$

3. _____

4. $2x^7$

4. _____

Express each polynomial in descending order. Give the degree of each polynomial.

5. $-3x + 5x^2 - 9$

5. _____

6. $2 - 4x^3 + 6x$

6. _____

7. 10

7. _____

Add.

8. $(8a^5 - 4) + (3a^5 + a - 2)$

8. _____

9. $(6m^5 + 1) + (2m^5 + 9m - 1)$

9. _____

10. $(3m^5 + 1) + (9m^5 + 3m - 2)$

10. _____

Add in columns.

11. $(-5x^2 - x + 4)$ and $(-3x^2 - 5x + 2)$

11. _____

12. $(-4x + 4x^3 + 7)$ and $(3x^3 - 9 - 3x)$

12. _____

13. $(3x^4 - 2x^2 + 1) + (-x^3 + 3x + 1)$

13. _____

Subtract.

14. $(-x^2 + x - 4) - (-3x^2 - 8x - 2)$

14. _____

15. $(8x^2 - 3x) - (5x - 5 - 8x^2)$

15. _____

16. $(-x^2 - 5x - 3) - (-7x^2 - 8x - 8)$

16. _____

17. $(-2x^3 + x - 3) - (7x - 3 + 7x^3)$

17. _____

Perform each subtraction using columns.

18. $5x^3 - 7x^2 + 6x - 9$ from $3x^3 + 3x^2 + 9$

18. _____

19. $6x^3 - 6x^2 + 8x - 5$ from $5x^3 + 5x^2 + 5$

19. _____

20. $7x^3 - 9x^2 + 8x - 5$ from $5x^3 + 3x^2 + 5$

20. _____

Additional Exercises 5.5

Name:_____

Date:_____

Multiply.

1. $(8x^3 y^2)(-3x^2 y^3)$

2. $(-9x^3 y)(-8x^2 y^3)$

3. $a^4(b^4 a^6)$

4. $2x^3 y(9x^2 + 5y)$

5. $5m^2(3m^3 + 5m^2 - 4m + 6)$

6. $-4x^2 y(x^2 + 7xy - 6y^3)$

7. $(x + 6)(x + 2)$

8. $(x - 6)(x + 9)$

9. $(4x - 3)(3x - 5)$

10. $(6a + 1)(5a + 2)$

11. $(5x + 4y)(2x + 5y)$

12. $(2x + y)(4x - 9y)$

13. $(6r - 5)(6r + 1)$

Multiply using a special products formula.

14. $(6c + 7)(6c - 7)$

15. $(3x + 5y)^2$

Multiply.

16. $(x - 2)(x^2 - x + 3)$

17. $(2x - 5)(5x^2 + 4x + 7)$

18. $(x - 1)(x^4 + x^3 + x^2 + x + 1)$

19. $(x + 2)^2(x + 3)^2$

20. A rectangular solid has a length of $3x + 2$, a width of $2x + 1$, and a height of $x + 4$. Find a formula for the volume of the solid and evaluate the volume when the height is 9.

1. _____

2. _____

3. _____

4. _____

5. _____

6. _____

7. _____

8. _____

9. _____

10. _____

11. _____

12. _____

13. _____

14. _____

15. _____

16. _____

17. _____

18. _____

19. _____

20. _____

Additional Exercises 5.6

Name:_____

Date:_____

Divide.

1. $\dfrac{9x-6}{3}$

2. $\dfrac{4x-7}{2}$

3. $\dfrac{x^2-3x+5}{x}$

4. $\dfrac{5x^2-25x+2}{-5x}$

5. $\dfrac{4x^{10}-5x^9-20x^4}{4x^2}$

6. $\dfrac{-x^6+x^5+7x^2-9}{x^4}$

7. $\dfrac{x^2+2x+6}{x}$

8. $\dfrac{3x^2-15x+5}{-3x^2}$

9. $\dfrac{2x^{11}-5x^7-10x^6}{2x^3}$

10. $\dfrac{-2x^6+5x^5+9x^2+2}{x^4}$

11. $(f^3+64)\div(f+4)$

1. _____

2. _____

3. _____

4. _____

5. _____

6. _____

7. _____

8. _____

9. _____

10. _____

11. _____

Additional Exercises 5.6 *(cont.)*

Name:_____

12. $\dfrac{4p-2+3p^2}{p-1}$

12. _____

13. $\dfrac{3m-4+2m^2}{m+5}$

13. _____

14. $(j^3-64)\div(j-4)$

14. _____

15. $(p^6+64)\div(p^2+4)$

15. _____

16. $\dfrac{4p+3p^2-1}{p+4}$

16. _____

17. $\dfrac{20x^2-13x+2}{2-5x}$

17. _____

18. $12x^2-6x^3-3-9x$ by $3x-3$

18. _____

19. $\dfrac{8x^2-2x-3}{2x+1}$

19. _____

20. $-2x^2+6x^3-4-x$ by $2x^2+1$

20. _____

Chapter 5 Test Form A

Name:_____

Date:_____

Simplify each expression.

1. $-2x^3 \cdot 5x^2$

1. _____

2. $(10x^4 y)^3$

2. _____

3. $\left(\dfrac{10x^2 y^0}{2x^5 y^2} \right)^2$

3. _____

4. $\dfrac{18x^7}{6x^5}$

4. _____

5. $(3x^{-3} y^{-3})^{-2}$

5. _____

Determine whether each expression is a polynomial. If the polynomial has a specific name, give that name.

6. $(x^2 + 1)^{-1}$

6. _____

7. $3x^3 + 4$

7. _____

Convert each number to scientific notation. Then calculate. Express your answer in scientific notation.

8. $(17,200)(3,250,000)$

8. _____

9. $\dfrac{0.000001}{2,500}$

9. _____

10. Write the following polynomial in descending order, and give its degree.
$-3x + 5 + 2x^3 + x^2$

10. _____

Chapter 5 Test Form A *(cont.)*

Perform the operations indicated.

11. $(3x+4)+(2x^2+4x-5)$

11. _____

12. $(2x^2+3x-1)-(5x^2+3x-6)$

12. _____

13. $-2x^2y\left(2x-2y^2\right)$

13. _____

14. $(-x+5)(2x-7)$

14. _____

15. $(2x-2)(2x^2+2x+1)$

15. _____

16. $\dfrac{3x^3+9x^2-6}{3x}$

16. _____

17. $\dfrac{10x^3-9x^2+8x}{x-1}$

17. _____

18. $\dfrac{2x^2+3x+35}{2x+5}$

18. _____

19. $\left(2x+1\right)^2+(x-3)(x+3)$

19. _____

20. Russia supplies about 5% of the world's cut diamonds. If all of Russia's supply is worth \$1.7 billion, what is the value of the total world production annually? Write your answer in scientific notation.

20. _____

Chapter 5 Test Form B

Simplify each expression.

1. $3x^2 \cdot 5x^5$

1. _____

2. $(-4x^2 y^3)^2$

2. _____

3. $\dfrac{6x^7}{3x^3}$

3. _____

4. $\left(\dfrac{5x^0 y^2}{10x^3 y}\right)^3$

4. _____

5. $(3x^{-3} y^3)^{-3}$

5. _____

Determine whether each expression is a polynomial. If the polynomial has a specific name, give that name.

6. $3x^3 + 6x + 4$

6. _____

7. $\dfrac{1}{2x+1}$

7. _____

Convert each number to scientific notation. Then calculate. Express your answer in scientific notation.

8. $(27{,}500{,}000)(0.1200)$

8. _____

9. $\dfrac{0.00035}{500{,}000}$

9. _____

10. Write the following polynomial in descending order, and give its degree.
    $3x^3 - 4x^5 + x^2 + 4$

10. _____

Chapter 5 Test Form B *(cont.)*

Perform the operations indicated.

11. $(3x^2 - 2) + (2x^2 - x + 5)$

11. _____

12. $(5x^2 + 4x - 3) - (2x^2 - x + 3)$

12. _____

13. $-3x^2 y(x^2 + 2y^2)$

13. _____

14. $(3x + 6)(-x + 2)$

14. _____

15. $(x - 2)(x^2 + 7x - 13)$

15. _____

16. $\dfrac{9x^2 - 18x - 81}{9x}$

16. _____

17. $\dfrac{-3x^3 + x^2 - 5x}{x - 2}$

17. _____

18. $\dfrac{8x^2 - 26x + 11}{2x - 1}$

18. _____

19. $(2x - 1)(2x + 1) - (x + 2)^2$

19. _____

20. If a computer can perform an operation in 0.0000007 seconds, how many seconds would it take the computer to do 10,000,000,000,000 calculations? Use scientific notation.

20. _____

Chapter 5 Test Form C

Simplify each expression.

1. $16x^3 2x^3$

1._____

2. $(2x^2 y)^3$

2._____

3. $\dfrac{5x^6}{15x^2}$

3._____

4. $\left(\dfrac{6xy^3}{2x^2 y^0}\right)^3$

4._____

5. $(2x^3 y^{-3})^{-2}$

5._____

Determine whether each expression is a polynomial. If the polynomial has a specific name, give that name.

6. $-x - x^{-3} + 3x^3$

6._____

7. $x^5 + 2$

7._____

Convert each number to scientific notation. Then calculate. Express your answer in scientific notation.

8. $(3.2750)(440,000)$

8._____

9. $\dfrac{0.003}{12,000,000}$

9._____

10. Write the following polynomial in descending order, and give its degree.
    $3x + 2x^5 - 4 - 5x^3$

10._____

Name:_____

Perform the operations indicated.

11. $(x^2 + 3x) + (2x^2 - 5x - 10)$

11. _____

12. $(3x^2 - 3x + 1) - (2x^2 - 4x - 1)$

12. _____

13. $-5xy^2(x^2 - 4xy)$

13. _____

14. $(2x - 7)(4x + 1)$

14. _____

15. $(x - 3)(-2x^2 + 7x - 1)$

15. _____

16. $\dfrac{16x^2 - 8x + 2}{8x}$

16. _____

17. $\dfrac{4x^3 - 3x^2 - x - 18}{x - 2}$

17. _____

18. $\dfrac{3x^2 - 7x + 6}{3x - 1}$

18. _____

19. $\dfrac{(2x + 1)^2}{(x - 2)(x + 2)}$

19. _____

20. The distance from earth to the moon is 2.38×10^5 miles. If a light year is approximately 5.87×10^{12} miles, calculate the distance in light years. Use scientific notation.

20. _____

Chapter 5 Test Form D

Name:_____

Date:_____

Simplify each expression.

1. $3x^3 \cdot 4x^4$

1. _____

2. $(2x^2 \cdot y^3)^3$

2. _____

3. $\dfrac{12x^5}{4x}$

3. _____

4. $\left(\dfrac{21x^0 y^2}{3x^3 y^3} \right)^2$

4. _____

5. $(8x^{-2} y^{-3})^{-2}$

5. _____

Determine whether each expression is a polynomial. If the polynomial has a specific name, give that name.

6. $3x^3 + 4x^2 + 5$

6. _____

7. $-2^{-1} + 3x^3$

7. _____

Convert each number to scientific notation. Then calculate. Express your answer in scientific notation.

8. $(16,200)(4,550,000)$

8. _____

9. $\dfrac{12,000}{0.003}$

9. _____

10. Write the following polynomial in descending order, and give its degree.
    $x - 3x^2 + 5x^3 + 12$

10. _____

Chapter 5 Test Form D *(cont.)*

Perform the operations indicated.

11. $(-2x^2 + 1) + (5x^2 + 6x - 10)$

11. _____

12. $(-3x^2 + 4x - 6) - (4x^2 + 3x - 1)$

12. _____

13. $-9x^2y(xy - 3y^2)$

13. _____

14. $(3x - 6)(x - 1)$

14. _____

15. $(-x^2 + 5x - 1)(x - 1)$

15. _____

16. $\dfrac{12x^2 + 2x - 6}{6x}$

16. _____

17. $\dfrac{2x^3 - 5x^2 - 8x + 15}{x - 3}$

17. _____

18. $\dfrac{8x^2 + 6x - 5}{4x + 5}$

18. _____

19. $(3x - 1)(3x + 1)(x - 2)^2$

19. _____

20. The Sombrero Galaxy is 40 million light years away. Given that a light year is approximately 5.87×10^{12} miles, calculate the distance in miles.

20. _____

Chapter 5 Test Form E

Name:_____

Date:_____

Simplify each expression.

1. $-3x \cdot 5x^4$

2. $(3x^2 \cdot 2y^3)^2$

3. $\dfrac{14x^5}{7x^3}$

4. $\left(\dfrac{2xy^2}{3x^3 y^0}\right)^3$

5. $(3x^3 \cdot 2y^{-4})^3$

1. _____

2. _____

3. _____

4. _____

5. _____

Determine whether each expression is a polynomial. If the polynomial has a specific name, give that name.

6. $(y^{-3} + 4)^2$

7. x^3

6. _____

7. _____

Convert each number to scientific notation. Then calculate. Express your answer in scientific notation.

8. $(3220)(2,550,000)$

9. $\dfrac{0.0004}{50,000}$

10. Write the following polynomial in descending order, and give its degree.
$5 - 3x + 7x^3 + x^4$

8. _____

9. _____

10. _____

Chapter 5 Test Form E *(cont.)*

Name:_____

Perform the operations indicated.

11. $(8x + 4) + (2x^2 - 6x - 6)$

11. _____

12. $(5x^2 - 1) - (3x^2 - 4x - 11)$

12. _____

13. $-7x^2 y(x^2 - y^3)$

13. _____

14. $(2x - 1)(4x + 5)$

14. _____

15. $(4x - 3)(4x^2 - 2x - 4)$

15. _____

16. $\dfrac{18x^5 - 6x^2 - 12x}{6x}$

16. _____

17. $\dfrac{-4x^3 + 3x^2 - 18}{x + 2}$

17. _____

18. $\dfrac{12x^2 + 5x - 3}{3x - 1}$

18. _____

19. $(3x - 1)(3x + 1) - (x - 1)^2$

19. _____

20. Insurance adjusters estimated a total damage of $1.163 billion to approximately 5,200 businesses in Oklahoma and 9,500 in Kansas. What is the average amount of damage on a single home or business? Use scientific notation.

20. _____

Chapter 5 Test Form F

Simplify each expression.

1. $5x^2 \cdot 3x^3$

1. _____

2. $(4xy^2)^4$

2. _____

3. $\dfrac{15x^3}{3x}$

3. _____

4. $\left(\dfrac{2xy^3}{6x^0 y^2}\right)^2$

4. _____

5. $(3x^2 y^{-3})^{-2}$

5. _____

Determine whether each expression is a polynomial. If the polynomial has a specific name, give that name.

6. $3x^2 + x - \dfrac{1}{x}$

6. _____

7. $12x^3 - 9x^2 + 3$

7. _____

Convert each number to scientific notation. Then calculate. Express your answer in scientific notation.

8. $(25{,}000{,}000)(0.034)$

8. _____

9. $\dfrac{0.004}{2{,}000}$

9. _____

10. Write the following polynomial in descending order, and give its degree.

    $5x^2 - 2 + 4x^4 - 7x$

10. _____

Perform the operations indicated.

11. $(x^2 + 3x) + (2x^2 - x + 3)$ 11. _____

12. $(x^2 + 4) - (x^2 - 4x - 5)$ 12. _____

13. $(x - y)(x + 2xy + y)$ 13. _____

14. $(4x + 3)(3x - 1)$ 14. _____

15. $(3x - 5)(-x^2 + x + 3)$ 15. _____

16. $\dfrac{15x^3 - 5x^2 + 10}{5x}$ 16. _____

17. $\dfrac{3x^3 + 8x^2 - 4x + 155}{x + 5}$ 17. _____

18. $\dfrac{2x^2 + 7x - 15}{2x - 3}$ 18. _____

19. $\dfrac{(2x^2 + 3)^2}{(x - 1)(x + 1)}$ 19. _____

20. The Whirlpool Galaxy is 13 million light years away. If a light year   20. _____
is approximately 5.87×10^{12} miles, calculate the distance in miles.
Write your answer in scientific notation.

Chapter 5 Test Form G

Name:_____

Date:_____

Simplify each expression.

1. $3x^3 \cdot 2x$

    (a) $5x^2$ **(b)** $6x^2$ **(c)** $5x^4$ **(d)** $6x^4$

2. $(4xy^3)^3$

    (a) $64x^3y^9$ **(b)** $12x^3y^6$ **(c)** $64xy^3$ **(d)** $12x^3y^9$

3. $\dfrac{27x^5}{6x^2}$

    (a) $21x^3$ **(b)** $\dfrac{9}{2}x^3$ **(c)** $\dfrac{9}{3}x^2$ **(d)** $3x^3$

4. $\left(\dfrac{8x^3y^0}{2xy^{-5}}\right)^2$

    (a) $4\dfrac{x^3}{y^5}$ **(b)** $8x^2y^5$ **(c)** $16x^4y^{10}$ **(d)** $4x^2y^5$

5. $(7x^2y^{-2})^{-3}$

    (a) $-21\dfrac{x^2}{y^2}$ **(b)** $\dfrac{y^6}{343x^6}$ **(c)** $343x^{-1}y^{-5}$ **(d)** $\dfrac{21}{xy^5}$

Determine whether each expression is a polynomial. If the polynomial has a specific name, give that name.

6. $3x^4 - 2x^3 + 12x^2$

    (a) monomial   **(b)** not a polynomial   **(c)** binomial   **(d)** trinomial

7. $(x-3)^{-2}$

    (a) monomial   **(b)** not a polynomial   **(c)** binomial   **(d)** trinomial

Convert each number to scientific notation. Then calculate. Express your answer in scientific notation.

8. $(0.000027)(0.002)$

    (a) -1.35×10^6 **(b)** 2.9×10^{-12} **(c)** 5.4×10^{-8} **(d)** -5.4×10^8

9. $\dfrac{0.00042}{0.02}$

    (a) 2.1×10^{-2} **(b)** 2.1×10^4 **(c)** 21×10^{-3} **(d)** 2.1×10^2

10. Write the following polynomial in descending order, and give its degree.
$2x^2 - 3 - 3x^3 + 5x$

    (a) $-3x^3 + 2x^2 + 5x - 3$; fourth   **(b)** $-3x^3 + 2x^2 + 5x - 3$; third

    (c) $5x - 3x^3 - 3 + 2x^2$; third   **(d)** $-3x + 5x + 2x^2 - 3x^3$; fourth

Chapter 5 Test Form G *(cont.)*

Name:_____

Perform the operations indicated.

11. $(6x^3 + 8) + (x^2 - 3x - 12)$

   (a) $7x^3 - 3x - 4$ (b) $7x^3 - x^2 - 4$ (c) $6x^3 + x^2 + 3x - 4$ (d) $6x^3 + x^2 - 3x - 4$

12. $(3x^2 - 4x + 1) - (x^2 + 3x - 5)$

   (a) $4x^2 - x - 6$ (b) $2x^2 - x - 4$ (c) $4x^2 - x - 6$ (d) $2x^2 - 7x + 6$

13. $-2x^2 y^2 (3x - 4y)$

   (a) $-6x^3 y^2 + 8x^2 y^3$ (b) $-5xy - 2x^2 y^2 - 4y$ (c) $-6x^3 - 8y$ (d) $3x^2 - 2x^2 y - 4y^3$

14. $(2x - 3)(x + 5)$

   (a) $2x^2 - 3x - 15$ (b) $2x^2 + 13x + 15$ (c) $2x^2 + 7x - 15$ (d) $2x^2 + 13x - 5$

15. $(5x - 7)(x^2 - 3x - 1)$

   (a) $5x^3 - 22x^2 + 16x + 7$ (b) $5x^3 - 7x^2 - 8x + 7$ (c) $5x^3 - 15x - 7x^2 - 7$ (d) $5x^3 - 15x + 16x + 7$

16. $\dfrac{12x^3 + 4x^2 - 2x}{2x}$

   (a) $6x^2 - 2x - 1$ (b) $14x^2 + 6x - 4$ (c) $6x^2 + 2x - 1$ (d) $12x^2 + 4x - 1$

17. $\dfrac{6x^3 + 4x^2 - 7x}{x + 2}$

   (a) $6x^2 - 8x + 9$ (b) $6x^2 - 8x + 9 - \dfrac{18}{x + 2}$ (c) $3x^3 + 8x^2 - \dfrac{3}{2}x - \dfrac{3}{2}$ (d) $6x^2 - 8x + 9 + \dfrac{9}{x + 2}$

18. $\dfrac{6x^2 - 13x + 6}{3x - 2}$

   (a) $2x + 3$ (b) $3x^2 - 3$ (c) $2x - 3$ (d) $3x - 3$

19. $(3x + 2)^2 - (2x - 3)(2x + 3)$

   (a) $2x^2 + 6x + 13$ (b) $5x^2 + 12x + 13$ (c) $5x^2 + 6x + 13$ (d) $5x^2 - 6x - 5$

20. If FedEx handles 5.5 million packages a day at 42,200 drop off locations, what is the average number of packages deposited at each drop off location? Use scientific notation.

   (a) 2.2×10^0 (b) 1.03×10^3 (c) 7.2×10^1 (d) 1.30×10^2

Chapter 5 Test Form H

Name:_____

Date:_____._____

Simplify each expression.

1. $-2x^3 \cdot 4x^4$

   (a) $-8x^{12}$ (b) $8x^{12}$ (c) $-8x^7$ (d) $8x^7$

2. $(2x^2 y^3)^4$

   (a) $16x^8 y^{12}$ (b) $16x^6 y^7$ (c) $8x^6 y^7$ (d) $8x^8 y^{12}$

3. $\dfrac{12x^6}{2x^2}$

   (a) $6x^3$ (b) $10x^4$ (c) $10x^3$ (d) $6x^4$

4. $\left(\dfrac{6x^4 y^0}{2x^2 y^2}\right)^3$

   (a) $\dfrac{9x^5}{y^5}$ (b) $\dfrac{6x^2}{y^6}$ (c) $\dfrac{18x^9}{y^3}$ (d) $\dfrac{27x^6}{y^6}$

5. $(5x^{-2} y^3)^{-3}$

   (a) $\dfrac{-15x^6}{y^{-9}}$ (b) $\dfrac{x^6}{125 y^9}$ (c) $\dfrac{-125x^5}{y^6}$ (d) $-15x^6 y^9$

Determine whether each expression is a polynomial. If the polynomial has a specific name, give that name.

6. $4x + 5$

   (a) monomial   (b) not a polynomial   (c) binomial   (d) trinomial

7. $(x+3)^{-2}$

   (a) monomial   (b) not a polynomial   (c) binomial   (d) trinomial

Convert each number to scientific notation. Then calculate. Express your answer in scientific notation.

8. $(13,000,000)(200,000)$

   (a) 26×10^{12} (b) $260,000$ (c) 2.6×10^{12} (d) 2.6×10^{10}

9. $\dfrac{0.00096}{300,000}$

   (a) 3.2×10^{-9} (b) -2.3×10^{-9} (c) 32×10^8 (d) 2.3×10^7

10. Write the following polynomial in descending order, and give its degree.
    $7 + 3x - 2x^3 + 4x^4 - 5x^2$

    (a) $-5x^2 + 4x^4 - 2x^3 + 7$; fifth   (b) $7 + 3x - 5x^2 - 2x^3 + 4x^4$; first
    (c) $4x^4 - 2x^3 - 5x^2 + 3x + 7$; fourth   (d) $7 - 5x^2 + 4x^4 + 3x - 2x^3$; fourth

Chapter 5 Test Form H *(cont.)*

Perform the operations indicated.

11. $(3x^3 + 3) + (x^3 - x^2 + 4)$

    (a) $2x^3 - x^2 + 12$ **(b)** $4x^3 - 3x^2 + 7$ **(c)** $4x^3 - x^2 - 7$ **(d)** $4x^3 - x^2 + 7$

12. $(2x^2 + x - 4) - (x^2 + 3x + 8)$

    (a) $x^2 - 2x - 12$ **(b)** $3x^2 + 4x + 4$ **(c)** $3x^2 + 4x + 4$ **(d)** $x^2 + 2x + 12$

13. $-5x^2 y(2xy + y)$

    (a) $-5x^2 y + 2xy + y$ **(b)** $5x^3 y^2 + 2x + 5x^2 y$ **(c)** $-10x^3 y^2 + 5x^2 y^2$ **(d)** $-10x^3 y^2 - 5x^2 y^2$

14. $(3x - 2)(2x - 3)$

    (a) $6x^2 - 13x + 6$ **(b)** $6x^2 - 4x - 6$ **(c)** $6x^2 - 5x - 5$ **(d)** $6x^2 - x - 6$

15. $(4x - 5)(2x^2 - x + 1)$

    (a) $8x^3 - 10x^2 - 5$ **(b)** $8x^3 - 6x^2 - 5x - 5$ **(c)** $8x^3 - 14x^2 - 5$ **(d)** $8x^3 - 14x^2 + 9x - 5$

16. $\dfrac{20x^3 + 10x^2 - 15x}{5x}$

    (a) $20x^2 + 10x - 3$ **(b)** $4x^2 + 2x - 3$ **(c)** $4x^2 - 10x - 3$ **(d)** $4x^2 - 2x - 3$

17. $\dfrac{6x^2 + 4x + 3}{x + 2}$

    (a) $6x + 4 + \dfrac{3}{x + 2}$ **(b)** $6x^2 + 4x + \dfrac{3}{2}$ **(c)** $6x - 8 + \dfrac{19}{x + 2}$ **(d)** $6x - 5 + \dfrac{13}{x + 2}$

18. $\dfrac{-3x^2 + 12}{3x + 6}$

    (a) $x^2 - 2$ **(b)** $-x + 2$ **(c)** $x^2 + 2$ **(d)** $x + 2$

19. $\dfrac{(x^2 - 3)^2}{(x - 2)(x + 2)}$

    (a) $x^2 - 6 + \dfrac{9}{x^2 - 4}$ **(b)** $x^4 - 7x^2 + 13$ **(c)** $x^2 - 2 + \dfrac{1}{x^2 - 4}$ **(d)** $x^2 - 5 + \dfrac{29}{x^2 - 4}$

20. The maize crop in Zimbabwe in the 1998/1999 agricultural season was 1.54×10^6 tons, which was 2.6×10^5 tons short of local requirements. How much maize should Zimbabwe farmers produce next year to make up for the difference? Give your answer in scientific notation.

    (a) 2.06×10^6 tons   **(b)** 1.8×10^6 tons   **(c)** 5.2×10^5 tons   **(d)** 7.26×10^6 tons

Chapter 6 Pretest Form A

Name:_____

Date:_____

1. Find the GCF of the set of terms: $12x^2, 4x^3, 8x$

1. _____

Factor completely.

2. $10x^3 - 25x^5$

2. _____

3. $20x^4 - 12x^3 + 28x^2$

3. _____

4. $4x(x+3) + 5(x+3)$

4. _____

5. $x^2 + 3x - 40$

5. _____

6. $x^2 + 2x - 15$

6. _____

7. $3x^3 - 15x^2 + 18x$

7. _____

8. $2x^2 + 11x - 21$

8. _____

9. $12x^2 + x - 20$

9. _____

10. $3x^4 - 24x$

10. _____

11. $4x^2 - 20x + 25$

11. _____

12. $5x^3 - 20x$

12. _____

Solve.

13. $7x^2 + 10x = 8$

13. _____

14. $x^2 + 30 = 17x$

14. _____

15. $2x^3 + 3x^2 - 8x = 12$

15. _____

16. The area of a rectangle is 60 square centimeters. Find its length, width, and diagonal if the length is 7 centimeters greater than its width.

16. _____

Chapter 6 Pretest Form B

Name:_____

Date:_____

1. Find the GCF of the set of terms: $4x, 2x^2$, and $6x^8$.

1. _____

Factor completely.

2. $9x^2 - 3x^3$

2. _____

3. $4x^3 - 10x^2 + 18x$

3. _____

4. $5x(x-7) + 2(x-7)$

4. _____

5. $x^2 + 3x + 2$

5. _____

6. $x^2 + 2x - 48$

6. _____

7. $4x^3 - 24x^2 + 20x$

7. _____

8. $3x^2 + 11x - 4$

8. _____

9. $12x^2 + 8x - 15$

9. _____

10. $5x^4 + 5x$

10. _____

11. $25x^3 - 36x$

11. _____

12. $3x^3 - 24x^2 + 48x$

12. _____

Solve.

13. $9x^2 + 22x = 15$

13. _____

14. $x^3 - 25x = 0$

14. _____

15. The legs of a right triangle are $x-4$ and $x+3$, and the hypotenuse is $x+4$. Find x.

15. _____

16. If each side of a square is increased by 5 meters, the area becomes 169 square meters. Find the length of a side of the original square.

16. _____

Mini-Lecture 6.1
Factoring a Monomial from a Polynomial

Learning Objectives:

1. Identify factors.
2. Determine the greatest common factor of two or more numbers.
3. Determine the greatest common factor of two or more terms.
4. Factor a monomial from a polynomial.
5. Key Vocabulary: *factors, factor an expression, greatest common factor, GCF, prime number, composite number, unit, prime factorization*

Examples:

1. List the factors of each expression.

   a) $8m^4$ b) $4xy^2$

2. a) Write 76 as the product of prime numbers.

   b) Determine the GCF of 42 and 60.   c) Determine the GCF of 64 and 80.

3. Determine the GCF of the terms.

   a) m^2n^4, m^5n^3, mn^8 b) $6x^2y$, $12x^3y^2$, $3xy^4$

   c) $-18a^5b^2$, $24a^2b^3$, $30a^3b$ d) $4a+7b$, $6(4a+7b)$

   e) $(2x+3)(x+5)$, $(2x+3)(x+9)$ f) $(x-2)^3$, $3(x-2)^2$

4. Factor.

   a) $3c+9$ b) $12h^3+15h^4$ c) $14x^2+35x-7$

   d) $2x^2y-20xy^3+6y^2$ e) $x(5x+1)-3(5x+1)$ f) $3x(x-2)+(x-2)$

Teaching Notes:

- Encourage students to simplify the factoring process by factoring out the GCF first.
- Students may have trouble seeing the GCF at first. Encourage those having trouble to start with a common factor and continue pulling out common factors until no common factors remain.
- Remind students that variables in the GCF will have the smallest exponent that occurs for that variable in any of the terms.
- Remind students that they can check their work by multiplying.

Answers: 1a) $1,2,4,8,m,2m,4m,8m,m^2,2m^2,4m^2,8m^2,m^3,2m^3,4m^3,8m^3,m^4,2m^4,4m^4,8m^4$; 1b) $1,2,4,x,2x,4x,y,2y,4y,y^2,2y^2,4y^2,xy,2xy,4xy,xy^2,2xy^2,4xy^2$; 2a) $2\cdot2\cdot19$; 2b) 6; 2c) 16; 3a) mn^3; 3b) $3xy$; 3c) $6a^2b$; 3d) $4a+7b$; 3e) $2x+3$; 3f) $(x-2)^2$; 4a) $3(c+3)$; 4b) $3h^3(4+5h)$; 4c) $7(2x^2+5x-1)$; 4d) $2y(x^2-10xy^2+3y)$; 4e) $(5x+1)(x-3)$; 4f) $(x-2)(3x+1)$

Mini-Lecture 6.2
Factoring by Grouping

Learning Objectives:

1. Factor a polynomial containing four terms by grouping.
2. Key vocabulary: *factoring by grouping, FOIL*

Examples:

1. Factor by grouping.

   a) $y^2 + 4y + 6y + 24$

   b) $3x^2 + 6x + 5x + 10$

   c) $4a^2 - 6a + 10a - 15$

   d) $10x^3 + 20x^2 + 40x^2 + 80x$

   e) $4x^2 - 2x + 2x - 1$

   f) $6b^4 + 4b^2 + 12b^2 + 8$

   g) $4x^2 - 20x - 7x + 35$

   h) $3x^3 - 15x^2 + 6x^2 - 30x$

   i) $8mn + 4m - 6n - 3$

   j) $6p^2q^3 - 4pq^2 + 15pq - 10$

Teaching Notes:

- Remind students to factor out a negative if the first term of the polynomial is negative.

- When factoring by grouping, students sometimes have trouble when factoring out a negative. Remind them to check their work by multiplying.

- Remind students to always factor out the GCF first.

- Remind students that certain groupings may not yield a common factor. If this happens, a different grouping may be required.

- Remind students that the order of multiplication does not matter. The form of their answer may be different, but the results are the same. For example, $(2x + 5)(x - 1)$ is equivalent to $(x - 1)(2x + 5)$.

Answers: *1a)* $(y + 4)(y + 6)$; *1b)* $(x + 2)(3x + 5)$; *1c)* $(2a - 3)(2a + 5)$; *1d)* $10x(x + 2)(x + 4)$; *1e)* $(2x - 1)(2x + 1)$; *1f)* $2(b^2 + 2)(3b^2 + 2)$; *1g)* $(x - 5)(4x - 7)$; *1h)* $3x(x - 5)(x + 2)$; *1i)* $(4m - 3)(2n + 1)$; *1j)* $(3pq - 2)(2pq^2 + 5)$

Mini-Lecture 6.3
Factoring Trinomials of the Form $ax^2 + bx + c$, $a = 1$

<u>**Learning Objectives:**</u>

1. Factor trinomials of the form $ax^2 + bx + c$, $a = 1$.
2. Remove a common factor from a trinomial.
3. Key vocabulary: *factor, product, sum, binomials, prime polynomial, trial and error*

<u>**Examples:**</u>

1. Factor. If the polynomial is prime, so state.

   a) $m^2 + 5m + 6$ b) $a^2 + 2a - 3$ c) $x^2 + 5x - 14$

   d) $p^2 - 3p + 6$ e) $q^2 - 7q + 10$ f) $c^2 - 5c + 4$

   g) $x^2 - 13x - 30$ h) $x^2 + 15x - 60$ i) $12 - 8w + w^2$

   j) $7y + y^2 - 44$ k) $x^2 + 12xy - 64y^2$ l) $a^2 + 10ab + 24b^2$

2. Factor completely.

   a) $5x^2 + 40x + 75$ b) $3a^3 + 12a^2 + 9a$ c) $-r^2 - 5r + 6$

   d) $-n^2 + 3n - 4$ e) $3x^4 + 42x^3 + 135x^2$ f) $-6x^3 - 18x^2 + 168x$

<u>**Teaching Notes:**</u>

- Remind students to always factor out the GCF first.
- It is often helpful for students to make a table where one column contains all possible factor pairs for c and another column contains the sum of the factors.
- Remind students that one or both of the factors for c can be negative.
- Remind students that when c is positive, the two factors will have the same sign (the sign of b) and when c is negative, the two factors will have opposite signs.
- Remind students that when the two factors of c have opposite signs, the factor with the larger absolute value has the same sign as b.
- Remind students to always check their result by multiplying.

Answers: 1a) $(m+2)(m+3)$; 1b) $(a-1)(a+3)$; 1c) $(x-2)(x+7)$; 1d) prime; 1e) $(q-5)(q-2)$; 1f) $(c-4)(c-1)$; 1g) $(x-15)(x+2)$; 1h) prime; 1i) $(w-6)(w-2)$; 1j) $(y-4)(y+11)$; 1k) $(x+16y)(x-4y)$; 1l) $(a+4b)(a+6b)$; 2a) $5(x+3)(x+5)$; 2b) $3a(a+1)(a+3)$; 2c) $-(r-1)(r+6)$; 2d) $-\left(n^2 - 3n + 4\right)$; 2e) $3x^2(x+5)(x+9)$; 2f) $-6x(x-4)(x+7)$

Mini-Lecture 6.4
Factoring Trinomials of the Form $ax^2 + bx + c$, $a \neq 1$

Learning Objectives:

1. Factor trinomials of the form $ax^2 + bx + c$, $a \neq 1$, by trial and error.
2. Factor trinomials of the form $ax^2 + bx + c$, $a \neq 1$, by grouping.
3. Key vocabulary: *factoring by trial and error, factoring by grouping, GCF, coefficient, pairs of factors*

Examples:

1. Factor completely using the trial and error method. If the polynomial is prime, so state.

   a) $2x^2 + x - 10$

   b) $12c^2 + 23c + 5$

   c) $10n^2 - 9n - 8$

   d) $7x^2 - 27x - 4$

   e) $6m^2 - 17m - 45$

   f) $35x^2 + 29xy + 6y^2$

   g) $5y^2 + 20y + 12$

   h) $12p^3 + 14p^2 - 10p$

   i) $-24a^3 - 18a^2 + 81a$

2. Factor completely using the grouping method. If the polynomial is prime, so state.

   a) $4n^2 - 31n - 8$

   b) $60b^2 + 43b - 10$

   c) $18x^2 + 45x + 28$

   d) $24k^2 - 17k - 20$

   e) $6h^2 - 5h - 56$

   f) $6x^2 + 3x + 10$

   g) $6d^2 + 34d - 12$

   h) $24x^3 + 14x^2y - 24xy^2$

   i) $-12m^3 + 72m^2 - 105m$

Teaching Notes:

- A quick review of factoring by grouping will be helpful.
- For the trial and error method, some students may wish to make a table listing all possible factor pair combinations for a and c, and then use these to form all possible binomial factor pairs for the polynomial.
- Remind students that if the original expression has no common factor, then neither of the two factors can have a common factor.
- Remind students to put the polynomial in standard form before attempting to factor.
- Students often prefer the grouping method because it gives them a structured approach to factoring.
- Remind students to check their results by multiplying.

Answers: 1a) $(x-2)(2x+5)$; 1b) $(3c+5)(4c+1)$; 1c) prime; 1d) $(7x+1)(x-4)$; 1e) $(2m-9)(3m+5)$; 1f) $(7x+3y)(5x+2y)$; 1g) prime; 1h) $2p(3p+5)(2p-1)$; 1i) $-3a(2a-3)(4a+9)$; 2a) $(n-8)(4n+1)$; 2b) prime; 2c) $(6x+7)(3x+4)$; 2d) $(8k+5)(3k-4)$; 2e) $(2h-7)(3h+8)$; 2f) prime; 2g) $2(3d-1)(d+6)$; 2h) $2x(4x-3y)(3x+4y)$; 2i) $-3m(2m-5)(2m-7)$

Mini-Lecture 6.5
Special Factoring Formulas and a General Review of Factoring

<u>**Learning Objectives:**</u>

1. Factor the difference of two squares.
2. Factor the sum and difference of two cubes.
3. Learn the general procedure for factoring a polynomial.
4. Key vocabulary: *perfect squares, perfect cubes, difference of two squares, difference of two cubes, sum of two cubes*

<u>**Examples:**</u>

1. Factor each difference of squares.

   a) $x^2 - 64$

   b) $121 - m^2$

   c) $16a^2 - 81b^2$

   d) $16m^4 - 256n^4$

2. Factor.

   a) $y^3 - 64$

   b) $125 + 8x^3$

   c) $64a^3 + 343b^3$

   d) $27p^3 - 8q^3$

3. Factor completely.

   a) $6m^3 + 8m^2 - 6m$

   b) $5n^2 + 30n + 45$

   c) $3p^4 - 24pq^3$

   d) $12x^2 - 3x + 8x - 2$

   e) $4w^3 - 2w^2 - 64w + 32$

   f) $5x^5 - 45x^3y^2$

   g) $10c^3 + 10,000$

   h) $30 - 15x + 24x - 12x^2$

<u>**Teaching Notes:**</u>

- It is worthwhile to spend a little extra time showing students how to write expressions as perfect squares or perfect cubes.
- Encourage students to learn to recognize special forms, particularly the sum/difference of cubes, as they will encounter these situations in later applications.
- Remind students that there is no special form for the sum of two squares.
- Emphasize that factoring often involves trial and error, but experience can help students determine a first course of action. For example, if there are only two terms, students should try special forms for squares or cubes. If there are more than three terms, students should try factoring by grouping.

Answers: 1a) $(x-8)(x+8)$; 1b) $(11-m)(11+m)$; 1c) $(4a-9b)(4a+9b)$;

1d) $16(m-2n)(m+2n)(m^2+4n^2)$; 2a) $(y-4)(y^2+4y+16)$; 2b) $(5+2x)(25-10x+4x^2)$;

2c) $(4a+7b)(16a^2-28ab+49b^2)$; 2d) $(3p-2q)(9p^2+6pq+4q^2)$; 3a) $2m(3m^2+4m-3)$; 3b) $5(n+3)^2$;

3c) $3p(p-2q)(p^2+2pq+4q^2)$; 3d) $(4x-1)(3x+2)$; 3e) $2(2w-1)(w-4)(w+4)$; 3f) $5x^3(x-3y)(x+3y)$;

3g) $10(c+10)(c^2-10c+100)$; 3h) $3(2-x)(5+4x)$

Mini-Lecture 6.6
Solving Quadratic Equations Using Factoring

<u>**Learning Objectives:**</u>

   1. Recognize quadratic equations.
   2. Solve quadratic equations using factoring.
   3. Key vocabulary: *quadratic equation, standard form, zero-factor property*

<u>**Examples:**</u>

   1. Determine whether the following are quadratic equations.

     a) $(x+3)(x-5) = 6$ b) $x^2 - 5x + 6$

     c) $x + 3 = 6$ d) $15 - 7x + 3x^2 = 4$

   2. Solve the equations.

     a) $3x(x+4) = 0$ b) $(x+9)(x-2) = 0$

     c) $10x^2 + 15x = 0$ d) $3x^2 - 6x - 45 = 0$

     e) $3x^2 = 16 - 8x$ f) $9x^2 = 100$

     g) $x(x+10) = -21$ h) $(x-6)(x-10) = 5$

<u>**Teaching Notes:**</u>

- Remind students to always put the equation into standard form before solving.
- Remind students that the zero-factor property can only be used if the equation is in the standard form $ax^2 + bx + c = 0$ (that is, the expression to be factored must be equal to 0).
- Remind students to check their answers in the original equation.
- Remind students about the difference between expressions and equations. Point out how in previous sections they *factored a polynomial expression* while in this section they are *using factoring to solve an equation* which involves factoring an expression but takes it one step further.

Answers: *1a) yes; 1b) no; 1c) no; 1d) yes; 2a) 0, −4; 2b) −9, 2; 2c) 0, $-\dfrac{3}{2}$; 2d) −3, 5; 2e) −4, $\dfrac{4}{3}$;*

2f) $-\dfrac{10}{3}$, $\dfrac{10}{3}$; 2g) −3, −7; 2h) 5, 11

Mini-Lecture 6.7
Applications of Quadratic Equations

Learning Objectives:

1. Solve applications by factoring quadratic equations.
2. Learn the Pythagorean Theorem.
3. Key vocabulary: *quadratic equation, right triangle, legs, hypotenuse, Pythagorean Theorem, consecutive integers, consecutive even integers, consecutive odd integers*

Examples:

1. a) The product of two positive numbers is 112. Determine the two numbers if one number is 6 more than the other.

   b) A picture frame in the shape of a rectangle has an area of 360 square inches. Determine the length and width if the length is 6 inches less than twice the width.

   c) The area of a circle is 16π square centimeters. Determine the radius of the circle.

   d) The product of two consecutive odd integers is 255. Find the numbers.

2. a) The legs of a right triangle are given by $a = 9$ and $b = 12$. Determine the length of the hypotenuse.

   b) The hypotenuse of a right triangle measures 40 units and one leg measures 32 units. What is the length of the other leg?

   c) One leg of a right triangle is 4 meters more than twice the other. The hypotenuse is 26 meters. Find the lengths of the legs of the triangle.

   d) A signal tower is supported by wires. One wire goes from the top of the tower to a point on the ground. The height of the tower is 10 feet less than twice the distance between the base of the tower and the wire anchored on the ground. The length of the wire is 10 feet more than twice the distance between the base of the tower and the wire anchored on the ground. Find the height of the tower.

Teaching Notes:

- Students may need a review on translating word statements to mathematical statements.
- Have students review formulas for basic geometric shapes. Encourage them to memorize the basic formulas.
- Remind students that the hypotenuse of a right triangle is always opposite the right angle and is always the longest side.
- Students sometimes have trouble understanding the consecutive even/odd integer problems.

Answers: *1a) 8, 14; 1b) width: 15 inches, length: 24 inches; 1c) 4 centimeters; 1d) 15, 17; 2a) $c = 15$; 2b) 24 units; 2c) 10 meters, 24 meters; 2d) 150 feet*

Additional Exercises 6.1

Name:_____

Date:_____

Write each number as a product of prime numbers.

1. 300

2. 360

3. 98

1. _____

2. _____

3. _____

Find the greatest common factor for the two given numbers.

4. 28, 42

5. 105, 120

6. 100, 57

4. _____

5. _____

6. _____

Find the greatest common factor for each set of terms.

7. x^5, x^2, x^4

8. $16x, 8x^2$, and $24x^8$

9. $x^3 y^9, x^9 y^7$, and $x^7 y^3$

10. $12x^2, 4x^3, 8x$

11. $20x, 7x^3, -10x^2$

12. $2x^3, -2x^2, 6$

7. _____

8. _____

9. _____

10. _____

11. _____

12. _____

Factor the GCF from each term in the expression. If an expression cannot be factored, so state.

13. $8x - 6$

14. $11x + 2y^2$

15. $12x^4 - 42x^6$

16. $21x^2 y - 14x^2 + 42x$

17. $5x^3 + 5x^2 + 15x$

18. $12x^5 - 9x^4 + 15x^3 + 9x^2$

19. $24x^3 y^2 - 28x^2 y^2 + 12xy^3$

20. $3x^3 + 3x^2 + 6x$

13. _____

14. _____

15. _____

16. _____

17. _____

18. _____

19. _____

20. _____

Additional Exercises 6.2

Name:_____

Date:_____

Factor by grouping.

1. $x^2 - 2x + 6x - 12$

2. $x^2 - 4x + 5x - 20$

3. $x^2 - 4x + 7x - 28$

4. $x^2 + 3x + 8x + 24$

5. $5x^2 + 5x + 3x + 3$

6. $3x^2 + 18x + 5x + 30$

7. $x^2 + 2x - 2x - 4$

8. $2x^3 + 4x^2 + x + 2$

9. $9x^2 + 3x + 3x + 1$

10. $2x^3 + 2x^2 + x + 1$

11. $4x^2 + 20x + 3x + 15$

12. $3x^3 - 6x^2 + x - 2$

13. $3x^2 + 18x + 4x + 24$

14. $2x^2 - 4xy + xy - 2y^2$

15. $3x^2 - xy + 15xy - 5y^2$

16. $2x^2 - 8xy + xy - 4y^2$

17. $y^2 + ay + 2by + 2ab$

18. $5x^2 + 4xy - 5xy - 4y^2$

19. $4x^2 - 8x - 8x + 16$

20. $30x^2 + 20xy + 45xy + 30y^2$

1. _____

2. _____

3. _____

4. _____

5. _____

6. _____

7. _____

8. _____

9. _____

10. _____

11. _____

12. _____

13. _____

14. _____

15. _____

16. _____

17. _____

18. _____

19. _____

20. _____

Additional Exercises 6.3

Name:_____

Date:_____

Factor each expression. If it cannot be factored, so state.

1. $x^2 - 6x - 16$

2. $x^2 - 10xy + 24y^2$

3. $x^2 + 5x + 6$

4. $x^2 - 3x + 2$

5. $x^2 - x - 30$

6. $x^2 + 7x - 44$

7. $x^2 - x + 2$

8. $x^2 - 5xy + 6y^2$

9. $x^2 + 10x + 16$

10. $x^2 + x - 72$

11. $x^2 - 8x - 9$

12. $x^2 + 4x + 4$

13. $x^2 - 13xy + 42y^2$

14. $x^2 + 8x + 12$

Factor completely.

15. $4x^3 - 8x^2 - 12x$

16. $2x^3 - 2x^2 - 4x$

17. $2x^3 - 4x^2 - 6x$

18. $2x^3 - 4x^2y - 6xy^2$

19. $5x^3y - 35x^2y + 50xy$

20. $3x^3y + 18x^2y - 21xy$

1. _____

2. _____

3. _____

4. _____

5. _____

6. _____

7. _____

8. _____

9. _____

10. _____

11. _____

12. _____

13. _____

14. _____

15. _____

16. _____

17. _____

18. _____

19. _____

20. _____

Additional Exercises 6.4

Factor completely. If an expression cannot be factored, so state.

1. $4x^2 + 1 - 4x$

2. $15x^2 + 12 + 29x$

3. $6r^2 + 5r - 6$

4. $35a^2 + 3a - 20$

5. $25x^2 + 8 + 30x$

6. $12x^2 + 3 + 13x$

7. $9x^2 - 27xy + 20y^2$

8. $25u^2 - 15u - 18$

9. $12f^2 - f - 6$

10. $5z^2 + 3z + 4$

11. $4x^2 - 1$

12. $20x^2 + 6 + 23x$

13. $6x^2 - 19xy + 10y^2$

14. $35p^2 + 13p - 4$

15. $50x^2 + 10x - 12$

16. $-30x^2 - 25x + 30$

17. $-18x^2 + 18x + 20$

18. $3x^3 - 22x^2 + 7x$

19. $15x^2 - 18x - 24$

20. $3x^3 + 4x^2 + 5x$

1. _____

2. _____

3. _____

4. _____

5. _____

6. _____

7. _____

8. _____

9. _____

10. _____

11. _____

12. _____

13. _____

14. _____

15. _____

16. _____

17. _____

18. _____

19. _____

20. _____

Additional Exercises 6.5

Factor completely.

1. $49x^2 - 42xy + 9y^2$

2. $36x^2 + 84xy + 49y^2$

3. $81r^2 - 16s^2$

4. $12x^2 - 75y^2$

5. $64x^2 - 9y^2$

6. $x^2 - 64$

7. $27m^3 + 64$

8. $64p^3 + 125$

9. $9x^2 + 12xy + 4y^2$

10. $49x^2 - 84xy + 36y^2$

11. $81a^2 - 16b^2$

12. $x^2y^2 - 64y^2$

13. $25x^2 - 81y^2$

14. $x^3 + 3x^2 - 4x - 12$

15. $27a^3 + 8$

16. $c^3 + 125$

17. $75x^2 - 48$

18. $x^5 - x^2 - x^3 + 1$

19. $x^4 + 4x^3 + 4x^2 - 9x^2 - 36x - 36$

20. $5c^3 + 40$

1. _____

2. _____

3. _____

4. _____

5. _____

6. _____

7. _____

8. _____

9. _____

10. _____

11. _____

12. _____

13. _____

14. _____

15. _____

16. _____

17. _____

18. _____

19. _____

20. _____

Additional Exercises 6.6

Solve.

1. $3x(x-6)=0$

2. $(x+2)(x-11)=0$

3. $(x+7)(x+12)=0$

4. $x^2-19x=0$

5. $x^2-9x=0$

6. $x^2-5x+6=0$

7. $4x^2+15x=25$

8. $x^2-x-30=0$

9. $x^2-3x=0$

10. $x^2-9x+18=0$

11. $2x^3-50x+1=1$

12. $x^2=x+6$

13. $3x^2+13x=10$

14. $20x^2-3=11x$

15. $6x^2+13x=28$

16. $x^2+8x+13=22x-36$

17. $x(x+5)=104$

18. $3x^2-4x+5=37$

19. $(x-1)(x-2)=x+47$

20. $162-9x=x^2$

1. _____

2. _____

3. _____

4. _____

5. _____

6. _____

7. _____

8. _____

9. _____

10. _____

11. _____

12. _____

13. _____

14. _____

15. _____

16. _____

17. _____

18. _____

19. _____

20. _____

Additional Exercises 6.7

Name:_____

Date:_____

Determine whether the following are the sides of a right triangle.

1. $a = 4, b = 5, c = 6$

    1. _____

2. $a = 12, b = 9, c = 16$

    2. _____

3. $a = 4.5, b = 6, c = 7.5$

    3. _____

In each of the following, a, b, c are the sides of a right triangle; solve for x

4. $a = x, b = 16, c = 20$

    4. _____

5. $a = 9, b = x, c = 41$

    5. _____

6. $a = 16, b = 30, c = x$

    6. _____

7. $a = 9, b = 2x, c = 15$

    7. _____

8. $a = 20 - x, b = 12, c = 20$

    8. _____

9. $a = x - 5, b = x + 9, c = 26$

    9. _____

10. $a = 12, b = x + 5, c = 2x - 2$

    10. _____

11. $a = x + 5, b = x - 2, c = 2x - 1$

    11. _____

12. Find two positive integers where one is 3 more than twice the other, and their product is 90.

    12. _____

13. Find two consecutive odd integers whose product is 323.

    13. _____

14. Find three consecutive integers whose product is 336.

    14. _____

15. Find two consecutive integers, the sum of whose squares is 61.

    15. _____

16. Find the dimensions of a rectangle if the length is twice the width and the area is 98.

    16. _____

17. Find the dimensions of a rectangle if the length is twice the width and (numerically) 11 times the perimeter is 3 times the area.

    17. _____

18. An aircraft flying at 2500 feet jettisons a fuel tank. How long will it take the fuel tank to hit the water? ($d = 16t^2$)

    18. _____

19. The sum of the first n consecutive even integers is $s = n^2 + n$. If the sum of the even integers from 2 to x is 600, what is x?

    19. _____

20. The width of a rectangular garden is $\frac{2}{3}$ its length and its area is 4374. A walkway x ft. wide is put all around the garden and the new area is 5670. Find the width of the walkway.

    20. _____

Chapter 6 Test Form A

Determine the greatest common factor.

1. $9x^3, 27x^4, 54x^2$

      1. _____

2. $15x^3y^2, 25x^2y^4, 20x^3y^2$

      2. _____

Factor completely.

3. $6x^3y^3 + 9x^2y$

      3. _____

4. $10a^2b^2 - 5ab^3 - 15ab^2$

      4. _____

5. $x^2 - x + 6$

      5. _____

6. $2x^2 + 9x - 5$

      6. _____

7. $a^2 - 6ab - 16b^2$

      7. _____

8. $3x^2 + 2xy - 8y^2$

      8. _____

9. $4a^2 - 4az + z^2$

      9. _____

10. $3x^2 + 27x + 60$

     10. _____

11. $t^2 - w^2$

     11. _____

12. $x^5 + 8x^2 - x^3 - 8$

     12. _____

13. $a^3 + 64$

     13. _____

Solve.

14. $7x^2 + 7x = 2x + 2$

     14. _____

15. $x^3 - 9x = 0$

     15. _____

16. $x^2 + 18x + 81 = 0$

     16. _____

17. $x^2(x + 2) - 2x(x + 2) - 3(x + 2) = 0$

     17. _____

18. $3x^2 - 23x + 30 = 0$

     18. _____

19. The product of two consecutive positive integers is 210. Determine the two integers.

     19. _____

20. The area of a rectangle is 168 square feet. Determine the length, width, and diagonal of the rectangle if the length is 3 feet greater than 3 times the width.

     20. _____

Chapter 6 Test Form B

Name:_____

Date:_____

Determine the greatest common factor.

1. $5x^3, 20x^2, 50x^4$

2. $28x^2y, 14x^3y^2, 35x^2y^2$

Factor completely.

3. $8x^3y^3 + 12x^2y^4$

4. $3a^3b - 6a^2b^2 - 3ab$

5. $x^2 + 3x + 4$

6. $2a^2 + 11a - 6$

7. $2x^2 + 9xy - 5y^2$

8. $x^2 + 8x + 15$

9. $9a^2 + 6a + 1$

10. $t^2 - z^2$

11. $4a^3 + 10a^2 - 24a$

12. $x^3 - 8y^3$

13. $x^5 - x^3 - x^2 + 1$

Solve.

14. $2x^2 + 10x = 3x + 15$

15. $x^4 = 8x$

16. $x^2 - 12x + 36 = 0$

17. $x^2(x+3) - 3x(x+3) = 10(x+3)$

18. $x^2 + 3x = 10$

19. The product of two consecutive negative integers is 132. Determine the two integers.

20. The area of a rectangle is 240 square feet. Determine the length, width, and diagonal of the rectangle if the length is 4 feet greater than twice the width.

1. _____

2. _____

3. _____

4. _____

5. _____

6. _____

7. _____

8. _____

9. _____

10. _____

11. _____

12. _____

13. _____

14. _____

15. _____

16. _____

17. _____

18. _____

19. _____

20. _____

Chapter 6 Test Form C

Determine the greatest common factor.

1. $36y^3, 12y^4, 18y^7$

1. _____

2. $16x^3y^3, 20x^4y^3, 4x^2y^4$

2. _____

Factor completely.

3. $13x^3y^2 - 5xy$

3. _____

4. $6a^3b - 6a^3b^3 + 24a^2b^2$

4. _____

5. $a^2 + 7a + 15$

5. _____

6. $3x^2 - 5x - 2$

6. _____

7. $3x^2 + 7xy + 2y^2$

7. _____

8. $3t^2 + ut - 2u^2$

8. _____

9. $a^2 - 4b^2$

9. _____

10. $4x^4 + 4x^3 - 8x^2$

10. _____

11. $8 - a^3$

11. _____

12. $4x^5 + 4x^2 - x^3 - 1$

12. _____

13. $9a^2 + 12a + 4$

13. _____

Solve.

14. $12x^3 = 48x^2 - 48x$

14. _____

15. $x^2(x-2) + 2x(x-2) = 15(x-2)$

15. _____

16. $x^2 - 10x + 25 = 0$

16. _____

17. $18x^3 = 2x$

17. _____

18. $3x^2 - 12x - 63 = 0$

18. _____

19. The product of two consecutive positive integers is 306. Determine the two integers.

19. _____

20. The legs of a right triangle are $x+11$, and $2x$, and the hypotenuse is $2x+9$. If $x > 5$, find x.

20. _____

Chapter 6 Test Form D

Determine the greatest common factor.

1. $12a^3, 8a^2, 4a^4$

    1. _____

2. $18x^2y^4, 3x^3y^3, 9x^4y^4$

    2. _____

Factor completely.

3. $10x^3y^2 + 5xy$

    3. _____

4. $4x^3y - 6x^3y^4 + 10x^2y$

    4. _____

5. $x^2 - 4x + 45$

    5. _____

6. $6x^2 - 5x + 1$

    6. _____

7. $2a^2 + 13a + 20$

    7. _____

8. $2x^2 - 19x - 33$

    8. _____

9. $x^2 - 4y^2$

    9. _____

10. $t^2 - 4ut + 4u^2$

    10. _____

11. $3x^2 - 21x - 54$

    11. _____

12. $8x^3 + 1$

    12. _____

13. $x^5 - 4x^3 + 27x^2 - 108$

    13. _____

Solve.

14. $x^3 = 4x^2 + 77x$

    14. _____

15. $x^3 = 16x$

    15. _____

16. $x^2 - 13x = -30$

    16. _____

17. $2x^3(x+3) = 3x(x+3) + 2(x+3)$

    17. _____

18. $7x^2 - 5x = 14x - 10$

    18. _____

19. The product of two consecutive positive even integers is 624. Determine the two integers.

    19. _____

20. Sally has a rectangular yard. Its length is 4 feet shorter than its diagonal and two feet less than twice its width. Find the area and the diagonal.

    20. _____

Chapter 6 Test Form E

Name:_____

Date:_____

Determine the greatest common factor.

1. $12x^4, 6x^6, 30x^3$

2. $10a^2b^3, 20a^2b^2, 25ab^4$

Factor completely.

3. $6a^3b - 2a^2b^3$

4. $3x^2y^2 + 15x^3y$

5. $3x^2 + 14x + 40$

6. $2t^2 - vt - 3v^2$

7. $2a^2 - 11a + 14$

8. $6x^2 + x - 15$

9. $10x^2 - 15x - 45$

10. $x^5 - x^3 + 8x^2 - 8$

11. $x^2 - 4ax + 4a^2$

12. $x^3 + 8$

13. $t^2 - 25$

Solve.

14. $x^2 = 9$

15. $x^2 - 13x + 42 = 0$

16. $x^3 = 2x^2 - x$

17. $24x^2 + 16x = 15x + 10$

18. $x^3 + 3x^2 = 25x + 75$

19. The product of two consecutive negative even integers is 288. Determine the two integers.

20. The area of a rectangle is 108 square inches. Determine the length, width, and diagonal of the rectangle if its length is 3 inches longer than its width.

1. _____

2. _____

3. _____

4. _____

5. _____

6. _____

7. _____

8. _____

9. _____

10. _____

11. _____

12. _____

13. _____

14. _____

15. _____

16. _____

17. _____

18. _____

19. _____

20. _____

Chapter 6 Test Form F

Determine the greatest common factor.

1. $10x^4, 15x^3, 55x^5$

2. $12x^4 y^4, 18x^2 y^5, 36x^3 y^2$

Factor completely.

3. $12a^3 x^3 + 21ax$

4. $8x^3 y^2 - 12x^2 y^3 + 4x^2 y^2$

5. $x^2 - 3x + 3$

6. $2a^2 - 9a - 18$

7. $x^2 - 5xy + 6y^2$

8. $6a^2 + 5ax - 25x^2$

9. $125a^2 - 125a + 30$

10. $4x^2 - 12xy + 9y^2$

11. $x^5 - x^2 - 4x^3 + 4$

12. $25t^2 - 4w^2$

13. $27x^3 - 1$

Solve.

14. $x^2(3x+2) - 6x(3x+2) = 7(3x+2)$

15. $3x^3 = 12x$

16. $x^2 - 22x + 121 = 0$

17. $2x^3 = 16$

18. $x^2 - 1 = 6x - 9$

19. The product of two consecutive negative odd integers is 483. Determine the two integers.

20. The legs of a right triangle are x and $2(x+1)$; the hypotenuse is $2x+3$. Find x.

1. _____

2. _____

3. _____

4. _____

5. _____

6. _____

7. _____

8. _____

9. _____

10. _____

11. _____

12. _____

13. _____

14. _____

15. _____

16. _____

17. _____

18. _____

19. _____

20. _____

Chapter 6 Test Form G

Name:_____

Date:_____

Choose the correct answer to each problem.

Determine the greatest common factor.

1. $8a^3, 12a^4, 20a^2$

   (a) $4a^2$ (b) $4a$ (c) $6a^3$ (d) $2a^2$

2. $6a^3x^2, 36a^2x, 3a^3x^3$

   (a) $6ax$ (b) $3a^2x$ (c) $3ax$ (d) $3ax^2$

Factor completely.

3. $10a^3b^4 + 8ab^3$

   (a) $4a^3b^3(6b + 2a^2)$ (b) $8ab^3(2a^2b + 1)$ (c) $8a^3b^3(2b + a^2)$ (d) $2ab^3(5a^2b + 4)$

4. $9x^3y - 3xy^2 - 21xy$

   (a) $9x^3y(1 - x^2y + x^2)$ (b) $3xy(3x^2 - y - 7)$ (c) $3x(3x^2y - xy - 7y)$ (d) $3y(3x^2 - y - 7)$

5. $a^2 + 4a + 45$

   (a) $(a + 5)(a + 9)$ (b) $(a + 5)(a - 9)$ (c) $(a - 5)(a - 9)$ (d) prime

6. $3x^2 + 11x + 6$

   (a) prime   (b) $(x - 3)(2x - 3)$ (c) $(x + 3)(2x + 3)$ (d) $(x + 3)(3x + 2)$

7. $a^2 + 2ab - 8b^2$

   (a) $(a - 2b)(a + 4b)$ (b) $(a - 2b)(a - 4b)$ (c) $(a + 2b)(a + 4b)$ (d) $(2a - b)(4a + b)$

8. $6x^2 + 19xy - 7y^2$

   (a) $(2x + 7y)(3x - y)$ (b) $(2x - 7)(3x + y)$ (c) $(2x - y)(3x - 7y)$ (d) $(x - 7y)(3x - 2y)$

9. $14a^2 - 49a - 105$

   (a) $14(a - 5)(a + 3)$ (b) prime   (c) $7(a - 5)(2a + 3)$ (d) $(7a + 35)(2a - 3)$

10. $x^5 + 8x^2 - x^3 - 8$

   (a) $(x + 1)^2(x + 2)(x^2 + 4)$ (b) $(x - 1)(x + 1)(x + 2)(x^2 - 2x + 4)$
   (c) $(x - 2)^3(x + 1)^2$ (d) $(x - 1)(x + 1)(x + 2)(x - 2)^2$

11. $4x^2 + 20x + 25$

(a) $2x(x-3)(2x+5)$ (b) $(4x+5)(x+5)$ (c) $4x(x+20)$ (d) $(2x+5)^2$

12. $64x^3 + a^3$

(a) $4x(x+4)(4x+1)$ (b) $(4x+a)(16x^2 - 4ax + a^2)$

(c) $x(4x+1)^2$ (d) $4x(x+1)(4x+1)$

13. $9a^2 - 25b^2$

(a) $(3a+5b)(3a-5b)$ (b) $(9a+5b)(a+5b)$ (c) $(3a+5b)^2$ (d) $(3a+5b)(5a-3b)$

Solve.

14. $2x^4 = 54x$

(a) 2, 54   (b) 0, 27   (c) 0, 3   (d) 3

15. $x^2 + 24x + 144 = 0$

(a) 12, 14   (b) –12   (c) 0, 24   (d) –12, 12

16. $3x^2 - 27 = 0$

(a) –9, 3   (b) 3, –9   (c) –3, 3   (d) –3

17. $x^2(13x+5) = (5x+14)(13x+5)$

(a) $-\dfrac{5}{13}, 2, -7$ (b) $2, \dfrac{13}{5}, 7$ (c) $-2, \dfrac{5}{13}, -7$ (d) $-2, -\dfrac{5}{13}, 7$

18. $x^2 + 15 = 11x - 13$

(a) 14   (b) 0, 14   (c) 5, 6   (d) 4, 7

19. The legs of a right triangle are x and $5x-1$, and the hypotenuse is $5x+1$. Find x.

(a) 20   (b) 10   (c) 5   (d) –10

20. The length of each side of a square is increased by 3 meters. If the area of the original square is 144 square meters, determine the length of a side of the resulting square.

(a) 9 m   (b) 15 m   (c) 14 m   (d) 10 m

Chapter 6 Test Form H

Name:_____

Date:_____

Choose the correct answer to each problem.

Determine the greatest common factor.

1. $6xy^2, 3y^3, 9y$

    (a) $6xy$ **(b)** $6y$ **(c)** $3y^2$ **(d)** $3y$

2. $6a^3b^2, 2a^3b, 14a^2b^3$

    (a) $2ab$ **(b)** $2a^2b$ **(c)** $4a^2b^2$ **(d)** $6a^2b$

Factor completely.

3. $6x^3y - 9x^2y^3$

    (a) $3xy(2x^2 - 3xy^2)$ **(b)** $6xy(x^2 - 3xy^2)$ **(c)** $6x^2y(x + 3y^2)$ **(d)** $3x^2y(2x - 3y^2)$

4. $10a^2b + 15ab^2 - 30ab$

    (a) $10a^2b(1 + 5a^2 + 20b)$ **(b)** $5a(2ab + 3b^2 - 6b)$ **(c)** $5ab(2a + 3b - 6)$ **(d)** $10ab(a + 5b - 20)$

5. $a^2 - 9a + 22$

    (a) prime   **(b)** $(a + 2)(a - 11)$ **(c)** $(a - 2)(a - 11)$ **(d)** $(a + 2)(a + 11)$

6. $3x^2 - x - 14$

    (a) $(x + 2)(x - 7)$ **(b)** $(x - 2)(x + 3)$ **(c)** $(x + 2)(3x - 7)$ **(d)** prime

7. $x^2 + 4ax - 21a^2$

    (a) $(x + 3a)(x - 7a)$ **(b)** $(x - 3a)(x + 7a)$ **(c)** $(ax - 3)(ax + 7)$ **(d)** $(x + 3a)(x - 7a)$

8. $6a^2 + 13ab + 6b^2$

    (a) $(2a - 3b)(3a + 2b)$ **(b)** $(2a - 3b)(3a - 2b)$ **(c)** $(2ab - 3)(3ab - 2)$ **(d)** $(2a + 3b)(3a + 2b)$

9. $x^5 - x^2 - 4x^3 + 4$

    (a) $(x^2 - 2)(x^2 + 2)(x + 1)$ **(b)** $(x - 1)(x + 2)(x + 1)^3$

    (c) $(x - 1)(x - 2)(x + 2)(x^2 + 1)$ **(d)** $(x - 2)(x + 2)(x - 1)(x^2 + x + 1)$

10. $24a^2 + 20a - 100$

    (a) $(6a - 25)(5a + 4)$ **(b)** $4(2a + 5)(3a - 5)$ **(c)** prime   **(d)** $4(2a + 5)^2$

11. $t^2 - 8wt + 16w^2$

    (a) $(t-4w)^2$ **(b)** $(t-w)(t-16w)$ **(c)** $(t-2w)(t+8w)$ **(d)** $w(t-4)(t+8)$

12. $x^3 - 64a^3$

    (a) $x(x-4a)(x+8a)$ **(b)** $(x-4)^2(x+16a)$

    (c) $(x-4a)(x^2+4ax+16a^2)$ **(d)** $4a(a^2-4x)(x-a)$

13. $4x^2 - 9y^2$

    (a) $(2x-3y)^2$ **(b)** $(2x+3y)(2x-3y)$

    (c) $4(x-3y)(x-2y)$ **(d)** $(2x+3y)(3x-2y)$

Solve.

14. $x^2(2x+5) = (x+6)(2x+5)$

    (a) $-2, 3, -\dfrac{5}{2}$ **(b)** $2, 3, \dfrac{5}{2}$ **(c)** $3, 0, \dfrac{5}{2}$ **(d)** $-2, 3, -5$

15. $x^4 = 8x$

    (a) $0, 2$ **(b)** $1, 8$ **(c)** 0 **(d)** $0, 8$

16. $3x^2 + 12x + 12 = 0$

    (a) -2 **(b)** $3, 4$ **(c)** $1, 12$ **(d)** $2, 6$

17. $3x^2 = 75$

    (a) $1, 75$ **(b)** 25 **(c)** $-5, 5$ **(d)** $0, 75$

18. $3x^2 + 12 = 10x + 9$

    (a) $3, 5$ **(b)** $\dfrac{1}{3}, 3$ **(c)** $3, 4\dfrac{1}{2}$ **(d)** $0, 3$

19. The product of two consecutive positive odd integers is 323. Determine the two integers.

    (a) 13, 23   **(b)** 11, 32   **(c)** 19, 21   **(d)** 17, 19

20. If the length of a rectangle is 1 yard less than twice the width, and the diagonal is 2 yards longer than the length, find the area and the diagonal.

    (a) A = 64 , d = 8   **(b)** A = 225, d = 15   **(c)** A = 100, d = 20.   **(d)** A = 120, d = 17

Chapters 1–6 Cumulative Test
Form A

1. Find: $-56 - 78$

    1. _____

2. Consider the set of numbers:

    $\{-17, 31, -\sqrt{7}, 0, \sqrt{3}, \pi, \frac{4}{7}, -\sqrt{\frac{4}{9}}, 1.75\}$

    Which of the numbers are rational numbers?

    2. _____

3. Evaluate the expression: $3 \cdot 2^2 - 6 \cdot 3^2$

    3. _____

4. Solve for x: $5(3 - x) = 9 - 5x$

    4. _____

5. Solve for d in the equation $S = 9c^2 d$.

    5. _____

6. If 10 gallons of insecticide can treat 2 acres of land, how many gallons of insecticide are needed to treat 40 acres of land?

    6. _____

7. If the product of a number and 6 is increased by 25, the result is 4. Find the number.

    7. _____

8. A triangle has a perimeter of 58 inches. Find the length of the three sides if one side is 18 inches larger than the smallest and the third side is three times the smallest.

    8. _____

9. Two trains, an express and a commuter, are 150 miles apart. Both start at the same time and travel toward each other. They meet 3 hours later. The speed of the express is 10 miles faster than the commuter. Find the speed of each train.

    9. _____

10. Find the slope of the line that passes through $(-2, 6)$ and $(-4, 4)$.

    10. _____

Chapters 1–6 Cumulative Test
Form A *(cont.)*

11. Graph the linear equation: $5x + 3y = 3$

11.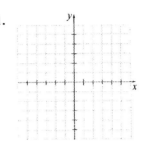

12. Simplify: $\left(\dfrac{4x^5 y}{12xy^3} \right)^5$

12. _____

13. Simplify and write with only positive exponents: $\left(\dfrac{2x^{-3} z^0}{3y^2 z^{-4}} \right)^{-2}$

13. _____

14. Add: $(2x-1)^2 + (x-3)(x+3)$

14. _____

15. Multiply: $(-5x^4 y^2)(6xy^4)$

15. _____

16. Multiply: $(6x+5)(4x^2 + 4x - 4)$

16. _____

17. Divide: $\dfrac{x^6 + x^4 + x - 1}{x^3 + 1}$

17. _____

18. Factor completely: $8x^2 - 6x - 35$

18. _____

19. Factor completely: $5m^3 - 80m$

19. _____

20. Solve: $3x^2 + 10x = 8$

20. _____

Chapters 1–6 Cumulative Test Form B

Name:_____

Date:_____

1. Subtract: $6 - 10$.

   (a) -16 (b) 16 (c) -4 (d) 4

2. What is the best classification for -0.9?

   (a) whole number   (b) integer   (c) rational number   (d) irrational number
       integer   rational number   real number   real number
       rational number   real number

3. Evaluate the expression: $-[10 - (-1 - 2)]^2$

   (a) 169 (b) 13 (c) -169 (d) -19

4. Solve for x: $8(x - 9) = -72 + 8x$

   (a) all real numbers   (b) -2 (c) no solution   (d) 0

5. Solve for a in $F = kma$.

   (a) $a = \dfrac{F}{km}$ (b) $a = \dfrac{km}{F}$ (c) $a = F - km$ (d) $a = Fkm$

6. A worker in an assembly line takes 4 hours to produce 29 parts. At that rate how many parts can she produce in 20 hours?

   (a) 290 parts   (b) 8 parts   (c) 145 parts   (d) 580 parts

7. If the sum of a number and 4 is decreased by 36, the result is 24. Find the number.

   (a) 15 (b) 56 (c) -3 (d) -16

8. The length of the base of an isosceles triangle is four times the length of one of its legs. If the perimeter of the triangle is 78 inches, what is the length of the base?

   (a) 18 in.   (b) 34 in.   (c) 13 in.   (d) 52 in.

9. A train A leaves a station traveling at 50 mph. Two hours later, train B leaves the same station traveling in the same direction at 60 mph. How long does it take for train B to catch up to train A?

   (a) 12 hr   (b) 10 hr   (c) 9 hr   (d) 11 hr

10. Determine the equation of a line that contains the points $(-3, 11)$ and $(2, -4)$.

   (a) $y = -\dfrac{1}{3}x + 10$ (b) $y = -3x + 2$ (c) $y = -3x - 10$ (d) $y = -\dfrac{1}{3}x - \dfrac{10}{3}$

Chapters 1–6 Cumulative Test Form B *(cont.)*

11. Graph: $x + 2y = 6$

(a)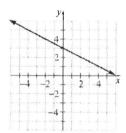

(b)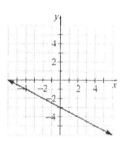

(c)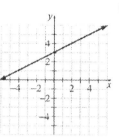

(d)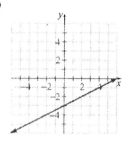

12. Simplify: $\left(\dfrac{3x^4 y}{9x^0 y^4}\right)^3$

(a) $\dfrac{3x^2}{y^3}$

(b) $\dfrac{y^9}{27x^{12}}$

(c) $\dfrac{3x^{12}}{y^9}$

(d) $\dfrac{x^{12}}{27y^9}$

13. Simplify and write with only positive exponents: $\left(\dfrac{4y^2}{5x^{-2}z^3}\right)^{-3}$

(a) $\dfrac{125z^9}{64x^6 y^6}$

(b) $\dfrac{64x^6 y^6}{125z^9}$

(c) $\dfrac{125x^6 z^9}{64y^6}$

(d) $\dfrac{125}{64xyz^6}$

14. Subtract: $(x^2 + 2)^2 - (2x - 3)(2x + 3)$

(a) $x^4 + 8x^2 - 5$

(b) $x^2 + 13$

(c) $4x^4 - 4x^2 + 5$

(d) $x^4 + 13$

15. Multiply: $\left(2x^3 y^3\right)\left(-7x^2 y^4\right)$

(a) $14x^5 y^7$

(b) $-14x^5 y^7$

(c) $-14x^3 y^4$

(d) $14x^3 y^4$

16. Multiply: $3x(x + 2)(x - 1)$

(a) $3x^3 + x^2 - 6x$

(b) $3x^3 + 3x^2 - 6$

(c) $3x^3 + 3x^2 - 6x$

(d) $3x^3 - 6x^2 - 2$

17. Divide: $\dfrac{5x^6 + 6x^4 - 9x^2}{5x^2 + 1}$

(a) $x^4 + x^2$

(b) $x^4 + x^2 - 2 + \dfrac{2}{5x^2 + 1}$

(c) $x^4 + \dfrac{7}{5}x^2 - \dfrac{8}{5} + \dfrac{2}{5x^2 + 1}$

(d) $x^4 + x^3 - 2x^2 + \dfrac{2}{5x^2 + 1}$

18. Factor completely: $8x^3z - 24x^2z + 4x^2z - 12xz$

    (a) $4xz(2x^2 - 6x + x - 3)$

    (b) $4xz(2x^2 - 5x - 3)$

    (c) $4xz(2x+1)(x-3)$

    (d) $(8x^2z + 4xz)(x-3)$

19. Factor completely: $24x^3z + 375y^3z$

    (a) $3z\left(8x^3 + 125y^3\right)$

    (b) $3z\left(2x - 5y\right)\left(4x^2 - 10xy + 25y^2\right)$

    (c) $3z\left(2x - 5y\right)\left(4x^2 + 10xy + 25y^2\right)$

    (d) $3z\left(2x + 5y\right)\left(4x^2 - 10xy + 25y^2\right)$

20. Solve: $2x^2 + 3x = 9$

    (a) $\left\{-\dfrac{2}{3}, 3\right\}$

    (b) $\left\{-\dfrac{3}{2}, 3\right\}$

    (c) $\left\{-3, \dfrac{2}{3}\right\}$

    (d) $\left\{-3, \dfrac{3}{2}\right\}$

Chapter 7 Pretest Form A

Determine the value or values of the variable for which the following expressions are defined.

1. $\dfrac{x+4}{4x+3}$

1. _____

2. $\dfrac{x-7}{x^2+5x-6}$

2. _____

Simplify to lowest terms.

3. $\dfrac{p+3}{p^2-9}$

3. _____

4. $\dfrac{x^2-6x-16}{8-x}$

4. _____

Perform the indicated operations.

5. $\dfrac{8x^2}{16x^3y^2} \div \dfrac{4x}{3y^5}$

5. _____

6. $\dfrac{x+1}{2x+y} \cdot \dfrac{4x^2-y^2}{4x^2-x-5}$

6. _____

7. $\dfrac{18}{y^2+3y-18} \div \dfrac{6y}{y-3}$

7. _____

8. $\dfrac{11}{5x} + \dfrac{4}{5x}$

8. _____

9. $\dfrac{7x+3}{x+7} - \dfrac{x-2}{x+7}$

9. _____

Find the least common denominator.

10. $\dfrac{5}{6x^2y^5} + \dfrac{1}{5x^3y^3}$

10. _____

11. $\dfrac{7}{x^2+5x-14} - \dfrac{1}{x^2-4}$

11. _____

Add or subtract.

12. $\dfrac{9}{8x^2 y^4} + \dfrac{1}{x^3 y^2}$

12. _____

13. $\dfrac{5}{x+4} + \dfrac{7}{(x+4)^2}$

13. _____

Simplify each complex fraction.

14. $\dfrac{2 - \frac{1}{2}}{2 - \frac{1}{4}}$

14. _____

15. $\dfrac{x + \frac{4x}{y}}{\frac{5}{3x}}$

15. _____

Solve.

16. $\dfrac{8}{32} = \dfrac{5}{x}$

16. _____

17. $\dfrac{x}{7} = \dfrac{x-5}{2}$

17. _____

18. $\dfrac{4}{x} - \dfrac{1}{8} = \dfrac{1}{x}$

18. _____

19. If y varies inversely as x, and $y = 10$ when $x = 3$, find y when $x = 15$.

19. _____

20. Audra can paint the living room in 2 hours, but it takes Marvin 3 hours. How long would it take them to paint the living room if they worked together?

20. _____

Chapter 7 Pretest Form B

Name:_____

Date:_____

Determine the value or values of the variable for which the following expressions are defined.

1. $\dfrac{x+2}{4x+12}$

1. _____

2. $\dfrac{x^2-8x-9}{x^2+6x+5}$

2. _____

Reduce to lowest terms.

3. $\dfrac{m+7}{m^2-49}$

3. _____

4. $\dfrac{x^2-3x-10}{5-x}$

4. _____

Perform the indicated operations.

5. $\dfrac{7y^2}{6}\cdot\dfrac{18x}{35y}$

5. _____

6. $\dfrac{a^2-8a+12}{a-2}\cdot\dfrac{a^2-4a-12}{(a-6)^2}$

6. _____

7. $\dfrac{20}{y^2+y-20}\div\dfrac{5y}{y-4}$

7. _____

8. $\dfrac{-4x+9}{3x}+\dfrac{5x-9}{3x}$

8. _____

9. $\dfrac{10x+1}{x+8}-\dfrac{6x-4}{x+8}$

9. _____

Find the least common denominator.

10. $\dfrac{7}{4x^3y^2}-\dfrac{9}{10x^2y^5}$

10. _____

11. $\dfrac{8x}{x^2-3x-4}+\dfrac{4}{x^2-2x-8}$

11. _____

Add or subtract.

12. $\dfrac{5}{14x^2y^3} - \dfrac{2}{7x^5y^2}$

12. _____

13. $\dfrac{9}{x+3} + \dfrac{4}{(x+3)^2}$

13. _____

Simplify each complex fraction.

14. $\dfrac{1+\frac{1}{5}}{1+\frac{1}{2}}$

14. _____

15. $\dfrac{1+\frac{2}{x}}{1-\frac{4}{x^2}}$

15. _____

Solve.

16. $\dfrac{6}{24} = \dfrac{7}{x}$

16. _____

17. $\dfrac{x}{9} = \dfrac{x-3}{6}$

17. _____

18. $\dfrac{5}{x} - \dfrac{1}{10} = \dfrac{1}{x}$

18. _____

19. If y varies inversely as x, and $y = 5$ when $x = 4$, find y when $x = 10$.

19. _____

20. A hose can fill a tank in 4 hours by itself. A smaller hose can fill the same tank in 6 hours by itself. How long would it take to fill fill the tank if both hoses are used together?

˙20. _____

Mini-Lecture 7.1
Simplifying Rational Expressions

Learning Objectives:

1. Determine the values for which a rational expression is defined.
2. Understand the three signs of a fraction.
3. Simplify rational expressions.
4. Factor a negative 1 from a polynomial.
5. Key vocabulary: *rational expression, numerator, denominator, factor, common factor, simplify, lowest terms*

Examples:

1. Determine the values for which each rational expression is defined.

   a) $\dfrac{5}{x+10}$

   b) $\dfrac{x-7}{3x+5}$

   c) $\dfrac{x+3}{x^2-2x-15}$

2. Write the fraction $\dfrac{5}{-8}$ in two other equivalent forms.

3. Simplify.

   a) $\dfrac{14x^3y^2}{21x^2y^5}$

   b) $\dfrac{12x^3+6x^2-30x}{12x^2}$

   c) $\dfrac{10x}{15x^3-25x^2-5x}$

   d) $\dfrac{x^2+9x+18}{x^2+8x+15}$

   e) $\dfrac{x^2-100}{x+10}$

   f) $\dfrac{3a^2+2a-8}{2a^2-a-10}$

4. Simplify.

   a) $\dfrac{5x-7}{7-5x}$

   b) $\dfrac{2n^2-7n+3}{3-n}$

   c) $\dfrac{x-9}{81-x^2}$

Teaching Notes:

- It may be a good idea to begin by reviewing simplification of numeric fractions such as $\dfrac{4}{6}$ before starting rational expressions.
- Remind students that only factors can be cancelled, not terms.
- Remind students to factor completely before simplifying.

Answers: 1a) $x \neq -10$; 1b) $x \neq -\frac{5}{3}$; 1c) $x \neq -3, x \neq 5$; 2) $-\frac{5}{8}$ or $\frac{-5}{8}$; 3a) $\frac{2x}{3y^3}$; 3b) $\frac{2x^2+x-5}{2x}$;

3c) $\frac{2}{3x^2-5x-1}$; 3d) $\frac{x+6}{x+5}$; 3e) $x-10$; 3f) $\frac{3a-4}{2a-5}$; 4a) -1; 4b) $-(2n-1)$ or $1-2n$; 4c) $-\frac{1}{x+9}$

Mini-Lecture 7.2
Multiplication and Division of Rational Expressions

Learning Objectives:

1. Multiply rational expressions.
2. Divide rational expressions.
3. Key vocabulary: *reciprocal*

Examples:

1. Multiply.

a) $\left(\dfrac{7}{10}\right)\left(\dfrac{-15}{14}\right)$

b) $\dfrac{8x^3}{3y^2} \cdot \dfrac{9y^3}{16x^4}$

c) $\left(x^2 - 4\right) \cdot \dfrac{7}{3x^2 + 6x}$

d) $\dfrac{9x^2 - 25y^2}{12xy^2} \cdot \dfrac{3xy}{10xy - 6x^2}$

e) $\dfrac{2x^2 + x - 3}{x^2 - x} \cdot \dfrac{5x^2 - 20x}{2x^2 - 5x - 12}$

f) $\dfrac{x^2 - 11x + 28}{x^2 - 11x + 10} \cdot \dfrac{x^2 - 16x + 60}{x^2 - 9x + 20}$

2. Divide.

a) $\dfrac{3}{14} \div \dfrac{5}{6}$

b) $\dfrac{15m^3}{6n^2} \div \dfrac{9m^4}{12n^4}$

c) $\dfrac{a^2 - 64}{a - 7} \div \dfrac{a + 8}{a - 7}$

d) $\dfrac{2x^2 - 3x - 2}{x^2} \div \left(x^2 - 4\right)$

e) $\dfrac{2x^2 - x - 15}{2x^2 - 3x - 20} \div \dfrac{x^2 - 9}{7x^2 - 28x}$

f) $\dfrac{t^2 + 10t + 25}{t^2 - t - 30} \div \dfrac{t^2 - 4t - 45}{t^2 + t - 42}$

Teaching Notes:

- It may be a good idea to begin by reviewing multiplication and division with numeric fractions before starting with rational expressions.
- Remind students to factor completely before simplifying.
- When dividing, remind students to change division problems to multiplication problems before factoring and cancelling.

Answers: 1a) $-\dfrac{3}{4}$; 1b) $\dfrac{3y}{2x}$; 1c) $\dfrac{7(x-2)}{3x}$; 1d) $-\dfrac{3x+5y}{8xy}$; 1e) 5; 1f) $\dfrac{(x-7)(x-6)}{(x-1)(x-5)}$; 2a) $\dfrac{9}{35}$; 2b) $\dfrac{10n^2}{3m}$;

2c) $a - 8$; 2d) $\dfrac{2x+1}{x^2(x+2)}$; 2e) $\dfrac{7x}{x+3}$; 2f) $\dfrac{t+7}{t-9}$

Mini-Lecture 7.3
Addition and Subtraction of Rational Expressions with a Common Denominator and Finding the Least Common Denominator

Learning Objectives:

1. Add and subtract rational expressions with a common denominator.
2. Find the least common denominator.
3. Key vocabulary: *least common denominator (LCD)*

Examples:

1. Add or subtract as indicated.

   a) $\dfrac{4}{9} + \dfrac{1}{9}$

   b) $\dfrac{7}{8} - \dfrac{1}{8}$

   c) $\dfrac{5}{x-4} + \dfrac{x-1}{x-4}$

   d) $\dfrac{8x^2 + 16x - 7}{2x+7} + \dfrac{2x^2 + 19x + 7}{2x+7}$

   e) $\dfrac{x^2 - 5x - 6}{(x+2)(x+5)} + \dfrac{9x+1}{(x+2)(x+5)}$

   f) $\dfrac{3x^2 + 4x - 15}{x^2 - 7x + 6} - \dfrac{2x^2 + 6x + 9}{x^2 - 7x + 6}$

   g) $\dfrac{4m}{m+3} - \dfrac{2m^2 + 7m - 9}{m+3}$

   h) $\dfrac{6n+5}{n^2 + 4n - 21} - \dfrac{5n-2}{n^2 + 4n - 21}$

   i) Add: $\dfrac{2}{5} + \dfrac{1}{6}$

   j) Subtract: $\dfrac{7}{12} - \dfrac{4}{9}$

2. Find the least common denominator (LCD).

   a) $\dfrac{1}{9} + \dfrac{3}{n}$

   b) $\dfrac{2}{x^3} - \dfrac{3}{8x}$

   c) $\dfrac{7}{20n^5} + \dfrac{5}{16n^4}$

   d) $\dfrac{7}{x} - \dfrac{3x}{x+2}$

   e) $\dfrac{1}{5x^2 - 20x} - \dfrac{2x^2}{x^2 - 8x + 16}$

   f) $\dfrac{3n}{n^2 - 3n - 10} - \dfrac{9n^2}{n^2 - 7n + 10}$

   g) $\dfrac{5}{m-9} - \dfrac{4m}{m+9}$

   h) $\dfrac{7x}{x^2 - 10x + 24} + x + 13$

Teaching Notes:

- Remind students to be sure to simplify final answers.
- Students often have difficulty finding the LCD. Review the process they use when finding the LCD for numeric fractions and parallel it with finding the LCD for rational expressions.

Answers: 1a) $\dfrac{5}{9}$; 1b) $\dfrac{3}{4}$; 1c) $\dfrac{x+4}{x-4}$; 1d) $5x$; 1e) $\dfrac{x-1}{x+2}$; 1f) $\dfrac{x+4}{x-1}$; 1g) $-(2m-3)$ or $-2m+3$;

1h) $\dfrac{1}{n-3}$; 1i) $\dfrac{17}{30}$; 1j) $\dfrac{5}{36}$; 2a) $9n$; 2b) $8x^3$; 2c) $80n^5$; 2d) $x(x+2)$; 2e) $5x(x-4)^2$;

2f) $(n+2)(n-5)(n-2)$; 2g) $(m-9)(m+9)$; 2h) $(x-4)(x-6)$

Mini-Lecture 7.4
Addition and Subtraction of Rational Expressions

Learning Objectives:

1. Add and subtract rational expressions.
2. Key vocabulary: *equivalent fractions*

Examples:

1. Add or subtract as indicated.

a) $\dfrac{3}{m} + \dfrac{5}{n}$

b) $\dfrac{3}{10x^3y} + \dfrac{1}{15xy^2}$

c) $\dfrac{8}{x+7} + \dfrac{4}{x}$

d) $\dfrac{k}{k-4} - \dfrac{8}{k-6}$

e) $\dfrac{x+3}{x-6} - \dfrac{x+4}{x+6}$

f) $\dfrac{6}{n-8} + \dfrac{n+4}{8-n}$

g) $\dfrac{3a-2}{4a-5} - \dfrac{a-11}{5-4a}$

h) $\dfrac{1}{2x^2-5x-12} + \dfrac{4}{x^2-2x-8}$

i) $\dfrac{10}{x^2-10x} - \dfrac{x}{10x-100}$

j) $\dfrac{5}{a+1} + \dfrac{20}{a^2-2a-3}$

Teaching Notes:

- Remind students that when subtracting, they must subtract the entire numerator of the fraction that follows the minus sign by applying the distributive property when appropriate.
- Remind students that when they multiply by expressions such as $\dfrac{k-6}{k-6}$, $\dfrac{x}{x}$, or $\dfrac{7}{7}$, they are applying the multiplicative identity.
- Remind students to be sure to simplify final answers.

Answers: 1a) $\dfrac{3n+5m}{mn}$; 1b) $\dfrac{9y+2x^2}{30x^3y^2}$; 1c) $\dfrac{12x+28}{x(x+7)}$; 1d) $\dfrac{k^2-14k+32}{(k-4)(k-6)}$; 1e) $\dfrac{11x+42}{(x-6)(x+6)}$; 1f) $\dfrac{-n+2}{n-8}$;

1g) $\dfrac{4a+9}{4a-5}$; 1h) $\dfrac{9x+14}{(2x+3)(x-4)(x+2)}$; 1i) $-\dfrac{x+10}{10x}$; 1j) $\dfrac{5}{a-3}$

Mini-Lecture 7.5
Complex Fractions

Learning Objectives:

1. Simplify complex fractions by combining terms.
2. Simplify complex fractions using multiplication first to clear fractions.
3. Key vocabulary: *complex fraction*

Examples:

1. Simplify each complex fraction by combining terms (using Method 1).

a) $\dfrac{\dfrac{9x^2}{25y^2z}}{\dfrac{3x^2z}{5y^4}}$

b) $\dfrac{2m - \dfrac{4}{n}}{n - \dfrac{2}{m}}$

c) $\dfrac{\dfrac{x}{4} - \dfrac{4}{x}}{\dfrac{x-4}{x}}$

2. Simplify each complex fraction using multiplication first to clear fractions (using Method 2).

a) $\dfrac{\dfrac{7}{8} - \dfrac{1}{6}}{\dfrac{5}{8} + \dfrac{1}{6}}$

b) $\dfrac{2m - \dfrac{4}{n}}{n - \dfrac{2}{m}}$

c) $\dfrac{a^3}{\dfrac{1}{a} - \dfrac{1}{b}}$

Teaching Notes:

- A brief review of multiplication, division, addition, and subtraction of fractions may be useful at the beginning of this lesson.
- Have students practice using both methods to simplify the same set of complex fractions. Then discuss the advantages and disadvantages of using one method over the other in specific cases.
- Remind students that when they multiply both the numerator and denominator of a complex fraction by the LCD, they are applying the multiplicative identity.

Answers: 1a) $\dfrac{3y^2}{5z^2}$; 1b) $\dfrac{2m}{n}$; 1c) $\dfrac{x+4}{4}$; 2a) $\dfrac{17}{19}$; 2b) $\dfrac{2m}{n}$; 2c) $\dfrac{a^4b}{b-a}$

Mini-Lecture 7.6
Solving Rational Equations

Learning Objectives:

1. Solve rational equations with integer denominators.
2. Solve rational equations where a variable appears in a denominator.
3. Key vocabulary: *rational equation, extraneous solution*

Examples:

1. Solve each equation.

a) $\dfrac{x}{9} - \dfrac{x}{8} = 1$

b) $\dfrac{3x}{10} + \dfrac{3}{2} = \dfrac{4x}{5}$

c) $\dfrac{n-2}{12} = \dfrac{1}{4} - \dfrac{n-5}{3}$

2. Solve each equation.

a) $4 - \dfrac{6}{x} = \dfrac{19}{5}$

b) $\dfrac{a-7}{a-5} = \dfrac{3}{5}$

. c) $\dfrac{8}{n-2} = \dfrac{2}{n+4}$

d) $x - \dfrac{6}{x} = 5$

e) $\dfrac{x^2 - 3}{x+2} = \dfrac{2x+5}{x+2}$

f) $\dfrac{2x}{x^2 - 25} - \dfrac{3}{x-5} = \dfrac{2}{x+5}$

Teaching Notes:

- Remind students to always determine the values for the variable that will cause the denominator to equal zero. This will aide in spotting extraneous solutions.

Answers: *1a)* $x = -72$; *1b)* $x = 3$; *1c)* $n = 5$; *2a)* $x = 30$; *2b)* $a = 10$; *2c)* $n = -6$; *2d)* $x = -1, x = 6$; *2e)* $x = 4$; *2f)* $x = -\dfrac{5}{3}$

Mini-Lecture 7.7
Rational Equations: Applications and Problem Solving

Learning Objectives:
1. Set up and solve applications containing rational expressions.
2. Set up and solve motion problems.
3. Set up and solve work problems.
4. Key vocabulary: *geometric problem, motion problem, work problem, reciprocal*

Examples:
1. Set up and solve each application problem.
   a) The area of a rectangle 120 square feet. Determine the length and width if the width is 2 feet less than two-thirds of the length.
   b) One number is 6 times another number. The sum of their reciprocals is $\frac{1}{2}$. Determine the numbers.

2. Set up and solve each motion problem.
   a) The speed of the current of a river is 1 mile per hour. Tom and Beth can canoe 12 miles down the river in the same time it takes them to canoe 10 miles up the river. How fast can Tom and Beth paddle the canoe in still water?
   b) John is riding a bike on a scenic trail that is 25 miles long. One part of the trail is on level ground, so John can average 12 miles per hour. Another part of the trail is hilly, so John can only average 8 miles per hour. If John traveled the entire length of the trail in 2.5 hours, how long was he on level ground and how long was he in the hills?
   c) George runs at 10 miles per hour and Bill runs at 6 miles per hour. If they start on a jogging course at the same time and George finishes it 1 hour ahead of Bill, how long is the course?

3. Set up and solve each work problem.
   a) A painter can paint a room in 6 hours. His apprentice can paint the same room in 9 hours. How long will it take them to paint the room if they work together?
   b) One pipe can fill a tank in 6 hours. Another pipe can drain the tank in 8 hours. If both pipes are open, how long will it take to fill the empty tank?
   c) It takes Karen 5 hours to stuff a case of envelopes for a mass mailing. When she works with Anna, it takes only 3 hours to stuff a case of envelopes. How long will it take Anna to stuff a case of envelopes if she works alone?
   d) Craig can mow his lawn by himself in 45 minutes. It takes his wife Diane 75 minutes to mow the lawn by herself. After Craig mows for 30 minutes, he stops and Diane takes over. How long does it take Diane to finish mowing the lawn?

Teaching Notes:
- Many students find these types of applications very difficult. Emphasize to students the importance of understanding the problems rather than trying to memorize procedures.
- When doing work problems, students often want to average the individual completion times in order to get the time for working together. Point out that the time for working together will be less than the times for working alone.

Answers: 1a) length = 15 ft, width = 8 ft; 1b) $\frac{7}{3}$ and 14; 2a) 11 mph; 2b) level = 1.25 hours, hilly = 1.25 hours; 2c) 15 miles; 3a) 3.6 hours; 3b) 24 hours; 3c) 7.5 hours; 3d) 25 minutes

Mini-Lecture 7.8
Variation

1. Solve direct variation problems.
2. Solve inverse variation problems.
3. Solve joint variation problems.
4. Solve combined variation problems.
5. Key vocabulary: *variation, direct variation, constant of proportionality, inverse variation, joint variation, combined variation*

Examples:

1. Set up and solve each direct variation problem.
   a) The circumference, C, of a circle varies directly with the diameter, d, of the circle. Write the variation as an equation. If the constant of proportionality is $k = 3.14$, find the circumference if the diameter is 12 inches.
   b) The cost to fill a car with gasoline varies directly with the number of gallons purchased. Suppose the cost for 15 gallons of gas is $49.35. Determine the cost for 12 gallons of gas.
   c) Suppose A varies directly as the square of d. If $A = 1256$ when $d = 40$, find A when $d = 50$.

2. Set up and solve each inverse variation problem.
   a) The time required to clean a basketball arena after a game is inversely proportional with the number of workers cleaning. If it takes 6 hours to clean the arena using 24 workers, how long will it take to clean the arena using 18 workers?
   b) In an electrical circuit, the resistance is inversely proportional to the square of the current. If the resistance in a specific circuit is 400 ohms when the current is 0.5 amps, determine the resistance if the current is 0.4 amps.

3. a) The volume of a cylinder varies jointly as its height and the square of its radius. If the volume of a cylinder is 471 cubic feet when its height is 6 feet and its radius is 5 feet. Find the volume of a cylinder whose height is 6 feet and radius is 4 feet.
   b) Suppose Q varies jointly as the cube of r and the square of s. If $Q = 3375$ when $r = 2.5$ and $s = 6$, find Q when $r = 1.5$ and $s = 8$.

4. a) The electrical resistance of a wire, R, varies directly as its length, L, and inversely as its cross-sectional area, A. If the resistance of a wire is 1.875 ohms when the length is 300 feet and the cross-sectional area is 0.08 square inches, find the resistance of a wire whose length is 1000 feet with a cross-sectional area of 0.05 square inches.
   b) Suppose M is jointly proportional to x and y and inversely proportional to the square of z. Express M in terms of x, y, and z.

Teaching Notes:

- Have students develop their own examples of direct and inverse variation.
- Point out that solving variation problems require two steps: (1) finding the constant of variation, and (2) solving the problem for the desired quantity.

Answers: *1a)* $C = kd$, $C = 37.68$ *inches; 1b)* $39.48; 1c) 1962.5; 2a) 8 hours; 2b) 625 ohms; 3a) 301.44 cubic feet; 3b) 1296; 4a) 10 ohms; 4b)* $M = \dfrac{kxy}{z^2}$

Additional Exercises 7.1

Determine the value or values of the variables for which the expression is defined.

1. $\dfrac{x+4}{x^2-16}$

    1. _____

2. $\dfrac{x-3}{5x-3}$

    2. _____

3. $\dfrac{x^2+5x-36}{x^2+7x+6}$

    3. _____

4. $\dfrac{64x^2-25}{-8x^2-10x}$

    4. _____

5. $\dfrac{x^2+5x-36}{x^2-8x+12}$

    5. _____

6. $\dfrac{16x^2-9}{x^2+9}$

    6. _____

Simplify.

7. $\dfrac{15a+9ab}{6a}$

    7. _____

8. $\dfrac{5p+pq}{8p}$

    8. _____

9. $\dfrac{8m}{6m+mn}$

    9. _____

10. $\dfrac{12x-24}{8-4x}$

    10. _____

Additional Exercises 7.1 *(cont.)*

11. $\dfrac{6x-24}{12-3x}$

11. _____

12. $\dfrac{x^2+2x-35}{5-x}$

12. _____

13. $\dfrac{x^2-5x-14}{7-x}$

13. _____

14. $\dfrac{x^2-5x-14}{x^2-49}$

14. _____

15. $\dfrac{u-1}{u^2-1}$

15. _____

16. $\dfrac{x^2-x-12}{x^2-16}$

16. _____

17. $\dfrac{2y+16}{y^2-64}$

17. _____

18. $\dfrac{8-x}{x^2-5x-24}$

18. _____

19. $\dfrac{x^2+7x-18}{x^2-4}$

19. _____

20. $\dfrac{x^2-9}{27-x^3}$

20. _____

Additional Exercises 7.2

Multiply.

1. $\dfrac{21y^2}{4x} \cdot \dfrac{8x^2}{7y}$

     1. _____

2. $\dfrac{2x^3 y}{3z^4} \cdot \dfrac{6xz^5}{10y^5}$

     2. _____

3. $\dfrac{9y^2}{8} \cdot \dfrac{32x}{27y}$

     3. _____

4. $\dfrac{2x^2 y}{3z^3} \cdot \dfrac{12xz^4}{6y^3}$

     4. _____

5. $\dfrac{3y^2}{5} \cdot \dfrac{10x}{15y}$

     5. _____

6. $\dfrac{4x^2 y}{2z^2} \cdot \dfrac{6xz^3}{20y^4}$

     6. _____

7. $\dfrac{y^2 - 16x^2}{9x^2 - 25y^2} \cdot \dfrac{3x - 5y}{4x - y}$

     7. _____

8. $\dfrac{a^2 - 10a + 21}{a - 7} \cdot \dfrac{a^2 + a - 12}{(a-3)^2}$

     8. _____

9. $\dfrac{x+1}{3x+y} \cdot \dfrac{9x^2 - y^2}{2x^2 + 3x + 1}$

     9. _____

10. $\dfrac{a^2 - 6a + 9}{a - 3} \cdot \dfrac{a^2 + 3a - 18}{(a-3)^2}$

     10. _____

Additional Exercises 7.2 *(cont.)*

Name:_____

Divide.

11. $28p^2q^4 \div \dfrac{4pq^4}{5r}$

11. _____

12. $\dfrac{r^3s}{t} \div \dfrac{rs^3}{t^3}$

12. _____

13. $24e^2d^4 \div \dfrac{3cd^4}{5f}$

13. _____

14. $\dfrac{u^5x}{y} \div \dfrac{ux^2}{y^4}$

14. _____

15. $\dfrac{x^2-9}{x-2} \div \dfrac{x+3}{x-3}$

15. _____

16. $\dfrac{3x^2+4x+1}{3x^2-5x-2} \div \dfrac{x^2-2x-3}{-5x^2+25x-30}$

16. _____

17. $\dfrac{2x^2+5x+3}{2x^2+7x+6} \div \dfrac{x^2+6x+5}{-5x^2-35x-50}$

17. _____

18. $\dfrac{50}{y^2+2y-15} \div \dfrac{10y}{y-3}$

18. _____

19. $\dfrac{15}{y^2+2y-8} \div \dfrac{5y}{y-2}$

19. _____

20. $\dfrac{x^2+3x-28}{x^2+4x+4} \div \dfrac{x^2-49}{x^2-5x-14}$

20. _____

Additional Exercises 7.3

Add or subtract.

1. $\dfrac{x-4}{3} + \dfrac{5x}{3}$

   1. _____

2. $\dfrac{2x+5}{7} - \dfrac{x}{7}$

   2. _____

3. $\dfrac{8}{x} + \dfrac{x+9}{x}$

   3. _____

4. $\dfrac{9x-11}{24x} + \dfrac{11-x}{24x}$

   4. _____

5. $\dfrac{3x-6}{24x} + \dfrac{3x+6}{24x}$

   5. _____

6. $\dfrac{2x+3}{x+5} - \dfrac{x-3}{x+5}$

   6. _____

7. $\dfrac{2x+3}{x+4} - \dfrac{x-7}{x+4}$

   7. _____

8. $\dfrac{8}{3(x+8)} + \dfrac{4}{3(x+8)}$

   8. _____

9. $\dfrac{3}{2(x-9)} + \dfrac{9}{2(x-9)}$

   9. _____

10. $\dfrac{x^2}{x+9} + \dfrac{9x}{x+9}$

   10. _____

Additional Exercises 7.3 *(cont.)*

11. $\dfrac{4x+7}{x+5} - \dfrac{x-6}{x+5}$

11. _____

12. $\dfrac{2x+4}{x^2-9} - \dfrac{x+5}{9-x^2}$

12. _____

13. $\dfrac{5x+1}{x^2-64} - \dfrac{4x-7}{x^2-64}$

13. _____

14. $\dfrac{2x^2+7x-3}{x^2+4x-12} - \dfrac{2x^2+6x-1}{x^2+4x-12}$

14. _____

Find the least common denominator.

15. $\dfrac{9}{4x^2y^6} + \dfrac{1}{7x^3y^4}$

15. _____

16. $\dfrac{7}{5x^4y^7} + \dfrac{1}{8x^5y^5}$

16. _____

17. $\dfrac{x^2-9x+10}{x^2-16} - \dfrac{3}{x-4}$

17. _____

18. $\dfrac{4x-7}{x^2-x} + \dfrac{9}{x}$

18. _____

19. $\dfrac{7x}{x^2+x-6} + \dfrac{6}{x^2-2x-8}$

19. _____

20. $\dfrac{5x}{x^2+5x+4} + \dfrac{8}{x^2-4x-5}$

20. _____

Additional Exercises 7.4

Add or subtract.

1. $\dfrac{2}{3x} + \dfrac{4}{x}$

1. _____

2. $\dfrac{6}{x} + \dfrac{7}{5x^2}$

2. _____

3. $\dfrac{3}{x} + \dfrac{6}{7x^2}$

3. _____

4. $\dfrac{4}{x} + \dfrac{7}{6x^2}$

4. _____

5. $\dfrac{6}{5x^3 y} - \dfrac{1}{2x^2 y^3}$

5. _____

6. $\dfrac{5}{xy} + \dfrac{7}{2x^2}$

6. _____

7. $2x + \dfrac{x}{y}$

7. _____

8. $\dfrac{5a - 3}{4a} - \dfrac{1}{6a}$

8. _____

9. $\dfrac{7}{3p} - \dfrac{4}{5p^2}$

9. _____

10. $\dfrac{2}{x^2 - 4} - \dfrac{1}{x^2 - 2x}$

10. _____

11. $\dfrac{x}{x-2}+\dfrac{x+2}{x}$

11. _____

12. $\dfrac{x-1}{x-2}-\dfrac{x^2+4x-4}{x^2+4x-12}$

12. _____

13. $\dfrac{x+2}{x-6}-\dfrac{x^2+5x+14}{x^2-2x-24}$

13. _____

14. $\dfrac{x+2}{x-7}-\dfrac{x^2+4x+13}{x^2-4x-21}$

14. _____

15. $\dfrac{x-4}{x^2+5x+6}+\dfrac{x-1}{x^2-4}$

15. _____

16. $\dfrac{x-4}{x^2+4x+3}-\dfrac{x-1}{x^2-9}$

16. _____

17. $\dfrac{2x}{x^2+5x+6}-\dfrac{x+1}{x^2+2x-3}$

17. _____

18. $-\dfrac{x-2}{x^2-2x-8}-\dfrac{x-1}{x^2-4}$

18. _____

19. $\dfrac{7}{3x^2+x-4}+\dfrac{9x+2}{3x^2-2x-8}$

19. _____

20. $\dfrac{x+1}{x^2+6x+9}-\dfrac{x-4}{x^2-9}$

20. _____

Additional Exercises 7.5

Name:_____

Date:_____

Simplify.

1. $\dfrac{3-\frac{3}{7}}{2+\frac{1}{4}}$

     1. _____

2. $\dfrac{1+\frac{3}{2}}{1+\frac{1}{4}}$

     2. _____

3. $\dfrac{2+\frac{4}{7}}{2+\frac{4}{3}}$

     3. _____

4. $\dfrac{1+\frac{2}{5}}{2+\frac{3}{8}}$

     4. _____

5. $-\dfrac{\frac{9x^9}{16y^4}}{\frac{6x^8}{4y^8}}$

     5. _____

6. $\dfrac{\frac{4}{x+5}}{\frac{8}{x^2-25}}$

     6. _____

7. $\dfrac{\frac{27x^7}{16y^3}}{\frac{18x^3}{2y^6}}$

     7. _____

8. $\dfrac{x+\frac{2x}{y}}{\frac{7}{3x}}$

     8. _____

9. $\dfrac{\frac{1}{5}+\frac{1}{x}}{\frac{x}{2}+\frac{5}{2}}$

     9. _____

10. $\dfrac{x+\frac{8}{x^2}}{x-\frac{16}{x^3}}$

     10. _____

Additional Exercises 7.5 *(cont.)*

11. $\dfrac{\frac{-2}{x+3}}{\frac{6}{x}+2}$

11. _____

12. $\dfrac{\frac{4}{x+2}}{\frac{1}{x}+4}$

12. _____

13. $\dfrac{x+\frac{8}{x^2}}{x+\frac{16}{x^3}}$

13. _____

14. $\dfrac{x+\frac{6x}{y}}{\frac{3}{7x}}$

14. _____

15. $\dfrac{x-\frac{5}{x^2}}{x-\frac{10}{x^3}}$

15. _____

16. $\dfrac{\frac{-1}{x-1}}{\frac{-1}{x}-2}$

16. _____

17. $\dfrac{\frac{2}{3x}+\frac{1}{4x}}{\frac{1}{x}+\frac{1}{2x}}$

17. _____

18. $\dfrac{\frac{1}{x}-\frac{1}{2x}}{\frac{2}{3x}-\frac{1}{4x}}$

18. _____

19. $\dfrac{\frac{1}{x}+\frac{2}{y}}{\frac{3}{xy}}$

19. _____

20. $\dfrac{\frac{x}{x+1}}{\frac{1}{x-1}+1}$

20. _____

Additional Exercises 7.6

Name:_____

Date:_____

Solve each equation and check your solution.

1. $\dfrac{5}{8} = \dfrac{x}{16}$

   1. _____

2. $\dfrac{6}{x} = -\dfrac{2}{9}$

   2. _____

3. $-\dfrac{1}{4} = \dfrac{7x}{10}$

   3. _____

4. $\dfrac{n}{6} = 5 - \dfrac{n}{3}$

   4. _____

5. $\dfrac{3}{5} + \dfrac{1}{10} = \dfrac{w}{10}$

   5. _____

6. $\dfrac{x}{6} - \dfrac{x}{9} = \dfrac{5}{18}$

   6. _____

7. $\dfrac{x}{2} - \dfrac{x}{9} = 7$

   7. _____

8. $\dfrac{x}{7} - \dfrac{x}{9} = 2$

   8. _____

9. $\dfrac{8}{x} + \dfrac{1}{4x} = \dfrac{11}{8}$

   9. _____

10. $\dfrac{4 - 3y}{5} = \dfrac{2 - 7y}{25}$

    10. _____

11. $\dfrac{2x-9}{x-7} = \dfrac{5}{x-7}$

11. _____

12. $\dfrac{6}{x^2-1} = \dfrac{5}{x-1} - \dfrac{3}{x+1}$

12. _____

13. $\dfrac{3x}{x-4} + 1 = \dfrac{2x}{x-4}$

13. _____

14. $x + \dfrac{4}{x} = \dfrac{20}{x}$

14. _____

15. $\dfrac{x+1}{x+3} + \dfrac{x-3}{x-2} = \dfrac{2x^2-15}{x^2+x-6}$

15. _____

16. $1 + \dfrac{2}{x-1} = \dfrac{4}{x^2-1}$

16. _____

17. $1 + \dfrac{4}{x-1} = \dfrac{8}{x^2-1}$

17. _____

18. $\dfrac{x}{x-8} + \dfrac{6}{x-4} = \dfrac{x^2}{x^2-12x+32}$

18. _____

19. $\dfrac{x}{x-16} + \dfrac{10}{x-8} = \dfrac{x^2}{x^2-24x+128}$

19. _____

20. $\dfrac{2}{x-3} + \dfrac{x-1}{x+5} = \dfrac{x^2-1}{x^2+2x-15}$

20. _____

Additional Exercises 7.7

Name:_____

Date:_____

Multiply.

1. Find two consecutive even integers such that the sum of their reciprocals is $\dfrac{7}{24}$.

    1. _____

2. The sum of the reciprocal of a number and the reciprocal of 4 less than the number is 6 times the reciprocal of the original number. Find the original number.

    2. _____

3. Bud travels 4 miles upstream in the same time he travels 6 miles downstream. If his boat will travel 14 miles per hour in still water what was the speed of the current?

    3. _____

4. One car travels 13 miles per hour faster than another. In the time it takes the slower car to travel 352 miles, the faster car travels 456 miles. Find the speed of both cars.

    4. _____

5. Two planes travel toward each other from cities that are 1280 km apart at a rate of 120 km/hr and 200 km/hr. They started at the same time. In how many hours will they meet?

    5. _____

6. Thelma's average driving speed is 16 miles per hour faster than Kirk's. In the same length of time it takes Thelma to drive 290 miles, Kirk only drives 210 miles. What is Thelma's average speed?

    6. _____

7. Crawford can paint a fence by himself in 15 hours and Bernice can paint the same fence in 16 hours. How long will it take them to paint the fence if they work together?

    7. _____

8. Alone, a large hose can fill a pool in half the time it takes a small hose to fill the pool alone. Together, the two hoses can fill the pool in 5 hours. How long does it take each hose to fill the pool alone?

    8. _____

9. Sandra can paint a kitchen in 4 hours and Roger can paint the same kitchen in 6 hours. How long would it take for both working together to paint the kitchen?

    9. _____

10. Ronald can mow his family's yard in 3 hours. His younger brother Randall can mow the yard in 6 hours. How long will it take the two to mow the yard together?

    10. _____

Additional Exercises 7.7 *(cont.)*

11. The sum of the reciprocal of a number and the reciprocal of 6 less than the number is 7 times the reciprocal of the original number. Find the original number.

11. _____

12. A plane flies 300 miles with a tail wind in 1 hour. It takes the same plane 2 hours to fly the 300 miles when flying against the wind. What is the plane's speed in still air?

12. _____

13. One car travels 4 miles per hour faster than another. In the time it takes the slower car to travel 192 miles, the faster car travels 208 miles. Find the speed of both cars.

13. _____

14. Doug's average driving speed is 7 miles per hour faster than Richard's. In the same length of time it takes Doug to drive 156 miles, Richard only drives 135 miles. What is Doug's average speed?

14. _____

15. Brandon can paint a fence by himself in 7 hours and Bernice can paint the same fence in 9 hours. How long will it take them to paint the fence if they work together?

15. _____

16. Mr. Johnson, on his tractor, can level a 1-acre field in 9 hours. Mr. Hackett, on his tractor, can level a 1-acre field in 18 hours. If they work together, how long will it take to level a 1-acre field?

16. _____

17. One integer is two more than twice the other. If the sum of their reciprocals is $\dfrac{7}{20}$, find the two integers.

17. _____

18. The sum of the reciprocal of a number and the reciprocal of 2 less than the number is 7 times the reciprocal of the original number. Find the original number.

18. _____

19. A plane flies 900 miles with a tail wind in 3 hours. It takes the same plane 4 hours to fly the 900 miles when flying against the wind. What is the plane's speed in still air?

19. _____

20. One car travels 1 mile per hour faster than another. In the time it takes the slower car to travel 368 miles, the faster car travels 376 miles. Find the speed of both cars.

20. _____

Additional Exercises 7.8

Name:_____

Date:_____

1. A variable x varies directly as y. If $x = 32$ when $y = 80$, find x when $y = 190$.

   1. _____

2. The wattage of an appliance, W, varies jointly as the square of the current, I, and the resistance, R. If the wattage is 18 watts when the current is 0.6 ampere and the resistance is 50 ohms, find the wattage when the current is 0.4 ampere and the resistance is 100 ohms.

   2. _____

3. W varies jointly as P and Q, and inversely as the square of T. If $W = 54$ when $P = 45$, $Q = 6$, and $T = 6$, find W if $P = 25$, $Q = 20$, and $T = 30$.

   3. _____

4. A variable a varies inversely as the square of b. If a is $\dfrac{5}{9}$ when b is 3, find a when b is 9.

   4. _____

5. A variable p varies jointly with m and the square of n. If k is the constant of proportionality, find the equation of the variation.

   5. _____

6. The amount a spring will stretch, S, varies directly with the force (or weight), F, attached to the spring. If a spring stretches 0.8 inches when 10 pounds is attached, how far will it stretch when 60 pounds is attached?

   6. _____

7. The intensity I, of light received at a source varies inversely as the square of the distance, d, from the source. If the light intensity is 25 foot-candles at 11 feet, find the light intensity at 12 feet. Round your answer to the nearest hundredth if necessary.

   7. _____

8. The electrical resistance of a wire, R, varies directly as its length, L, and inversely as its cross-sectional area, A. If the resistance of a wire is 12.5 ohms when the length is 500 feet and the cross-sectional area is 0.12 square inches, find the resistance of a wire whose length is 1200 feet with a cross-sectional area of 0.10 square inches.

   8. _____

9. The surface area, A, of a sphere varies directly as the radius, r, of the sphere. If the surface area is 5024 square inches when the radius is 20 inches, find the surface area when the radius is 15 inches.

   9. _____

10. A variable h varies jointly with f and the cube of g. If k is the constant of proportionality, find the equation of the variation.

    10. _____

Additional Exercises 7.8 *(cont.)*

11. A variable p varies inversely with the variable q. If k is the constant of proportionality, find the equation of the variation.

11. _____

12. A variable y varies directly with the square of x. If $x = 2$ when $y = 3$, find the constant of proportionality, k.

12. _____

13. If n varies jointly with w and the square of s, and n is 243 when w is 3 and s is 3, find n when w is 6 and s is 8.

13. _____

14. If $x = 0.4$ when $y = 625$ and y varies inversely as x, find y when $x = 2.5$.

14. _____

15. The variable A is directly proportional to the variable B and inversely proportional to the square of the variable C. If $A = 54$ when $B = 6$ and $C = 10$, find A when $B = 5$ and $C = 6$.

15. _____

16. The volume, V, of a gas varies inversely as its pressure, P. If the volume of the gas is 480 cubic centimeters when the pressure is 250 millimeters of mercury, find the volume when the pressure is 300 millimeters of mercury.

16. _____

17. An enclosed gas exerts a pressure P on the walls of the container. This pressure is directly proportional to the temperature T of the gas. If the pressure is 4 lb per in^2 when the temperature is 520°F, find the constant of proportionality, k.

17. _____

18. The horsepower that a rotating shaft can safely transmit varies jointly as the cube of its diameter, d, and the number, n, of revolutions it makes per minute. If a 3-inch shaft at a speed of 1800 revolutions per minute can safely transmit 900 horsepower, find the constant of proportionality, k.

18. _____

19. The variable A varies jointly as B and the square of C. If A is 625 when B is 5 and C is 5, find A when B is 6 and C is 2.

19. _____

20. The owners of a pizza parlor find that the number, N, of large pizzas sold weekly varies directly as their advertising budget, A, and inversely as the price, P. When the advertising budget is $500 and the price is $15, the number of large pizzas sold is 700. Find the number of pizzas sold when the advertising budget is $600 and the price is $12.

20. _____

Chapter 7 Test Form A

Simplify.

1. $\dfrac{x-3}{3-x}$

    1. _____

2. $\dfrac{x^2 - 2x + 1}{x - 1}$

    2. _____

Perform the indicated operations.

3. $\dfrac{a^2 - 10a + 21}{a - 2} \cdot \dfrac{a^2 + 7a - 18}{a^2 - 6a + 9}$

    3. _____

4. $\dfrac{x^2 + 6x + 9}{x^2 - 4} \cdot \dfrac{x^2 + 9x + 14}{x^2 - 2x - 15}$

    4. _____

5. $\dfrac{x^2 - 25}{x + 4} \cdot \dfrac{9x + 36}{75 - 3x^2}$

    5. _____

6. $\dfrac{x^2 - 16y^2}{4x + 12y} \div \dfrac{x + 4y}{x + 3y}$

    6. _____

7. $\dfrac{x^2 + 6x + 9}{x^2 + 3x - 10} \div \dfrac{x^2 + 5x + 6}{x^2 + 8x + 15}$

    7. _____

8. $\dfrac{3xy}{x + 9} + \dfrac{27y}{x + 9}$

    8. _____

9. $5 - \dfrac{3}{x + 3}$

    9. _____

10. $\dfrac{3x - 8}{x^2 + 2x} - \dfrac{4}{x + 2}$

    10. _____

11. $\dfrac{8}{4x} + \dfrac{x}{3x}$

    11. _____

12. $\dfrac{x - 3}{x^2 - x - 6} - \dfrac{x - 4}{x^2 - 4}$

    12. _____

Chapter 7 Test Form A *(cont.)*

Simplify.

13. $\dfrac{7 - \frac{1}{4}}{3 + \frac{3}{4}}$

13. _____

14. $\dfrac{\frac{5x}{y} - x}{\frac{y}{x} + 1}$

14. _____

Solve.

15. $\dfrac{x}{8} = \dfrac{x-5}{13}$

15. _____

16. $\dfrac{x-3}{5} = \dfrac{x+2}{6}$

16. _____

17. $\dfrac{x}{2x-2} - \dfrac{2}{3x+3} = \dfrac{5x^2 - 2x + 9}{12x^2 - 12}$

17. _____

18. If y varies directly as x, and $y = 9$ when $x = 75$, find y when $x = 100$.

18. _____

19. It takes Bob and Jane 2 hours to rake all the leaves in their yard. It takes Bob 5 hours to do the leaves by himself. How long does it take Jane to rake the leaves by herself?

19. _____

20. Officer Smith rides his bicycle around Oklahoma State University campus as part of his patrolling duties. During the first part of his ride he is peddling mostly uphill and his average speed is 3 miles an hour. After a certain point, he is traveling mostly downhill and averages 7 miles per hour. If the total distance he travels is 20 miles and the total time he rides is 4 hours, how long did he ride at each speed?

20. _____

Chapter 7 Test Form B

Name:_____

Date:_____

Simplify.

1. $\dfrac{x+4}{x^2+6x+8}$

                                              1. _____

2. $\dfrac{x^2-x-6}{3-x}$

                                              2. _____

Perform the indicated operations.

3. $\dfrac{(2xy)^2}{5z^2} \div \dfrac{10y^3}{15xz^3}$

                                              3. _____

4. $\dfrac{2m-2n}{m} \cdot \dfrac{m^2}{6m-6n}$

                                              4. _____

5. $\dfrac{4x^2-4x}{2x^2+3x} \cdot \dfrac{2x^2+7x+6}{x^2-x-6}$

                                              5. _____

6. $\dfrac{z^2+2z-15}{z^2-8z+15} \div \dfrac{z^2-25}{z^2-2z-15}$

                                              6. _____

7. $\dfrac{2x}{x^2-7x+10} - \dfrac{x+2}{x^2-7x+10}$

                                              7. _____

8. $\dfrac{2z}{z^2-9z+20} + \dfrac{3}{z^2-4z}$

                                              8. _____

9. $\dfrac{x}{3x^3y^4} - \dfrac{5}{9x^2y^6}$

                                              9. _____

10. $\dfrac{6x^2+35x}{x^3+6x^2+5x} + \dfrac{-2x^2-15x}{x^3+6x^2+5x}$

                                            10. _____

11. $\dfrac{m+2}{3m-6} + \dfrac{m}{3m^2-2m-8}$

                                            11. _____

Chapter 7 Test Form B *(cont.)*

Simplify.

12. $\dfrac{5-\frac{5}{8}}{1+\frac{3}{8}}$

12. _____

13. $\dfrac{\frac{z}{y}-z}{\frac{y}{z}+1}$

13. _____

Solve.

14. $\dfrac{x+3}{4}+\dfrac{1}{5}=\dfrac{x-7}{8}$

14. _____

15. $\dfrac{4}{x}=\dfrac{x}{3x+16}$

15. _____

16. $\dfrac{8}{x-4}+6=\dfrac{2x}{x-4}$

16. _____

17. $\dfrac{3}{x-4}+\dfrac{2}{x+4}=\dfrac{6}{x^2-16}$

17. _____

18. The pressure exerted by a liquid at a given point varies directly as the depth of the point beneath the surface of the liquid. If a certain liquid exerts a pressure of 40 pounds per square foot at a depth of 10 feet, what is the pressure at a depth of 30 feet?

18. _____

19. Antonio can paint a fence by himself in 12 hours. With Carlotta's help, it takes only 5 hours. How long would it take Carlotta to paint the fence by herself?

19. _____

20. Hilda can drive 600 miles in the same time it takes Elliott to drive 500 miles. If Hilda drives 10 miles per hour faster than Elliott, then how fast does Hilda drive?

20. _____

Chapter 7 Test Form C

Name:_____

Date:_____

Simplify.

1. $\dfrac{x+5}{-5x-25}$

1. _____

2. $\dfrac{4x+32}{x^2+10x+16}$

2. _____

Perform the indicated operations.

3. $\dfrac{a+2}{a^2+5a+4} \cdot \dfrac{a^2+7a+6}{a^2+9a+18}$

3. _____

4. $\dfrac{x^2-13x+22}{x^2+11x+10} \cdot \dfrac{x^2-8x-20}{x^2-9x-22}$

4. _____

5. $\dfrac{x^2-36}{x+3} \cdot \dfrac{16x+48}{144-4x^2}$

5. _____

6. $\dfrac{9x^2-4y^2}{32x+8y} \div \dfrac{3x+2y}{4x+y}$

6. _____

7. $\dfrac{x^2-3x-10}{x-1} \div \dfrac{x^2-10x+25}{x^2-1}$

7. _____

8. $\dfrac{5xy}{x-7} - \dfrac{35y}{x-7}$

8. _____

9. $10 - \dfrac{4}{x+2}$

9. _____

10. $\dfrac{4x+5}{3x^2+6x} - \dfrac{6}{x+2}$

10. _____

11. $\dfrac{9}{5x} + \dfrac{x}{3x}$

11. _____

12. $\dfrac{x+4}{x^2+3x+2} - \dfrac{x-3}{x^2-2x-3}$

12. _____

Simplify.

13. $\dfrac{6-\frac{3}{5}}{4-\frac{5}{2}}$

13. _____

14. $\dfrac{\frac{5x}{y}+x}{\frac{y}{x}-4}$

14. _____

Solve.

15. $\dfrac{x}{2}=\dfrac{x+4}{6}$

15. _____

16. $\dfrac{x-1}{x+1}=\dfrac{x+2}{x+1}$

16. _____

17. $\dfrac{x-1}{x+2}+\dfrac{x+4}{x-2}=\dfrac{2x^2+20x+18}{x^2-4}$

17. _____

18. If y varies inversely as x, and $y=18$ when $x=20$, find y when $x=45$.

18. _____

19. Frank can paint a mural in 10 days. With the help of an apprentice it would take only 6 days to paint the mural. How long would it take the apprentice to paint the mural working alone?

19. _____

20. A Model T can travel between Tulsa and Stillwater at an average rate of 30 miles per hour. A Porsche 900 can make the same trip at an average speed of 100 miles per hour. If the distance between Tulsa and Stillwater is 68 miles, how long does it take each vehicle to get to Stillwater from Tulsa?

20. _____

Chapter 7 Test Form D

Simplify.

1. $\dfrac{2x^2 + 2x}{x^2 + 2x + 1}$

1. _____

2. $\dfrac{2x^2 + 6x - 8}{x + 4}$

2. _____

Perform the indicated operations.

3. $\dfrac{a^2 + a - 6}{a + 3} \cdot \dfrac{a^2 + 5a + 4}{a^2 + 2a - 8}$

3. _____

4. $\dfrac{x^2 - 7x + 12}{x^2 - 5x - 14} \cdot \dfrac{x^2 - 2x - 8}{x^2 + x - 12}$

4. _____

5. $\dfrac{x^2 - 121}{x - 1} \cdot \dfrac{15x - 15}{-605 + 5x^2}$

5. _____

6. $\dfrac{x^2 - 16y^2}{6x + 24y} \div \dfrac{x - 4y}{2x + 6y}$

6. _____

7. $\dfrac{x^2 - 2x + 1}{x^2 - 4} \div \dfrac{x^2 + x - 2}{-x^2 - x + 2}$

7. _____

8. $\dfrac{-4xy}{x - 2} + \dfrac{8y}{x - 2}$

8. _____

9. $7 - \dfrac{4}{x + 2}$

9. _____

10. $\dfrac{2x - 7}{2x^2 + 2x} - \dfrac{9}{x + 1}$

10. _____

11. $\dfrac{7}{5x} + \dfrac{x}{12x}$

11. _____

12. $\dfrac{x + 2}{x^2 + 4x + 3} - \dfrac{x - 1}{x^2 + 2x - 3}$

12. _____

Chapter 7 Test Form D *(cont.)*

Simplify.

13. $\dfrac{2+\frac{3}{8}}{9-\frac{5}{8}}$

13. _____

14. $\dfrac{\frac{7x}{y}+x}{\frac{y}{x}-2}$

14. _____

Solve.

15. $\dfrac{2x}{7}=\dfrac{x+5}{3}$

15. _____

16. $\dfrac{x-2}{x+4}=\dfrac{x+2}{x+4}$

16. _____

17. $\dfrac{x+2}{x+4}+\dfrac{x-7}{x+3}=\dfrac{2x^2+5x+3}{x^2+7x+12}$

17. _____

18. If y varies inversely as x, and $y=18$ when $x=25$, find y when $x=60$.

18. _____

19. Sally and Mary can paint a house in 8 hours. If Mary paints by herself, it takes her 20 hours. How long would it take Sally to paint the house by herself?

19. _____

20. Carl likes to swim in the river near his house. One day he went swimming and the current was 3 miles per hour. If it took Carl the same amount of time to swim 15 miles downstream as 9 miles upstream, determine the rate Carl would swim in still water.

20. _____

Chapter 7 Test Form E

Name:_____

Date:_____

Simplify.

1. $\dfrac{3x-6}{2-x}$

        1. _____

2. $\dfrac{x^2-4}{x^2+x-6}$

        2. _____

Perform the indicated operations.

3. $\dfrac{18a^3b^4}{c^9} \cdot \dfrac{7c^2}{12ab^6}$

        3. _____

4. $\dfrac{x^2+7x+10}{x^2+x-12} \div \dfrac{x^2-3x-10}{x^2+2x-15}$

        4. _____

5. $\dfrac{4a+5}{3a+10} \cdot \dfrac{9a^2-100}{32a+40}$

        5. _____

6. $\dfrac{x^2-4y^2}{5x-15y} \div \dfrac{x+2y}{x-3y}$

        6. _____

7. $\dfrac{x^2+3x-4}{x^2-4} - \dfrac{x^2+2x-6}{x^2-4}$

        7. _____

8. $\dfrac{3xy}{x+2} + \dfrac{6y}{x+2}$

        8. _____

9. $3 - \dfrac{2}{x-7}$

        9. _____

10. $\dfrac{4x-3}{2x^2-10x} - \dfrac{2}{x-5}$

        10. _____

11. $\dfrac{x+5}{x^2-2x-3} - \dfrac{x-4}{x^2-3x-4}$

        11. _____

12. $\dfrac{6}{xy} + \dfrac{5}{xy^3}$

        12. _____

Chapter 7 Test Form E *(cont.)*

Simplify.

13. $\dfrac{5+\frac{1}{4}}{\frac{1}{x}-\frac{1}{4x}}$

13. _____

14. $\dfrac{\frac{4x}{y}-x}{\frac{y}{x}+2}$

14. _____

Solve.

15. $\dfrac{x+7}{4}-\dfrac{x}{6}=\dfrac{1}{2}$

15. _____

16. $\dfrac{x^2+2}{x+2}=\dfrac{x^2-2}{x+2}$

16. _____

17. $\dfrac{x-3}{x+1}+\dfrac{x-2}{x+2}=\dfrac{2x^2-4x+2}{x^2+3x+2}$

17. _____

18. If y varies inversely as x, and $y=30$ when $x=20$, find y when $x=15$.

18. _____

19. Fred likes to canoe the White River in Arkansas. One day he went canoeing and the river was flowing at a rate of 8 miles per hour. If it took Fred the same amount of time to go 25 miles downstream as 6 miles upstream, determine the canoe's speed in still water.

19. _____

20. Cathy can decorate a wedding cake in 4 hours. It takes Betty 6 hours to decorate a similar cake. How long would it take them to decorate a cake if they worked together?

20. _____

Chapter 7 Test Form F

Name:_____

Date:_____

Simplify.

1. $\dfrac{3x^2 - 6x}{4 - x^2}$

   1. _____

2. $\dfrac{x^2 - 5xy - 14y^2}{7x^2 + 13xy - 2y^2}$

   2. _____

Perform the indicated operations.

3. $\dfrac{a^2 + 5a + 6}{a + 1} \cdot \dfrac{a^2 + 8a + 7}{a^2 - 4}$

   3. _____

4. $\dfrac{3x^2 + 6x}{x^2 - 5x - 24} \cdot \dfrac{x^2 - 3x - 40}{x^2 + 7x + 10}$

   4. _____

5. $\dfrac{x^2 - 9}{x + 6} \cdot \dfrac{5x + 30}{6 - 2x}$

   5. _____

6. $\dfrac{12m - 6n}{2m + 2n} \div \dfrac{2m - n}{8m^2 - 8n^2}$

   6. _____

7. $\dfrac{m^2 - 2mn - 3n^2}{m^2 - mn - 6n^2} \div \dfrac{3m^2 + mn - 2n^2}{2m^2 + 3mn - 2n^2}$

   7. _____

8. $\dfrac{3xy}{x + 5} + \dfrac{15y}{x + 5}$

   8. _____

9. $9 - \dfrac{3}{x - 2}$

   9. _____

10. $\dfrac{5x - 7}{x^2 + 7x} - \dfrac{7}{x + 7}$

    10. _____

11. $\dfrac{3}{4x} + \dfrac{x}{2x}$

    11. _____

12. $\dfrac{x + 2}{x^2 + 6x + 5} - \dfrac{x + 2}{x^2 + 3x + 2}$

    12. _____

Simplify.

13. $\dfrac{5 + \frac{2}{9}}{\frac{1}{3} + \frac{1}{12}}$ 13. _____

14. $\dfrac{\frac{5x}{y} - 1}{\frac{y}{x} - 3}$ 14. _____

Solve.

15. $\dfrac{4x}{2} = \dfrac{x-8}{3}$ 15. _____

16. $\dfrac{x+1}{x-11} = \dfrac{x-1}{x-11}$ 16. _____

17. $\dfrac{x+1}{x+4} + \dfrac{x-3}{x+5} = \dfrac{2x^2 + 6x + 11}{x^2 + 9x + 20}$ 17. _____

18. If y varies inversely as x, and $y = 2.5$ when $x = 80$, find y when $x = 16$.   18. _____

19. Spencer likes to paddle-boat in a river. He was told the current of the river was 2 miles per hour. If it took Spencer the same amount of time to travel 10 miles downstream as 2 miles upstream, determine the speed of his paddle-boat in still water.   19. _____

20. Cindy can mow her lawn in 3 hours. If her husband Ken helps mow, it takes them 1.2 hours to mow the lawn. How long would it take Ken to mow the yard by himself?   20.

Chapter 7 Test Form G

Name:_____

Date:_____

Choose the correct answer to each problem.

Simplify.

1. $\dfrac{2x+4}{4x-4}$

(a) $\dfrac{x+2}{x-1}$　　　　(b) 2　　　　(c) $\dfrac{x+2}{2(x-1)}$　　　　(d) $2x+2$

2. $\dfrac{x^2-x-42}{x-7}$

(a) $x-7$　　　　(b) $\dfrac{x^2-x+42}{x-7}$　　　　(c) $x+6$　　　　(d) $x-6$

Perform the indicated operations.

3. $\dfrac{a^2-16a-63}{a+2}\cdot\dfrac{a^2-a-6}{a^2-10a+21}$

(a) $\dfrac{a^2-16a-63}{a-7}$　　(b) $\dfrac{2a^2-17a-69}{a^2-9a+23}$　　(c) $\dfrac{a^4+16a^2+378}{a^3-10a+42}$　　(d) $\dfrac{a^2-10a+21}{a^2-a-6}$

4. $\dfrac{x^2-1}{x^2-4x-12}\cdot\dfrac{x^2-12x+36}{x^2+x-2}$

(a) $\dfrac{x^2-5x-6}{x+2}$　　(b) $\dfrac{x^2-5x-6}{x^2+4x+4}$　　(c) 1　　(d) $\dfrac{x^2-12x+35}{x^2-3x-14}$

5. $\dfrac{x^2-16}{x-2}\cdot\dfrac{-7x+14}{336-21x^2}$

(a) -3　　　　(b) $\dfrac{1}{3}$　　　　(c) $\dfrac{x^2-16}{336-21x^2}$　　　　(d) x^2-24

6. $\dfrac{x^2-4y^2}{x-3y}\div\dfrac{x-2y}{5x-15y}$

(a) $5(x+2y)$　　(b) $5x^2-8xy-30y^2$　　(c) 5　　(d) $\dfrac{x^2-4y}{5x-15y}$

7. $\dfrac{x^2 - 3x - 10}{x^2 - 6x + 9} \div \dfrac{x^2 - 8x + 15}{x^2 - 5x + 6}$

(a) $\dfrac{x + 2}{x^2 - 6x + 9}$ (b) $\dfrac{x^2 - 4}{x^2 - 6x + 9}$ (c) $\dfrac{x^2 - 4}{x^2 - 8x + 15}$ (d) $\dfrac{x + 2}{x - 2}$

8. $\dfrac{5xy}{x + 6} + \dfrac{30y}{x + 6}$

(a) $\dfrac{5xy + 30y}{x^2 + 12x + 36}$ (b) $\dfrac{5xy + 30y}{2x + 12}$ (c) $5y$ (d) $\dfrac{5y}{x + 6}$

9. $7 - \dfrac{5}{x + 4}$

(a) $\dfrac{7x + 23}{x + 4}$ (b) $\dfrac{2}{x + 4}$ (c) $\dfrac{1}{7}$ (d) $\dfrac{2}{-x - 4}$

10. $\dfrac{5x + 3}{2x^2 + 6x} - \dfrac{3}{x + 3}$

(a) $\dfrac{5x}{2x^2 + 5x - 3}$ (b) $\dfrac{3 - x}{2x^2 + 6x}$ (c) $\dfrac{2x + 3}{2x^2 + 6x}$ (d) $\dfrac{5x}{2x^2 + 6x}$

11. $\dfrac{5}{6x} + \dfrac{x}{8}$

(a) $\dfrac{20 + 3x^2}{24x}$ (b) $\dfrac{5 + x}{6x + 8}$ (c) $\dfrac{5}{14}$ (d) $\dfrac{5x}{48x}$

12. $\dfrac{x + 1}{x^2 + 3x + 2} - \dfrac{x - 1}{x^2 + 5x + 6}$

(a) $\dfrac{4}{x^2 + 5x + 6}$ (b) $\dfrac{2}{-2x - 4}$ (c) $\dfrac{4x + 4}{(x + 2)(x + 1)(x + 3)}$ (d) $\dfrac{-2x + 2}{-2x^2 - 2x - 4}$

13. $\dfrac{10}{xy} + \dfrac{22}{xy^3}$

(a) $\dfrac{32}{x^2 y^4}$ (b) $\dfrac{2(y^2 + x)}{xy^3}$ (c) $\dfrac{10y^3 + 22}{xy^3}$ (d) $\dfrac{2(5y^2 + 11)}{xy^3}$

Chapter 7 Test Form G *(cont.)*

Simplify.

14. $\dfrac{5+\frac{1}{6}}{\frac{3}{4}-\frac{2}{3}}$

    (a) $\dfrac{\frac{31}{6}}{\frac{1}{12}}$ (b) $\dfrac{372}{6}$ (c) $\dfrac{77}{12}$ (d) 62

15. $\dfrac{\frac{2x}{y}+4x}{\frac{y}{x}}$

    (a) $\dfrac{6x^2}{y^2}$ (b) $\dfrac{2xy(1+2y)}{xy}$ (c) $\dfrac{2x^2(1+2y)}{y^2}$ (d) $\dfrac{2x^2(1+2y)}{y}$

Solve.

16. $\dfrac{3x}{7}=\dfrac{x-3}{2}$

    (a) $x=\dfrac{21}{4}$ (b) $x=-21$ (c) no solution   (d) $x=21$

17. $\dfrac{x+2}{2x-1}=\dfrac{2x+1}{2x-1}$

    (a) $x=1$ (b) $x=\dfrac{1}{2}$ (c) $x=0$ (d) no solution

18. $\dfrac{x+4}{x+5}+\dfrac{x-2}{x+1}=\dfrac{2x^2+7x+1}{x^2+6x+5}$

    (a) $x=7$ (b) $x=-7$ (c) no solution   (d) all real numbers

19. Sam can eat a whole pie in 2 minutes. Peter can eat a whole pie in $2\frac{1}{2}$ minutes. How long would it take them to eat the pie if they shared?

    (a) $4\frac{1}{2}$ minutes   (b) $1\frac{1}{9}$ minutes   (c) $1\frac{1}{2}$ minutes   (d) $2\frac{1}{4}$ minutes

20. If y varies inversely as x, and $y=30$ when $x=14$, find y when $x=35$.

    (a) 420   (b) 75   (c) $16\frac{1}{3}$ (d) 12

Chapter 7 Test Form H

Name: _____

Date: _____

Choose the correct answer to each problem.

Simplify.

1. $\dfrac{3n^2 - 9n}{n^2 - 9}$

   (a) $\dfrac{3n}{n+3}$ (b) 1 (c) $\dfrac{1}{2}$ (d) $3 - n$

2. $\dfrac{3x^2 - 12}{3x - 6}$

   (a) 2 (b) $x^2 - 2$ (c) $x + 2$ (d) x^2

Perform the indicated operations.

3. $\dfrac{a^2 + 5a + 6}{a + 1} \cdot \dfrac{a^2 + 3a + 2}{a^2 + 6a + 9}$

   (a) $\dfrac{1}{a+1}$ (b) $\dfrac{a^2 + 4a + 4}{a + 3}$ (c) $a^2 + 4a + 4$ (d) 1

4. $\dfrac{x^2 - 5x + 6}{x^2 + 6x + 9} \cdot \dfrac{x^2 - x - 12}{x^2 - x - 6}$

   (a) $\dfrac{x - 4}{x + 3}$ (b) $\dfrac{x^2 - 6x + 8}{x + 2}$ (c) $\dfrac{1}{x + 3}$ (d) $\dfrac{x^2 - 6x + 8}{x^2 + 5x + 6}$

5. $\dfrac{x + 5}{x^2 - 1} \cdot \dfrac{2 - 2x^2}{4x + 20}$

   (a) $-\dfrac{1}{2}$ (b) $\dfrac{1}{2}$ (c) $\dfrac{2x - 1}{4x^2 - 4}$ (d) $\dfrac{1}{x^2 + 4x - 5}$

6. $\dfrac{2x - y}{4x^2 - 4y^2} \div \dfrac{6(2x - y)}{2x + 2y}$

   (a) $\dfrac{6}{2x - 2y}$ (b) 3 (c) $\dfrac{1}{12(x - y)}$ (d) $\dfrac{1}{6}$

7. $\dfrac{x^2 + 4x + 4}{x^2 + x - 12} \div \dfrac{x^2 - x - 6}{x^2 + 6x + 8}$

(a) 1

(b) $\dfrac{x^2 + 4x + 4}{x^2 + 8x + 16}$

(c) $\dfrac{x^2 + 4x + 4}{x^2 - 6x + 9}$

(d) $\dfrac{x + 2}{x - 3}$

8. $\dfrac{-2xy}{x - 1} + \dfrac{2y}{x - 1}$

(a) $-2x$

(b) $\dfrac{-2xy + 2y}{2x - 2}$

(c) $\dfrac{2y}{x - 1}$

(d) $-2y$

9. $3 - \dfrac{4}{x - 4}$

(a) $-\dfrac{1}{x - 4}$

(b) $\dfrac{3x - 16}{x - 4}$

(c) $\dfrac{3x - 1}{x}$

(d) $3x - 8$

10. $\dfrac{2x - 5}{x^2 + x} - \dfrac{3}{x + 1}$

(a) $\dfrac{2x - 8}{x^2 - 1}$

(b) $\dfrac{2x^2 - 5x - 3}{x^2 + x}$

(c) $\dfrac{-x - 5}{x^2 + x}$

(d) $-\dfrac{5}{x + 1}$

11. $\dfrac{5}{7x} + \dfrac{x}{3x}$

(a) $\dfrac{15 + 7x}{21x}$

(b) $\dfrac{5 + x}{10x}$

(c) $\dfrac{5 + x}{7x + 3x}$

(d) $\dfrac{6x}{10x}$

12. $\dfrac{x - 3}{x^2 + 6x - 27} - \dfrac{x + 9}{x^2 + 10x + 9}$

(a) $\dfrac{6}{x^2 - 2x - 3}$

(b) 1

(c) $-\dfrac{8}{x^2 + 10x + 9}$

(d) 0

13. $\dfrac{3}{xy} + \dfrac{9}{xy^3}$

(a) $\dfrac{12}{x^2 y^4}$

(b) $\dfrac{3(y^2 + 3)}{xy^3}$

(c) $\dfrac{3y^3 + 9}{xy^3}$

(d) $\dfrac{12}{2xy^3}$

Simplify.

14. $\dfrac{7+\frac{1}{3}}{1-\frac{7}{18}}$

    (a) 12
    (b) $\dfrac{121}{27}$
    (c) $\dfrac{7}{3}$
    (d) $\dfrac{8}{9}$

15. $\dfrac{\frac{7x}{y}-x}{\frac{y}{x}-4}$

    (a) $\dfrac{6x}{y}-\dfrac{3y}{x}$
    (b) $\dfrac{7-y}{y-4}$
    (c) $\dfrac{7x-xy}{y-4x}$
    (d) $\dfrac{x^2(7-y)}{y(y-4x)}$

Solve.

16. $\dfrac{x+8}{5}=\dfrac{x-4}{3}$

    (a) $x=6$
    (b) $x=2$
    (c) $x=22$
    (d) no solution

17. $\dfrac{x-2}{x+3}=\dfrac{2x-1}{x+3}$

    (a) $x=-3$
    (b) $x=1$
    (c) $x=-1$
    (d) no solution

18. $\dfrac{x+2}{x-3}+\dfrac{x-6}{x+5}=\dfrac{2x^2+3x-12}{x^2+2x-15}$

    (a) $x=8$
    (b) $x=-40$
    (c) $x=-8$
    (d) $x=-5$

19. If y varies inversely as the square of x, and $y=16$ when $x=5$, find y when $x=2$.

    (a) 40
    (b) 100
    (c) 400
    (d) 2.56

20. Johnny can paint a car in 5 hours. If Johnny works with Larry they can paint the car in 3 hours. How long does it take Larry to paint the car by himself?

    (a) 8 hours
    (b) 4 hours
    (c) $2\frac{1}{2}$ hours
    (d) $7\frac{1}{2}$ hours

Chapter 8 Pretest Form A

Graph.

1. $y = 3x - 4$

1.

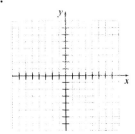

2. $y = -x^2$

2.

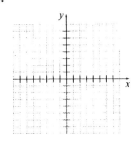

3. $f(x) = 2|x|$

3.

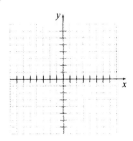

4. Write $\{x \mid x > -3.4 \text{ and } x \in I\}$ in roster form.

4. _____

5. Represent $\{x \mid -2 < x \le 7\}$ using interval notation.

5. _____

6. Graph using x- and y-intercepts. $2x - 5y = 10$

6.

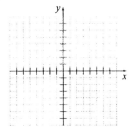

For problems 7–9, use the relation $\{(-7, 9), (4, -7), (8, 2), (2, 9)\}$.

7. Find the domain of the relation.

7. _____

8. Find the range of the relation.

8. _____

9. Is the relation a function?

9. _____

Chapter 8 Pretest Form A *(cont.)*

Determine whether the following relations are functions.

10.

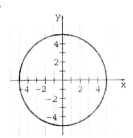

10. _____

11.

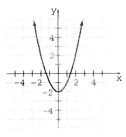

11. _____

12. If $f(x) = 3x^2 + 5$, find $f(-1)$.

12. _____

13. Determine the slope and y-intercept of the equation $2x + 3y = 12$.

13. _____

14. The profit of a skateboard manufacturer can be estimated by the function $p(x) = 3.7x - 12,250$, where x is the number of skateboards produced and sold. Use the function to estimate the profit on 10,000 skateboards.

14. _____

15. Determine the slope of the line through the points $(-2, 6)$ and $(3, -4)$.

15. _____

16. Determine whether or not the graphs of the two equations are parallel.
$$2x + y = -5$$
$$4x + 2y = 1$$

16. _____

17. Find the equation of the line through $(2, -1)$ with slope 3.

17. _____

For problems 18–19, let $f(x) = 3x^2 - 5$ and $g(x) = x + 9$.

18. Find $(f + g)(x)$.

18. _____

19. Find $(f + g)(-3)$.

19. _____

20. Find $\left(\dfrac{g}{f}\right)(-2)$.

20. _____

Chapter 8 Pretest Form B

Graph.

   1. $y = 4 - 2x$

1.
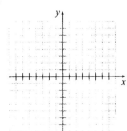

   2. $y = x^2 - 5$

2.
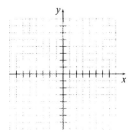

   3. $f(x) = -|x|$

3.
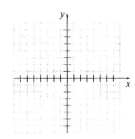

   4. Express in set builder notation.

          $\xleftarrow{\quad}\overset{-4}{|}\quad\overset{-2}{\circ}\quad\overset{0}{|}\quad\overset{2}{|}\quad\overset{4}{\bullet}\quad\overset{6}{|}\xrightarrow{\quad}$

4. _____

   5. Represent $\{x \mid 3 < x < 12\}$ in interval notation.

5. _____

   6. Graph using x- and y-intercepts. $3x - 5y = 15$

6.
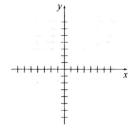

For problems 7–9, use the relation $\{(7, 2), (-4, 3), (5, 2), (-3, -1)\}$.

   7. Find the domain of the relation.

7. _____

   8. Find the range of the relation.

8. _____

Chapter 8 Pretest Form B *(cont.)*

9. Is the relation a function?

9. _____

Determine whether the following relations are functions.

10.

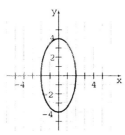

10. _____

11.

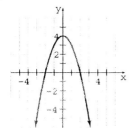

11. _____

12. If $f(x) = x^2 + 7x + 9$, find $f(-1)$.

12. _____

13. The yearly profit, p, for Grandma's Bakery on frozen pies can be estimated by the function $p(x) = 2.8x - 12,000$, where x is the number of pies sold. Use the function to estimate the profit on 15,000 pies.

13. _____

14. Determine the slope and y-intercept of the graph $7x - 6y = 12$.

14. _____

15. Determine the slope of the line through the points $(-2, 6)$ and $(3, -5)$.

15. _____

16. Determine whether or not the graphs of the two equations are parallel.
$4x - 9y = 1$
$8x = 18y + 4$

16. _____

17. Find the equation of the line through $(-2, 4)$ with slope -3.

17. _____

For problems 18–20, let $f(x) = 4x^2 - 10$ and $g(x) = x + 2$.

18. Find $(f + g)(x)$.

18. _____

19. Find $(f + g)(-2)$.

19. _____

20. Find $(f \cdot g)(2)$

20. _____

Mini-Lecture 8.1
More on Graphs

<u>Learning Objectives:</u>

1. Graph equations by plotting points.
2. Interpret graphs.
3. Key vocabulary: *ordered pair, coordinates, graph, first-degree equations, nonlinear equations*

<u>Examples:</u>

1. (review) Plot the following points on the same set of axes.
   a) A(4, 2)   b) B(3, –5)   c) C(–2, 4)

   d) D(–3, –1)   e) E(1, 0)   f) F(0, –4)

2. Determine whether the following ordered pairs are solutions to the equation $y = 3x - 2$.
   a) (4, 10)   b) (–2, –4)

3. Graph each equation.
   a) $y = 2x + 1$ b) $y = 1 - x^2$ c) $y = |x| - 1$

4. (review) Determine the quadrant in which each point is located.
   a) (–4, 7)   b) (5, –1)   c) (–20, –50)

5. Refer to Example 5 from Section 8.1 and decide which of the given graphs best illustrates the following situation.
   A plane sat on the runway for 20 minutes and then took off, increasing its speed to 600 mph over the next 20 minutes. The plane flew at about 600 mph for about 2 hours. For the next 20 minutes, the plane steadily slowed down until the speed reached about 400 mph. The plane then flew at 400 mph for 20 minutes. The speed of the plane was then quickly reduced in preparation to land. After landing, the plane was forced to hurry off the runway and, as a result, the speed of the plane increased to about 100 mph. Finally, the plane began slowing down as it approached the gate.

<u>Teaching Notes:</u>

- Students often place ordered pairs having 0 as one its coordinates on the wrong axis.
- Emphasis the importance of using parentheses when expressing the coordinates of an ordered pair. Indicate that (3, 5), {3, 5}, and 3, 5 all have different mathematical meanings.
- Remind students that a point on the graph of an equation represents an ordered-pair solution to the equation.

Answers: 1) see Chapter 8 Answers; 2a) yes; 2b) no; 3) see Chapter 8 Answers; 4a) II; 4b) IV; 4c) III; 5) (b)

Mini-Lecture 8.2
Functions

Learning Objectives:

1. Understand set builder notation and interval notation.
2. Understand relations.
3. Recognize functions.
4. Use the vertical line test.
5. Understand function notation.
6. Applications of functions in daily life.
7. Key vocabulary: *set builder notation, interval notation, roster form, endpoints, 'between', 'inclusive', dependent variable, independent variable, relation, function, domain, range, vertical line test, function notation*

Examples:

1. Express in set builder notation and interval notation.

   a)    b)

2. Determine if the following relations are functions. Give the domain and range of each relation or function.

   a) $\{(7, -4), (5, 1), (4, 8)\}$ b) $\{(3,8), (-2,8), (0,8)\}$ c) $\{(-6, 7), (-6, 2), (-6, 10)\}$

3. Use the vertical line test to determine whether the following graphs represent functions. Also give the domain and range of each function or relation.

   a) 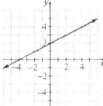   b)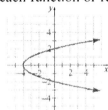

4. If $f(x) = -2x^2 + 3x - 1$, find

   a) $f(3)$ b) $f(-1)$ c) $f(0)$

5. Suppose that the total profit, in dollars, at local theater can be determined by the function $P(x) = -x^2 + 35x - 150$, where x is the cost, in dollars, of an individual ticket. Determine the value of $P(10)$ and interpret its meaning.

Teaching Notes:

- A short phrase to help remember whether a relation involving x-values and y-values is a function is "For each different x-value, there can only be one y-value."
- Elements of the domain can be thought of as input values; elements of the range can be thought of as output values.
- Emphasize to students that function notation is not to be confused with multiplication. The notation $f(x)$ does not mean to multiply f and x.

Answers: 1) a) $\{x \mid -3 \le x < 3\}$, $[-3,3)$ b) $\{x \mid x \le 7\}$, $(-\infty, 7]$; 2a) yes, D:{4, 5, 8}, R:{-4, 1, 8}; 2b) yes, D:{ -2, 0, 3}, R:{8}; 2c) no, D:{-6}, R:{2, 7, 10}; 3a) yes, D: $\mathbb{R}$, R: $\mathbb{R}$; 3b) no, D: $\{x \mid x \ge -4\}$, R: $\mathbb{R}$; 4a) -10; 4b) -6; 4c) -1; 5) $P(10) = 100$, When tickets are sold at $10 each, the total profit will be $100.

Mini-Lecture 8.3
Linear Functions

Learning Objectives:

1. Graph linear functions.
2. Graph linear functions using intercepts.
3. Study applications of functions.
4. Solve linear equations in one variable graphically.
5. Key vocabulary: *linear function, x-intercept, y-intercept, constant function*

Examples:

1. Graph each equation.

   a) $y = 2x - 3$ b) $f(x) = -\dfrac{1}{3}x$

2. Graph each equation using the *x*- and *y*-intercepts.

   a) $2x - 3y = 6$ b) $f(x) = -\dfrac{1}{2}x + 3$

3. Wayne is told that his total monthly salary in his new sales job will be $900 plus 8% commission on his monthly sales.
   a) Write a function expressing Wayne's monthly salary, *s*, in terms of his monthly sales, *x*.
   b) Draw a graph of Wayne's monthly salary versus his monthly sales, for up to and including $10,000 in sales.
   c) What is Wayne's monthly salary if his sales are $8000?
   d) If Wayne's monthly salary was $1420, what were his monthly sales?

4. Solve equation $3x + 5 = 2x + 9$ graphically as shown in Example 5 in Section 8.3

Teaching Notes:

- Remind students that $f(x)$ and *y* are interchangeable.
- Remind students that it is often difficult to determine the exact answer from a graph. Students should solve the problem algebraically and use the graph as a check to see if their answer is reasonable.

Answers: 1)   2)

3a) $s = 0.08x + 900$;   3b)

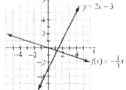

3c) $1540; 3d) $6500; 5) x = 4

262

Mini-Lecture 8.4
Slope, Modeling, and Linear Relationships

Learning Objectives:
1. Recognize slope as a rate of change.
2. Use the slope-intercept form to construct models from graphs.
3. Use the point-slope form to construct models from graphs.
4. Recognize vertical translations.
5. Use slope to identify and construct perpendicular lines.
6. Key vocabulary: *point-slope form of a line, parallel, perpendicular, negative reciprocal, translation, parallel, slope, positive slope, zero slope, negative slope, vertical line (undefined slope), rate of change, slope-intercept form*

Examples:
1. Find the slope of the line in each figure. Then write the equation of the given line.

   a) 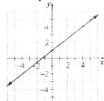   b)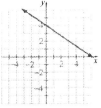

2. Refer to Example 2 from Section 8.4.
   a) Determine the slope of the line segment between 1990 and 2002. Round to two decimal places.
   b) Explain what the answer to part (a) means in terms of the U.S. public debt.

3. Kathy is making fixed monthly payments to her rich brother to pay back a loan that allowed her to have a swimming pool installed in her backyard. After three months, Kathy owed $6800 and after seven months, she still owed $5200.
   a) Using ordered pairs of the form (x, P) write a linear equation for Kathy's balance, P, as a function of the number of months, x.
   b) Using the function from part (a), determine Kathy's balance after 12 months.
   c) Using the function from part (a), determine how many months it will take until her loan is paid off with a balance of $0.

4. Use the point-slope form to find the equation of a line with the properties given. Then write the equation in slope-intercept form.
   a) Slope = 3, through $(2, -5)$ b) Through $(8, -5)$ and $(-5, 4)$
   c) Through $(4, 3)$ and perpendicular to $y = -\frac{1}{2}x + 5$.

5. Determine whether the two equations represent lines that are parallel, perpendicular, or neither.

   a) $6x - 9y = 15$ b) $-2x + 3y = 6$ c) $4x + 8y = 24$

   $-4x + 6y = -12$ $5x - 7y = 9$ $12x - 6y = 11$

Teaching Notes:

- Another common word definition for slope related to $\dfrac{\text{vertical change}}{\text{horizontal change}}$ is $\dfrac{\text{rise}}{\text{run}}$.

- Some students confuse the concept of slope with the plotting of ordered pairs. Since they are used to plotting points like (2, 3) by starting at the origin and moving 2 units right and 3 units up, they may show a slope of 2/3 by moving right 2 units and up 3 instead of going up 2 units and right 3.

Answers: 1a) $m = \frac{3}{4}$, $y = \frac{3}{4}x + 1$; 1b) $m = -\frac{2}{3}$, $y = -\frac{2}{3}x + 4$; 2a) 219.49; 2b) The U.S. public debt, from 1990 to 2002, increased by about $219.49 billion per year; *3a)* $P(x) = -400x + 8000$; *3b) $3200;*

3c) 20 months;; 4a) $y = 3x - 11$; 4b) $y = -\frac{9}{13}x + \frac{7}{13}$; 4c) $y = 2x - 5$; 5a) parallel; 5b) neither;

5c) perpendicular

Mini-Lecture 8.5
The Algebra of Functions

Learning Objectives:

1. Find the sum, difference, product and quotient of functions.
2. Graph the sum of functions.
3. Key vocabulary: *sum of functions, difference of functions, product of functions, quotient of functions*

Examples:

1. If $f(x) = x - 4$ and $g(x) = x^2 + 2x - 1$, find
   a) $(f + g)(x)$ b) $(f - g)(x)$ c) $(f \cdot g)(x)$ d) $(f / g)(x)$

2. If $f(x) = x^2 + 2$ and $g(x) = x - 3$, find
   a) $(f + g)(-1)$ b) $(g - f)(2)$ c) $(f \cdot g)(1)$ d) $(g / f)(4)$

3. Construct a line graph, including the total, for the following data.

	2003	2004	2005	2006
Average monthly home phone bill	$64	$60	$62	$67
Average monthly mobile phone bill	$41	$47	$56	$61

4. Construct a bar graph, similar to Example 4 from Section 8.5, for the above data.

5. Construct a stacked line graph for the data given in problem #3.

Teaching Notes:

- Once again, students are often confused by function notation. For instance, some students may think that $(f + g)(5)$ means to add the two functions f and g, and then multiply by 5 rather than substitute 5.

Answers: 1a) $x^2 + 3x - 5$; 1b) $-x^2 - x - 3$; 1c) $x^3 - 2x^2 - 9x + 4$; 1d) $\dfrac{x - 4}{x^2 + 2x - 1}$; 2a) -1; 2b) -7;

2c) -6; 2d) $\dfrac{1}{18}$; 3–5) see Chapter 8 Answers

Additional Exercises 8.1

1. What are the coordinates of point A?

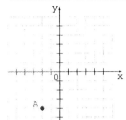

2. Graph the point $B(-3, 4)$.

2.

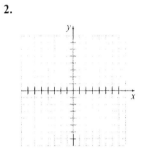

3. Name the coordinates of the points A, B, C, and D.

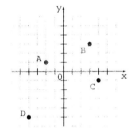

3. _____

4. Graph: $x - y = 5$

4.

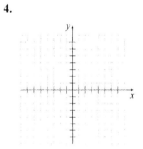

5. Graph: $-3x + 4y = -12$

5.

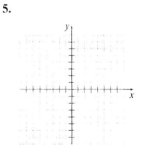

Additional Exercises 8.1 *(cont.)*

Name:_____

6. Graph: $8x - 3y = 24$

6.

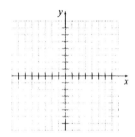

7. Graph: $y = -3x - 3$

7.

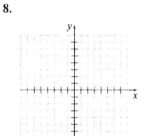

8. Graph the equation $2x + y = 2$.

8.

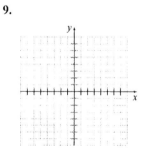

9. Graph: $5x - 6y = 30$

9.

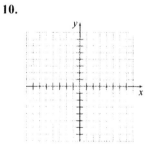

10. Graph: $-2x + y = 6$

10.

11. Graph: $-x - 3y = -3$

11.

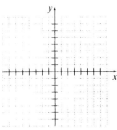

12. Graph: $5x - 7y = 35$

12.

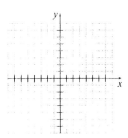

13. Graph: $y = -3x + 2$

13.

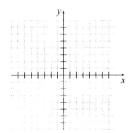

14. Graph the equation $-2x + y = 2$.

14.

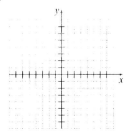

15. Graph: $5x - 3y = 15$

15.

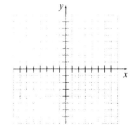

Additional Exercises 8.1 *(cont.)*

16. Graph: $y = -1 - |x|$

16.

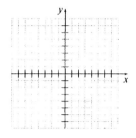

17. Graph: $y = 2 + |x|$

17.

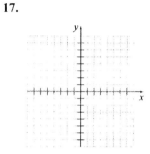

18. Graph: $y = -x^2 + 5$

18.

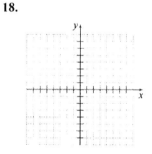

19. Graph the data in the following table using a broken-line graph.

Number of Rigs Drilling for Oil (Monthly Averages)			
1980	1982	1984	1986
2000	4000	3500	3000

19.

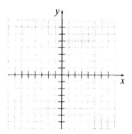

20. The double line graph below compares high temperatures in Honolulu and Miami in August. Use the graph below to determine the days that Honolulu's temperature was lower than Miami's.

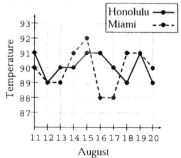

20. _____

Additional Exercises 8.2

1. Find the domain and range for the relation graphed below. Express both in set builder notation and interval notation.

    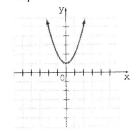

    1. _____

2. Find the domain of the relation $\{(1,-2),(-2,1),(-1,-5)\}$.

    2. _____

3. Find the range of the relation $\{(2,-5),(-1,7),(4,-5),(3,2)\}$.

    3. _____

4. Find the range of the relation $A=\{(x,y)|x^2+y^2=49\}$. Express your answer in set builder notation and interval notation.

    4. _____

5. Determine if the relation $\{(0, 7), (7, 8), (0, -6)\}$ is a function.

    5. _____

6. Determine if the following is a function: $\{(x,y)|x+y=5\}$

    6. _____

7. Determine if the relation $\{(-2, 4), (-4, 4), (-1, 2)\}$ is a function.

    7. _____

8. Is the relation $\{(5, -4), (5, 3), (3, -5)\}$ a function?

    8. _____

9. Find $f(-2)$ given $f(x)=x+2$.

    9. _____

10. Given the function $f(x)=\dfrac{4}{5}x+3$; find $f(-30)$.

    10. _____

11. Find $f(-2)$ given $f(x)=-x^2+5x+1$.

    11. _____

12. If $P(x)=x^2-3x-2$, find $P(-3)$.

    12. _____

13. The cost of a long-distance phone call from New York to Athens is defined by $C(t)=0.75(t-1)+1.45$, where the cost is $1.45 for the first minute and $0.75 for each additional minute. Find the cost of a 7-minute phone call.

    13. _____

Additional Exercises 8.2 *(cont.)*

14. The measure in degrees of an interior angle of a regular polygon with n sides is given by the function $f(n) = 180 - \dfrac{360}{n}$. Find the measure of an interior angle of a regular decagon (10 sides).

14. _____

15. The area of an equilateral triangle with sides of length s can be found by the function: $f(x) = \dfrac{1}{4}\sqrt{3}s^2$. Find the area of an equilateral triangle with sides of length 4.5. Round answers to the nearest 0.1.

15. _____

16. Determine if the relation $\{(x, y) \mid x^2 + y^2 = 1\}$ is a function.

16. _____

17. Is the relation $\{(5, 6), (-1, 6), (1, 6)\}$ a function?

17. _____

18. Find $f(-1)$ given $f(x) = 2x + 1$.

18. _____

19. Given the function $f(x) = -\dfrac{1}{6}x + 3$; find $f(-18)$.

19. _____

20. Find $f(2)$ given $f(x) = -x^3 - 2x^2 + 28$.

20. _____

Additional Exercises 8.3

1. Graph the linear equation by finding x- and y-intercepts.
   $3x + y - 3 = 0$

   1.
   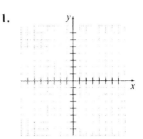

2. Graph the linear equation by finding x- and y-intercepts.
   $-y + 2x = -2$

   2.
   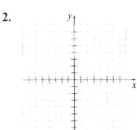

3. Graph: $f(x) = -x + 3$

   3.
   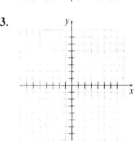

4. Graph: $f(x) = \dfrac{1}{2}x - 4$

   4.
   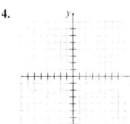

5. Graph: $g(x) = \dfrac{2}{3}x + 1$

   5.
   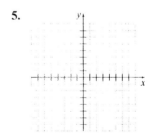

6. Graph: $g(x) = -\dfrac{2}{5}x - 4$

   6.
   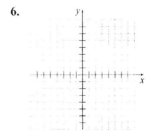

Additional Exercises 8.3 *(cont.)*

Name:_____

7. A real estate agent's initial monthly salary is $800 plus 1% of the total homes sales he has for that month. Write an equation expressing the relationship between the agent's monthly salary and his monthly homes sales. Then use this equation to determine his monthly salary if his total homes sales in one particular month is $680,000.

7. _____

For problems 8 – 10, the annual profit, p, of a bike manufacturer can be estimated by the formula $p = 18x - 10,000$ where x is the number of bikes sold per year.

8. Draw a graph of profits versus bikes that must be sold for up to 6000 bikes.

8.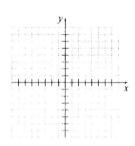

9. Estimate the number of bikes that must be sold for the company to break even.

9. _____

10. Estimate the number of bikes sold if the company has a $30,000 profit.

10. _____

11. Write the linear equation $y = \dfrac{2}{3}x + 5$ in standard form.

11. _____

12. Write the linear equation $y + 2 = -\dfrac{3}{4}(x - 5)$ in standard form.

12. _____

13. Graph the linear equation by finding x- and y-intercepts.
$3x - 2y + 6 = 0$

13.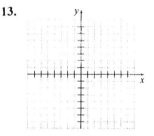

14. Graph the linear equation by finding x- and y-intercepts.
$-4y - 3x = 12$

14.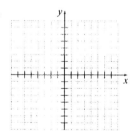

Additional Exercises 8.3 *(cont.)*

Name:_____

15. Graph: $h(x) = 4x - 3$

15.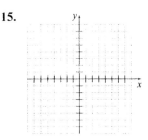

16. Graph: $h(x) = -\dfrac{7}{3}x + 4$

16.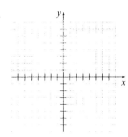

17. John's salary is $1400 plus 6% commission on monthly sales. Write a function expressing the relationship between John's salary and his monthly sales, and use it to find his sales for the month if his salary for the month is $4000.

17.

For problems 18 – 20, the annual profit, p, of a bike manufacturer can be estimated by the function $p(x) = 18x - 30{,}000$ where x is the number of bikes sold per year.

18. Draw a graph of profits versus bikes that must be sold for up to 6000 bikes.

18.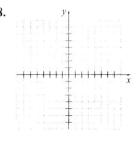

19. Estimate the number of bikes that must be sold for the company to break even.

19. _____

20. Estimate the number of bikes sold if the company has a $45,000 profit.

20. _____

Additional Exercises 8.4

Name:_____

Date:_____

1. Determine the slope of the line graphed below.

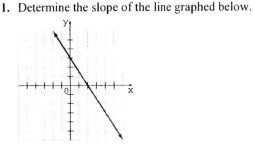

2. Find the slope of the line going through the points (2, 7) and (9, 10).

3. Find the slope and the *y*-intercept of the line $9x + 3y = -54$.

4. Draw the graph of a line with *y*-intercept 3 and slope of $-\dfrac{3}{2}$.

5. Graph: $y = 4x + 8$

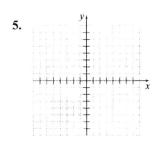

6. Find the slope of the line going through the points $(6, -5)$ and $(-3, -1)$.

7. Find the slope and *y*-intercept of $7x - 4y = 2$.

8. Find an equation of the line having slope 4 and *y*-intercept 5.

9. Graph: $y = -2x - 6$

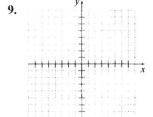

10. Are the two given lines parallel? (Answer yes or no.)
$9x + 2y = -10$

$y = -\dfrac{9}{2}x - 6$

1. _____

2. _____

3. _____

4.

5.

6. _____

7. _____

8. _____

9.

10. _____

Additional Exercises 8.4 *(cont.)*

11. Are the two given lines perpendicular? (Answer yes or no.)

    $-8x - 10y = -11$

    $-4x + 5y = -2$

    11. _____

12. Two points on line 1 are (3, 1) and (–2, –2). Two points on line 2 are (–5, 4) and (–8, 9). Determine if line 1 and line 2 are parallel lines, perpendicular lines, or neither.

    12. _____

13. Find an equation of the line passing through the point (–7, –2) with slope $m = -5$.

    13. _____

14. Find an equation of the line that passes through the point (–3, 6) and is parallel to the line $2x + y = -6$.

    14. _____

15. Write the equation of the line (in slope-intercept form) passing through the point (–5, –1) and perpendicular to $7x + 6y = 6$.

    15. _____

16. Find an equation of the line that passes through the point (2, –5) and is parallel to the line $2x + 3y = 7$.

    16. _____

17. Write the equation of the line (in slope-intercept form) passing through the point (–1, –8) and perpendicular to $4x - 5y = 6$.

    17. _____

18. Are the two given lines parallel?

    $-2x + 3y = 21$

    $y = -2x + 7$

    18. _____

19. Are the two given lines perpendicular?

    $5x - y = 9$

    $-2x - 10y = 3$

    19. _____

20. Two points on line 1 are (2, –7) and (–6, –10). Two points on line 2 are (–6, 2) and (–9, 10). Determine if line 1 and line 2 are parallel lines, perpendicular lines, or neither.

    20. _____

Additional Exercises 8.5

1. Let $f(x) = 16 - x^2$, $g(x) = 4 - x$. Find $\left(\dfrac{f}{g}\right)(x)$.

    1. _____

2. Let $f(x) = x^2 + 5x - 3$, $g(x) = 3x + 7$. Find $(f + g)(x)$.

    2. _____

3. Given $f(x) = 4x^2 - 9x + 3$ and $g(x) = x^3$, find $\left(\dfrac{f}{g}\right)(x)$.

    3. _____

4. Given $f(x) = x^2 + 9x + 6$ and $g(x) = x + 6$, find $(f + g)(x)$.

    4. _____

5. Given $f(x) = x^3$ and $g(x) = 4 + 3x$, find $(g \cdot f)(x)$.

    5. _____

6. Let $f(x) = 1 - x^2$, $g(x) = 1 - x$. Find $(fg)(x)$.

    6. _____

7. Let $f(x) = 9 - x^2$, $g(x) = 3 - x$. Find $(f - g)(x)$.

    7. _____

8. Given $f(x) = 5x^2 - 6x + 5$ and $g(x) = x^4$, find $\left(\dfrac{f}{g}\right)(x)$.

    8. _____

9. Given $f(x) = x^2 + 8x - 8$ and $g(x) = x + 9$, find $(f + g)(x)$.

    9. _____

10. Given $f(x) = x^3$ and $g(x) = 7x - 4$, find $(g \cdot f)(x)$.

    10. _____

11. Let $f(x) = 4x - 9x^2 + 3$, $g(x) = 2x^2 - 3x$. Find $(f - g)(x)$.

    11. _____

12. Let $f(x) = 1 - x^2$, $g(x) = 1 + x$. Find $(fg)(x)$.

    12. _____

13. Given $f(x) = 9x^2 - 4x + 7$ and $g(x) = x^3$, find $\left(\dfrac{f}{g}\right)(x)$.

    13. _____

14. Given $f(x) = x^2 - 2x + 3$ and $g(x) = x - 7$, find $(f + g)(x)$.

    14. _____

15. Given $f(x) = x^3$ and $g(x) = 5 - 3x^2$, find $(f \cdot g)(x)$.

    15. _____

16. If $f(x) = x + 3$ and $g(x) = x$, find the domain of $\left(\dfrac{g}{f}\right)(x)$.

    16. _____

17. If $f(x) = x^2$ and $g(x) = 8$, find the range of $(f + g)(x)$.

    17. _____

18. If $f(x) = -1 + 3x^2$ and $g(x) = 2x$, find the domain of $(f \cdot g)(x)$.

    18. _____

19. If $f(x) = \dfrac{x + 3}{x}$ and $g(x) = -2$, find the domain of $(f \cdot g)(x)$.

    19. _____

20. If $f(x) = |x|$ and $g(x) = 2$, find the range of $(f - g)(x)$.

    20. _____

Chapter 8 Test Form A

Graph.

1. $y = x - 2$

1.

2. $f(x) = x^2 + 2$

2.

3. $g(x) = |x| - 1$

3.

4. Represent the interval $(3, 17]$ using set builder notation.

4. _____

5. Represent $\{x \mid x > -3\}$ using interval notation.

5. _____

Problems 6 and 7 refer to the following relation:
$\{(-1, 4), (2, 7), (3, 6), (-1, -5)\}$

6. Determine if the relation is a function.

6. _____

7. Give the domain and range of the function or relation.

7. _____

Problems 8 and 9 refer to the following relation:

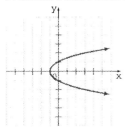

8. Determine if the relation is a function.

8. _____

9. Give the domain and range of the function or relation.

9. _____

Name:_____

10. Graph $y = \frac{1}{3}x + 1$ using the x and y intercepts.

10.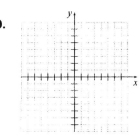

11. Graph $f(x) = -2$.

11.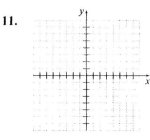

12. Determine the slope of the line through the points $(3, -2)$ and $(7, 3)$.

12. _____

13. Determine the slope and y-intercept of the graph of the equation $11x - 2y = 12$.

13. _____

14. Determine if the graphs of the two equations are parallel, perpendicular, or neither. Explain your answer. $2x + y = -3$ and $x - 2y = -4$

14. _____

15. Find the equation of the line through $(-8, -1)$ that is parallel to the graph of $y = \frac{3}{2}x - 4$ in slope-intercept form.

15. _____

16. Find the equation of the line through $(1, -2)$ that is perpendicular to the graph of $6x - 3y = 9$ in standard form.

16. _____

17. The yearly profit, p, for Waylan Publishing Company on the sales of a particular book can be estimated by the function $p(x) = 11.4x - 45,000$, where x is the number of books produced and sold. Use the function to estimate the number of books that the company must sell to make a profit of at least \$100,000.

17. _____

If $f(x) = x^2 + x - 6$ and $g(x) = 2x - 4$, find:

18. $(f \cdot g)(3)$

18. _____

19. $(f + g)(x)$

19. _____

20. $\left(\frac{f}{g}\right)(x)$

20. _____

Chapter 8 Test Form B

1. What is a relation?

2. Express the interval $\left[-10, -5\right)$ in set builder notation.

Graph.

3. $y = 2x - 3$

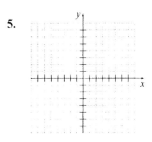

4. $f(x) = x^2 - 1$

5. $g(x) = |x| + 2$

Problems 6 and 7 refer to the following relation:
$\{(2, 6), (3, 6), (4, 6), (5, 6)\}$

6. Determine if the relation is a function.

7. Give the domain and range of the function or relation.

Problems 8 and 9 refer to the following relation:

8. Determine if the relation is a function.

9. Give the domain and range of the function or relation.

10. Determine the slope and y-intercept of the graph of the equation $4x + 5y = -4$.

1. _____

2. _____

3.

4.

5.

6. _____

7. _____

8. _____

9. _____

10. _____

Name:_____

11. Graph $y = \dfrac{-4}{5}x + 4$ using the x and y intercepts.

11.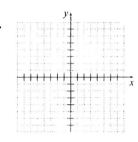

12. Graph $f(x) = 5$.

12.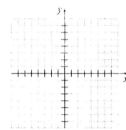

13. Determine the slope of the line through the points $(-1, 7)$ and $(4, 8)$.

13. _____

14. Determine if the graphs of the two equations are parallel, perpendicular, or neither.
$3x - 5y = -5$ and $5x - 3y = 6$

14. _____

15. Find an equation of the line shown in the following graph.

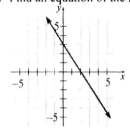

15. _____

16. Find the equation of the line through $(1, -4)$ that is perpendicular to the graph of $2x - 3y = 5$ in standard form.

16. _____

17. The yearly profit, p, for Waylan Publishing Company on the sales of a particular book can be estimated by the function
$p(x) = 12.1x - 40,000$, where x is the number of books produced
and sold. Use the function to estimate the number of books that the company must sell to make a profit of at least \$125,000.

17. _____

If $f(x) = x^2 + 2x$ and $g(x) = 3x - 9$, find:

18. $(f \cdot g)(-1)$

18. _____

19. $(f + g)(x)$

19. _____

20. $\left(\dfrac{f}{g}\right)(x)$

20. _____

Chapter 8 Test Form C

Name:_____

Date:_____

1. What is a function?

1. _____

2. Does x usually represent the dependent variable, or the independent variable?

2. _____

Graph.

3. $y = 3x - 2$

3.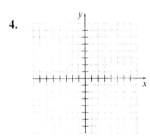

4. $y = x^3 - 1$

4.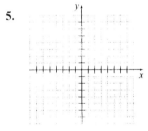

5. $g(x) = -|x|$

5.

Problems 6 and 7 refer to the following relation:
$\{(2, 5), (-2, 3), (1, 5), (0, -1)\}$

6. Determine if the relation is a function.

6. _____

7. Give the domain and range of the function or relation.

7. _____

Problems 8 and 9 refer to the following relation:

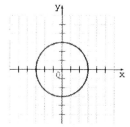

8. Determine if the relation is a function.   8. _____

9. Give the domain and range of the function or relation.   9. _____

Graph each equation using the *x*- and *y*-intercepts.

10. $y = -2x + 4$ 10.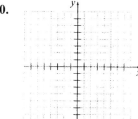

11. $x - 6y = 6$ 11.

12. Graph $f(x) = 0$.   12.

13. Write $\{x \mid x \leq -3 \text{ and } x \in I\}$ in roster form.   13. _____

14. Determine the slope of the line through the points   14. _____
    $(-2, -1)$ and $(0, 7)$.

15. Determine the slope and y-intercept of the graph of the equation $-4x - 3y = 7$.

15. _____

16. Determine if the graphs of the two equations are parallel, perpendicular, or neither. Explain your answer.
$x + 2y = 16$ and $x + 2y = -10$

16. _____

17. Find the equation of the line through $(-4, 0)$ that is parallel to the graph of $y = -3x + 6$ in slope-intercept form.

17. _____

18. Find the equation of the line through $(-7, -1)$ that is perpendicular to the graph of $x - y = 3$ in standard form.

18. _____

If $f(x) = x - 11$ and $g(x) = 2x^2 - 4$, find:

19. $(f - g)(-1)$

19. _____

20. $(f \cdot g)(-1)$

20. _____

21. $\left(\dfrac{f}{g}\right)(2)$

21. _____

If $f(x) = \dfrac{1}{2}x + 2$ and $g(x) = x^2 - 1$, find:

22. $(f \cdot g)(x)$.

22. _____

23. The domain of $(f \cdot g)(x)$.

23. _____

24. The domain of $\left(\dfrac{f}{g}\right)(x)$.

24. _____

25. The graph shows the monthly mean rainfall for Fort Ogelthorpe.

25. _____

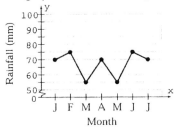

Between which two months did the monthly mean rainfall increase the most?

Chapter 8 Test Form D

1. Does y usually represent the dependent variable or the independent variable?

2. Describe the vertical line test.

Graph.

3. $y = -2x + 1$

4. $y = -x^3 + 1$

5. $f(x) = -|x| + 1$

1. _____

2. _____

3.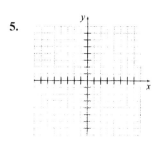

4.

5.

Problems 6 and 7 refer to the following relation:
{(−4, 0), (−3, 2), (−2, 4), (−3, 6)}

6. Determine if the relation is a function.

7. Give the domain and range of the function or relation.

6. _____

7. _____

Problems 8 and 9 refer to the following relation:

Lunch

Sam • ⟶ Bagel

Will • Pasta

Rita • Grapes

8. Determine if the relation is a function.

9. Give the domain and range of the function or relation.

8. _____

9. _____

Chapter 8 Test Form D *(cont.)*

Graph each equation using the *x*- and *y*-intercepts.

10. $y = x + 5$

10.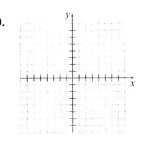

11. $3x - 2y = 6$

11.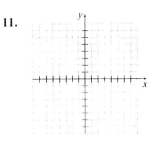

12. Graph $f(x) = 7$.

12.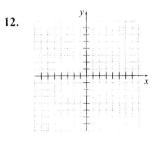

13. Express $\{x \mid x \leq 6 \text{ and } x \in N\}$ in roster form.

13. _____

14. Determine the slope of the line through the points (0, 5) and (3, 1).

14. _____

15. Determine the slope and *y*-intercept of the graph of the equation $7x + 9y = 10$.

15. _____

16. Determine if the graphs of the two equations are parallel, perpendicular, or neither. Explain your answer.
$2x - 5y = 5$ and $5x + 2y = -2$

16. _____

17. Find the equation of the line through (3, 0) that is parallel to the graph of $y = 2x + 1$ in slope-intercept form.

17. _____

18. Find the equation of the line through (0, 4) that is perpendicular to the graph of $y = \frac{1}{3}x + 8$ in standard form.

18. _____

If $f(x) = \frac{1}{2}x + 4$ and $g(x) = 3 - x^2$, find:

19. $(f \cdot g)(4)$

19. _____

20. $(f + g)(-2)$

20. _____

If $f(x) = 2x + 5$ and $g(x) = x^2 - 9$, to find:

21. $(f + g)(x)$

21. _____

22. The domain of $(f + g)(x)$

22. _____

23. The domain of $\left(\dfrac{f}{g}\right)(x)$

23. _____

24. Represent $\{x \mid x \geq 5\}$ using interval notation.

24. _____

25. The double line graph below compares rainfall in inches in Oregon and Washington in March 1993. Use the graph to determine the combined rainfall on March 7th.

25. _____

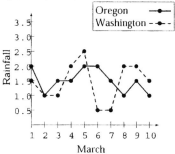

Chapter 8 Test Form E

1. True or False: If a horizontal line intersects a graph more than once, the graph is not a function.

2. Write the equation $y = 2x - 7$ in function notation.

1. _____

2. _____

Graph.

3. $y = \dfrac{1}{2}x + 1$

3.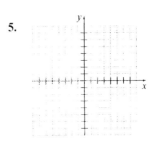

4. $y = -x^2 - 1$

4.

5. $f(x) = -|x| + 2$

5.

Problems 6 and 7 refer to the following relation:
$\{(8, -2), (6, -3), (4, -4), (2, -5)\}$

6. Determine if the relation is a function.

7. Give the domain and range of the function or relation.

6. _____

7. _____

Problems 8 and 9 refer to the following relation:

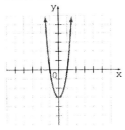

8. Determine if the relation is a function.

9. Give the domain and range of the function or relation.

8. _____

9. _____

Graph each equation using the *x*- and *y*-intercepts.

10. $y = -\dfrac{3}{4}x + 3$

10.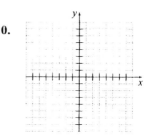

11. $5x - 2y = -10$

11.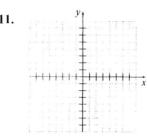

12. Graph $y = -8$.

12.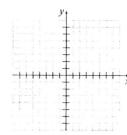

13. Express $\{x \mid x < -9\}$ using interval notation.

13. _____

14. Determine the slope of the line through the points $\left(\frac{2}{3}, 4\right)$ and $\left(-\frac{7}{3}, -1\right)$.

14. _____

15. Determine the slope and *y*-intercept of the graph of the equation $7x + 2y = 8$.

15. _____

16. Determine if the graphs of the two equations are parallel, perpendicular, or neither. Explain your answer.
$y = \dfrac{-2}{3}x + \dfrac{5}{3}$ and $y = \dfrac{3}{2}x - 7$

16. _____

17. Find the equation of the line through (5, –3) that is parallel to the graph of $y = 6x - 8$ in slope-intercept form.

17. _____

18. Find the equation of the line through (–3, 4) that is perpendicular to the graph of $9x - 3y = -12$ in standard form.

18. _____

If $f(x) = 6x - 12$ and $g(x) = x^2 - 1$, find:

19. $(f - g)(2)$

19. _____

20. $(f \cdot g)(-2)$

20. _____

21. $f(5) - g(-2)$

21. _____

22. The yearly profit, p, of a bagel company can be estimated by the function $p(x) = 0.12x - 6000$, where x is the number of bagels sold per year. Estimate the number of bagels sold if the company has a profit of $24,000.

22. _____

If $f(x) = x^2 - 2x + 7$ and $g(x) = x$, find:

23. The domain of $(f \cdot g)(x)$

23. _____

24. The domain of $\left(\dfrac{f}{g}\right)(x)$

24. _____

25. The double line graph below compares rainfall in inches in Oregon and Washington in March 1993. Use the graph to determine which state had the most rain on March 4th.

25. _____

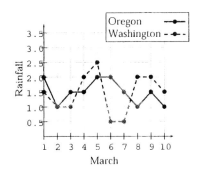

Chapter 8 Test Form F

1. Name the test used to determine whether a graph represents a function?

   1. _____

2. True or false? To find the y-intercept of a function, you set the x-value equal to 0 and solve for y.

   2. _____

3. Determine an equation for the given graph.

   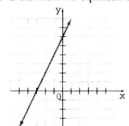

   3. _____

4. Determine the range.

   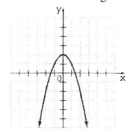

   4. _____

5. Determine the domain.

   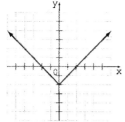

   5. _____

6.

   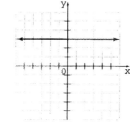

   6. _____

7. Represent $\{x \mid -0.5 < x \leq 4.25\}$ using interval notation.

   7. _____

8. Determine if $\{(0,1),(1,2),(2,3),(1,4)\}$ is a function.

   8. _____

Chapter 8 Test Form F *(cont.)*

9. Determine if $\{(9,0),(10,0),(11,0),(12,0)\}$ is a function.

9. _____

10. Determine the domain and range of $y = x + 1$.

10. _____

11. Determine the domain and range of $y = \dfrac{1}{x}$.

11. _____

Determine the x- and y-intercepts of each equation:

12. $5x - y = 3$

12. _____

13. $y = \dfrac{1}{2}x + 4$

13. _____

14. Determine the slope of the line through the points $(-2, 4)$ and $(-6, 12)$.

14. _____

15. Determine the slope and y-intercept of the graph of the equation $5x - 4y = 7$.

15. _____

16. Determine if these two equations have graphs that are perpendicular. $y = 2x + 3; \; y = -2x - 6$

16. _____

17. Find the equation of the line through $(-4, 3)$ that is parallel to the graph of $y = \dfrac{2}{3}x - 5$ in slope-intercept form.

17. _____

18. Find the equation of the line through $(0, 2)$ that is perpendicular to the graph of $4x - 2y = 8$ in standard form.

18. _____

If $f(x) = 3x - 1$ and $g(x) = x^2 - 2$, find:

19. $(f + g)(-1)$

19. _____

20. $\left(\dfrac{f}{g}\right)(1)$

20. _____

If $f(x) = 2x - 11$ and $g(x) = x^2 - 4$, find:

21. The domain of $(f \cdot g)(x)$

21. _____

22. The domain of $\left(\dfrac{f}{g}\right)(x)$

22. _____

23. The graph of $y = 3x + 5$ is translated down 8 units. Find the slope and y-intercept of the translated graph.

23. _____

24. The yearly profit, *p*, of a bagel company can be estimated by the function $p(x) = 0.13x - 6500$, where *x* is the number of bagels sold per year. Estimate the number of bagels sold if the company has a profit of $26,000.

24. _____

25. The graph shows the monthly mean rainfall for Fort McCoy.

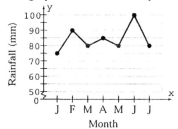

Between which two months did the monthly mean rainfall increase the most?

25. _____

Chapter 8 Test Form G

Name:_____

Date:_____

1. Which of the following is true for a function?
   (a) each y value corresponds to exactly one x value
   (b) each x value corresponds to exactly one y value
   (c) each y value corresponds to more than one x value
   (d) each x value corresponds to more than one y value

2. Which is the standard form of a linear equation?

   (a) $y = mx + b$ (b) $y - y_0 = m(x - x_0)$ (c) $ax + by = c$ (d) $ay + bx = c$

For problems 3–7, determine which equation describes the given graph.

3.

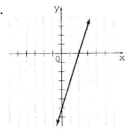

   (a) $y = 3x - 6$ (b) $y = 3x + 6$ (c) $y = -6x + 2$ (d) $y = -6x - 2$

4.

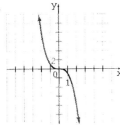

   (a) $y = x^2$ (b) $y = -x^2$ (c) $y = x^3$ (d) $y = -x^3$

5.

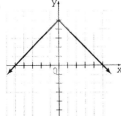

   (a) $f(x) = |x| + 5$ (b) $f(x) = -|x| - 5$ (c) $f(x) = -|x| + 5$ (d) $f(x) = -|x| - 5$

6.

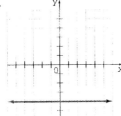

   (a) $y = -4$ (b) $x = -4$ (c) $y = -4x$ (d) $y = -\dfrac{1}{4}x$

Chapter 8 Test Form G *(cont.)* Name:_____

7. Express in set builder notation.

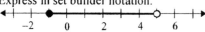

   (a) $\{x\,|\,-1 < x < 5\}$ (b) $\{x\,|\,-1 \le x < 5\}$ (c) $\{x\,|\,-1 \le x \le 5\}$ (d) $\{x\,|\,-1 < x \le 5\}$

8. Determine which of the following relations is a function.

   (a) $\{(1, 2), (-1, 2), (1, 3), (-1, 3)\}$ (b) $\{(1, 2), (-1, 2), (-3, 2), (1, 0)\}$

   (c) $\{(0, 1), (0, 2), (0, 3), (0, 4)\}$ (d) $\{(1, 0), (2, 0), (3, 0), (4, 0)\}$

9. Determine which of the following relations is a function.

   (a) $\{(1, 7), (2, 7), (3, 7), (4, 7)\}$ (b) $\{(2, 3), (3, 4), (5, 6), (7, 8)\}$

   (c) both are functions   (d) neither are functions

10. Determine the domain and range of $y = -\dfrac{1}{2}x - 2$.

   (a) D: $\mathbb{R}$, R: $\mathbb{R}$ (b) D: $\left\{x \middle| x > -\dfrac{1}{2}\right\}$, R: $\mathbb{R}$

   (c) D: $\left\{x \middle| x \le -\dfrac{1}{2}\right\}$, R: $\mathbb{R}$ (d) D: $\mathbb{R}$, R: $\left\{y \middle| y \ge -\dfrac{1}{2}\right\}$

11. Determine the domain and range of $y = |x| - 1$.

   (a) D: $\mathbb{R}$, R: $\{y\,|\,y \le -1\}$ (b) D: $\mathbb{R}$, R: $\{y\,|\,y \ge -1\}$

   (c) D: $\{x\,|\,x \le -1\}$, R: $\mathbb{R}$ (d) D: $\{x\,|\,x \ge -1\}$, R: $\mathbb{R}$

For problems 12 – 13, determine the slope and *y*-intercept of each equation:

12. $3x - 6y = -9$

   (a) slope $= \dfrac{1}{2}$; *y*-intercept $\left(0, \dfrac{3}{2}\right)$ (b) slope $= -\dfrac{1}{2}$; *y*-intercept $\left(0, -\dfrac{3}{2}\right)$

   (c) slope $= 3$; *y*-intercept $(0, -9)$ (d) slope $= 3$; *y*-intercept $\left(0, \dfrac{3}{2}\right)$

13. $y = 4x - 8$

   (a) slope $= -\dfrac{1}{4}$; *y*-intercept $(0, -8)$ (b) slope $= \dfrac{1}{4}$; *y*-intercept $(0, -8)$

   (c) slope $= -4$; *y*-intercept $(0, -8)$ (d) slope $= 4$; *y*-intercept $(0, -8)$

14. Determine the slope of the line through the points $(3, 7)$ and $(-2, -3)$.

   (a) -4 (b) -2 (c) 2 (d) 4

Chapter 8 Test Form G *(cont.)*   Name:_____

15. Determine the slope and y-intercept of the graph of the equation $x - 3y = -6$.

    (a) slope $= \dfrac{1}{3}$; y-intercept $(0, 2)$ **(b)** slope $= \dfrac{1}{3}$; y-intercept $(0, -2)$

    (c) slope $= 1$; y-intercept $(0, 6)$ **(d)** slope $= 1$; y-intercept $(0, -2)$

16. Determine which two equations have graphs that are parallel.

    (a) $y = 2x + 3$; $y = 2x - 5$ **(b)** $y = 2x + 3$; $y = -2x - 5$

    (c) $y = 2x + 3$; $y + \dfrac{1}{2}x = 5$ **(d)** $y = 2x + 3$; $y = -\dfrac{1}{2}x + 3$

17. Find the equation of the line through $(0, 3)$ that is parallel to the graph of $y = -2x - 7$ in slope-intercept form.

    (a) $y = 3x - 7$ **(b)** $y = -2x + 3$ **(c)** $y = \tfrac{1}{2}x + 3$ **(d)** $y = \tfrac{1}{2}x - 7$

18. Find the equation of the line through $(5, -2)$ that is perpendicular to the graph of $8x - 2y = 4$ in standard form.

    (a) $y = -\dfrac{1}{4}x + \dfrac{5}{4}$ **(b)** $y = -\dfrac{1}{4}x - \dfrac{3}{4}$ **(c)** $x + 4y = 5$ **(d)** $x + 4y = -3$

For problems 19 – 20, if $f(x) = 3x$ and $g(x) = 2 - x^2$, find:

19. $(f + g)(3)$

    (a) 9   **(b)** 3   **(c)** 2   **(d)** 5

20. $\left(\dfrac{f}{g}\right)(0)$

    (a) 0   **(b)** 1.5   **(c)** 3   **(d)** undefined

For problems 21 – 22, if $f(x) = 3x^2 + 1$ and $g(x) = x^2 - 9$, find:

21. The domain of $(f + g)(x)$

    (a) $\mathbb{R}$ **(b)** $\{x \mid x \neq 3\}$ **(c)** $\{x \mid x \neq \pm 3\}$ **(d)** $\{x \mid x \neq 9\}$

22. The domain of $\left(\dfrac{f}{g}\right)(x)$

    (a) $\mathbb{R}$ **(b)** $\{x \mid x \neq 3\}$ **(c)** $\{x \mid x \neq \pm 3\}$ **(d)** $\{x \mid x \neq 9\}$

23. Express $\{x \mid -15 < x \leq 22\}$ in interval notation.

    (a) $(-15, 22)$ **(b)** $(-15, 22]$ **(c)** $[-15, 22)$ **(d)** $[-15, 22]$

24. If the graph of $y = -\dfrac{2}{3}x + 5$ is translated up 6 units, what are the slope and y-intercept of the translated graph?

(a) $m = \dfrac{2}{3}$; $(0, -1)$ (b) $m = \dfrac{16}{3}$; $(0, 5)$ (c) $m = -\dfrac{2}{3}$; $(0, 11)$ (d) $m = -\dfrac{2}{3}$; $(0, 5)$

25. The double line graph below compares high temperatures in Honolulu and Miami in August. From the days listed, use the graph below to determine the day that Honolulu's temperature was the same as Miami's.

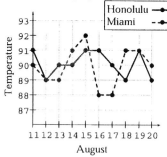

(a) August 15 (b) August 19 (c) August 16 (d) August 18

Chapter 8 Test Form H

1. The point where the graph crosses the *x*-axis is called

   (a) the *x*-intercept   **(b)** the *y*-intercept   **(c)** the origin   **(d)** the slope

2. To find the *x*-intercept of an equation, you

   (a) set $y = 0$ and solve for *x*   **(b)** set $x = 0$ and solve for *y*

   (c) set $x = 0$ and $y = 0$ and solve for the slope   **(d)** use the vertical line test

For problems 3–7, determine which equation describes the given graph.

3.

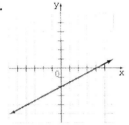

   (a) $y = 2x - 2$ **(b)** $y = 2x + 4$ **(c)** $y = \frac{1}{2}x - 2$ **(d)** $y = \frac{1}{2}x + 4$

4.

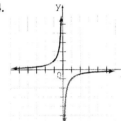

   (a) $y = \frac{1}{x}$ **(b)** $y = -\frac{1}{x}$ **(c)** $y = x^3$ **(d)** $y = -x^3$

5.

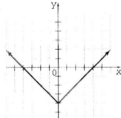

   (a) $f(x) = |x| + 4$ **(b)** $f(x) = -|x| - 4$ **(c)** $f(x) = -|x| + 4$ **(d)** $f(x) = |x| - 4$

6.

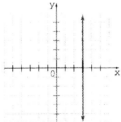

   (a) $x = 3$ **(b)** $x = -3$ **(c)** $y = 3$ **(d)** $y = -3$

7. Express in set builder notation.

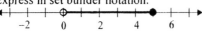

   (a) $\{x \mid 0 < x < 5\}$ (b) $\{x \mid 0 \le x < 5\}$ (c) $\{x \mid 0 < x \le 5\}$ (d) $\{x \mid 0 \le x \le 5\}$

8. Determine which of the following relations is a function.

   (a) $\{(2, -1), (2, 0), (3, -1), (3, 0)\}$ (b) $\{(2. -1), (0, 2), (3, -1), (0, 3)\}$
   (c) $\{(2, -1), (0, 2), (3, -1), (2, 4)\}$ (d) $\{(2, -1), (0, 2), (3, -1), (4, 2)\}$

9. Determine which of the following relations is a function.

   (a) $\{(1, 4), (2, 4), (3, 4), (4, 4)\}$ (b) $\{(1, 4), (2, -4), (3, 0), (1, 5)\}$
   (c) $\{(4, 1), (4, 2), (4, 3), (4, 4)\}$ (d) none of these are functions

10. Determine the domain and range of $y = x^2$.

    (a) D: $\mathbb{R}$, R: $\mathbb{R}$ (b) D: $\mathbb{R}$, R: $\{y \mid y \le 0\}$

    (c) D: $\mathbb{R}$, R: $\{y \mid y \ge 0\}$ (d) D: $\{x \mid x \ge 0\}$, R: $\mathbb{R}$

11. Determine the domain and range of $y = x + 3$.

    (a) D: $\mathbb{R}$, R: $\mathbb{R}$ (b) D: $\mathbb{R}$, R: $\{y \mid y > 3\}$

    (c) D: $\mathbb{R}$, R: $\{y \mid y \ge 3\}$ (d) D: $\{x \mid x \ge 3\}$, R: $\mathbb{R}$

For problems 12 – 13, determine the *x*- and *y*-intercepts of each equation:

12. $y = \frac{1}{2}x + 11$

    (a) $(11, 0); (0, \frac{1}{2})$ (b) $(\frac{1}{2}, 0); (0, -\frac{11}{2})$ (c) $(-22, 0); (0, 11)$ (d) $(\frac{11}{2}, 0); (0, 11)$

13. $5x - 10y = 30$

    (a) $\left(\frac{1}{2}, 0\right); (0, 30)$ (b) $\left(-\frac{1}{2}, 0\right); (0, 30)$ (c) $(-6, 0); (0, -3)$ (d) $(6, 0); (0, -3)$

14. Determine the slope of the line through the points $(-7, -11)$ and $(2, 7)$.

    (a) $-\frac{1}{2}$ (b) $\frac{1}{2}$ (c) -2 (d) 2

15. Determine the slope and *y*-intercept of the graph of the equation $x - 10y = 5$.

    (a) slope = 1; *y*-intercept $(0, 5)$ (b) slope = 1; *y*-intercept $\left(0, -\frac{1}{2}\right)$

    (c) slope $= \frac{1}{10}$; *y*-intercept $\left(0, -\frac{1}{2}\right)$ (d) slope $= -\frac{1}{10}$; *y*-intercept $\left(0, \frac{1}{2}\right)$

16. Determine which two equations have graphs that are perpendicular.

    (a) $y = 3x + 2$; $y = 3x - 7$ (b) $y = 3x + 2$; $y = -3x - 7$

    (c) $y = 3x + 2$; $y = \frac{1}{3}x - 7$ (d) $y = 3x + 2$; $y = -\frac{1}{3}x - 7$

17. Find the equation of the line through (–4, –4) that is parallel to the graph of $y = 2x + 9$ in slope-intercept form.

(a) $y = 2x + 4$ (b) $y = 2x + 8$ (c) $y = 2x - 4$ (d) $y = -\dfrac{1}{2}x + 4$

18. Find the equation of the line through (7, 7) that is perpendicular to the graph of $2x - y = 3$ in standard form.

(a) $x + 2y = 7$ (b) $x + 2y = 21$ (c) $2x - y = 7$ (d) $2x - y = -7$

For problems 19 – 20, if $f(x) = x^2 + x - 3$ and $g(x) = x^2 - 16$, find:

19. $(f - g)(0)$

(a) -19 (b) -13 (c) 13 (d) 19

20. $\left(\dfrac{f}{g}\right)(-4)$

(a) 15 (b) $-\dfrac{15}{12}$ (c) -15 (d) undefined

For problems 21 – 22, if $f(x) = x^3 + 2x^2 - 1$ and $g(x) = x^3 - 1$, find:

21. The domain of $(f \cdot g)(x)$

(a) $\mathbb{R}$ (b) $\{x \mid x \neq 1\}$ (c) $\{x \mid x \neq \pm 1\}$ (d) $\{x \mid x \geq -1\}$

22. The domain of $\left(\dfrac{f}{g}\right)(x)$

(a) $\mathbb{R}$ (b) $\{x \mid x \neq 1\}$ (c) $\{x \mid x \neq \pm 1\}$ (d) $\{x \mid x > -1\}$

23. Express $\{x \mid 4 \leq x \leq 17\}$ in interval notation.

(a) $(4, 17)$ (b) $[4, 17]$ (c) $(4, 17]$ (d) $[4, 17)$

24. If the graph of $y = -\dfrac{2}{3}x + 5$ is translated up 6 units, what are the slope and y-intercept of the translated graph?

(a) $m = \dfrac{2}{3}; (0, -1)$ (b) $m = \dfrac{16}{3}; (0, 5)$ (c) $m = -\dfrac{2}{3}; (0, 11)$ (d) $m = -\dfrac{2}{3}; (0, 5)$

25. What is the combined monthly average number of rigs drilling for oil for the years 1982 and 1984?

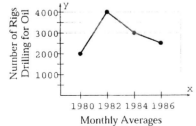

(a) 4000   (b) 4500   (c) 5500   (d) 7000

Cumulative Review 1 – 8
Test Form A

Name:_____

Date:_____

1. Evaluate $3x^2 - 9xy + 6y^2$ when $x = -2$ and $y = 5$.

    1. _____

2. Evaluate $10 - \left\{ 2 \left[4 - 3 \left(8^2 \div 4 \right) \right] \right\}$.

    2. _____

3. Solve $3x + 5(x - 7) = -2 \left[x - (3x - 2) \right]$.

    3. _____

4. Solve the formula $M = \dfrac{4R - 2V}{S}$ for V.

    4. _____

5. Write the equation $7x - 2y = 15$ in slope-intercept form.

    5. _____

6. Simplify $\left(5x^2 - 2x + 3 \right) - \left(-4x^2 + 3x - 1 \right)$.

    6. _____

7. Simplify $\left(\dfrac{12x^3 y^2}{4xy^4} \right)^4$.

    7. _____

8. Multiply $\left(5x^2 - 2x + 7 \right)(2x - 3)$.

    8. _____

9. Factor $4w^2 - 8w - 5w + 10$.

    9. _____

10. Solve $3x^2 = 2x + 5$.

    10. _____

11. Multiply $\dfrac{x^2 - 25}{2x^2 - 7x - 4} \cdot \dfrac{x^2 - 2x - 8}{x^2 + 7x + 10}$.

    11. _____

12. Subtract $\dfrac{2x}{x - 2} - \dfrac{5}{x + 4}$.

    12. _____

13. Add $\dfrac{3}{x^2 + x - 30} + \dfrac{7}{x^2 - 2x - 15}$.

    13. _____

14. Solve $\dfrac{x}{12} - \dfrac{x}{8} = \dfrac{1}{6}$.

    14. _____

15. Solve $\dfrac{2}{x + 4} - \dfrac{5}{x - 2} = \dfrac{4}{x^2 + 2x - 8}$.

    15. _____

16. Determine whether the graphs of the two given equations are parallel, perpendicular, or neither. Explain.
    $3x - 5y = 11$
    $10x + 6y = 15$

    16. _____

Cumulative Review 1 – 8
Test Form A *(cont.)*

17. Determine whether the following graph represents a function. Find the domain and range of the graph.

17. _____

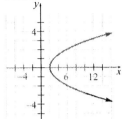

In 18 and 19, use $f(x) = x^2 - 4x + 7$ and $g(x) = 3x - 2$.

18. Find $(f + g)(x)$.

18. _____

19. Find $(f \cdot g)(4)$.

19. _____

20. The total consumption of natural gas in 2003 was 21.8 trillion cubic feet (2.18×10^{13}). The following pie chart shows the breakdown of consumption by sector. What was the natural gas consumption by the Industrial sector in 2003?

20. _____

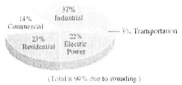

Natural Gas Consumption by Sector
(2.18 x 10¹³ cubic feet)

37% Industrial
14% Commercial
3% Transportation
23% Residential
22% Electric Power

(Total is 99% due to rounding.)

Source: Energy Information Administration

Cumulative Review 1 – 8 Test Form B

Name:_____

Date:_____

1. Evaluate $2x^2 + 7xy - y^2$ when $x = 3$ and $y = -2$.

   (a) 64   (b) 56   (c) -20 (d) -28

2. Evaluate $7 + \left\{3\left[10 - 3\left(12 - 6^2 \div 3\right)\right]\right\}$.

   (a) 7   (b) 100   (c) 109   (d) 37

3. Solve $2x - 4(x+1) = -8\left[4x - (3x+1)\right]$.

   (a) 8   (b) 2   (c) $\dfrac{2}{5}$ (d) $\dfrac{2}{7}$

4. Solve the formula $W = \dfrac{R+3}{2T}$ for R.

   (a) $R = 2TW - 3$ (b) $R = \dfrac{W-3}{2T}$ (c) $R = \dfrac{2T}{3W}$ (d) $R = \dfrac{2TW}{3}$

5. Write the equation $4x + 3y = 21$ in slope-intercept form.

   (a) $y = \dfrac{3}{4}x - \dfrac{21}{4}$ (b) $y = \dfrac{4}{3}x - 7$ (c) $y = -\dfrac{3}{4}x + \dfrac{21}{4}$ (d) $y = -\dfrac{4}{3}x + 7$

6. Simplify $\left(2x^2 + 5x - 3\right) - \left(3x^2 - 4x + 6\right)$.

   (a) $-x^2 + x + 3$ (b) $x^2 + 9x + 3$ (c) $-x^2 + 9x - 9$ (d) $x^2 - x - 9$

7. Simplify $\left(\dfrac{32x^2 y^5}{16x^3 y^2}\right)^3$.

   (a) $\dfrac{8y^9}{x^3}$ (b) $\dfrac{2y^6}{x^3}$ (c) $\dfrac{2y^3}{x}$ (d) $\dfrac{8x^6}{y}$

8. Multiply $\left(x^2 + x - 4\right)(3x - 2)$.

   (a) $3x^3 - 5x^2 - 10x + 8$ (b) $3x^3 + x^2 - 2x - 8$ (c) $3x^3 + x^2 - 14x + 8$ (d) $3x^3 + x^2 + 10x - 8$

9. Factor $6w^2 - 15w + 14w - 35$.

   (a) $(2w-7)(3w+5)$ (b) $(2w-5)(3w+7)$ (c) $(2w+5)(3w-7)$ (d) $(2w+7)(3w-5)$

10. Solve $x^2 = 4x + 12$.

    (a) $\{-2, 6\}$ (b) $\{-3, 12\}$ (c) $\{-6, 3\}$ (d) no solution

Cumulative Review 1 – 8
Test Form B *(cont.)*

Name:_____

11. Multiply $\dfrac{2x-1}{2x^2-9x-5} \cdot \dfrac{x+3}{6x^2+x-2}$.

    (a) $\dfrac{x+3}{(9x-5)(3x+2)(2x-1)}$ **(b)** $\dfrac{x+3}{(3x+2)(x-5)(2x+1)}$ **(c)** $\dfrac{2x-1}{(x-5)(2x+1)}$ **(d)** $\dfrac{x+3}{(2x+3)(x+5)(2x-1)}$

12. Subtract $\dfrac{3x}{x^2-4} - \dfrac{5}{x^2+x-2}$.

    (a) $\dfrac{-2x}{x^2+x-6}$ **(b)** $\dfrac{3x-5}{x-6}$ **(c)** $\dfrac{3x^2+11x-10}{x^2-4}$ **(d)** $\dfrac{3x^2+x+10}{x^2-4}$

13. Add $\dfrac{2x}{x-3} + \dfrac{x-1}{x+2}$.

    (a) $\dfrac{3x-1}{2x-1}$ **(b)** $\dfrac{2x-1}{x+5}$ **(c)** $\dfrac{3x^2+3}{x^2-x-6}$ **(d)** $\dfrac{3x^2+x+10}{x^2-4}$

14. Solve $\dfrac{4x}{5} - \dfrac{3x}{10} = -\dfrac{3}{2}$.

    (a) $x = -\dfrac{15}{2}$ **(b)** $x = -3$ **(c)** $x = -\dfrac{3}{2}$ **(d)** $x = -15$

15. Solve $\dfrac{6x-12}{x^2-4x-5} - \dfrac{2}{x-5} = \dfrac{2-x}{x+1}$.

    (a) $\{4\}$ **(b)** $\{3,4\}$ **(c)** $\{5,-1\}$ **(d)** $\{2\}$

16. Determine whether the graphs of the two given equations are parallel, perpendicular, or neither. Explain.
$$4x+2y=7$$
$$6x+3y=8$$

    (a) cannot determine   **(b)** neither   **(c)** perpendicular   **(d)** parallel

17. Determine whether the following graph represents a function. Find the domain and range of the graph.

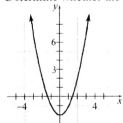

    (a) Function   **(b)** Not a Function   **(c)** Function   **(d)** Not a Function

        $D:\{x\,|\,x\geq-2\}$ $D:\mathbb{R}$ $D:\mathbb{R}$ $D:\{x\,|\,x\geq-2\}$

        $R:\mathbb{R}$ $R:\{y\,|\,y\geq-2\}$ $R:\{y\,|\,y\geq-2\}$ $R:\mathbb{R}$

In 18 and 19, use $f(x) = 2x^2 - 5x + 9$ and $g(x) = x - 4$.

18. Find $(f - g)(x)$.

(a) $2x^2 - 6x + 5$ (b) $2x^2 - 4x + 5$,   (c) $2x^2 - 6x + 13$ (d) $2x^2 - 4x - 13$

19. Find $\left(\dfrac{f}{g}\right)(3)$.

(a) 0   (b) −12   (c) 2   (d) −6

20. The total consumption of natural gas in 2003 was 21.8 trillion cubic feet $\left(2.18 \times 10^{13}\right)$. The following pie chart shows the breakdown of consumption by sector. How much more natural gas was consumed by the Residential sector than by the Commercial sector in 2003?

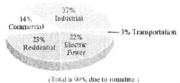

Natural Gas Consumption by Sector
$(2.18 \times 10^{13}$ cubic feet)

37% Industrial
14% Commercial
3% Transportation
23% Residential
22% Electric Power

(Total is 99% due to rounding.)

Source: Energy Information Administration

(a) 3.052×10^{12} ft^3 (b) 5.014×10^{12} ft^3 (c) 1.744×10^{12} ft^3 (d) 1.962×10^{12} ft^3

Chapter 9 Pretest Form A

Name:_____

Date:_____

1. Determine which, if any, of the ordered pairs satisfy the system of equations.

    $5x - y = -30$

    $3x + y = -10$

    (a) $(-4, 10)$ **(b)** $(-5, 5)$ **(c)** $(-2, 4)$ **(d)** $(-3, 15)$

    1. _____

2. Identify the system as consistent, inconsistent, or dependent. State whether the system has exactly one solution, no solution, or an infinite number of solutions.

    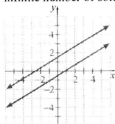

    2. _____

For questions 3 – 5, write each question in slope-intercept form. Then determine, without solving the system, whether the system has exactly one solution, no solution, or an infinite number of solutions.

3. $3x + 4y = -7$

    $3x - 4y = -7$

    3. _____

4. $2x - 5y = 8$

    $5y - 2x = -10$

    4. _____

5. $3x + 4y = -7$

    $12x + 16y = -28$

    5. _____

For questions 6 – 7, solve each system of equations graphically.

6. $x - 2y = 4$

    $x + y = -5$

    6. _____

    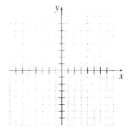

7. $2x + 3y = 12$

    $x - 3y = -3$

    7. _____

    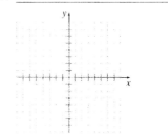

For questions 8 – 9, solve each system of equations using substitution.

8. $3x - y = -3$
   $2x + 5y = -2$

8. _____

9. $x + 6y = 2$
   $4x + 5y = -11$

9. _____

For questions 10 – 11, solve each system of equations using the addition method.

10. $x + y = -9$
    $x - y = 1$

10. _____

11. $8x + 5y = 1$
    $3x + 2y = 2$

11. _____

12. Solve the system of equations using the method of your choice.
    $x + 2y \quad = -1$
    $\quad y + 3z = 7$
    $2x \quad + 5z = 21$

12. _____

13. Consider the augmented matrix below. Show the results obtained by multiplying the elements in the first row by −2 and adding the products to their corresponding elements in the second row.
    $$\begin{bmatrix} 1 & 1 & 3 & | & 5 \\ 2 & 3 & 2 & | & 0 \\ 1 & 2 & 1 & | & -1 \end{bmatrix}$$

13. _____

For questions 14 – 15, refer to the following system of equations:
$$-4x + 3y = 11$$
$$5x + 2y = -8$$

14. Write the augmented matrix for the system.

14. _____

15. Solve the system by using matrices.

15. _____

For questions 16 – 17, evaluate each determinant.

16. $\begin{vmatrix} 14 & -5 \\ 2 & -1 \end{vmatrix}$

16. _____

17. $\begin{vmatrix} 3 & 6 & -4 \\ 2 & 0 & 3 \\ -1 & -5 & 2 \end{vmatrix}$

17. _____

18. Solve the system of equations using determinants and Cramer's Rule.
$$x + y = -11$$
$$x - 2y = 10$$

18. _____

19. The Modern Grocery has cashews that sell for $5.00 a pound and peanuts that sell for $2.75 a pound. How much of each must Albert, the grocer, mix to get 90 pounds of a mixture that he can sell for $3.00 per pound?

19. _____

20. How many liters each of a 60% chlorine solution and a 20% chlorine solution must be mixed in order to obtain 10 liters of a 50% chlorine solution?

20. _____

Chapter 9 Pretest Form B

Name:_____

Date:_____

1. Determine which, if any, of the ordered pairs satisfy the system of equations.

   $2x - 5y = 29$

   $5x - 4y = -4$

   (a) $(5, 4)$ **(b)** $(-8, -9)$ **(c)** $(17, 1)$ **(d)** $(7, -3)$

   1. _____

2. Identify the system as consistent, inconsistent, or dependent. State whether the system has exactly one solution, no solution, or an infinite number of solutions.

   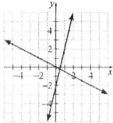

   2. _____

For questions 3 – 5, write each question in slope-intercept form. Then determine, without solving the system, whether the system has exactly one solution, no solution, or an infinite number of solutions.

3. $3x + 4y = -7$

   $3x + 4y = -6$

   3. _____

4. $3x - y = -3$

   $6x - 2y = -6$

   4. _____

5. $3x + 4y = -7$

   $x - 4y = -6$

   5. _____

For questions 6 – 7, solve each system of equations graphically.

6. $4x - 2y = -8$

   $3x + y = -1$

   6. _____

   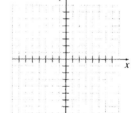

7. $2x + y = 1$

   $x - y = 5$

   7. _____

   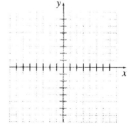

Chapter 9 Pretest Form B *(cont.)*

For questions 8 – 9, solve each system of equations using substitution.

8. $2x - y = -5$
   $x + 3y = 1$

8. _____

9. $14x + 7y = 6$
   $2x + y = -10$

9. _____

For questions 10 – 11, solve each system of equations using the addition method.

10. $x + 2y = -5$
   $x - y = 4$

10. _____

11. $8x - 7y = -16$
   $5x - 2y = -10$

11. _____

12. Solve the system of equations using the method of your choice.
   $x + y + z = 6$
   $x + 2y \quad = 3$
   $\quad 3y - z = -7$

12. _____

13. Consider the augmented matrix below. Show the results obtained by multiplying the elements in the first row by -4 and adding the products to their corresponding elements in the second row.

$$\begin{bmatrix} 1 & 2 & -1 & | & 0 \\ 4 & 1 & -2 & | & -3 \\ 2 & -1 & 3 & | & 5 \end{bmatrix}$$

13. _____

For questions 14 – 15, refer to the following system of equations:
   $x + 3y = 15$
   $4x - 3y = 10$

14. Write the augmented matrix for the system.

14. _____

15. Solve the system by using matrices.

15. _____

For questions 16 – 17, evaluate each determinant.

16. $\begin{vmatrix} 2 & 3 \\ -6 & -8 \end{vmatrix}$

16. _____

17. $\begin{vmatrix} 1 & 3 & 9 \\ 1 & 0 & 2 \\ 4 & 1 & 2 \end{vmatrix}$

17. _____

18. Solve the system of equations using determinants and Cramer's Rule.
$4x + y = -3$
$6x - 2y = 13$

18. _____

19. Mrs. Wong invests a total of $10,005 in two savings accounts. One account yields 9% simple interest and the other 10% simple interest. She would like to find the amount placed in each account if a total of $968.54 in interest is received after one year.

19. _____

20. Hearts Rent a Car Agency charges $50 per day plus 13 cents per mile to rent a midsized car. Mavis Rentals charges $59 per day plus 10 cents per mile to rent the same sized car. How many miles will have to be driven for the cost of the Harts car to equal the cost of the Mavis car?

20. _____

Mini-Lecture 9.1
Solving Systems of Equations Graphically

Learning Objectives:

1. Determine if an ordered pair is a solution to a system of equations.
2. Determine if a system of equations is consistent, inconsistent, or dependent.
3. Solve a system of equations graphically.
4. Key vocabulary: *system of linear equations, solution to a system of equations, consistent, inconsistent, dependent, independent*

Examples:

1. Determine which of the following ordered pairs satisfies the system of equations.

   $y = 3x - 8$

   $4x - 3y = 9$

   a) $(3, 1)$ b) $(5, 7)$

2. Determine whether each of the following systems has exactly one solution, no solution, or an infinite number of solutions. Identify each system as consistent, inconsistent, or dependent.

   a) $x - 2y = 10$ b) $10x - 15y = -3$ c) $3x - 4y = -20$

   $2y = x + 8$ $\dfrac{10}{3}x = 5y - 1$ $4y = 20 - 3x$

3. Solve each system of equations graphically.

   a) $x - 2y = -8$ b) $y = -3x + 5$ c) $x - \dfrac{2}{3}y = 1$

   $2x + y = -1$ $6x + 2y = -8$ $y = \dfrac{3}{2}x - 2$

   d) The U-Move rental company charges a daily fee of \$20, plus \$0.25 per mile to rent a moving van. The Pack-It rental company charges a daily fee of \$30, plus \$0.15 per mile. For how many miles will the cost for the two companies be the same? If Randy will have to drive 150 miles, which company will be the better choice for him?

Teaching Notes:

- Point out that the solution to a system of equations is an ordered pair.
- If graphing calculators are available, show students how to use the INTERSECT command.
- Remind students to check their solutions by substituting them into the original equations.

Answers: *1a) yes; 1b) no; 2a) no solution, inconsistent; 2b) infinite number of solutions, dependent; 2c) one solution, consistent; 3a) $(-2, 3)$; 3b) no solution, inconsistent; 3c) infinite number of solutions, dependent; 3d) 100 miles, Pack-It*

Mini-Lecture 9.2
Solving Systems of Equations by Substitution

Learning Objectives:

1. Solve systems of equations by substitution.
2. Key vocabulary: *substitution*

Examples:

1. Solve each system of equations by substitution.

   a) $x + 3y = 11$
      $2x + y = 2$

   b) $-2x + y = 10$
      $6x - 3y = 9$

   c) $y = 4x - 8$
      $x - \dfrac{1}{4}y = 2$

   d) $4x - 2y = 6$
      $3x + 5y = -15$

   e) $3x - 8y = 4$
      $6x - 4y = 5$

   f) There are 27 more children than adults at a party. There are 75 people in total at the party. How many children and how many adults are at the party?

Teaching Notes:

- Emphasize to students that they should solve for a convenient variable such as one with a coefficient of 1 or -1. For example, in Example 1b) above, it is most convenient to solve the first equation for y.
- Discuss with students that there can be more than one way to solve a problem.
- Students sometimes fail to find the entire solutions by stopping after finding the value of the first variable of choice. Remind students that the solution should be an ordered pair.
- Remind students to check their solutions by substituting them into the original equations.

Answers: 1a) $(-1, 4)$; 1b) no solution, inconsistent; 1c) infinite number of solutions, dependent; 1d) $(0, -3)$; 1e) $\left(\dfrac{2}{3}, -\dfrac{1}{4}\right)$; 1f) 51 children, 24 adults

Mini-Lecture 9.3
Solving Systems of Equations by the Addition Method

Learning Objectives:

1. Solve systems of equations by the addition method.
2. Key vocabulary: *addition (or elimination) method*

Examples:

1. Solve each system of equations by using the addition method.

a) $x + 5y = -7$
$3x - 5y = 19$

b) $x + 2y = 1$
$x + 4y = 5$

c) $3x + y = 10$
$x + 6y = -8$

d) $2x = -6y + 3$
$3y = 4x + 4$

e) $-2x + 5y = 6$
$3x - 7y = -9$

f) $-4x + 2y = 5$
$6x - 3y = 4$

g) $5x + 15y = 10$
$2x + 6y = 4$

h) $5x - 3y = 3$
$9x + 5y = 7$

i) The sum of a number and twice a second number is 27. When the first number is subtracted from four times the second number, the result is 3. Find the two numbers.

Teaching Notes:

- Emphasize to students that the purpose of this method is to obtain two equations whose sum will be an equation containing only one variable.
- Encourage students to contemplate which variable will be easiest to eliminate and what multiplications will be needed to make the elimination possible.
- Discuss with students that there can be more than one way to solve a problem.
- Students sometimes fail to find the entire solutions by stopping after finding the value of the first variable of choice. Remind students that the solution should be an ordered pair.
- Remind students to check their solutions by substituting them into the original equations.

Answers: *1a) (3,−2); 1b) (−3, 2); 1c) (4,−2); 1d) $\left(-\frac{1}{2}, \frac{2}{3}\right)$; 1e) (−3,0); 1f) no solution,*

inconsistent; 1g) infinite number of solutions, dependent; 1h) $\left(\frac{9}{13}, \frac{2}{13}\right)$; 1i) 17 and 5

Mini-Lecture 9.4
Solving Systems of Linear Equations in Three Variables

<u>Learning Objectives</u>:

1. Solve systems of linear equations in three variables.
2. Learn the geometric interpretation of a system of equations in three variables.
3. Recognize inconsistent and dependent systems.
4. Key vocabulary: *system of three linear equations, ordered triple*

<u>Examples</u>:

1. Solve the following system by substitution.
$$x = 5$$
$$x + y = 3$$
$$x + y - z = 2$$

2. Solve the following system of equations using the addition method.
$$3x - y + 2z = 39$$
$$2x - 2y - 3z = 18$$
$$4x + 3y + z = 15$$

3. Solve the following systems of equations using any method.

   a) $2a + 3b + 2c = 1$ $\qquad$ b) $2a + 3b + c = -4$

   $\quad 3a + 2c = 10$ $\qquad\qquad\qquad$ $3a + 2b + 4c = 19$

   $\quad\; a + b = 1$ $\qquad\qquad\qquad\quad$ $a - b - 3c = -19$

4. Determine whether the following systems are inconsistent, dependent or neither.

   a) $\quad a + 2b + 3c = 2$ $\qquad$ b) $\quad 2a + b + c = 3$

   $\quad 4a + 5b + 6c = 3$ $\qquad\qquad\qquad$ $a + 2b - 2c = 2$

   $\quad\; a + b + c = 1$ $\qquad\qquad\qquad\;$ $a - b + 3c = 1$

<u>Teaching Notes</u>:

- Have students use construction paper and scissors to illustrate the various ways that three planes can intersect with each other.

<u>Answers</u>: 1) (5, –2, 1); 2) (8, –7, 4); 3a) (3, –2, 0.5); 3b) (–1, –3, 7); 4a) inconsistent; 4b) dependent

Mini-Lecture 9.5
Systems of Linear Equations: Applications and Problem Solving

Learning Objectives:

1. Use systems of equations to solve applications.
2. Use linear systems in three variables to solve applications.

Examples:

1. Use systems of equations in two variables to solve the following applied problems.

   a) There were a total of 28 students in a philosophy class. The number of men was six more than the number of women. How many men and women were in the class?

   b) A boat traveling with the current can go 24 miles in 2 hours. Against the current, it takes 3 hours to go the same distance. Find the rate of the boat in calm water and the rate of the current.

   c) Benito receives a weekly salary plus a commission, which is a percentage of his sales. One week, with sales of $2000, Benito's gross pay was $980. On another week with sales of $3500, his gross pay was $1115. Find his weekly salary and his commission rate.

   d) Two trains are 330 miles apart, and their speeds differ by 20 mph. They travel toward each other and meet in 3 hours. Find the speed of each train.

   e) How many gallons of a 3% salt solution must be mixed with 50 gallons of a 7% solution to obtain a 5% solution?

2. Use systems of equations in three variables to solve the following applied problems.

   a) Melissa works at Fitz's Pizza. Her last three orders were 5 slices of pizza, 2 salads, and 2 sodas for $9.75; 3 slices of pizza, 2 salads and 1 soda for $7.15; and 2 slices of pizza, 1 salad and 1 soda for a total of $4.35. What is the price of 1 slice of pizza, 1 salad and 1 soda sold individually?

   b) The sum of the measures of the three angles of a triangle is $180°$. The middle-sized angle measures $2°$ more than 3 times the smallest angle. The middle-sized angle measures $43°$ less than the largest angle. Find the measure of the middle-sized angle.

Teaching Notes:

- In applied problems involving money, remind students to be consistent when writing their equations; express coefficients as decimals (dollars) or whole numbers (pennies), but do not mix the two coefficient forms.

Answers: 1a) 17 men, 11 women; 1b) boat's rate in calm water = 10 mph, current's rate = 2 mph; 1c) weekly salary = $800, rate of commission = 9%; 1d) 65 mph and 45 mph; 1e) 50 gallons; 2a) each slice of pizza = $1.05, each salad = $1.75, each soda = $0.50; 2b) 59°

Mini-Lecture 9.6
Solving Systems of Equations Using Matrices

Learning Objectives:

1. Write an augmented matrix.
2. Solve systems of linear equations.
3. Solve systems of linear equations in three variables.
4. Recognize inconsistent and dependent systems.
5. Key vocabulary: *matrix, matrices, elements, square matrix, augmented matrix, triangular form, row transformations*

Examples:

1. Write an augmented matrix for the following system of equations.

$$4x - 3y = 1$$
$$5x + 2y = 4$$

2. Solve the following systems of equations in two variables using matrices.

a) $x + y = 1$
$2x - y = 8$

b) $3x + 2y = 4$
$3x + y = 8$

c) $x + 7y = -68$
$4x - 6y = 34$

3. Solve the following systems of equations in three variables using matrices.

a) $x + y + z = 6$
$-2x - y + z = -2$
$x - 2y - z = 4$

b) $x + y + z = 0$
$-2x - y + z = -12$
$x - 2y - z = 1$

4. Use matrices to determine if the following systems are dependent or inconsistent.

a) $3x - 2y = 8$
$-3x + 2y = -12$

b) $2x - 5y = 19$
$-0.2x + 0.5y = -1.9$

c) $2x + y - 3z = 4$
$-4x + y - z = 5$
$3y - 7z = -2$

Teaching Notes:

- Using row transformations on augmented matrices is a very tedious task. Encourage students to check their arithmetic after every step.
- When practicing the new procedure, also encourage students to have other students check for errors when they themselves cannot identify them.

Answers: 1) $\begin{bmatrix} 4 & -3 & | & 1 \\ 5 & 2 & | & 4 \end{bmatrix}$; *2a) (3, –2); 2b) (4, –4); 2c) (–5, –9); 3a) (4, –2, 4); 3b) (2, 3, –5); 4a) inconsistent; 4b) dependent; 4c) inconsistent*

Mini-Lecture 9.7
Solving Systems of Equations Using Determinants and Cramer's Rule

Learning Objectives:

1. Evaluate a determinant of a 2×2 matrix.
2. Use Cramer's rule.
3. Evaluate a determinant of a 3×3 matrix.
4. Use Cramer's rule with systems in three variables.
5. Key vocabulary: *determinant, Cramer's rule, minor determinant, expansion by minors*

Examples:

1. Evaluate the following determinants.

   a) $\begin{vmatrix} 5 & -2 \\ 4 & 3 \end{vmatrix}$

   b) $\begin{vmatrix} 3 & -7 \\ -2 & 5 \end{vmatrix}$

   c) $\begin{vmatrix} -6 & 3 \\ -1 & -4 \end{vmatrix}$

2. Solve the following systems of equations using Cramer's Rule.

   a) $x - 4y = -7$

   $3x + y = -8$

   b) $x + y = 3$

   $2x - y = 5$

   c) $2x + 4y = 8$

   $x + 2y = 6$

3. Evaluate the following determinants.

   a) $\begin{vmatrix} -1 & -2 & -3 \\ 3 & 4 & 2 \\ 0 & 1 & 2 \end{vmatrix}$

   b) $\begin{vmatrix} 1 & 2 & 2 \\ 2 & 1 & 0 \\ 3 & 3 & 1 \end{vmatrix}$

   c) $\begin{vmatrix} -1 & 2 & 1 \\ 2 & 1 & -3 \\ 1 & 1 & 1 \end{vmatrix}$

4. Solve the following systems of equations using Cramer's Rule.

   a) $2x + y - z = -8$

   $4x - y + 2z = -3$

   $-3x + y + 2z = 5$

   b) $3x - 2y + 4z = 15$

   $x - y + z = 3$

   $x + 4y - 5z = 0$

   c) $x - 4y - 3z = -1$

   $-3x + 12y + 9z = 3$

   $2x - 10y - 7z = 5$

Teaching Notes:

- Emphasize that $\begin{bmatrix} 5 & -2 \\ 4 & 3 \end{bmatrix}$ does not have a numerical value but $\begin{vmatrix} 5 & -2 \\ 4 & 3 \end{vmatrix}$ does.

- Indicate that Cramer's Rule is especially useful when the solution to a system of equations involves fractional values.

Answers: 1a) 23; 1b) 1; 1c) 27; 2a) (–3, 1); 2b) $\left(\dfrac{8}{3}, \dfrac{1}{3}\right)$; 2c) no solution; 3a) –3; 3b) 3; 3c) –13;

4a) (–2, –3, 1); 4b) (3, 3, 3); 4c) infinite number of solutions

Additional Exercises 9.1

Name:_____

Date:_____

Decide which ordered pair, if any, is a solution of each system of equations.

1. $y = 6x - 25$
   $3x + y = 11$

    (a) $(3, -7)$ **(b)** $(4, -1)$ **(c)** $(5, -4)$

1. _____

2. $y = 4x + 9$
   $3x - 5y = 6$

    (a) $(-1, 5)$ **(b)** $(2, 0)$ **(c)** $(-3, -3)$

2. _____

3. $y = 2x - 10$
   $9x + 4y = -6$

    (a) $(2, -6)$ **(b)** $(2, -6)$ **(c)** $(5, 0)$

3. _____

4. Does $(7, -2)$ satisfy the system of linear equations? Answer yes or no.
   $3x + 8y = 5$
   $y = x + 9$

4. _____

5. Does $(-3, 4)$ satisfy the system of linear equations? Answer yes or no.
   $2x + 5y = 14$
   $y = x + 1$

5. _____

6. Does $\left(\frac{2}{3}, -\frac{1}{4}\right)$ satisfy the system of linear equations? Answer yes or no.
   $9x + 8y = 4$
   $3x + 12y = -1$

6. _____

Identify each system as consistent, inconsistent, or dependent. State whether the system has exactly one solution, no solution or an infinite number of solutions.

7.

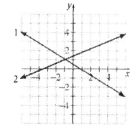

7. _____

8.

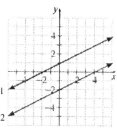

8. _____

9.

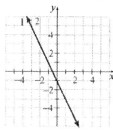

9. _____

318

Additional Exercises 9.1 *(cont.)*

Name:_____

Express each equation in slope-intercept form. Then determine, without graphing the equations, whether the system has exactly one solution, no solution or an infinite number of solutions.

10. $2x + 3y = -11$
$-3x - 4y = 16$

10. _____

11. $3x - 4y = 3$
$3x - 4y = 10$

11. _____

12. $-4x + 2y = 12$
$x - 2y = -9$

12. _____

13. $3x + 2y = 0$
$9x + 6y = 0$

13. _____

14. $3x + 4y = -9$
$3x + 4y = -1$

14. _____

15. $-3x + 2y = 5$
$-4x + y = 10$

15. _____

Determine the solution to each system graphically. If the system is dependent or inconsistent, so state.

16. $4x + y = 4$
$x = 1 - \dfrac{1}{4}y$

16. _____

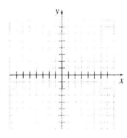

17. $x + y = -1$
$y = 2x + 8$

17. _____

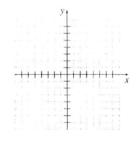

Additional Exercises 9.1 *(cont.)*

18. $x + y = 4$

$\quad\quad y = 3x - 4$

18. _____

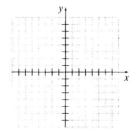

19. $2y = x$

$\quad\quad y = \dfrac{1}{2}x - 3$

19. _____

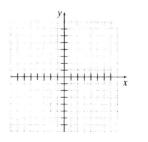

20. $x - 2y = -2$

$\quad\quad x = -2 + 2y$

20. _____

Additional Exercises 9.2

Find the solution to each system of equations using substitution.

1. $2x + 5y = -10$
   $y = -3x + 11$

2. $5x - 3y = 19$
   $y = 2x - 7$

3. $6x + 5y = 14$
   $2x - 3y = 42$

4. $3x + 6y = 15$
   $x = -2y + 1$

5. $-4x - y = -26$
   $-7x - 7y = -35$

6. $x + 2y = -5$
   $y = 3x + 1$

7. $9x - 5y = 2$
   $-4x - 9y = -8$

8. $6x + 6y = -3$
   $y = -x$

9. $6x - 4y = 16$
   $7x + 7y = 112$

10. $3x + 2y = 0$
    $x - 2y = 3$

11. $3x + 4y = 6$
    $y = 4x + 11$

12. $2x + 3y = 22$
    $x - 3y = -7$

13. $3x - 4y = 8$
    $y = -x + 5$

14. $-5x - y = 1$
    $4x + 8y = -5$

15. $4x + 4y = -3$
    $y = -x$

16. $7x + 5y = -24$
    $-6x - 5y = 27$

17. $-4x - y = 4$
    $x = -1 - \dfrac{1}{4}y$

18. $x + 4y = 11$
    $y = 4x + 7$

19. $x - 2y = -8$
    $x - 2y = 7$

20. $2x + y = 2$
    $-x - y = 1$

1. _____

2. _____

3. _____

4. _____

5. _____

6. _____

7. _____

8. _____

9. _____

10. _____

11. _____

12. _____

13. _____

14. _____

15. _____

16. _____

17. _____

18. _____

19. _____

20. _____

Additional Exercises 9.3

Solve each system of equations using the addition method.

1. $x - 2y = 13$
   $3x + 2y = 15$

2. $-2x + y = 8$
   $3x + 4y = -12$

3. $-4x + 2y = 14$
   $2x + y = 3$

4. $4x + y = 2$
   $10x - 5y = -1$

5. $2x - y = 4$
   $x + y = 5$

6. $4x - 3y = 1$
   $3x - 4y = 4$

7. $4x + 7y = -49$
   $-5x - 5y = 50$

8. $2x - 3y = -21$
   $3x + 3y = 6$

9. $2x + 3y = 3$
   $x - 4y = -4$

10. $8x + 3y = -16$
    $-4x + 4y = -36$

11. $3x - 4y = 3$
    $5x + 4y = 10$

12. $4x - 3y = 11$
    $3x + 3y = -18$

13. $3x - 4y = 3$
    $9x - 12y = 9$

14. $3x + 2y = 0$
    $3x + 2y = 3$

15. $7x - 3y = -3$
    $6x + 4y = 3$

16. $4x - 3y = 3$
    $5x - 2y = 4$

17. $8x - 2y = 70$
    $-5x + 9y = -67$

18. $x - 2y = -8$
    $2x - 4y = -16$

19. $y = -2x + 13$
    $3x + y = 20$

20. $3x + 2y = 0$
    $3x - 2y = 0$

1. _____

2. _____

3. _____

4. _____

5. _____

6. _____

7. _____

8. _____

9. _____

10. _____

11. _____

12. _____

13. _____

14. _____

15. _____

16. _____

17. _____

18. _____

19. _____

20. _____

Additional Exercises 9.4

Solve by substitution.

1. $3x - y + z = 9$
$x + y + 2z = 14$
$3x + 2y - z = 5$

2. $x = 5$
$6x + 6y = 24$
$-x + 6y - 7z = -46$

3. $x - y + z = 6$
$2x - 2y - 2z = 12$
$x + 2y - z = 0$

4. $x = 3$
$x - 2y = 7$
$x + y - 4z = -3$

5. $x + 2y + z = -9$
$3x + 2y + z = -15$
$x - 2y - z = 3$

6. $2y + z = 7$
$2x - z = 3$
$x - y = 3$

Solve using the addition method.

7. $9x - 6y + 6z = 1$
$6x + 3y - 3z = -4$
$3x - 3y + 6z = 2$

8. $2x - y + 3z = 9$
$x + 4y + 4z = 5$
$3x + 2y + 2z = 5$

9. $3x + y - 2z = -2$
$10x + 5y - 5z = -6$
$5x - 5y + 10z = -2$

10. $x + y + z = 0$
$-2x - y + z = -12$
$x - 2y - z = 1$

11. $3x - y + z = 11$
$x + 4y - 2z = -12$
$2x + 2y - z = -3$

12. $x + y + z = -9$
$-2x - y + z = 13$
$x - 2y - z = -1$

1. _____

2. _____

3. _____

4. _____

5. _____

6. _____

7. _____

8. _____

9. _____

10. _____

11. _____

12. _____

Additional Exercises 9.4 *(cont.)*   Name:_____

Determine whether the following systems are inconsistent, dependent, or neither.

13. $2x + y + 2z = -5$
$-4x + y + z = 2$
$3y + 5z = 5$

13. _____

14. $2x + y - 2z = -1$
$-4x + y - 3z = -6$
$3y - 7z = -4$

14. _____

15. $2x + y + z = 4$
$-x + 2y + z = 3$
$9x + 2y + 3z = 13$

15. _____

16. $2x + y + z = 4$
$-x + 2y + z = 3$
$7x + 6y + 5z = 19$

16. _____

17. $2x + y - z = -5$
$-4x + y - z = 4$
$3y - 3z = 3$

17. _____

18. $2x + y - 3z = 4$
$-4x + y - z = 5$
$3y - 7z = -2$

18. _____

19. $2x + y + z = 4$
$-x + 2y + z = 3$
$7x + 6y + 5z = 19$

19. _____

20. $2x + y + z = 4$
$-x + 2y + z = 3$
$9x + 2y + 3z = 13$

20. _____

Additional Exercises 9.5

Name:_____

Date:_____

1. Tickets to a local movie were sold at $5.00 for adults and $3.50 for students. If 150 tickets were sold for a total of $615.00, how many adult tickets were sold?

   1. _____

2. Max has cashews that sell for $4.75 a pound and peanuts that sell for $2.75 a pound. How much of each must he mix to get 80 pounds of a mixture that he can sell for $3.00 per pound?

   2. _____

3. A collection of quarters and dimes contains 44 coins and has a total value of $6.50. How many coins of each kind are in the collection?

   3. _____

4. Clare has cashews that sell for $3.50 a pound and peanuts that sell for $2.00 a pound. How much of each must she mix to obtain 60 pounds of a mixture that she can sell for $3.00 per pound?

   4. _____

5. Tickets for a band concert cost $8 for the main floor and $6 for the balcony. If 1125 tickets were sold and the ticket sales totaled $7700, how many tickets of each type were sold?

   5. _____

6. A physician invests $24,000 in two bonds. If one bond yields 6% and the other yields 12%, how much is invested in each if the annual income from both bonds is $1980?

   6. _____

7. An actuary invests $26,000 in two bonds. If one bond yields 12% and the other yields 7%, how much is invested in each if the annual income from both bonds is $2420?

   7. _____

8. An engineer invests $27,000 in two bonds. If one bond yields 7% and the other yields 9%, how much is invested in each if the annual income from both bonds is $2130?

   8. _____

9. Linda's Bakery sells three kinds of cookies: chocolate chip cookies at 15 cents each, oatmeal cookies at 20 cents each, and peanut butter cookies at 25 cents each. Charles buys some of each kind and chooses three times as many peanut butter cookies as chocolate chip cookies. If he spends $4.10 on 19 cookies, how many oatmeal cookies did he buy?

   9. _____

10. The sum of three numbers is 161. The second number is 7 more than the first number, and the third number is 5 times the first number. Find the second number.

    10. _____

11. The sum of the measures of the three angles of a triangle is 180°. The middle-sized angle measures 89° less than the largest angle. The middle-sized angle measures 14° less than 3 times the measure of the smallest angle. Find the measure of the middle-sized angle.

    11. _____

Additional Exercises 9.5 *(cont.)*

12. The sum of three numbers is 20. The first number is the sum of second and third. The third number is three times the first. What are the three numbers?

12. _____

13. Tasty Bakery sells three kinds of cookies: chocolate chip cookies at 30 cents each, oatmeal cookies at 35 cents each, and peanut butter cookies at 40 cents each. Candace buys some of each kind and chooses twice as many peanut butter cookies as chocolate chip cookies. If she spends $5.75 on 16 cookies, how many oatmeal cookies did she buy?

13. _____

14. The sum of three numbers if 159. The second number is 4 more than the first number, and the third number is 3 times the first number. Find the numbers.

14. _____

15. An investor has $70,000 invested in mutual funds, bonds, and a fast food franchise. She has twice as much invested in bonds as in mutual funds. Last year, the mutual funds paid a 2% dividend, the bonds paid 10%, and the fast food franchise paid 6%; her total dividend income was $4800. How much is invested in each of the three investments?

15. _____

16. The sum of the measures of the three angles of a triangle is 180°. The middle-sized angle measures 8° more than 2 times the smallest angle. The middle-sized angle measures 59° less than the largest angle. Find the measure of the middle-sized angle.

16. _____

17. Tasty Bakery sells three kinds of cookies: chocolate chip cookies at 20 cents each, oatmeal cookies at 25 cents each, and peanut butter cookies at 30 cents each. Melissa buys some of each kind and chooses twice as many peanut butter cookies as chocolate chip cookies. If she spends $2.85 on 11 cookies, how many peanut butter cookies did she buy?

17. _____

18. The sum of three number is 286. The second number is 7 more than the first number, and the third number is 7 times the first number. Find the numbers.

18. _____

19. The sum of the measures of the three angles of a triangle is 180°. The middle-sized angle measures 4° more than 4 times the smallest angle. The middle-sized angle measures 64° less than the largest angle. Find the measure of the middle-sized angle.

19. _____

20. The sum of the measures of the three angles of a triangle is 180°. The middle-sized angle measures 1° more than 2 times the smallest angle. The middle-sized angle measures 38° less than the largest angle. Find the measure of the middle-sized angle.

20. _____

Additional Exercises 9.6

Solve the system using matrices.

1. $2x - 4y = -16$
   $3x + 4y = 6$

2. $x - 2y = -2$
   $y = 4x - 6$

3. $4x + 4y = 2$
   $8x + 8y = -5$

4. $2x + 5y = -5$
   $4x + 10y = 4$

5. $3x + 2y = 4$
   $3x + y = 8$

6. $2x + 5y = -15$
   $x - 5y = 30$

7. $3x + 4y = 14$
   $y = 2x - 2$

8. $4x - 3y = 6$
   $2x - 5y = -4$

9. $3x - 9y = 12$
   $2x - 6y = 8$

10. $4x - 3y = 3$
    $x - 4y = 2$

1. _____

2. _____

3. _____

4. _____

5. _____

6. _____

7. _____

8. _____

9. _____

10. _____

Solve the system using matrices.

11. $6x + 4y + 7z = 4$
    $-x + 9y - 2z = 51$
    $5x - y + z = 1$

12. $x + y + z = 6$
    $-2x - y + z = -2$
    $x - 2y - z = 4$

11. _____

12. _____

Additional Exercises 9.6 *(cont.)*

13. $-3x - 4y + z = -19$
$-x - 3y - z = -1$
$-x - 2y - 3z = 7$

13. _____

14. $x = 7$
$7x + 3y = 40$
$-x + 7y - 3z = -22$

14. _____

15. $-3x + 4y + 4z = -22$
$3x + 3y - 3z = 3$
$-4x - 4y - 3z = 31$

15. _____

16. $-3x - 3y + 4z = -1$
$x - 3y + 3z = 4$
$-4x - 3y - 3z = 5$

16. _____

17. $-4x + 2y - 2z = 18$
$2x + y + 4z = 15$
$x + 3y + 3z = 21$

17. _____

18. $5x - 2y - 5z = -42$
$-x - 8y + 9z = -10$
$3x - y + z = -8$

18. _____

19. $3x - 2y + z = 9$
$6x - 4y + 2z = 18$
$x - \dfrac{2}{3}y + \dfrac{1}{3}z = 3$

19. _____

20. $4x - 8y + 6z = 2$
$6x - 12y + 9z = 5$
$3x + 8y - 4z = -1$

20. _____

Additional Exercises 9.7

Name:_____

Date:_____

Evaluate the determinants.

1. $\begin{vmatrix} 4 & 3 \\ -2 & 1 \end{vmatrix}$

1. _____

2. $\begin{vmatrix} 4 & 7 \\ 2 & -3 \end{vmatrix}$

2. _____

3. $\begin{vmatrix} 1 & 2 & -5 \\ 3 & 2 & 1 \\ 2 & 2 & 4 \end{vmatrix}$

3. _____

4. $\begin{vmatrix} 2 & 5 & 4 \\ 4 & 2 & 3 \\ 1 & 1 & 3 \end{vmatrix}$

4. _____

5. $\begin{vmatrix} 2 & -3 & 2 \\ 1 & 3 & -1 \\ 0 & -2 & 2 \end{vmatrix}$

5. _____

6. $\begin{vmatrix} 3 & 3 & 1 \\ 4 & 3 & 2 \\ 1 & 1 & 2 \end{vmatrix}$

6. _____

7. $\begin{vmatrix} 1 & -7 & -1 \\ -2 & 6 & 2 \\ 9 & 4 & -9 \end{vmatrix}$

7. _____

8. $\begin{vmatrix} 7 & 12 \\ 7 & 12 \end{vmatrix}$

8. _____

9. $\begin{vmatrix} -1 & 4 \\ -2 & 6 \end{vmatrix}$

9. _____

10. $\begin{vmatrix} 1 & 4 & -3 \\ 0 & -2 & 8 \\ 3 & 6 & 0 \end{vmatrix}$

10. _____

Additional Exercises 9.7 *(cont.)*

Name:_____

Solve the system of equations using determinants.

11. $x - 5y = -5$
    $4x + 5y = -4$

11. _____

12. $5x + 3y = -5$
    $5x + y = -1$

12. _____

13. $8x + y = 36$
    $5x + y = 21$

13. _____

14. $x = 3y + 4$
    $-2x + 6y = -8$

14. _____

15. $x + 3y - z = 8$
    $3x + 3y + 3z = -6$
    $2x + 3y + 3z = -3$

15. _____

16. $x + 2y + z = 5$
    $x + y - z = 6$
    $5x + 8y + z = 7$

16. _____

17. $x + 2y + z = 8$
    $x + y - z = 9$
    $9x + 14y + z = 10$

17. _____

18. $2x + y - 3z = -2$
    $x - 2y - 3z = -13$
    $3x - 2y - z = -3$

18. _____

19. $2x - y - 3z = 17$
    $3x + y + z = -2$
    $x + y + 2z = -9$

19. _____

20. $2x - y + 3z = 2$
    $x + 2y - 3z = -10$
    $3x - 3y - 3z = -21$

20. _____

Chapter 9 Test Form A

Name:_____

Date:_____

1. Determine which, if any, of the ordered pairs satisfy the system of equations.
   $2x + y = 9$
   $2x - 5y = 15$

   (a) $(1, -5)$ **(b)** $(-5, 1)$ **(c)** $(5, -1)$ **(d)** $(-1, -5)$

   1. _____

2. Identify the system as consistent, inconsistent, or dependent. State whether the system has exactly one solution, no solution, or an infinite number of solutions.

   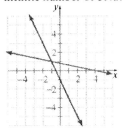

   2. _____

For questions 3–5, write each question in slope-intercept form. Then determine, without solving the system, whether the system has exactly one solution, no solution, or an infinite number of solutions.

3. $x - 2y = 14$
   $2x - 4y = 3$

   3. _____

4. $2x + 6y = -4$
   $-3x - 9y = 6$

   4. _____

5. $x - 3y = 9$
   $5x + 2y = 11$

   5. _____

For questions 6 – 7, solve each system of equations graphically.

6. $2x + y = 5$
   $x - y = 4$

   6. _____

   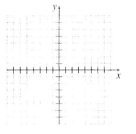

7. $3x - y = -6$
   $x - 2y = -2$

   7. _____

   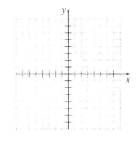

Chapter 9 Test Form A *(cont.)*

Name:_____

For questions 8 – 9, solve each system of equations using substitution.

8. $y - x = 4$
   $x - 5y = -12$

   8. _____

9. $2x - 4y = 0$
   $x + 2y = 5$

   9. _____

For questions 10 – 11, solve each system of equations using the addition method.

10. $3x + 4y = -10$
    $2x + 3y = -6$

    10. _____

11. $4x + 2y = 8$
    $6x - 2y = -13$

    11. _____

12. Solve the system of equations using the method of your choice.
    $x + y + z = 2$
    $-3y + 4z = 11$
    $3x - 4y + 2z = 11$

    12. _____

13. Write the augmented matrix for the system of equations.
    $x - y + z = 6$
    $2x + 3y + 2z = 2$
    $5y + 4z = -2$

    13. _____

14. Consider the augmented matrix below. Show the results obtained by multiplying the elements in the first row by –1 and adding the products to their corresponding elements in the second row.
    $$\begin{bmatrix} 1 & 1 & 1 & | & 4 \\ 1 & -2 & -1 & | & 1 \\ 2 & -1 & -2 & | & -1 \end{bmatrix}$$

    14. _____

15. Solve the system of equations using matrices.
    $x + y = 9$
    $2x - y = -3$

    15. _____

Name:_____

For questions 16 – 17, evaluate each determinant.

16. $\begin{vmatrix} 5 & -2 \\ -1 & 7 \end{vmatrix}$

16. _____

17. $\begin{vmatrix} 0 & 2 & 0 \\ 3 & -1 & 1 \\ 1 & -2 & 2 \end{vmatrix}$

17. _____

18. Solve the system using determinants and Cramer's Rule.
$3x - y = 4$
$7x + 2y = 5$

18. _____

For questions 19 – 20, express the problem as a system of linear equations and use the method of your choice to find the solution to the problem.

19. A mechanic has 2% and 6% solutions of alcohol. How much of each solution should she mix to obtain 60 liters of a 3.2 solution?

19. _____

20. The sum of three numbers is 5. The first number minus the second plus the third is 1. The first minus the third is 3 more than the second. Find the numbers.

20. _____

Chapter 9 Test Form B

1. Determine which, if any, of the ordered pairs satisfy the system of equations.
   $2x - y = 7$
   $3x + 4y = 5$

   (a) $(-1, -2)$ **(b)** $(1, 0)$ **(c)** $(3, -1)$ **(d)** $(-1, 3)$

   1. _____

2. Identify the system as consistent, inconsistent, or dependent. State whether the system has exactly one solution, no solution, or an infinite number of solutions.

   2. _____

For questions 3–5, write each question in slope-intercept form. Then determine, without solving the system, whether the system has exactly one solution, no solution, or an infinite number of solutions.

3. $4x - 3y = 2$
   $12x - 9y = 6$

   3. _____

4. $3x - 2y = 4$
   $2y - 3x = 5$

   4. _____

5. $x + 1 = y$
   $y - 3x = -1$

   5. _____

For questions 6 – 7, solve each system of equations graphically.

6. $x + y = 2$
   $2x + 3y = 9$

   6. _____

   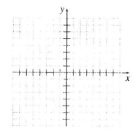

7. $x + 2y = 6$
   $x - 2y = -2$

   7. _____

   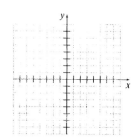

Name:_____

For questions 8 – 9, solve each system of equations using substitution.

8. $3x - 2y = -3$
   $y - 2x = 1$

8. _____

9. $x - 2y = 6$
   $x + y = 0$

9. _____

For questions 10 – 11, solve each system of equations using the addition method.

10. $2x - 4y = -6$
    $-x + 2y = 3$

10. _____

11. $2x - 3y = -4$
    $2x + y = -4$

11. _____

12. Solve the system of equations using the method of your choice.
    $2x + 3y \quad = 4$
    $3x + 7y - 4z = -3$
    $x - y + 2z = 9$

12. _____

13. Write the augmented matrix for the system of equations.
    $2x - 3y + z = 5$
    $x + 3y + 8z = 22$
    $3x - y + 2z = 12$

13. _____

14. Consider the augmented matrix below. Show the results obtained by multiplying the elements in the first row by -1 and adding the products to their corresponding elements in the second row.

$$\begin{bmatrix} 1 & 1 & 1 & | & 2 \\ 1 & -2 & -1 & | & 2 \\ 3 & 2 & 1 & | & 2 \end{bmatrix}$$

14. _____

15. Solve the system of equations using matrices.
    $x + y = 5$
    $2x + 6y = 22$

15. _____

For questions 16 – 17, evaluate each determinant.

16. $\begin{vmatrix} 2 & -5 \\ 6 & 7 \end{vmatrix}$

16. _____

17. $\begin{vmatrix} 1 & 2 & 1 \\ 2 & -2 & -2 \\ 1 & 6 & 3 \end{vmatrix}$

17. _____

18. Solve the system using determinants and Cramer's Rule.
$2x - y = 6$
$x + 3y = 4$

18. _____

For questions 19 – 20, express the problem as a system of linear equations and use the method of your choice to find the solution to the problem.

19. The sum of three numbers is 57. The second is 3 more than twice the first. The third is 6 more than the first. Find the numbers.

19. _____

20. A pharmacy has 5% and 25% saline solutions. How much of each must be mixed to obtain 6 liters of a 20% solution?

20. _____

Chapter 9 Test Form C

1. Determine which, if any, of the ordered pairs satisfy the system of equations.
   $3x - 2y = -3$
   $-2x + y = 1$

       (a) $(1, 3)$ **(b)** $(0, -5)$ **(c)** $(3, -2)$ **(d)** $(5, -3)$

1.

2. Identify the system as consistent, inconsistent, or dependent. State whether the system has exactly one solution, no solution, or an infinite number of solutions.

2. _____

For questions 3–5, write each question in slope-intercept form. Then determine, without solving the system, whether the system has exactly one solution, no solution, or an infinite number of solutions.

3. $5x + 3y = 6$
   $10x + 6y = 2$

3. _____

4. $x + 1 = y$
   $y - 3x = -1$

4. _____

5. $-x + 2y = 3$
   $2x - 4y = -6$

5. _____

For questions 6 – 7, solve each system of equations graphically.

6. $x + y = -5$
   $x - 2y = 4$

6. _____

   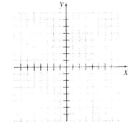

7. $x - 2y = 8$
   $3x - 4y = 20$

7. _____

   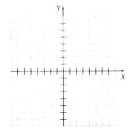

For questions 8 – 9, solve each system of equations using substitution.

8. $3y - 3x = -2$
   $3x - 3y = 2$

8. _____

9. $2x - y = 1$
   $3x + 2y = 12$

9. _____

For questions 10 – 11, solve each system of equations using the addition method.

10. $2x = 3y - 1$
    $6y - 4x = 5$

10. _____

11. $5x + 3y = -8$
    $2x - 4y = 2$

11. _____

12. Solve the system of equations using the method of your choice.
    $$x + y + z = -1$$
    $$y + 3z = 10$$
    $$2x - y + z = 0$$

12. _____

13. Write the augmented matrix for the system of equations.
    $$6x - 4y + 5z = 31$$
    $$5x + 2y + 2z = 13$$
    $$x + y + z = 2$$

13. _____

14. Consider the augmented matrix below. Show the results obtained by multiplying the elements in the second row by 3 and adding the products to their corresponding elements in the third row.
    $$\begin{bmatrix} 1 & 1 & 1 & | & 6 \\ 0 & 1 & 1 & | & 5 \\ 0 & -3 & 1 & | & -3 \end{bmatrix}$$

14. _____

15. Solve the system of equations using matrices.
    $$x + 2y = 0$$
    $$2x + 5y = -1$$

15. _____

Chapter 9 Test Form C *(cont.)*

Name:_____

For questions 16 – 17, evaluate each determinant.

16. $\begin{vmatrix} -4 & 5 \\ 1 & 7 \end{vmatrix}$

16. _____

17. $\begin{vmatrix} 3 & 4 & 0 \\ 0 & 1 & 2 \\ 1 & -1 & 5 \end{vmatrix}$

17. _____

18. Solve the system using determinants and Cramer's Rule.
    $3x - 2y = 7$
    $3x + 2y = 9$

18. _____

Express the problem as a system of linear equations and use the method of your choice to find the solution to the problem.

19. A 150-foot rope is cut into two pieces. If one piece is 2 feet more than three times the other piece, find the length of the two pieces.

19. _____

20. Ann invested $9000 in two accounts. One account gave 10% interest and the other 8%. Find the amount placed in each account if she received $840 in interest after one year.

20. _____

Chapter 9 Test Form D

Name:_____

Date:_____

1. Determine which, if any, of the ordered pairs satisfy the system of equations.
   $-x - 4y = 2$
   $4x - y = 9$
       (a) $(2, 3)$ (b) $(1, -1)$ (c) $(-2, 1)$ (d) $(2, -1)$

   1. _____

2. Identify the system as consistent, inconsistent, or dependent. State whether the system has exactly one solution, no solution, or an infinite number of solutions.

   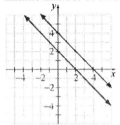

   2. _____

For questions 3–5, write each question in slope-intercept form. Then determine, without solving the system, whether the system has exactly one solution, no solution, or an infinite number of solutions.

3. $3x - 4y = -10$
   $x - y = -2$

   3. _____

4. $x - 2y = 7$
   $2x - 4y = 14$

   4. _____

5. $2y + x = -3$
   $2x + 4y = 1$

   5. _____

For questions 6 – 7, solve each system of equations graphically.

6. $3x + y = 6$
   $3x - y = 6$

   6. _____

   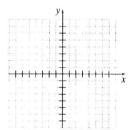

7. $-5x + 2y = 4$
   $3x - 2y = 0$

   7. _____

   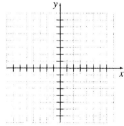

Name:_____

For questions 8 – 9, solve each system of equations using substitution.

8. $2x - y = 1$
$\quad 3x + 2y = 12$

8. _____

9. $x - y = -1$
$\quad x + y = 9$

9. _____

For questions 10 – 11, solve each system of equations using the addition method.

10. $2x - 4y = 6$
$\quad\; 5x + 3y = -11$

10. _____

11. $2x + y = 4$
$\quad\; 4x + 2y = 8$

11. _____

12. Solve the system of equations using the method of your choice.
$\quad x - 2y - z = -6$
$\quad 5x + y \quad\;\; = 6$
$\quad\quad 3y + z = 4$

12. _____

13. Write the augmented matrix for the system of equations.
$\quad 3x - 4y + z = 7$
$\quad 5x + y - 3z = 11$
$\quad\quad 7y + 2z = -5$

13. _____

14. Consider the augmented matrix below. Show the results obtained by multiplying the elements in the second row by –5 and adding the products to their corresponding elements in the third row.

$$\begin{bmatrix} 1 & -2 & -1 & -2 \\ 5 & 1 & 0 & 6 \\ 0 & 3 & 1 & 4 \end{bmatrix}$$

14. _____

15. Solve the system of equations using matrices.
$\quad x + y = 12$
$\quad 3x + y = 20$

15. _____

Chapter 9 Test Form D *(cont.)*

For questions 16 – 17, evaluate each determinant.

16. $\begin{vmatrix} -7 & 6 \\ -3 & 2 \end{vmatrix}$

16. _____

17. $\begin{vmatrix} 1 & 0 & 1 \\ 1 & 3 & 0 \\ 1 & -2 & 4 \end{vmatrix}$

17. _____

18. Solve the system using determinants and Cramer's Rule.
$x - y = -2$
$x + 2y = 10$

18. _____

For questions 19 – 20, express the problem as a system of linear equations and use the method of your choice to find the solution to the problem.

19. A collection of nickels and dimes has a value of $3.70. If there are a total of 52 coins, how many nickels and dimes are there?

19. _____

20. A gardener has 5% and 15% solutions of fertilizer. How much of each should he mix to obtain 100 liters of a 12% solution?

20. _____

Chapter 9 Test Form E

Name:_____

Date:_____

1. Determine which, if any, of the ordered pairs satisfy the system of equations.
   $-3x + y = -3$
   $2x - 3y = -5$

   (a) $(1, -3)$ **(b)** $(2, 3)$ **(c)** $(2, -2)$ **(d)** $(2, -3)$

   1. _____

2. Identify the system as consistent, inconsistent, or dependent. State whether the system has exactly one solution, no solution, or an infinite number of solutions.

   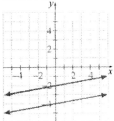

   2. _____

For questions 3–5, write each question in slope-intercept form. Then determine, without solving the system, whether the system has exactly one solution, no solution, or an infinite number of solutions.

3. $3x - 3y = 2$
   $3y - 3x = -2$

   3. _____

4. $2x - 3y = -1$
   $6y - 4x = 5$

   4. _____

5. $x - 2y = 7$
   $x - 3y = 5$

   5. _____

Solve each system of equations graphically.

6. $x + y = -4$
   $y = 3x + 4$

   6. _____

   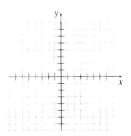

7. $2x - 3y = 3$
   $x + 3y = 6$

   7. _____

Name:_____

For questions 8 – 9, solve each system of equations using substitution.

8. $x - 2y = 7$
    $2x - 4y = 14$

8. _____

9. $2x + y = 14$
    $-2x + y = -6$

9. _____

For questions 10 – 11, solve each system of equations using the addition method.

10. $3x - 2y = 4$
    $2y - 3x = 5$

10. _____

11. $5x + 4y = -2$
    $-3x - 5y = 9$

11. _____

12. Solve the system of equations using the method of your choice.
    $2x + 8y + 3z = 1$
    $6x - 9y \quad\;\; = 5$
    $\quad\;\; -3y + z = 2$

12. _____

13. Write the augmented matrix for the system of equations.
    $-2x - y - z = -3$
    $3x - 2y - 2z = -5$
    $-x + y \quad\;\;\; = 0$

13. _____

14. Consider the augmented matrix below. Show the results obtained by multiplying the elements in the second row by –6 and adding the products to their corresponding elements in the third row.

$$\begin{bmatrix} 1 & 3 & -2 & | & 2 \\ 0 & 1 & 4 & | & 5 \\ 0 & 6 & -1 & | & 5 \end{bmatrix}$$

14. _____

15. Solve the system of equations using matrices.
    $x + y = 3$
    $2x - 3y = 6$

15. _____

Name:_____

For questions 16 – 17, evaluate each determinant.

16. $\begin{vmatrix} 7 & -2 \\ 9 & 2 \end{vmatrix}$

16. _____

17. $\begin{vmatrix} 1 & -3 & 7 \\ 1 & 1 & 1 \\ 1 & -2 & 3 \end{vmatrix}$

17. _____

18. Solve the system using determinants and Cramer's Rule.

$$x + y = -4$$
$$2x - 3y = 12$$

18. _____

For questions 19 – 20, express the problem as a system of linear equations and use the method of your choice to find the solution to the problem.

19. A boat can travel 20 miles per hour with the current and 12 miles per hour against the current. Find the speed of the current and the speed of the boat in still water.

19. _____

20. The sum of three numbers is 26. Twice the first minus the second is 2 less than the third. The third is the second minus 3 times the first. Find the numbers.

20. _____

Chapter 9 Test Form F

Name:_____

Date:_____

1. Determine which, if any, of the ordered pairs satisfy the system of equations.
   $2x + y = 2$
   $3x + y = 4$

   (a) $(-2, 2)$ (b) $(2, -2)$ (c) $(3, 1)$ (d) $(1, 0)$

   1. _____

2. Identify the system as consistent, inconsistent, or dependent. State whether the system has exactly one solution, no solution, or an infinite number of solutions.

   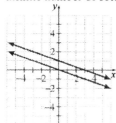

   2. _____

For questions 3–5, write each question in slope-intercept form. Then determine, without solving the system, the number of solutions the system has.

3. $2x = 3y - 1$
   $6y - 4x = 5$

   3. _____

4. $x + y = 2$
   $x - y = 2$

   4. _____

5. $-x + 2y = 3$
   $2x - 4y = -6$

   5. _____

For questions 6 – 7, solve each system of equations graphically.

6. $x - y = -2$
   $3x - y = 0$

   6. _____

   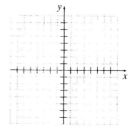

7. $x - 4y = 8$
   $x - 2y = 2$

   7. _____

   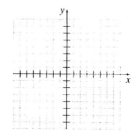

Name:_____

For questions 8 – 9, solve each system of equations using substitution.

8. $4x + 5y = 15$
  $-2x + y = 17$

8. _____

9. $x - 2y = 5$
  $3x + 2y = -13$

9. _____

For questions 10 – 11, solve each system of equations using the addition method.

10. $4x - y = 4$
  $3x + y = 10$

10. _____

11. $x - 2y = 7$
  $2x - 4y = 14$

11. _____

12. Solve the system of equations using the method of your choice.
  $x + 2y + z = 1$
  $3x - y \quad = 18$
  $\quad y - 3z = -9$

12. _____

13. Write the augmented matrix for the system of equations.
  $x + y \quad = 6$
  $x - y + z = 3$
  $x \quad - z = 5$

13. _____

14. Consider the augmented matrix below. Show the results obtained by multiplying the elements in the first row by –4 and adding the products to their corresponding elements in the third row.

$$\begin{bmatrix} 1 & 5 & 0 & | & 6 \\ 0 & 2 & 1 & | & 3 \\ 4 & 0 & 3 & | & 7 \end{bmatrix}$$

14. _____

15. Solve the system of equations using matrices.
  $5x + 2y = -23$
  $2x + y = -10$

15. _____

Chapter 9 Test Form F *(cont.)*

For questions 16 – 17, evaluate each determinant.

16. $\begin{vmatrix} 2 & 8 \\ -1 & 7 \end{vmatrix}$

16. _____

17. $\begin{vmatrix} 1 & -2 & 3 \\ 3 & 1 & 1 \\ 2 & -1 & -2 \end{vmatrix}$

17. _____

18. Solve the system using determinants and Cramer's Rule.

$$3x - y = 4$$
$$-2x + 5y = -1$$

18. _____

For questions 19 – 20, express the problem as a system of linear equations and use the method of your choice to find the solution to the problem.

19. John picked a total of 87 quarts of strawberries in three days. Tuesday's yield was 15 quarts more than Monday's. Wednesday's yield was 3 less than Tuesday's. How many quarts did he pick on Monday?

19. _____

20. Admission for a school play was $10 for adults and $5 for children. A total of 300 tickets were sold. How many adults tickets were sold if a total of $2500 was collected?

20. _____

Chapter 9 Test Form G

1. Determine which, if any, of the ordered pairs satisfy the system of equations.
   $4x - 3y = -5$
   $10x - y = 7$

   (a) $(2, -3)$ **(b)** $(-1, 3)$ **(c)** $(1, 3)$ **(d)** $(1, 2)$

2. Which of the following best describes the system of equations shown in the graph?

   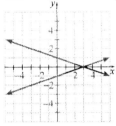

   (a) Consistent with one solution   **(b)** Inconsistent with an infinite number of solutions
   (c) Inconsistent with no solution   **(d)** Dependent with an infinite number of solutions

For questions 3–5, write each question in slope-intercept form. Then determine, without solving the system whether the system has exactly one solution, no solution, or an infinite number of solutions.

3. $4x - 3y = 2$
   $12x - 9y = 6$

   (a) One solution   **(b)** No solution
   (c) Infinite number of solutions   **(d)** Exactly two solutions

4. $5x + 3y = 6$
   $10x + 6y = 2$

   (a) One solution   **(b)** No solution
   (c) Infinite number of solution   **(d)** Exactly two solutions

5. $x + 1 = y$
   $y - 3x = -1$

   (a) One solution   **(b)** No solution
   (c) Infinite number of solutions   **(d)** Exactly two solutions

6. The following system is shown in the graph. What is the solution to the system?
   $x + y = -4$
   $-x + y = 0$

   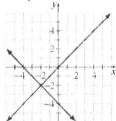

   (a) $(2, 2)$ **(b)** $(2, -2)$ **(c)** $(0, 4)$ **(d)** $(-2, -2)$

7. The following system is shown in the graph. What is the solution of the system?
    $2x + 3y = 7$
    $-x + 4y = 2$

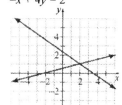

(a) $\left(0, \dfrac{7}{3}\right)$ (b) $(-2, 0)$ (c) $(2, 1)$ (d) $(1, 2)$

For questions 8–14, solve each system of equations.

8. $4x - y = -5$
    $-2x + \dfrac{1}{2}y = 1$

(a) $(2, -13)$ (b) inconsistent system   (c) $(1, -9)$ (d) $(-1, 4)$

9. $4x - 3y = 8$
    $2x + y = 14$

(a) $\left(4, -\dfrac{8}{3}\right)$ (b) dependent system   (c) $(5, -4)$ (d) $(5, 4)$

10. $5x - 6y = -1$
    $4x - 3y = 1$

(a) $(-1, -1)$ (b) $\left(-2, -\dfrac{3}{2}\right)$ (c) inconsistent system   (d) $(1, 1)$

11. $10x - 5y = -15$
    $6x - 3y = -9$

(a) $(2, 7)$ (b) dependent system   (c) $(0, 3)$ (d) $(-2, -1)$

12. $x + 3y = 4$
    $x + 2y = 2$

(a) dependent system   (b) $(2, -2)$ (c) $(-2, 2)$ (d) $(-2, -2)$

13. $-x + 3y = 2$
    $x - 5y = 14$

(a) $(-26, -8)$ (b) inconsistent system   (c) $(7, 3)$ (d) $(-8, -2)$

14. $x + 3y - z = 8$

$x + y + z = -2$

$2x + 3y + 3z = -3$

(a) $(3, -1, -8)$ (b) dependent system   (c) $(-4, 2, 0)$ (d) $(-3, 3, -2)$

15. Write the augmented matrix for the system of equations.

$x - 3y \quad = -2$

$x + y + z = 3$

$-2y + 3z = 1$

(a) $\begin{bmatrix} 1 & -3 & 0 & | & 2 \\ 1 & 1 & 1 & | & 3 \\ 1 & -2 & 3 & | & 1 \end{bmatrix}$ (b) $\begin{bmatrix} 0 & -3 & 0 & | & -2 \\ 0 & 0 & 0 & | & 3 \\ 0 & -2 & 3 & | & 1 \end{bmatrix}$ (c) $\begin{bmatrix} 1 & -3 & 1 & | & -2 \\ 1 & 1 & 1 & | & 3 \\ 1 & -2 & 3 & | & 1 \end{bmatrix}$ (d) $\begin{bmatrix} 1 & -3 & 0 & | & -2 \\ 1 & 1 & 1 & | & 3 \\ 0 & -2 & 3 & | & 1 \end{bmatrix}$

16. Consider the augmented matrix. Determine the results obtained by multiplying the elements in the second row by -5 and adding the products to their corresponding elements in the third row.

$\begin{bmatrix} 1 & 0 & 4 & | & 5 \\ 0 & 1 & 3 & | & 4 \\ 0 & 5 & -2 & | & 3 \end{bmatrix}$

(a) $\begin{bmatrix} 1 & 0 & 4 & | & 5 \\ 0 & 1 & 3 & | & 4 \\ 0 & -5 & -15 & | & -20 \end{bmatrix}$ (b) $\begin{bmatrix} 1 & 0 & 4 & | & 5 \\ 0 & -5 & -15 & | & -20 \\ 0 & 5 & -2 & | & 3 \end{bmatrix}$ (c) $\begin{bmatrix} 1 & 0 & 4 & | & 5 \\ 0 & 1 & 3 & | & 4 \\ 0 & 0 & -17 & | & -17 \end{bmatrix}$ (d) $\begin{bmatrix} 1 & 0 & 4 & | & 5 \\ 0 & 0 & -17 & | & -17 \\ 0 & 5 & -2 & | & 3 \end{bmatrix}$

For questions 17 – 18, evaluate each determinant.

17. $\begin{vmatrix} 6 & -9 \\ 2 & 3 \end{vmatrix}$

(a) 0   (b) 36   (c) –36   (d) –15

18. $\begin{vmatrix} 1 & -1 & 2 \\ 2 & 1 & -2 \\ 1 & -2 & 6 \end{vmatrix}$

(a) 6   (b) –13   (c) 1   (d) 3

For questions 19 – 20, use the method of your choice to find each solution.

19. The sum of three numbers is 6. Twice the first plus 3 times the second is 9. The third number is one less than the first. Find the numbers.

(a) 3, 1, 2   (b) –1, 4, 3   (c) 2, 3, 1   (d) 5, –3, 4

20. Chad sold 30 decorated sweatshirts at a craft show. White ones cost $10.00 and black ones cost $15.00. He received a total of $390 for the shirts. How many black shirts did he sell?

(a) 24   (b) 20   (c) 18   (d) 15

Chapter 9 Test Form H

1. Determine which, if any, of the ordered pairs satisfy the system of equations.
   $3x - 2y = -3$
   $-2x + y = 1$

   (a) $(2, 3)$ **(b)** $(0, -5)$ **(c)** $(1, -3)$ **(d)** $(1, 3)$

2. Which of the following best describes the system of equations shown in the graph?

   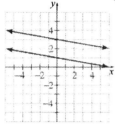

   (a) Inconsistent with no solution   **(b)** Inconsistent with an infinite number of solutions
   (c) Consistent with one solution   **(d)** Dependent with an infinite number of solutions

For questions 3–5, write each question in slope-intercept form. Then determine, without solving the system whether the system has exactly one solution, no solution, or an infinite number of solutions.

3. $x - 2y = -7$
   $x + y = 2$

   (a) One solution   **(b)** No solution
   (c) Infinite number of solutions   **(d)** Exactly two solutions

4. $2x + y = 4$
   $4x + 2y = 8$

   (a) One solution   **(b)** No solution
   (c) Infinite number of solutions   **(d)** Exactly two solutions

5. $3x - 2y = 4$
   $2y - 3x = 5$

   (a) One solution   **(b)** No solution
   (c) Infinite number of solutions   **(d)** Exactly two solutions

6. The following system is shown in the graph. What is the solution of the system?
   $2x - 3y = -1$
   $x - 4y = -8$

   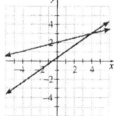

   (a) $\left(0, \dfrac{1}{3}\right)$ **(b)** $(4, 3)$ **(c)** $(0, 2)$ **(d)** $(3, 4)$

7. The following system is shown in the graph. What is the solution of the system?

$y = -x + 3$

$y = x - 5$

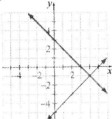

   (a) $(-1, 4)$ **(b)** $(0, 3)$ **(c)** $(0, -5)$ **(d)** $(4, -1)$

For questions 8–14, solve each system of equations.

8. $4x + 3y = 14$

   $6x + 3y = 24$

   (a) $(5, -2)$ **(b)** $(5, 2)$ **(c)** dependent system   **(d)** $(-2, 1)$

9. $2x = 3y - 1$

   $6y - 4x = 5$

   (a) $(4, 3)$ **(b)** $(5, -3)$ **(c)** inconsistent system   **(d)** $(1, 1)$

10. $2x - y = 1$

    $3x + 2y = 12$

   (a) $(-2, -3)$ **(b)** $(2, 3)$ **(c)** inconsistent system   **(d)** $(-1, 1)$

11. $5x - 6y = -1$

    $4x - 3y = 1$

   (a) $(1, 1)$ **(b)** $(1, -1)$ **(c)** inconsistent system   **(d)** $(-1, 1)$

12. $3x - y = 2$

    $5x - y = -14$

   (a) $(1, 1)$ **(b)** dependent system   **(c)** $(-2, -8)$ **(d)** $(-8, -26)$

13. $3x - 12y = 9$

    $5x - 20y = 15$

   (a) $(3, 0)$ **(b)** $(-1, -1)$ **(c)** dependent system   **(d)** $(7, 1)$

14. $2x - y + 3z = 2$

$x + 2y - 3z = -10$

$x - y - z = -7$

(a) $(3, 1, -1)$ **(b)** $(-3, 1, 3)$ **(c)** dependent system   **(d)** inconsistent system

15. Write the augmented matrix for the system of equations.

$a \quad - 3c = 6$

$b + 2c = 2$

$7a - 3b \quad = 9$

(a) $\begin{bmatrix} 1 & -3 & 6 \\ 1 & 2 & 2 \\ 7 & -3 & 9 \end{bmatrix}$ **(b)** $\begin{bmatrix} 1 & 1 & -3 & 6 \\ 1 & 1 & 2 & 2 \\ 7 & -3 & 1 & 9 \end{bmatrix}$ **(c)** $\begin{bmatrix} 1 & 0 & -3 & 6 \\ 0 & 1 & 2 & 2 \\ 7 & -3 & 0 & 9 \end{bmatrix}$ **(d)** $\begin{bmatrix} 0 & 0 & -3 & 6 \\ 0 & 0 & 2 & 2 \\ 7 & -3 & 0 & 9 \end{bmatrix}$

16. Consider the augmented matrix. Show the results obtained by multiplying the elements in the first row by -5 and adding the products to the corresponding elements in the second row.

$\begin{bmatrix} 1 & -1 & 2 & 2 \\ 5 & 0 & -3 & 2 \\ 0 & -4 & 6 & 2 \end{bmatrix}$

(a) $\begin{bmatrix} -5 & 5 & -10 & -10 \\ 5 & 0 & -3 & 2 \\ 0 & -4 & 6 & 2 \end{bmatrix}$ **(b)** $\begin{bmatrix} 1 & -1 & 2 & 2 \\ 0 & 5 & -13 & -8 \\ 0 & -4 & 6 & 2 \end{bmatrix}$ **(c)** $\begin{bmatrix} 0 & 5 & -15 & -8 \\ 5 & 0 & -3 & 2 \\ 0 & -4 & 6 & 2 \end{bmatrix}$ **(d)** $\begin{bmatrix} 1 & -1 & 2 & 2 \\ 5 & 0 & -3 & 2 \\ 0 & 5 & -15 & -8 \end{bmatrix}$

For questions 17 – 18, evaluate each determinant.

17. $\begin{vmatrix} 3 & -2 \\ 3 & 2 \end{vmatrix}$

(a) 0   **(b)** -12 **(c)** 12   **(d)** 5

18. $\begin{vmatrix} -1 & -2 & -3 \\ 3 & 4 & 2 \\ 0 & 1 & 2 \end{vmatrix}$

(a) -3 **(b)** 3   **(c)** 9   **(d)** -5

For questions 19 – 20, use the method of your choice to find each solution.

19. The sum of the angles in a triangle is $180°$. In triangle ABC, the measure of angle B is $2°$ more than three times the measure of angle A. The measure of angle C is $8°$ more than the measure of angle A. Find the measure of angle A.

(a) $34°$ **(b)** $42°$ **(c)** $104°$ **(d)** $114°$

20. Two investments are made totaling $8800. In one year, these investments yield $663 in simple interest. Part of the $8800 was invested at 7% and part at 8%. Find the amount invested at 7%.

(a) $2,700   **(b)** $3,000   **(c)** $5,000   **(d)** $4,100

Chapter 10 Pretest Form A

For problems 1 – 2, use $A = \{1, 2, 3\}$ and $B = \{4, 5, 6\}$.

1. Find $A \cup B$.

1. _____

2. Find $A \cap B$.

2. _____

For problems 3 – 4, solve each inequality and graph the solution on a number line.

3. $4x - 7 \geq -15$

3. _____

4. $8 - 5x < -7$

4. _____

5. Write $-2 \leq x < 7$ in interval notation.

5. _____

6. Solve the inequality and write the solution in set-builder notation.
$x - 3 \geq 7$ or $x + 1 > 1$

6. _____

7. Solve the inequality and write the solution in interval notation.
$x - 4 > -9$ and $3x - 5 \leq 7$

7. _____

For problems 8 – 11, find the solution set to each equation or inequality.

8. $|2x + 5| = 13$

8. _____

9. $|x + 1| > 4$

9. _____

10. $|2x + 6| - 6 < 5$

10. _____

11. $|4x - 6| = |2x + 1|$

11. _____

12. *Write an inequality that can be used to solve the following problem, then solve the inequality and answer the question:* The parking garage at the Broward Convention Center charges $1.25 for the first hour and $0.75 for each addition hour. What is the maximum length of time you can park in the garage if you wish to pay no more than $4.25?

12. _____

For problems 13 – 15, graph each inequality.

13. $y \geq 4x - 3$

13.

14. $x - 4y < -4$

14.

15. $x + y \leq 1$

15.

For problems 16 – 20, determine the solution to each system of inequalities.

16. $x - 2y > 6$
$3x - 4y \leq 8$

16.

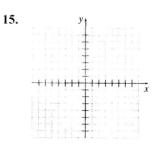

17. $5x - 3y < 15$
$x + y \leq 4$

17.

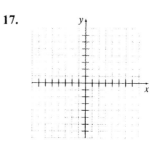

Name:_____

18. $4x - 2y \geq -8$
$\qquad 3x + 4y < 12$

18.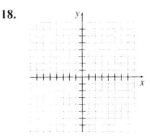

19. $|x - 1| < 2$
$\qquad x + y < 2$

19.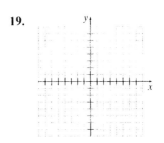

20. $\qquad x \geq -2$
$\qquad\quad y < 3$
$\qquad x + 3y > 3$

20.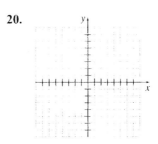

Chapter 10 Pretest Form B

Name:_____

Date:_____

For problems 1 – 2, use $A = \{1,3,5,7,9\}$ and $B = \{4,5,6,7,8\}$.

1. Find $A \cup B$.

1. _____

2. Find $A \cap B$.

2. _____

For problems 3 – 4, solve each inequality and graph the solution on a number line.

3. $3x - 11 \leq 16$

3. _____

4. $-4 - 5x < -9$

4. _____

5. Write $-4 \leq x < 5$ in interval notation.

5. _____

6. Solve the inequality and write the solution in set-builder notation.
$2x \leq 6$ or $x + 1 > 5$

6. _____

7. Solve the inequality and write the solution in interval notation.
$x - 5 < 12$ and $2x + 9 \geq 3$

7. _____

For problems 8 – 11, find the solution set to each equation or inequality.

8. $|2x - 3| = 7$

8. _____

9. $|x + 4| \leq 6$

9. _____

10. $|x + 2| + 3 > 8$

10. _____

11. $|6x + 1| = |2x - 5|$

11. _____

12. *Write an inequality that can be used to solve the following problem, then solve the inequality and answer the question:* To receive an "A" in the course Mei Ling must obtain an average score of 90 or higher on five exams. If Mei's first four exam scores are 90, 87, 96, and 79, what is the minimum score that she can receive on the fifth exam in order to get an "A" in the course?

12. _____

Chapter 10 Pretest Form B *(cont.)* Name:_____

For problems 13 – 15, graph each inequality.

13. $y > x - 4$

13.

14. $2x - 5y > -10$

14.

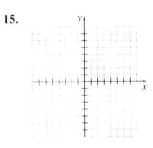

15. $3x - y \geq 1$

15.

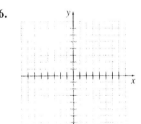

For problems 16 – 20, determine the solution to each system of inequalities.

16. $5x - 3y \geq -15$
$\quad$ $2x + 3y < -6$

16.

17. $\quad 4x + y > 1$
$\quad$ $2x - 5y \leq 5$

17.

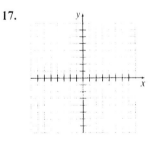

18. $4x + 3y < -6$

       $x \geq -3$

18.

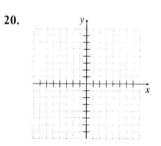

19. $|y - 1| < 3$

       $x - y \leq -2$

19.

20. $x > -1$

         $y \geq -2$

    $2x + y < 1$

20.

Mini-Lecture 10.1
Solving Linear Inequalities in One Variable

Learning Objectives:
1. Solve inequalities.
2. Graph solutions on a number line, in interval notation, and as a set.
3. Find the union and intersection of sets.
4. Solve compound inequalities involving *and*.
5. Solve compound inequalities involving *or*.
6. Key vocabulary: *inequality, order (or sense) of an inequality, compound inequality, intersection, union*

Examples:
1. Solve each inequality, stating the solution set and graphing the solution on the number line.
   a) $2x + 3 > 7$ b) $5x + 4 \geq 6x - 16$ c) $-x - 27 > 11$

2. Solve each inequality and give the solution in interval notation.

   a) $12 - 7x > 6x - 7(2x + 3)$ b) $\frac{2}{3}x - \frac{2}{3} \geq \frac{3}{4}x - \frac{7}{2}$

   c) $4(x + 2) < 4x - 10$ d) $2x + (x + 4) \geq 3x + 4$

3. If $A = \{0, 2, 4, 6, 8, 10, 12\}$ and $B = \{0, 3, 6, 9, 12\}$, find $A \cap B$ and $A \cup B$.

4. Solve each inequality and give the solution in interval notation.

   a) $-2 < 3x - 5 \leq 7$ b) $-2 \leq \frac{5 - 3x}{2} \leq 2$ c) $4x - 5 > 3 \text{ and } 4 - \frac{1}{2}x < 1$

5. Solve each inequality and given the solution set.
   a) $5x + 3 > 4x \text{ or } 3(x + 2) - 1 < 2x$ b) $-4(x + 2) \geq 12 \text{ or } 3x + 8 < 11$

6. A rental car agency charges a daily rate of $37.50 per day plus $0.21 per mile. Sam is budgeted $150 per day for business expenses related to car rentals. To the nearest tenth of a mile, determine how far Sam can travel per day to stay within his budget.
   a) Write an inequality that could be used to solve this problem.
   b) Solve the inequality and use the solution to answer the given question.

7. Ella wants to make a B in her algebra class. The grading scale for a B is 85 to 93, inclusive. Her first four test scores were 93, 100, 88, and 87. Determine the range of grades she must score on her fifth test in order to make a B.

Teaching Notes:
- Some students change the direction of the inequality symbol anytime the operation on the inequality somehow involves a negative number.
- Encourage students to always end up with the variable on the left side of the solution statement. Otherwise, errors in expressing the solution graphically or with interval notation are likely to occur.

Answers: 1a) $\{x | x > 2\}$ 1b) $\{x | x \leq 20\}$

1c) $\{x | x < -38\}$ 2a) $(-33, \infty)$; 2b) $(-\infty, 34]$; 2c) no solution;

2d) $(-\infty, \infty)$; 3) $A \cap B = \{0, 6, 12\}$, $A \cup B = \{0, 2, 3, 4, 6, 8, 9, 10, 12\}$; 4a) $(-1, 4]$; 4b) $\left[\frac{1}{3}, 3\right]$; 4c) $(6, \infty)$;

5a) $\{x | x < -5 \text{ or } x > -3\}$; 5b) $\{x | x < 1\}$; 6a) $37.5 + 0.21x \leq 150$; 6b) 535.7 miles; 7) 57 to 97, inclusive

Mini-Lecture 10.2
Solving Equations and Inequalities Containing Absolute Values

Learning Objectives:

1. Understand the geometric interpretation of absolute value.
2. Solve equations of the form $|x| = a,\ a > 0$.
3. Solve inequalities of the form $|x| < a,\ a > 0$.
4. Solve inequalities of the form $|x| > a,\ a > 0$.
5. Solve inequalities of the form $|x| > a$ or $|x| < a,\ a < 0$.
6. Solve inequalities of the form $|x| > 0$ or $|x| < 0$.
7. Solve equations of the form $|x| = |y|$.

Examples:

1. Find the solution set for each equation.
   a) $|x| = 7$ b) $|x| = -7$ c) $|2x - 3| = 5$

2. Find the solution set for each inequality.
   a) $|3 - 4x| < 13$ b) $|4 - \frac{1}{3}x| + 2 < 8$ c) $|x - 2| > 4$

   d) $|3 - \frac{1}{2}x| \geq 4.5$ e) $|9 - 5x| + 6 \leq 4$ f) $|7x - 6| > -1$

   g) $|x - 8| > 0$ h) $|4x - 8| \leq 0$

3. Find the solution set for each equation.
   a) $|x + 2| = |x - 3|$ b) $|5x + 3| = |3x + 25|$

Teaching Notes:

- Some students write answers to equations having two solutions using parentheses instead of braces. Distinguish between the mathematical meanings of $\{2, 6\}$ and $(2, 6)$.

- With absolute value equations and inequalities, students typically ignore the absolute value symbols and think of $|2x - 7| = 12$ and $2x - 7 = 12$ as the same problem.

- Remind students to isolate the absolute value expression before removing the absolute value symbols and writing two equations or inequalities.

Answers: 1a) $\{-7, 7\}$ 1b) *no solution* 1c) $\{-1, 4\}$; 2a) $\{x | -2.5 < x < 4\}$ 2b) $\{x | -6 < x < 30\}$ 2c) $\{x | x < -2\ or\ x > 6\}$ 2d) $\{x | x \leq -3\ or\ x \geq 15\}$ 2e) *no solution* 2f) *all real numbers* 2g) $\{x | x < 8\ or\ x > 8\}$ 2h) $\{2\}$; 3a) $\{\frac{1}{2}\}$ 3b) $\{-\frac{7}{2}, 11\}$

Mini-Lecture 10.3
Graphing Linear Inequalities in Two Variables
and Systems of Linear Inequalities

Learning Objectives:

1. Graph linear inequalities in two variables.
2. Solve systems of linear inequalities.
3. Solve linear programming problems.
4. Solve systems of linear inequalities containing absolute value.
5. Key vocabulary: *linear inequality, half plane, boundary, systems of linear inequalities, linear programming, constraints*

Examples:

1. Graph each inequality.

   a) $y < 2x - 4$

   b) $y \geq -\dfrac{2}{3}x + 1$

   c) $3x + 4y \leq 12$

2. Determine the solution to each system of inequalities.

   a) $x \leq 5$
   $x + y \geq 3$

   b) $2x + y < 4$
   $-2x + y > 2$

   c) $x > -2$
   $y < 4$

3. Determine the solution to each system of inequalities.

   a) $x \geq 0$
   $y \geq 0$
   $x + 2y \leq 6$

   b) $x \geq 0$
   $y \geq 0$
   $x + y \leq 5$
   $y \geq 2x - 4$

4. Determine the solution to each inequality.

   a) $|x| \leq 4$

   b) $|y - 3| > 2$

5. Determine the solution to each system of inequalities.

   a) $|y| \leq 4$
   $y \leq -2x + 4$

   b) $|x - 1| > 3$
   $|y| > 2$

Teaching Notes:

- When finished graphing linear inequalities in which the variable y is already isolated on the left side, the following guidelines can be used to double-check your solution. For inequalities involving $<$ or $\leq$, shading should occur below the boundary line. For inequalities involving $>$ or $\geq$, shading should occur above the boundary line.
- Students can easily confuse solving absolute value inequalities in one variable with graphing absolute value inequalities in two variables.

Answers: See answer pages.

Additional Exercises 10.1

Name:_____

Date:_____

Find both $A \cup B$ and $A \cap B$. Be sure to identify which is which.

1. $A = \{7, 8, 9, 10, 17\}$ and $B = \{3, 7, 10, 12\}$

1. _____

2. $A = \{7, 9, 11, 13, \ldots\}$ and $B = \{9, 11, 13, 15\}$

2. _____

3. $A = \{e, h, i, k, m\}$ and $B = \{e, i, m, o\}$

3. _____

Express each inequality (a) using the number line, (b), in interval notation, and (c) as a solution set.

1. $x \geq 2$

1. (a)

    (b) _____

    (c) _____

2. $x < -\dfrac{4}{3}$

2. (a)

    (b) _____

    (c) _____

3. $-5 \leq x < \dfrac{2}{3}$

3. (a)

    (b) _____

    (c) _____

Solve the inequality and graph the solution on the number line.

4. $2x < 29$

4. _____

5. $5x - 10 \leq 10$

5. _____

6. $\dfrac{2x - 8}{4} > 2$

6. _____

Solve the inequality and give the solution in interval notation.

7. $3x \leq 15$

7. _____

8. $6 - 4x < 14$

8. _____

Additional Exercises 10.1 *(cont.)* Name:_____

9. $1 < -7 - \dfrac{1}{3}x < 2$ 9. _____

10. $4 < \dfrac{7 - 4x}{8} \le 6$ 10. _____

Solve the inequality and indicate the solution set.

11. $-9 \le -9x - 18 < 9$ 11. _____

12. $-12 < \dfrac{6x + 6}{3} < 5$ 12. _____

13. $-6 \le \dfrac{3(x+5)}{2} < 3$ 13. _____

14. $-4 \le 4x$ and $x \le 3$ 14. _____

15. $x > -12$ or $x > 7$ 15. _____

16. $3x + 4 > 10$ and $2x - 2 < 18$ 16. _____

17. $-5x + 15 > -15$ or $3x < 24$ 17. _____

18. The width of a rectangle is 39 centimeters. Find all possible   18. _____
values for the length of the rectangle if the perimeter
is at least 784 centimeters.

19. The width of a rectangle is 38 centimeters. Find all possible   19. _____
values for the length of the rectangle if the perimeter
is at least 456 centimeters.

20. Karl's grades on his first five exams are 71, 65, 72, 80 and 77.   20. _____
What range of scores can Karl earn on his sixth exam so that his
final grade is a C? Assume that all of the exams have a maximum
score of 100 and also assume that a C is the final grade given when
the test average is greater than or equal to 70 and less than 80.

Additional Exercises 10.2

Name:_____

Date:_____

Find the solution set for each equation or inequality.

1. $|2x - 6| = 4$

2. $|3x + 3| = 6$

3. $|x - 7| < 1$

4. $\left|\dfrac{8x - 3}{5}\right| \le \dfrac{2}{5}$

5. $|4x + 2| \le 6$

6. $|3x - 4| < 5$

7. $|x - 2| > 8$

8. $|2x - 9| + 9 > 10$

9. $|3x + 5| > 7$

10. $|2 - 5x| > 6$

11. $|x + 8| \le -7$

12. $|x - 8| \ge -4$

13. $|12x - 3| \ge -5$

14. $|3x + 7| + 6 < 2$

15. $\left|\dfrac{3}{7}x + 6\right| + 3 = 8$

16. $|x - 3| = 9$

17. $|x - 2| = |x - 4|$

18. $|2x - 11| = |x + 5|$

19. Find the solution set for the equation.
$\left|\dfrac{2}{5}x + 4\right| = \left|\dfrac{1}{4}x + 2\right|$

20. Find the solution set for the equation.
$\left|\dfrac{3}{5}x + 2\right| = \left|\dfrac{1}{3}x + 3\right|$

1. _____

2. _____

3. _____

4. _____

5. _____

6. _____

7. _____

8. _____

9. _____

10. _____

11. _____

12. _____

13. _____

14. _____

15. _____

16. _____

17. _____

18. _____

19. _____

20. _____

Additional Exercises 10.3

Name:_____

Date:_____

1. Graph: $y > 3x + 2$

1.

2. Graph: $y \le -4x$

2.

3. Graph: $y < \dfrac{1}{3}x - 1$

3.

4. Graph: $3x \ge -y + 3$

4.

5. Graph: $2x < -3y - 12$

5.

6. Graph: $y > 1 - \dfrac{3}{2}x$

6.

Additional Exercises 10.3 *(cont.)* Name:_____

Determine the solution to each system of inequalities.

7. $y \geq x + 7$
    $3x + y \leq 3$

7.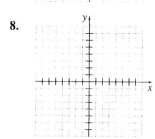

8. $2x + y \geq -2$
    $6x + 3y \leq 6$

8.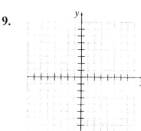

9. $y \leq -\frac{1}{2}x + 3$
    $y \geq x - 3$

9.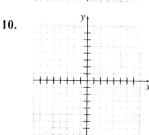

10. $y > 2x - 6$
    $x + y < 0$

10.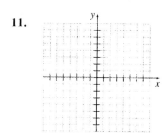

11. $2x + 3y \leq 6$
    $x - 3y \geq -9$

11.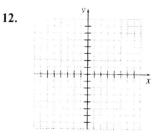

12. $y \geq -2x - 2$
    $y \leq x + 2$

12.

Additional Exercises 10.3 *(cont.)*

13.
$$x \geq 0$$
$$y \geq 0$$
$$3x + 5y \leq 19$$
$$x + 4y \leq 11$$

13.

14.
$$x \geq 0$$
$$y \geq 0$$
$$x + y \leq 9$$
$$2x + y \leq 15$$

14.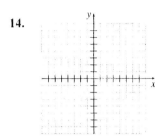

15.
$$x \geq 0$$
$$y \geq 0$$
$$4x + 3y \leq 23$$
$$x + 3y \leq 17$$

15.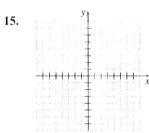

16.
$$3x + 5y \geq 15$$
$$x \leq y$$
$$x \leq 6$$

16.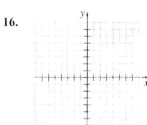

17.
$$x \geq 0$$
$$y \geq 0$$
$$x + y \leq 11$$
$$3x + y \leq 21$$

17.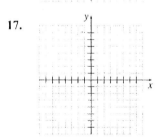

18.
$$|x - 4| \geq 1$$
$$x + y \leq 4$$

18.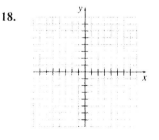

Additional Exercises 10.3 *(cont.)*

19. $|x - 2| \geq 3$

$x + y \leq 2$

19.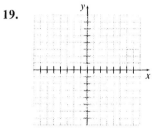

20. $|x| < 6$

$|y| \geq 3$

20.

Chapter 10 Test Form A

For problems $1-2$, find $A \cup B$ and $A \cap B$. Be sure to identify which is which.

1. $A = \{2,4,6\}$, $B = \{2,4,6,8,11\}$

   1. _____

2. $A = \{0,2,4,6,8\}$, $B = \{0,3,6,9\}$

   2. _____

For problems $3-4$, solve each inequality and graph the solution on a number line.

3. $\dfrac{5x-15}{5} \leq 5$

   3. _____

   <+-+-+-+-+-+-+-+-+-+-+-+-+-+->

4. $3x+5 > -10$ and $2x+6 < 12$

   4. _____

   <+-+-+-+-+-+-+-+-+-+-+-+-+-+->

For problems $5-7$, solve each inequality. Write each solution in interval notation.

5. $1 \leq \dfrac{6-2x}{9} \leq 2$

   5. _____

6. $x > 4$ or $x \geq -4$

   6. _____

7. $5x-3 > 10$ and $5-3x < -3$

   7. _____

For problems $8-11$, find the solution set to each equation or inequality.

8. $|x+4| = 6$

   8. _____

9. $\left|\dfrac{7x-2}{5}\right| \leq \dfrac{4}{3}$

   9. _____

10. $|3x+6| > 9$

    10. _____

11. $|4x-7| = |3x+5|$

    11. _____

12. *Write an inequality that can be used to solve the following problem, then solve the inequality and answer the question:*
    The width of a rectangle is 12 cm. Find all possible values for the length of the rectangle if the perimeter is at least 416 cm.

    12. _____

Chapter 10 Test Form A *(cont.)*

Name:_____

For problems 13 – 15, graph each inequality.

13. $y < 2x - 2$

13.

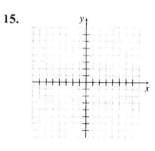

14. $x \geq -3$

14.

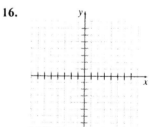

15. $3x - 2y \geq 12$

15.

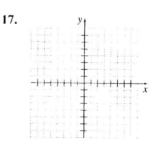

For problems 16 – 20, determine the solution to each system of inequalities.

16. $y \leq -2x$
$\quad\ y \geq x$

16.

17. $3x + 2y < 6$
$\quad\ 2x - y \leq 4$

17.

Name:_____

18. $2x + y > -5$
     $2x - 3y \geq 3$

18.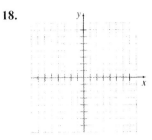

19. $|x| < 4$
     $|y| \leq 4$

19.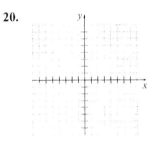

20. $y > 1$
     $x < -2$
     $y > x + 3$

20.

Chapter 10 Test Form B

Name:_____

Date:_____

For problems 1 – 2, find $A \cup B$ and $A \cap B$. Be sure to identify which is which.

1. $A = \{-1, 0, 1, e, i, \pi\}$, $B = \{-1, 0, 1\}$

     1. _____

2. $A = \{1, 2, 4, 8, 16\}$, $B = \{2, 4, 6, 8, 10\}$

     2. _____

For problems 3 – 4, solve each inequality and graph the solution on a number line.

3. $-5x + 20 > -20$

     3. _____

     ←+·+·+·+·+·+·+·+·+·+·+·+·+→

4. $x + 3 \leq 6$ and $-2x < 8$

     4. _____

     ←+·+·+·+·+·+·+·+·+·+·+·+·+→

For problems 5 – 7, solve each inequality. Write each solution in interval notation.

5. $-2 \leq \dfrac{7 - x}{7} \leq 3$

     5. _____

6. $4 - x < -2$ or $3x - 1 < -1$

     6. _____

7. $-2 \leq 2x$ and $x \leq 2$

     7. _____

For problems 8 – 11, find the solution set to each equation or inequality.

8. $|x + 4| + 5 = 14$

     8. _____

9. $|2x - 1| \leq 3$

     9. _____

10. $|2x - 1| + 4 > 9$

     10. _____

11. $\left| \dfrac{1}{2}x + 5 \right| = \left| \dfrac{2}{3}x - 4 \right|$

     11. _____

12. *Write an inequality that can be used to solve the following problem, then solve the inequality and answer the question:* An elevator can carry a maximum load of 1200 pounds. How many 70 pound boxes can Harold take on the elevator if he weighs 150 pounds?

     12. _____

For problems 13 – 15, graph each inequality.

13. $y \geq -\dfrac{1}{2}x - 2$

13.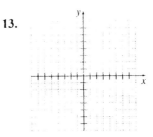

14. $x < 4$

14.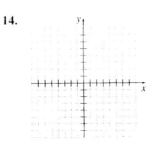

15. $6x - 5y \geq -30$

• **15.**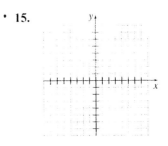

For problems 16 – 20, determine the solution to each system of inequalities.

16. $y > 2x - 3$
  $y < -2x + 3$

16.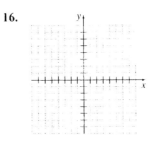

17. $5x + 2y < 0$
  $4x - y \leq -3$

17.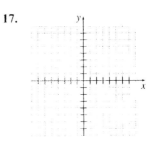

Name:_____

18. $3x - 2 \geq 6$

    $x < 2$

18.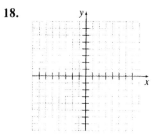

19. $|x - 2| \leq 3$

    $x - 2y < 8$

19.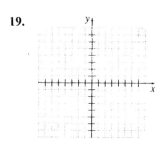

20. $x \geq 1$

    $y \leq 2$

    $2x + 3y \geq 6$

20.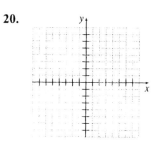

Chapter 10 Test Form C

For problems 1 – 2, find $A \cup B$ and $A \cap B$. Be sure to identify which is which.

1. $A = \{-3, -1, 1, 3, 5\}$, $B = \{1, 3, 5, 7, 9\}$

1. _____

2. $A = \{2, 4, 6, 8, ...\}$, $B = \{..., -3, -2, -1, 0, 1, 2, 3, ...\}$

2. _____

For problems 3 – 4, solve each inequality and graph the solution on a number line.

3. $-3x - 12 > -6$

3. _____

                                         ←++−+−++−+−++−+−++−+−++−→

4. $x + 2 \leq 4$ and $-4x < 12$

4. _____

                                         ←++−+−++−+−++−+−++−+−++−→

For problems 5 – 7, solve each inequality. Write each solution in interval notation.

5. $6 \leq -3(2x - 4) < 12$

5. _____

6. $x < -7$ or $x \leq 6$

6. _____

7. $x \geq -1$ and $2x + 1 > 5$

7. _____

For problems 8 – 11, find the solution set to each equation or inequality.

8. $|x - 3| = 7$

8. _____

9. $|x + 1| \leq -5$

9. _____

10. $|7 - 3x| > 5$

10. _____

11. $|2x + 7| = \left|\frac{2}{3}x - 1\right|$

11. _____

12. *Write an inequality that can be used to solve the following problem, then solve the inequality and answer the question:*
Tracy's first three exam scores are 83, 71, and 96. If an average of 80 or greater and less than 90 is needed for a B average, what score does Tracy need to make on her fourth exam to have a B average? Assume a maximum score of 100.

12. _____

For problems 13 – 15, graph each inequality.

13. $y \leq 3$

13.

14. $y > -3x - 2$

14.

15. $3x - 4y > -12$

15.

For problems 16 – 20, determine the solution to each system of inequalities.

16. $y < x - 3$
$\quad\ y > -2x$

16.

17. $x + y < -3$
$\quad\ \ y \leq 3x + 1$

17.

18. $3x - 2y \geq 0$

$\quad\quad x + 4y < 8$

18.

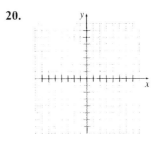

19. $|y + 1| \leq 3$

$\quad\quad\quad y > 3x$

19.

20. $y \leq 3$

$\quad x \geq -4$

$\quad y \geq \dfrac{1}{2}x + 2$

20.

Chapter 10 Test Form D

For problems 1 – 2, find $A \cup B$ and $A \cap B$. Be sure to identify which is which.

1. $A = \{3, 4, 6, 9\}, \quad B = \{1, 4, 5, 8\}$

      1. _____

2. $A = \{0, 1, 2, 3, 4\}, \quad B = \{3, 4, 5, 6, 7\}$

      2. _____

For problems 3 – 4, solve each inequality and graph the solution on a number line.

3. $\dfrac{4x + 4}{6} \geq -4$

      3. _____

                                      ◄++++++++++++++++++++►

4. $x + 1 \leq 4$ and $-3x < 12$

      4. _____

                                        ◄++++++++++++++++++++►

For problems 5 – 7, solve each inequality. Write each solution in interval notation.

5. $-3 < -3x - 9 \leq 3$

      5. _____

6. $x > 3$ or $x \geq -2$

      6. _____

7. $-x - 1 \leq 3$ and $2x < 6$

      7. _____

For problems 8 – 11, find the solution set to each equation or inequality.

8. $|x - 5| = 2$

      8. _____

9. $|2x + 3| \leq 5$

      9. _____

10. $|14x - 6| \geq -2$

      10. _____

11. $\left| \dfrac{5}{4}x + 1 \right| = |x - 10|$

      11. _____

12. *Write an inequality that can be used to solve the following problem, then solve the inequality and answer the question:*
     Angie can rent a car for $40 per day plus $0.10 per mile. If she needs the car for 5 days, how many miles can she drive without spending more than $300 on the rental car?

      12. _____

Chapter 10 Test Form D *(cont.)*

Name:_____

For problems 13 – 15, graph each inequality.

13. $y \le 5$

13.

14. $y > -\dfrac{1}{3}x + 2$

14.

15. $5x - 3y < -15$

15.

For problems 16 – 20, determine the solution to each system of inequalities.

16. $y < -\dfrac{1}{2}x + 1$
$\quad\;\; y > x$

16.

17. $2x + y < 4$
$\quad\;\; x + y \ge 1$

17.

Name:_____

18. $x \geq -2$
$x - 3y > 0$

18.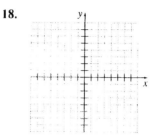

19. $|x+1| \leq 4$
$|y| < 3$

19.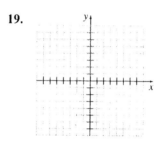

20. $y \geq -2$
$x \geq 1$
$y \leq -x + 4$

20.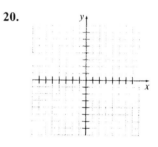

Chapter 10 Test Form E

For problems 1 – 2, find $A \cup B$ and $A \cap B$. Be sure to identify which is which.

1. $A = \{0, 2, 4, 6, ...\}$, $B = \{1, 3, 5, 7, ...\}$

1. _____

2. $A = \{-7, -4, -1, 2, 5\}$, $B = \{-7, -5, -3, -1, 1, 3, 5, 7\}$

2. _____

For problems 3 – 4, solve each inequality and graph the solution on a number line.

3. $\dfrac{2x-8}{6} < -2$

3. _____

     ⟵┼┼┼┼┼┼┼┼┼┼┼┼┼┼┼┼┼⟶

4. $2x + 2 > -6$ and $3x + 2 \leq 11$

4. _____

     ⟵┼┼┼┼┼┼┼┼┼┼┼┼┼┼┼┼┼⟶

For problems 5 – 7, solve each inequality. Write each solution in interval notation.

5. $-1 \leq \dfrac{3-x}{4} \leq 5$

5. _____

6. $x < -5$ or $x \leq 2$

6. _____

7. $4x < 16$ and $-3x \leq 12$

7. _____

For problems 8 – 11, find the solution set to each equation or inequality.

8. $|2x - 4| = 2$

8. _____

9. $|2x + 3| \leq 7$

9. _____

10. $|3x - 4| + 10 \geq 6$

10. _____

11. $|7 - 3x| = \left|\dfrac{1}{2}x + 1\right|$

11. _____

12. *Write an inequality that can be used to solve the following problem, then solve the inequality and answer the question*:
Carol receives a base wage of $200 per week plus 5% commission on all sales. What amount of sales does she need in order to have a weekly salary of at least $400?

12. _____

Name:_____

For problems 13 – 15, graph each inequality.

13. $x < -3$

13.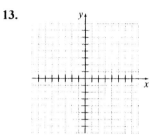

14. $y \geq 2x - 5$

14.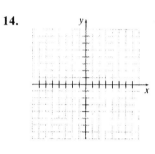

15. $5x - 2y > -10$

15.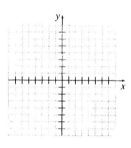

For problems 16 – 20, determine the solution to each system of inequalities.

16. $y < x + 1$
   $y > x - 1$

16.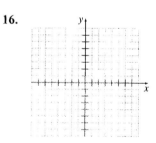

17. $y \geq 3x - 5$
   $y < 4$

17.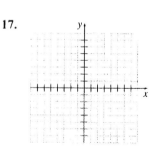

Name:_____

18. $3x + 4y < 8$
$3x - 2y \leq -2$

18.

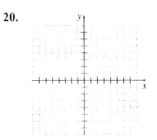

19. $|x + 2| \leq 3$
$x - 3y < 0$

19.

20. $y \geq 1$
$x \geq 1$
$x + y \leq 4$

20.

Chapter 10 Test Form F

Name:_____

Date:_____

For problems 1 – 2, find $A \cup B$ and $A \cap B$. Be sure to identify which is which.

1. $A = \{0, 2, 4, 6, 8\}$, $B = \{1, 3, 5, 7, 9\}$

1. _____

2. $A = \{-3, -2, -1, 0, 1\}$, $B = \{0, 1, 2, 3\}$

2. _____

For problems 3 – 4, solve each inequality and graph the solution on a number line.

3. $2x - 10 < 40$

3. _____

4. $-4 \leq 2x$ and $5x \leq 15$

4. _____

For problems 5 – 7, solve each inequality. Write each solution in interval notation.

5. $1 \leq \dfrac{9-x}{8} \leq 8$

5. _____

6. $x < 3$ or $x \leq -2$

6. _____

7. $5x - 3 < 7$ and $x \leq 5$

7. _____

For problems 8 – 11, find the solution set to each equation or inequality.

8. $|6x - 5| = 2$

8. _____

9. $|x - 3| \leq -2$

9. _____

10. $|5x + 1| \geq 9$

10. _____

11. $|5x - 8| = |4x - 19|$

11. _____

12. *Write an inequality that can be used to solve the following problem, then solve the inequality and answer the question:*
The width of a rectangle is 38 cm. Find all possible values for the length of the rectangle if the perimeter is at least 300 cm.

12. _____

Name:_____

For problems 13 – 15, graph each inequality.

13. $y > 2$

13.

14. $2x + y \leq 0$

14.

15. $2x - 3y \geq -15$

15.

For problems 16 – 20, determine the solution to each system of inequalities.

16. $y < 2x - 4$
 $x + y > 4$

16.

17. $2x - y \geq 3$
 $x + y < 3$

17.

Name:_____

18. $3x + 2y > -8$

$y - x \le 0$

18.

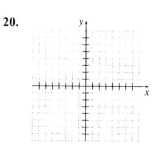

19. $x + y > -3$

$x + y < 2$

$|y| < 1$

19.

20. $x \ge 1$

$y \le 3$

$y \ge -x$

20.

Chapter 10 Test Form G

Name:_____

Date:_____

Choose the correct answer to each problem.

For problems 1 – 2, consider the sets $A = \{-3, 0, 2\}$ and $B = \{0, 1, 2, 3\}$.

1. Find $A \cup B$.
   (a) $\{0, 2\}$ (b) $\{-3, 0, 2\}$ (c) $\{0, 1, 2, 3\}$ (d) $\{-3, 0, 1, 2, 3\}$

2. Find $A \cap B$.
   (a) $\{0, 2\}$ (b) $\{-3, 0, 2\}$ (c) $\{0, 1, 2, 3\}$ (d) $\{-3, 0, 1, 2, 3\}$

For problems 3 – 4, solve each inequality and graph the solution on a number line.

3. $\dfrac{4x + 20}{3} > 8$

   (a)
   (b)
   (c)
   (d)

4. $-5x - 5 > 10$ or $2x < 6$

   (a)
   (b)
   (c)
   (d)

For problems 5 – 7, solve each inequality. Write each solution in interval notation.

5. $3 \le \dfrac{7 - 2x}{9} \le 7$
   (a) $[-10, 28]$ (b) $[-28, 10]$ (c) $[-28, -10]$ (d) $[10, 28]$

6. $x < -1$ or $x \le 2$
   (a) $(-\infty, 2]$ (b) $(-1, 2]$ (c) $(-1, \infty)$ (d) $[2, \infty)$

7. $x < 0$ and $x - 2 \ge -5$
   (a) $(-\infty, 0)$ (b) $(-\infty, -7)$ (c) $[-3, \infty)$ (d) $[-3, 0)$

For problems 8 – 11, find the solution set to each equation or inequality.

8. $|x - 7| = |x - 3|$
   (a) $\{-4\}$ (b) $\{5\}$ (c) $\{-4, 5\}$ (d) $\{-5\}$

9. $|x - 8| < 5$
   (a) $\{x | 3 < x < 13\}$ (b) $\{x | -3 < x < 13\}$ (c) $\{x | x < 13\}$ (d) $\varnothing$

10. $|8x - 4| \ge -3$
    (a) $\left\{x \middle| x \le -\dfrac{1}{8}\right\}$ (b) $\left\{x \middle| x \ge \dfrac{1}{8}\right\}$ (c) $\varnothing$ (d) all real numbers

11. $|5x - 9| = |3x - 1|$

    (a) $\{4\}$ **(b)** $\{-4\}$ **(c)** $\left\{\dfrac{5}{4}, 4\right\}$ **(d)** $\left\{-4, -\dfrac{5}{4}\right\}$

12. Fran has 300 feet of fence. She wants to build a rectangular pen with a length of at least 60 feet. What are all possible values for the width of the pen?

    (a) $w \geq 180$ ft   **(b)** $w \leq 180$ ft   **(c)** $w \geq 90$ ft   **(d)** $w \leq 90$ ft

For problems 13–15, determine which inequality describes the given graph.

13.

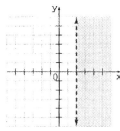

    (a) $x \geq 2$ **(b)** $x > 2$ **(c)** $y \geq 2$ **(d)** $y > 2$

14.

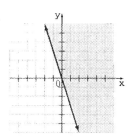

    (a) $y \geq 3x$ **(b)** $y > 3x$ **(c)** $y \geq -3x$ **(d)** $y > -3x$

15.

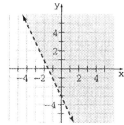

    (a) $2x + y < -3$ **(b)** $2x + y \leq -3$ **(c)** $2x + y > -3$ **(d)** $2x + y \geq -3$

For problems 16 – 19, determine the solution to the system of inequalities.

16. $x - y \leq 3$
    $2x + y < 4$

    (a)   **(b)**   **(c)**   **(d)**

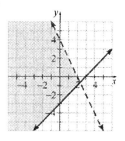
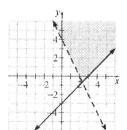
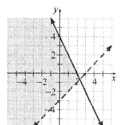
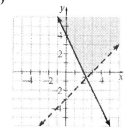

17. $y < x + 3$

$y > -\dfrac{3}{4}x + 3$

(a)

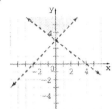

(b)

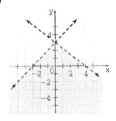

(c)

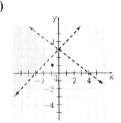

(d)

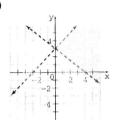

18. $3x - y > 4$
$x + y \leq 4$

(a)

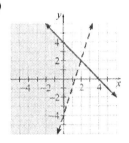

(b)

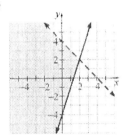

(c)

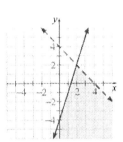

(d)

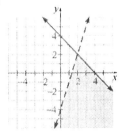

19. $|x| \leq 3$

$|y| < 2$

(a)

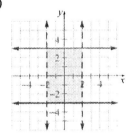

(b)

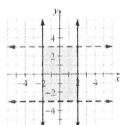

(c)

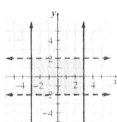

(d)

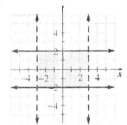

20. Identify the system of inequalities shown by the graph

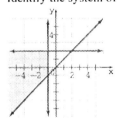

(a) $x \leq -1$
    $y \leq 2$
    $y \leq x$

(b) $x \leq -1$
    $y \leq 2$
    $y \geq x$

(c) $y \leq -1$
    $y \leq 2$
    $y \leq x$

(d) $y \leq -1$
    $x \leq 2$
    $y \geq x$

Chapter 10 Test Form H

Name:_____

Date:_____

Choose the correct answer to each problem.

For problems 1 – 2, consider the sets $A = \{1, 3, 5\}$ and $B = \{0, 1, 2, 3, 4\}$.

1. Find $A \cup B$.
   (a) $\{1, 2, 3, 4, 5\}$ (b) $\{0, 1, 2, 3, 4, 5\}$ (c) $\{1, 3, 5\}$ (d) $\{1, 3\}$

2. Find $A \cap B$.
   (a) $\{1, 2, 3, 4, 5\}$ (b) $\{0, 1, 2, 3, 4, 5\}$ (c) $\{1, 3, 5\}$ (d) $\{1, 3\}$

For problems 3 – 4, solve each inequality and graph the solution on a number line.

3. $-3x + 15 \leq 9$

   (a)
   (b)
   (c)
   (d)

4. $-3x + 6 > -9$ or $-3x < -24$

   (a)
   (b)
   (c)
   (d)

For problems 5 – 7, solve each inequality. Write each solution in interval notation.

5. $0 \leq \dfrac{4 - 2x}{3} \leq 8$
   (a) $[-10, -2]$ (b) $[-10, 2]$ (c) $[-2, 10]$ (d) $[2, 10]$

6. $x \geq -1$ or $x > 5$
   (a) $(-\infty, -1)$ (b) $(5, \infty)$ (c) $[-1, 5)$ (d) $[-1, \infty)$

7. $x < 2$ and $2x - 3 \geq -7$
   (a) $[-2, 2)$ (b) $[-2, \infty)$ (c) $(-\infty, -2)$ (d) $(-\infty, 2]$

For problems 8 – 11, find the solution set to each equation or inequality.

8. $|2x - 8| = |x - 6|$

   (a) $\{-2\}$ (b) $\{2\}$ (c) $\left\{ 2, \dfrac{14}{3} \right\}$ (d) $\{-6, -8\}$

9. $|x + 3| + 8 \leq 5$
   (a) $\{x | -16 \leq x \leq 13\}$ (b) $\{x | x \leq 10\}$ (c) $\varnothing$ (d) all real numbers

10. $|x + 5| < 7$
    (a) $\{x | -2 < x < 12\}$ (b) $\{x | -12 < x < 2\}$ (c) $\{x | -12 < x < -2\}$ (d) $\varnothing$

11. $|2x-13|>5$

   (a) $\{x|x<4 \text{ or } x>9\}$ (b) $\{x|x>9\}$ (c) all real numbers   (d) $\{x|x>-4\}$

12. Morgan's scores on four exams were 80, 95, 100 and 85. What range can the score for the fifth exam fall into for Morgan to have an average of 90 or above. Assume a maximum score of 100.

   (a) $x \geq 95$ (b) $x \geq 85$ (c) $x \geq 80$ (d) $x \geq 90$

For problems 13–15, determine which inequality describes the given graph.

13.

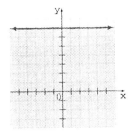

   (a) $y \leq 7$ (b) $y < 7$ (c) $x \leq 7$ (d) $x < 7$

14.

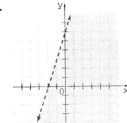

   (a) $y < -3x+6$ (b) $y \leq -3x+6$ (c) $y < 3x+6$ (d) $y \leq 3x+6$

15.

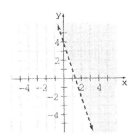

   (a) $y > 3x-4$ (b) $y > 4-3x$ (c) $y < 3x-4$ (d) $y < 4-3x$

For problems 16 – 19, determine the solution to the system of inequalities.

16. $x-y>-2$

   $y<-x$

   (a)   (b)   (c)   (d)

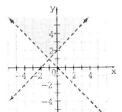 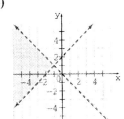 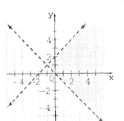 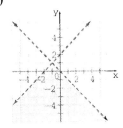

17. $x + y < 3$
    $2x - y \le -2$

(a)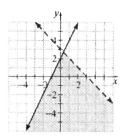

(b)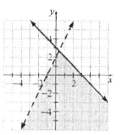

(c)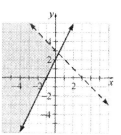

(d)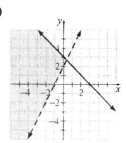

18. $y - 3x > 0$
    $2x + 3y \le 6$

(a)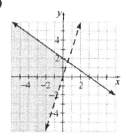

(b)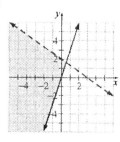

(c)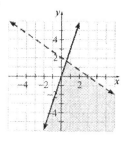

(d)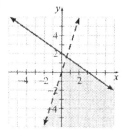

19. $|y| \le 3$

    $|x| < 2$

(a)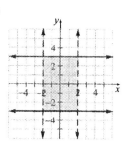

(b)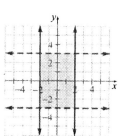

(c)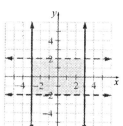

(d)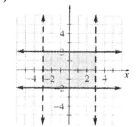

20. Identify the system of inequalities shown by the graph.

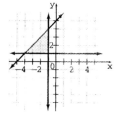

(a) $x \ge 1$
     $y \le -1$
     $y \le x - 5$

(b) $x \ge 1$
     $y \le -1$
     $y \ge x - 5$

(c) $y \ge 1$
     $x \le -1$
     $y \le x + 5$

(d) $y \ge 1$
     $x \le -1$
     $y \ge x + 5$

Cumulative Review Test 1–10 Form A

Name:_____

Date:_____

1. Illustrate the set $\left\{ x \left| -\frac{7}{2} \le x < \frac{7}{5} \right. \right\}$.

1. ←+++++++++++++++++++++→

2. Evaluate $(x-4)^2 + 3xy^2 - 7$ when $x = 4$ and $y = -2$.

2. _____

3. Simplify and leave no negative or zero exponents in the answer:
$$\left(\frac{-3a^3b^4c^0}{2a^4b^5c^3} \right)^{-2}$$

3. _____

4. Express in scientific notation: 0.000294

4. _____

The double line graph below compares high temperatures in Honolulu and Miami in August. Use the graph for questions 5 through 7. Let $H(t)$ represent Honolulu's temperature as a function of time and let $M(t)$ represent Miami's temperature as a function of time.

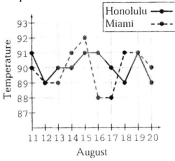

5. On which two days in the period shown did the two cities have the same high temperature?

5. _____

6. Find $(H + M)(17)$.

6. _____

7. Find $(M - H)(18)$.

7. _____

8. Evaluate $\dfrac{-b - \sqrt{b^2 - 4ac}}{2a}$ when $a = 2$, $b = -10$, and $c = 8$.

8. _____

9. Solve the equation $4[3x - 2(4x - 1)] = -[5(x + 1) + 4(3x - 1)]$.

9. _____

10. Graph $y = x^2 - 5$.

10.

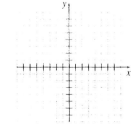

Cumulative Review Test 1–10 Form A *(cont.)*

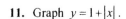

Name:_____

11. Graph $y = 1 + |x|$.

11.

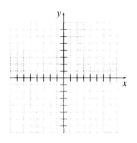

12. Determine whether or not the graph shown is a function.

12. _____

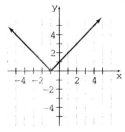

13. Write the equation of the line (in slope-intercept form) passing through the point $(4, 2)$ and perpendicular to $2x + 3y = 12$.

13. _____

14. The position on a line of an object after t seconds is given by the function: $P(t) = -\frac{7}{3}t^3 + \frac{9}{2}t^2 + 2t - 9$. Find the position of the object after 3 seconds. Round the answer to the nearest 0.1.

14. _____

15. Solve the system of equations:
$$3x - 3y - 3z = 33$$
$$2x - 2y - \ z = 18$$
$$x + \ y - 3z = 11$$

15. _____

16. How much of a 12% sulfuric acid solution should be mixed with a 20% sulfuric acid solution to get 5 liters of a 15% sulfuric acid solution?

16. _____

17. Given $f(x) = x^2 + 9x + 6$ and $g(x) = x + 6$, find $(f + g)(x)$.

17. _____

18. Factor completely: $16x^4 - 81y^4$

18. _____

19. Solve for x: $\dfrac{x}{x-1} - \dfrac{3}{x+2} = \dfrac{9}{x^2 + x - 2}$

19. _____

20. Solve for r: $\dfrac{5}{r+4} + \dfrac{2}{r+2} = \dfrac{4}{r^2 + 6r + 8}$

20. _____

Cumulative Review Test 1–10 Form B

Name:_____

Date:_____

1. Which of the following illustrates the set $\{x \mid x > -4 \text{ or } x \geq 4\}$.

   (a)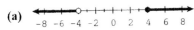

   (b)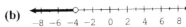

   (c)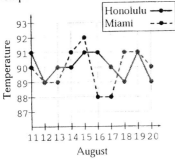

   (d)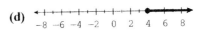

2. Evaluate $2y^2(x+y)$ when $x = 6$ and $y = 5$.

   (a) 1100   (b) 550   (c) 305   (d) 605

3. Simplify and leave no negative or zero exponents in the answer:

   $$\left(\frac{-3a^2b^4c^0}{2a^5b^6c^2}\right)^{-4}$$

   (a) $-\dfrac{16a^{12}}{81b^8c^8}$ (b) $\dfrac{16b^8c^8}{81a^{12}}$ (c) $-\dfrac{81a^{12}}{16b^8c^8}$ (d) $\dfrac{16a^{12}b^8c^8}{81}$

4. Express in scientific notation: 125,000,000

   (a) 125×10^6 (b) 125×10^{-6} (c) 1.25×10^8 (d) 1.25×10^{-8}

The double line graph below compares high temperatures in Honolulu and Miami in August. Use the graph for questions 5 through 7. Let $H(t)$ represent Honolulu's temperature as a function of time and let $M(t)$ represent Miami's temperature as a function of time.

5. For which of the days listed was Honolulu's temperature higher than Miami's?

   (a) August 14   (b) August 13   (c) August 18   (d) August 12

6. Find $(M+N)(11)$.

   (a) 181°   (b) 178°   (c) 2°   (d) 1°

7. Find $(H-M)(16)$.

   (a) 180°   (b) 179°   (c) 1°   (d) 3°

8. Evaluate $\dfrac{-b - \sqrt{b^2 - 4ac}}{2a}$ when $a = 1$, $b = -12$ and $c = 11$.

   (a) 1   (b) 2   (c) 10   (d) 12

9. Solve the equation $3\left[2x-4\left(3x+1\right)\right]=-\left[\left(10x-1\right)+9\left(5x+7\right)\right]$.

   (a) $x=2$ **(b)** $x=-2$ **(c)** $x=25$ **(d)** $x=-25$

10. Which equation matches the graph?

    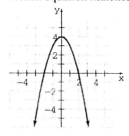

    (a) $y=4-x^2$ **(b)** $y=x^2-4$ **(c)** $y=x^2+4$ **(d)** $y=4x^2$

11. Which equation matches the graph?

    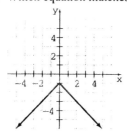

    (a) $y=\left|x\right|+1$ **(b)** $y=1-\left|x\right|$ **(c)** $y=\left|x\right|-1$ **(d)** $y=-1-\left|x\right|$

12. Determine which of the following graphs does NOT represent a function.

    (a) 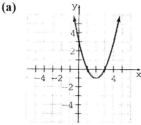   **(b)**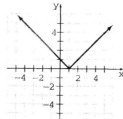

    (c) 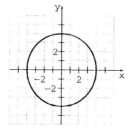   **(d)**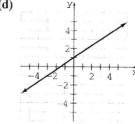

13. Find an equation of the line that passes through the point $(8,-5)$ and is parallel to the line $3x+4y=-8$.

    (a) $3x+4y=4$ **(b)** $4x-3y=47$ **(c)** $4x+3y=17$ **(d)** $3x-4y=44$

14. The area of an equilateral triangle with sides of length s can be found by the function: $f(s) = \dfrac{\sqrt{3}}{4}s^2$. Find the area of an equilateral triangle with sides of length 8.5. Round answers to the nearest 0.1.

    (a) 58.9   **(b)** 3.7   **(c)** 10.4   **(d)** 31.3

15. Solve for z:
$$3x + 2y + 3z = 5$$
$$2x + 3y - 3z = -10$$
$$x + 3y - 3z = -11$$

    (a) –2   **(b)** 2   **(c)** –6   **(d)** 1

16. Mr. Simon invests a total of $6046 in two savings accounts. One account yields 8.5% simple interest and the other yields 9% simple interest. He would like to find the amount placed in each account if a total of $522.14 interest is received after one year. Which system of linear equations expresses this problem?

    (a)
$$x + y = 522.14$$
$$0.085x + 0.09y = 6046$$

    (b)
$$x + y = 522.14$$
$$8.5x + 9y = 6046$$

    (c)
$$x + y = 6046$$
$$0.085x + 0.09y = 522.14$$

    (d)
$$x + y = 6046$$
$$8.5x + 9y = 522.14$$

17. Given $f(x) = 3x^2 - 8x + 7$ and $g(x) = x^2 - 6x - 2$, find $(f - g)(x)$.

    (a) none of these   **(b)** $2x^2 - 14x + 5$ **(c)** $2x^2 - 14x + 9$ **(d)** $2x^2 - 2x + 9$

18. Factor completely: $20x^4 - 56x^3y - 12x^2y^2$

    (a) $4x^2\left(5x^2 + 14xy - 3y^2\right)$ **(b)** $4x^2(5x + y)(x - 3y)$ **(c)** $4x^2\left(5x^2 - 14xy - 3y^2\right)$ **(d)** $4x^2(5x - y)(x + 3y)$

19. Solve for x: $\dfrac{x - 9}{x + 5} = \dfrac{x + 4}{x + 8}$

    (a) $x = -\dfrac{46}{5}$ **(b)** $x = \dfrac{26}{5}$ **(c)** $x = -\dfrac{5}{18}$ **(d)** $x = 2$

20. Solve for r: $\dfrac{4}{r + 3} + \dfrac{3}{r + 2} = \dfrac{-4}{r^2 + 5r + 6}$

    (a) $r = -3$ **(b)** $r = -2$ **(c)** $r = -4$ **(d)** no solution

Chapter 11 Pretest Form A

Name:_____

Date:_____

Evaluate.

1. $\sqrt{121}$

1. _____

2. $\sqrt[5]{-32}$

2. _____

3. Use absolute value to evaluate $\sqrt{m^2 + 8m + 16}$.

3. _____

4. Write $\sqrt{(-5b)^2}$ as an absolute value.

4. _____

Write in exponential form. Assume all variables represent positive real numbers.

5. $\sqrt{53}$

5. _____

6. $\sqrt[4]{3a}$

6. _____

Write in radical form. Assume all variables represent positive real numbers.

7. $(3x)^{1/4}$

7. _____

8. $\left(\dfrac{1}{2}x^3 y\right)^{1/4}$

8. _____

Simplify. Assume all variables represent positive real numbers.

9. $\sqrt[5]{\dfrac{3}{32}}$

9. _____

10. $\sqrt[3]{18x^{12}y^3}$

10. _____

11. $\sqrt{80c}$

11. _____

12. $\dfrac{\sqrt{180ab^4}}{\sqrt{5ab^2}}$

12. _____

13. $\sqrt{98} - \sqrt{50} - \sqrt{72}$

13. _____

14. $\left(\sqrt{2}+1\right)\left(\sqrt{2}-3\right)$ 14. _____

15. $\dfrac{\sqrt{3}}{\sqrt{3}-2}$ 15. _____

16. $\left(8-\sqrt{-1}\right)\left(-2+\sqrt{-16}\right)$ 16. _____

17. $\dfrac{-12}{7-\sqrt{-1}}$ 17. _____

Solve.

18. $2\sqrt{x}=\sqrt{5x-16}$ 18. _____

19. $2+\sqrt{u}=\sqrt{2u+7}$ 19. _____

20. $\sqrt{x}+3=\sqrt{x-3}$ 20. _____

Chapter 11 Pretest Form B

Name:_____

Date:_____

Evaluate.

1. $\sqrt{81}$

1. _____

2. $\sqrt[3]{-1000}$

2. _____

3. Use absolute value to evaluate $\sqrt{(-17)^2}$.

3. _____

4. Write $\sqrt{x^2 + 6x + 9}$ as an absolute value.

4. _____

Write in exponential form. Assume all variables represent positive real numbers.

5. $\sqrt{x^9}$

5. _____

6. $\left(\sqrt[5]{m}\right)^4$

6. _____

Write in radical form. Assume all variables represent positive real numbers.

7. $a^{1/6}$

7. _____

8. $(5m - n)^{8/5}$

8. _____

Simplify each radical expression by changing the expression to exponential form. Write the answer in radical form.

9. $\sqrt[5]{6^{10}}$

9. _____

10. $\sqrt[18]{x^3}$

10. _____

Simplify.

11. $\sqrt{45x^{10}y^{31}}$

11. _____

12. $\sqrt{\dfrac{x^5 y^6}{12z}}$

12. _____

13. $\sqrt{24} - 3\sqrt{6} + \sqrt{252}$

13. _____

14. $\left(\sqrt{6} - \sqrt{2}\right)\left(\sqrt{3} - \sqrt{2}\right)$

14. _____

15. $\dfrac{\sqrt{6}}{6 + \sqrt{30}}$

15. _____

16. $(5 + 7i) - (4 - 3i)$

16. _____

17. $\sqrt{-16}$

17. _____

Solve.

18. $\sqrt{7x - 3} = 5$

18. _____

19. $\sqrt{x - 8} = \sqrt{x} - 2$

19. _____

20. A new bypass is being constructed to relieve congestion at an intersection. The road will run from 45 feet north of the perpendicular intersection to 60 feet east of it. How long is the new road?

20. _____

Mini-Lecture 11.1
Roots and Radicals

Learning Objectives:

1. Find square roots.
2. Find cube roots.
3. Understand odd and even roots.
4. Evaluate radicals using absolute value.
5. Key Vocabulary: *radical sign, radicand, radical expression, index, square root, principal square root, cube root, even root, odd root*

Examples:

1. For the function $f(x) = \sqrt{4x+9}$, find each of the following:

   a) $f(10)$ b) $f(-2)$ c) $f(-5)$

2. For the function $f(x) = \sqrt[3]{7x-8}$, find each of the following:

   a) $g(5)$ b) $g(-5)$ c) $g(19)$

3. Indicate whether or not each radical expression is a real number. If the expression is a real number, find its value.

   a) $\sqrt[4]{-625}$ b) $-\sqrt[4]{625}$ c) $\sqrt[5]{-243}$ d) $-\sqrt[5]{-243}$

4. Use absolute value to evaluate.

   a) $\sqrt{5^2}$ b) $\sqrt{0^2}$ c) $\sqrt{(2.6)^2}$

   d) $\sqrt{(-10)^2}$ e) $\sqrt{(2x-5)^2}$ f) $\sqrt{36y^2}$

   g) $\sqrt{64x^6}$ h) $\sqrt{400m^4}$ i) $\sqrt{n^2-20n+100}$

Teaching Notes:

- Remind students that the result of a square root is always nonnegative and that the result is only rational if the radicand is a perfect square.
- Students may need extra practice to remember the difference between expressions like $-\sqrt{5}$ and $\sqrt{-5}$.
- A quick review of factoring real numbers may be helpful as students begin simplifying radical expressions.
- Encourage students to memorize as many perfect squares and perfect cubes as they can.
- Stress that calculators only give approximate values for irrational numbers. The radical form is the exact value.
- The textbook often assumes variables and radicands represent positive numbers. Remind students to pay attention to this detail in the instructions.

Answers: 1a) 7; 1b) 1; 1c) not a real number; 2a) 3; 2b) -4; 2c) 5; 3a) not a real number; 3b) real number, 5; 3c) real number, -3; 3d) real number, -3; 4a) 5; 4b) 0; 4c) 2.6; 4d) 10; 4e) $|2x-5|$; 4f) $|6y|$ or $6|y|$; 4g) $|8x^3|$ or $8|x^3|$; 4h) $20m^2$; 4i) $|n-10|$

Mini-Lecture 11.2
Rational Exponents

Learning Objectives:

1. Change a radical expression to an exponential expression.
2. Simplify radical expressions.
3. Apply the rules of exponents to rational and negative exponents.
4. Factor expressions with rational exponents.
5. Key vocabulary: *rational exponent, exponential expression, radical expression*

Examples:

1. Write each expression in exponential form.

   a) $\sqrt{13}$

   b) $\sqrt[3]{6xy}$

   c) $\sqrt[5]{x^2}$

   d) $\sqrt[6]{\dfrac{2m^5}{3n}}$

2. Write each expression in radical form.

   a) $16^{1/4}$

   b) $(-64)^{1/3}$

   c) $\left(5x^3y\right)^{1/5}$

   d) $6ab^{1/3}$

3. Write each expression in exponential form and then simplify.

   a) $\sqrt[3]{x^{15}}$

   b) $\left(\sqrt[4]{a}\right)^8$

   c) $\sqrt[5]{m^{10}n^{15}}$

4. Write each expression in radical form.

   a) $x^{3/8}$

   b) $(2ab)^{2/7}$

   c) $\left(\dfrac{x}{7}\right)^{3/4}$

5. Simplify.

   a) $9^{3/2}$

   b) $\sqrt[6]{4^3}$

   c) $\sqrt[10]{m^5}$

   d) $\left(\sqrt[9]{a}\right)^6$

6. Evaluate.

   a) $81^{-3/4}$

   b) $16^{-5/4}$

   c) $\left(\dfrac{4}{9}\right)^{-1/2}$

   d) $\left(-\dfrac{125}{8}\right)^{-1/3}$

7. Simplify each expression and write the answer without negative exponents.

   a) $x^{1/4} \cdot x^{-2/3}$

   b) $\left(3x^4y^{-2}\right)^{-1/2}$

   c) $\left(\dfrac{9x^{-2}z^{1/3}}{z^{-2/3}}\right)^{1/4}$

   d) $2.1x^{1/2}\left(1.6x^{1/3} + x^{-1/4}\right)$

8. Simplify.

   a) $\sqrt[18]{(13n)^6}$

   b) $\left(\sqrt[5]{xy^3z^4}\right)^{15}$

   c) $\sqrt[3]{\sqrt[5]{a}}$

9. Factor $r^{3/8} + r^{-5/8}$.

Teaching Notes:

- Demonstrate for students the convenience of using rational exponents to evaluate radical expressions on a calculator. For example, it is convenient to evaluate $16^{1/4}$ for $\sqrt[4]{16}$.

Answers: 1a) $13^{1/2}$; 1b) $(6xy)^{1/3}$; 1c) $x^{2/5}$; 1d) $\dfrac{2m^5}{3n}$; 2a) $\sqrt[4]{16} = 2$; 2b) $\sqrt[3]{-64} = -4$; 2c) $\sqrt[5]{5x^3y}$; 2d) $6a\sqrt[3]{b}$; 3a) $x^{15/3} = x^5$; 3b) $a^{8/4} = a^2$; 3c) $m^{10/5}n^{15/5} = m^2n^3$; 4a) $\sqrt[8]{x^3}$ or $\left(\sqrt[8]{x}\right)^3$; 4b) $\sqrt[7]{(2ab)^2}$ or $\left(\sqrt[7]{2ab}\right)^2$; 4c) $\sqrt[4]{\left(\dfrac{x}{7}\right)^3}$ or $\left(\sqrt[4]{\dfrac{x}{7}}\right)^3$; 5a) 27; 5b) 2; 5c) $\sqrt{m}$ or $m^{1/2}$; 5d) $\sqrt[3]{a^2}$ or $a^{2/3}$; 6a) $\dfrac{1}{27}$; 6b) $\dfrac{1}{32}$; 6c) $\dfrac{3}{2}$; 6d) $-\dfrac{2}{5}$; 7a) $\dfrac{1}{x^{5/12}}$; 7b) $\dfrac{y}{3^{1/2}x^2}$; 7c) $\dfrac{3^{1/2}z^{1/4}}{x^{1/2}}$; 7d) $3.36x^{5/6} + 2.1x^{1/4}$; 8a) $\sqrt[3]{13n}$; 8b) $x^3y^9z^{12}$; 8c) $\sqrt[15]{a}$; 9) $\dfrac{r+1}{r^{5/8}}$

Mini-Lecture 11.3
Simplifying Radicals

Learning Objectives:

1. Understand perfect powers.
2. Simplify radicals using the product rule for radicals.
3. Simplify radicals using the quotient rule for radicals.
4. Key vocabulary: *product rule for radicals, quotient rule for radicals, perfect power, perfect square, perfect cube*

Examples:

1. Simplify. Assume all variables represent non-negative real numbers.

   a) $\sqrt{400}$ b) $\sqrt[3]{1000}$ c) $\sqrt[4]{x^{20}}$ d) $\sqrt[3]{x^{14}}$

2. Simplify. Assume all variables and expressions in radicands are non-negative.

   a) $\sqrt{50}$ b) $\sqrt{63}$ c) $\sqrt[3]{80}$ d) $\sqrt[4]{162}$

   e) $\sqrt[3]{x^{18}}$ f) $\sqrt{y^5}$ g) $\sqrt[4]{a^{35}}$ h) $\sqrt[3]{m^{16}}$

   i) $\sqrt{x^{10}y^{13}}$ j) $\sqrt[4]{m^{10}n^{21}}$ k) $\sqrt{48a^7b^{10}c^5}$ l) $\sqrt[3]{125p^{14}q^{10}}$

3. Simplify. Assume all variables and expressions in radicands are non-negative.

   a) $\dfrac{\sqrt{63}}{\sqrt{7}}$ b) $\dfrac{\sqrt[3]{108x}}{\sqrt[3]{4x^4}}$ c) $\dfrac{\sqrt[3]{a^{-2}b^{11}}}{\sqrt[3]{a^4b^{-4}}}$

   d) $\sqrt{\dfrac{49x^4}{144}}$ e) $\sqrt[3]{\dfrac{27x^5y}{64x^2y^{16}}}$ f) $\sqrt[4]{\dfrac{20xy^{10}}{4x^{13}y^2}}$

Teaching Notes:

- Warn students to be careful to pay attention to the type of root being taken.
- Encourage students to memorize as many perfect powers as they can.

Answers: 1a) 20; 1b) 10; 1c) x^5; 1d) x^5; 2a) $5\sqrt{2}$; 2b) $3\sqrt{7}$; 2c) $2\sqrt[3]{10}$; 2d) $3\sqrt[4]{2}$; 2e) x^6; 2f) $y^2\sqrt{y}$; 2g) $a^8\sqrt[4]{a^3}$; 2h) $m^5\sqrt[3]{m}$; 2i) $x^5y^6\sqrt{y}$; 2j) $m^2n^5\sqrt[4]{m^2n}$; 2k) $4a^3b^5c^2\sqrt{3ac}$; 2l) $5p^4q^3\sqrt[3]{p^2q}$; 3a) 3; 3b) $\dfrac{3}{x}$; 3c) $\dfrac{b^5}{a^2}$; 3d) $\dfrac{7x^2}{12}$; 3e) $\dfrac{3x}{4y^5}$; 3f) $\dfrac{y^2\sqrt[4]{5}}{x^3}$

Mini-Lecture 11.4
Adding, Subtracting, and Multiplying Radicals

Learning Objectives:

1. Add and subtract radicals.
2. Multiply radicals.
3. Key vocabulary: *like radicals, unlike radicals*

Examples:

1. Simplify. Assume all variables represent non-negative real numbers.

   a) $8 + 5\sqrt{7} - \sqrt{7} + 6$
   b) $4\sqrt[3]{x} - 6 + 5\sqrt[3]{x} + 7x$
   c) $\sqrt{18} + \sqrt{72}$

   d) $10\sqrt{12} + 3\sqrt{75}$
   e) $3\sqrt{63} - 4\sqrt{28} + \sqrt{175}$
   f) $\sqrt[3]{125} + \sqrt[3]{40} - 9\sqrt[3]{5}$

   g) $\sqrt{32} + \sqrt{12} - \sqrt{18}$
   h) $\sqrt{m^6} - \sqrt{m^4 n} + m^2\sqrt{n}$
   i) $\sqrt[3]{a^{11}b} - \sqrt[3]{a^5 b^7}$

2. Multiply and simplify. Assume all variables represent non-negative real numbers.

   a) $\sqrt{15x^5}\,\sqrt{20x^4}$
   b) $\sqrt[3]{9x}\,\sqrt[3]{3x^5}$

   c) $\sqrt[4]{9ab^9}\,\sqrt[4]{36a^{10}b^{12}}$
   d) $\sqrt{7x}\left(\sqrt{14x} + \sqrt{28}\right)$

   e) $\left(\sqrt{a} + \sqrt{b}\right)\left(\sqrt{a} + b\right)$
   f) $\left(2\sqrt{10} - \sqrt{5}\right)^2$

   g) $\left(\sqrt[3]{m} - \sqrt[3]{6n}\right)\left(\sqrt[3]{m^2} + \sqrt[3]{9n^2}\right)$
   h) $\left(5 + \sqrt{7}\right)\left(5 - \sqrt{7}\right)$

3. If $f(x) = \sqrt[3]{x}$ and $g(x) = \sqrt[3]{x^5} + \sqrt[3]{x^4}$, find each of the following:

   a) $(f \cdot g)(x)$
   b) $(f \cdot g)(3)$

4. Simplify. Assume that variables may be any real number.

   a) $f(x) = \sqrt{x-2}\,\sqrt{x-2},\ x \geq 2$
   b) $g(x) = \sqrt{5x^2 - 30x + 45}$

Teaching Notes:

- A quick review of the previous discussion of collecting like terms may be helpful.

- When adding or subtracting like radicals, students sometimes mistakenly multiply the like radicals.

Answers: 1a) $14 + 4\sqrt{7}$; 1b) $9\sqrt[3]{x} + 7x - 6$; 1c) $9\sqrt{2}$; 1d) $35\sqrt{3}$; 1e) $12\sqrt{7}$; 1f) $5 - 7\sqrt[3]{5}$;
1g) $2\sqrt{3} + \sqrt{2}$; 1h) m^3; 1i) $(a^3 - ab^2)\sqrt[3]{a^2 b}$; 2a) $10x^4\sqrt{3x}$; 2b) $3x^2$; 2c) $3a^2 b^5\sqrt[4]{4a^3 b}$;
2d) $7x\sqrt{2} + 14\sqrt{x}$; 2e) $a + b\sqrt{a} + \sqrt{ab} + b\sqrt{b}$; 2f) $45 - 10\sqrt{2}$; 2g) $m + \sqrt[3]{9mn^2} - \sqrt[3]{6m^2 n} - 3n\sqrt[3]{2}$;
2h) 18; 3a) $x^2 + x\sqrt[3]{x^2}$; 3b) $9 + 3\sqrt[3]{9}$; 4a) $f(x) = x - 2$; 4b) $g(x) = \sqrt{5}\,|x - 3|$

Mini-Lecture 11.5
Dividing Radicals

Learning Objectives:

1. Rationalize denominators.
2. Rationalize a denominator using the conjugate.
3. Understand when a radical is simplified.
4. Use rationalizing the denominator in an addition problem.
5. Divide radical expressions with different indices
6. Key Vocabulary: *rationalizing a denominator, conjugate*

Examples:

1. Simplify. Assume all variables represent non-negative real numbers.

   a) $\dfrac{1}{\sqrt{11}}$

   b) $\dfrac{x}{3\sqrt{7}}$

   c) $\dfrac{15}{\sqrt{3x}}$

   d) $\dfrac{\sqrt[3]{56m^5}}{\sqrt[3]{n}}$

   e) $\dfrac{5y}{\sqrt[3]{2x^2}}$

   f) $\sqrt{\dfrac{3}{5}}$

   g) $\sqrt[3]{\dfrac{3m^2}{4n}}$

   h) $\sqrt[4]{\dfrac{405a^6b^9}{4c^3}}$

2. Simplify. Assume all variables represent non-negative real numbers.

   a) $\dfrac{23}{5-\sqrt{2}}$

   b) $\dfrac{12}{\sqrt{7}+\sqrt{3}}$

   c) $\dfrac{\sqrt{x}+y}{\sqrt{x}-y}$

3. Simplify each expression. Assume all variables represent non-negative real numbers.

   a) $\sqrt{150m^9}$

   b) $\sqrt{\dfrac{1}{7}}$

   c) $\dfrac{1}{\sqrt{10}}$

4. Simplify: $2\sqrt{5}+\dfrac{6}{\sqrt{5}}-\sqrt{45}$

5. Simplify. Assume all variables and expressions in radicands are non-negative.

   a) $\dfrac{\sqrt{x}}{\sqrt[6]{x}}$

   b) $\dfrac{\sqrt[5]{(a+b)^4}}{\sqrt[6]{(a+b)^3}}$

   c) $\dfrac{\sqrt[3]{m^4n^2}}{\sqrt[4]{m^3n}}$

Teaching Notes:

- Remind students that when they form the binomial conjugate, they only change the sign in the middle.

- Point out that, when we multiply by an expression such as $\dfrac{\sqrt{2}+\sqrt{3}}{\sqrt{2}+\sqrt{3}}$, we are really multiplying by 1.

Answers: 1a) $\dfrac{\sqrt{11}}{11}$; 1b) $\dfrac{x\sqrt{7}}{21}$; 1c) $\dfrac{5\sqrt{3x}}{x}$; 1d) $\dfrac{2m\sqrt[3]{7m^2n^2}}{n}$; 1e) $\dfrac{5y\sqrt[3]{4x}}{2x}$; 1f) $\dfrac{\sqrt{15}}{5}$; 1g) $\dfrac{\sqrt[3]{6m^2n^2}}{2n}$;

1h) $\dfrac{3ab^2\sqrt[4]{20a^2bc}}{2c}$; 2a) $5+\sqrt{2}$; 2b) $3\sqrt{7}-3\sqrt{3}$; 2c) $\dfrac{x+2y\sqrt{x}+y^2}{x-y^2}$; 3a) $5m^4\sqrt{6m}$; 3b) $\dfrac{\sqrt{7}}{7}$; 3c) $\dfrac{\sqrt{10}}{10}$;

4) $\dfrac{\sqrt{5}}{5}$; 5a) $\sqrt[6]{x}$; 5b) $\sqrt[20]{a+b}$; 5c) $\sqrt[12]{m^7n^5}$

Mini-Lecture 11.6
Solving Radical Equations

Learning Objectives:

1. Solve equations containing one radical.
2. Solve equations containing two radicals.
3. Solve equations containing two radical terms and a nonradical term.
4. Solve applications using radical equations.
5. Solve for a variable in a radicand.
6. Key vocabulary: *radical equation, isolating a radical, extraneous solutions*

Examples:

1. Solve.

   a) $\sqrt{x} = 9$

   b) $\sqrt{x+5} - 8 = 0$

   c) $\sqrt[3]{x+9} = 7$

   d) $\sqrt{x} + 4 = 0$

   e) $\sqrt{2x+31} = x - 2$

   f) $x - 4\sqrt{x} - 5 = 0$

2. Solve.

   a) $\sqrt{25x^2 + 15} = 5\sqrt{x^2 + x - 8}$

   b) $4(x-3)^{1/3} = (4x+48)^{1/3}$

3. Solve.

   a) $\sqrt{x-1} + \sqrt{x} = 2$

   b) $\sqrt{5x-1} = 1 + \sqrt{4x-3}$

4. Solve each application.

   a) Find the length of the diagonal of a rectangle that is 10 feet long and 4 feet wide.

   b) Find the period of a pendulum if its length is 3 feet. The formula $T = 2\pi\sqrt{\dfrac{L}{32}}$, where L is the length in feet, and T is the time in seconds.

5. Solve $V = \sqrt{\dfrac{2e}{m}}$ for e.

Teaching Notes:

- Remind students that solving an equation means to isolate the variable of interest and that for radical equations we are first trying to isolate the radical so we can square both sides.

- Emphasize the importance of checking solutions in the original equation.

- Be sure to explain clearly what an extraneous root is and why they might come about when solving a radical equation.

Answers: 1a) 81; 1b) 59; 1c) −8; 1d) no real solution; 1e) 9; 1f) 25; 2a) $\dfrac{43}{5}$; 2b) 4; 3a) $\dfrac{25}{16}$;

3b) 1, 13; 4a) 10.77 feet or $2\sqrt{29}$ feet; 4b) ≈ 1.92 seconds; 5) $e = \dfrac{v^2 m}{2}$

Mini-Lecture 11.7
Complex Numbers

Learning Objectives:

1. Recognize a complex number.
2. Add and subtract complex numbers.
3. Multiply complex numbers.
4. Divide complex numbers.
5. Find powers of i.
6. Key vocabulary: *imaginary numbers, the imaginary unit, complex numbers, conjugate of a complex number*

Examples:

1. Write each complex number in the form $a + bi$.

   a) $11 - \sqrt{-81}$ b) $6 + \sqrt{-54}$ c) 12

   d) $\sqrt{-72}$ e) $\sqrt{99}$ f) $3 + \sqrt{5}$

2. Add or subtract.

   a) $(-5 + 7i) + (8 - 12i) + 4$ b) $\left(6 - \sqrt{-20}\right) - \left(-7 + \sqrt{-45}\right)$

3. Multiply.

   a) $4i(7 - 3i)$ b) $\sqrt{-36}\left(\sqrt{-7} + 5\right)$ c) $\left(3 - \sqrt{-54}\right)\left(\sqrt{-6} + 4\right)$

4. Divide.

   a) $\dfrac{10 - 3i}{i}$ b) $\dfrac{2 + 5i}{5 - 3i}$

5. Evaluate.

   a) i^{46} b) i^{121}

6. Let $f(x) = x^2$. Find each of the following:

   a) $f(8i)$ b) $f(3 - 4i)$

7. Using the formula $Z = \dfrac{V}{I}$, find the impedance, Z, when $V = 2.1 + 0.4i$ and $I = 0.5i$.

Teaching Notes:

- Remind students that the imaginary unit i is not a variable. Emphasize $i = \sqrt{-1}$.

Answers: 1a) $11 - 9i$; 1b) $6 + 3i\sqrt{6}$; 1c) $12 + 0i$; 1d) $0 + 6i\sqrt{2}$; 1e) $3\sqrt{11} + 0i$; 1f) $\left(3 + \sqrt{5}\right) + 0i$; 2a) $7 - 5i$; 2b) $13 - 5i\sqrt{5}$; 3a) $12 + 28i$; 3b) $-6\sqrt{7} + 30i$; 3c) $30 - 9i\sqrt{6}$; 4a) $-3 - 10i$; 4b) $\dfrac{-5 + 31i}{34}$ or $-\dfrac{5}{34} + \dfrac{31}{34}i$; 5a) -1; 5b) i; 6a) -64; 6b) $-7 - 24i$; 7) $0.8 - 4.2i$

Additional Exercises 11.1

Name:_____

Date:_____

Evaluate the radical expression if it is a real number. If it is not real, indicate so.

1. $\sqrt{144}$

2. $\sqrt{\dfrac{25}{81}}$

3. $\sqrt{0.36}$

4. $\sqrt{2.25}$

5. $\sqrt[4]{-81}$

6. Evaluate: $\sqrt[3]{-64}$

Use absolute value to evaluate.

7. $\sqrt{(29)^2}$

8. $\sqrt{(-0.14)^2}$

Write as an absolute value.

9. $\sqrt{\left(5x^2 - y\right)^2}$

10. $\sqrt{\left(4x + 2\right)^2}$

Use absolute value to simplify. You may need to factor first.

11. $\sqrt{y^{16}}$

12. $\sqrt{x^2 - 10x + 25}$

13. $\sqrt{4a^2 - 28ab + 49b^2}$

1. _____

2. _____

3. _____

4. _____

5. _____

6. _____

7. _____

8. _____

9. _____

10. _____

11. _____

12. _____

13. _____

Additional Exercises 11.1 *(cont.)* Name:_____

Find the value of the function. Use your calculator to approximate
irrational numbers to the nearest thousandth.

14. If $f(x) = \sqrt{x+1}$, find $f(15)$.

14. _____

15. If $f(x) = \sqrt{25+5x}$, find $f(-3)$.

15. _____

16. If $f(x) = \sqrt[3]{7x^2 - 3}$, find $f(2)$.

16. _____

17. If $f(x) = \sqrt[4]{4x^2 - 3x + 9}$, find $f(-1)$.

17. _____

18. Find the domain of $\dfrac{\sqrt[3]{x-5}}{\sqrt{x+1}}$

18. _____

19. The velocity, v, of an object, in feet per second, after it has
fallen a distance, h, in feet, can be found by the formula
$v = \sqrt{64.4h}$. With what velocity will an object hit the ground
if it falls from 30 feet?

19. _____

20. Graph $f(x) = \sqrt{x-3}$.

20.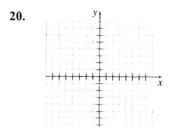

Additional Exercises 11.2

Assume all variables represent positive real numbers. Write in exponential form.

1. $\left(\sqrt[9]{x}\right)^4$

1. _____

2. $\sqrt[5]{y^3}$

2. _____

Write in radical form.

3. $\left(12b^5\right)^{2/3}$

3. _____

4. $\dfrac{1}{3^{-1/2}}$

4. _____

Simplify each radical expression by changing the expression to exponential form. Write the answer in radical form, when appropriate.

5. $\left(\sqrt[3]{x^2}\right)^3$

5. _____

6. $\sqrt[4]{x^{28}}$

6. _____

7. $\left(\sqrt[3]{a^2b}\right)^6$

7. _____

8. $\sqrt[5]{\sqrt{x^7}}$

8. _____

Evaluate if possible. If the expression is not a real number, so state.

9. $(25)^{-1/2} + (27)^{-1/3}$

9. _____

10. $\left(\dfrac{121}{4}\right)^{-1/2}$

10. _____

Simplify. Write the answer in exponential form without negative exponents.

11. $x^{5/4} \cdot x^{5/2}$

11. _____

12. $x^{-1/2} \cdot x^{-1/2}$

12. _____

13. $\left(\dfrac{16x^{3/5}}{2x^{1/3}}\right)^2$

13. _____

14. Multiply $-4y^{-3/8}\left(5y^{1/8}-y^2\right)$

14. _____

15. Use a calculator to evaluate $\sqrt[5]{209}$. Round to the nearest hundredth.

15. _____

Factor. Write the answer without negative exponents.

16. $x^{1/2}+x^{3/2}$

16. _____

17. $x^{-5}+x^{-6}$

17. _____

The formula used for carbon dating is $P=P_0\,2^{-t/5600}$ where P_0 represents the original amount of carbon 14 $\left(C_{14}\right)$ present and P represents the amount of C_{14} present after t years.

18. If 12 milligrams of C_{14} is present in a fossil now, how many milligrams will be present in 4000 years?

18. _____

19. If 25 milligrams of C_{14} is present in a piece of bone now, how much will be present in 7000 years.

19. _____

20. Find the domain of $f(x)=(x-5)^{1/2}(x+3)^{1/2}$.

20. _____

Additional Exercises 11.3

Simplify. Assume all variables represent non-negative, real numbers.

1. $\sqrt{176}$

2. $\sqrt[4]{405}$

3. $\sqrt[3]{192a^7b^3}$

4. $\sqrt[3]{x^5y^8}$

5. $\sqrt{216x^2y^5}$

6. $\sqrt[5]{1215a^6b^8}$

7. $\sqrt{100a^{100}}$

8. $\sqrt{x^4y^9}$

9. $\sqrt{1575}$

10. $\sqrt{108x^4y^5}$

11. $\sqrt[3]{375x^{14}y^{13}}$

12. $\sqrt[3]{49x^8y^4}$

13. $\sqrt[5]{128a^{13}b^{23}}$

14. $\sqrt{8x^6y^5}$

15. $\sqrt{\dfrac{32}{2}}$

16. $\sqrt{\dfrac{16}{81}}$

17. $\dfrac{\sqrt{75x^2}}{\sqrt{3}}$

18. $\sqrt{\dfrac{28x^3y^3}{7x^3y^5}}$

19. $\sqrt[3]{\dfrac{5xy}{27x^{10}}}$

20. $\sqrt[3]{\dfrac{3x^6y^9}{48x^2y}}$

1. _____

2. _____

3. _____

4. _____

5. _____

6. _____

7. _____

8. _____

9. _____

10. _____

11. _____

12. _____

13. _____

14. _____

15. _____

16. _____

17. _____

18. _____

19. _____

20. _____

Additional Exercises 11.4

Name:_____

Date:_____

Simplify. Assume all variables represent positive real numbers.

1. $5\sqrt[4]{9} - 8\sqrt[4]{9}$

2. $19\sqrt[3]{y} - 8\sqrt[3]{y} - 4\sqrt[3]{y}$

3. $6\sqrt{x} - 17 - 10\sqrt{x} + 24$

4. $\sqrt{45} - \sqrt{320}$

5. $\sqrt{63x^2 y} + x\sqrt{28y}$

6. $\sqrt[3]{640} - \sqrt[3]{80}$

7. $\sqrt[3]{-8x} - 4\sqrt[3]{x^4} + 9\sqrt[3]{x} + 3x\sqrt[3]{x}$

8. $\sqrt[3]{64x} - 9\sqrt[3]{x^4} - 6\sqrt[3]{x} + 6x\sqrt[3]{x}$

9. $\sqrt{12x^3 y^3}\sqrt{8xy^2}$

10. $\sqrt[5]{16a^7 b^{14}}\sqrt[5]{8a^6 b^9}$

11. $\left(\sqrt[3]{7x^4 y^2}\right)^2$

12. $\sqrt{3}\left(\sqrt{75} + \sqrt{27}\right)$

13. $\sqrt{6y}\left(\sqrt{12y} - \sqrt{y^3}\right)$

14. $\left(\sqrt{2} + 5\right)\left(\sqrt{2} - 4\right)$

15. $\left(-6\sqrt{x} + \sqrt{y}\right)\left(\sqrt{x} + 8\sqrt{y}\right)$

16. $\left(\sqrt{7} - 5\right)^2$

17. $\left(\sqrt[3]{4} - \sqrt[3]{5}\right)\left(\sqrt[3]{2} - \sqrt[3]{25}\right)$

18. $\left(\sqrt[3]{x} - 2\right)\left(\sqrt[3]{x^2} - 7\right)$

1. _____

2. _____

3. _____

4. _____

5. _____

6. _____

7. _____

8. _____

9. _____

10. _____

11. _____

12. _____

13. _____

14. _____

15. _____

16. _____

17. _____

18. _____

19. Find the perimeter and area of the rectangle below. Write both in simplified radical form.

19. _____

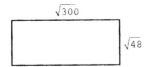

20. Find the perimeter and area of the triangle below. Write both in simplified radical form.

[·] 19. _____

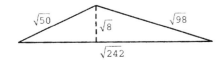

Additional Exercises 11.5

Simplify. Assume all variables represent non-negative, real numbers.

1. $\dfrac{h}{\sqrt{6}}$

2. $\dfrac{5\sqrt{6}}{\sqrt{3}}$

3. $\sqrt{\dfrac{3}{32}}$

4. $\sqrt{\dfrac{6x^4}{2x^5}}$

5. $\dfrac{10}{\sqrt{10}}$

6. $\sqrt{\dfrac{18x^3}{3x^4}}$

7. $\sqrt{\dfrac{2x^4y^5}{8z}}$

8. $\sqrt[6]{\dfrac{r^4}{9s^{20}}}$

9. $\sqrt[3]{\dfrac{16x^5y^9}{2x^6}}$

10. $\dfrac{1}{\sqrt{7}}+\dfrac{\sqrt{7}}{7}$

11. $\sqrt{\dfrac{1}{8}}+\sqrt{72}$

12. $\dfrac{\sqrt{5}}{2}+3\sqrt{\dfrac{1}{5}}+\sqrt{45}$

13. $\dfrac{3}{5-\sqrt{6}}$

1. _____

2. _____

3. _____

4. _____

5. _____

6. _____

7. _____

8. _____

9. _____

10. _____

11. _____

12. _____

13. _____

14. $\dfrac{\sqrt{x}}{\sqrt{x}+\sqrt{2}}$

14. _____

15. $\dfrac{5-\sqrt{3}}{5+\sqrt{3}}$

15. _____

16. $\dfrac{6-\sqrt{3}}{6+\sqrt{3}}$

16. _____

17. $\dfrac{\sqrt[3]{x}}{\sqrt[4]{x}}$

17. _____

18. $\dfrac{\sqrt[4]{x^{10}y^{6}}}{\sqrt[6]{x^{2}y^{4}}}$

18. _____

The radius, r, of a sphere with volume v can be found by the formula $r=\sqrt[3]{\dfrac{3v}{4\pi}}$.

19. At a state fair, a spherical decoration is to have a volume of 1436 cubic feet. Find the approximate radius of the decoration.

19. _____

20. Find the radius of a ball for a circus act, if the ball must have a volume of 113,040 cubic inches.

20. _____

Additional Exercises 11.6

Name:_____

Date:_____

Solve and check your solution(s). If the equation has no real solution, so state.

1. $\sqrt[3]{x-5} = 2$

1. _____

2. $\sqrt{x-11} + 2 = 9$

2. _____

3. $\sqrt[3]{x-9} = 6$

3. _____

4. $\sqrt{x^2 + 5x - 10} = x$

4. _____

5. $\sqrt{x^2 - x - 30} = x + 5$

5. _____

6. $\sqrt{4-x} + 4 = x$

6. _____

7. $\sqrt{m+9} + 3 = m$

7. _____

8. $(4a-9)^{1/4} = (a-3)^{1/4}$

8. _____

9. $\sqrt{x} + 2 = \sqrt{4+9x}$

9. _____

10. $\sqrt{x-11} - \sqrt{x} = -1$

10. _____

11. $\sqrt{x+4} = \sqrt{x} - 4$

11. _____

12. $5 + \sqrt{x} = \sqrt{25 + 10x}$

12. _____

Given $f(x)$ and $g(x)$, find all real values of x where $f(x) = g(x)$.

13. $f(x) = \sqrt{x+5}$, $g(x) = \sqrt{2x-3}$

13. _____

14. $f(x) = \sqrt[3]{x^2 - 3x + 5}$, $g(x) = \sqrt[3]{x^2 + 7x - 15}$

14. _____

15. $f(x) = (20x - 11)^{\frac{1}{2}}$, $g(x) = 3(4x-3)^{\frac{1}{2}}$

15. _____

Solve each formula for the variable indicated.

16. $u = \sqrt{\dfrac{HR}{a}}$ for H

16. _____

17. $r = \sqrt{\dfrac{qX}{y}}$ for y

17. _____

The formula for the period of a pendulum on the moon is $T = \dfrac{\pi\sqrt{3\ell}}{2}$, where T is the period in seconds and ℓ is the length in feet.

18. Find the period of a pendulum on the moon whose length is 12 feet.

18. _____

19. Solve the formula for the period of a pendulum on the moon for ℓ.

19. _____

20. Find the length of a pendulum that has a period of 5 seconds on the moon.

20. _____

Additional Exercises 11.7

Name:_____

Date:_____

Write each expression as a complex number in the form $a + bi$.

1. $-7 + \sqrt{-100}$

2. $3i - \sqrt{-64}$

Perform the operations as indicated.

3. $(-6 - 9i) - (8 + 5i)$

4. $(-13 + 7i) - (-4 - 11i)$

5. $\sqrt{-108} + \sqrt{-75}$

6. $\sqrt{-98} + \sqrt{-32}$

7. $\left(14 + \sqrt{-63}\right) - \left(\sqrt{49} - \sqrt{-28}\right)$

8. $-4(3 - 5i)$

9. $3i(11 - 4i)$

10. $(5 + i)(-5 - 6i)$

11. $(8 + i)(-4 + 9i)$

12. $\left(8 - \sqrt{-18}\right)\left(\frac{1}{4} + \sqrt{-2}\right)$

13. $\dfrac{7}{4i}$

14. $\dfrac{3 + 8i}{2i}$

15. $\dfrac{1 - 4i}{9 + i}$

16. $\dfrac{\sqrt{2}}{7 - \sqrt{-343}}$

Indicate whether the value is i, -1, $-i$, or 1.

17. i^{203}

18. i^{112}

19. If $f(x) = x^2$, find $f(2 - i)$.

20. If $f(x) = x^2 - 5x$, find $f(3 + i)$.

1. _____

2. _____

3. _____

4. _____

5. _____

6. _____

7. _____

8. _____

9. _____

10. _____

11. _____

12. _____

13. _____

14. _____

15. _____

16. _____

17. _____

18. _____

19. _____

20. _____

Chapter 11 Test Form A

Name:_____

Date:_____

1. Evaluate: $-\sqrt[3]{-125}$

   1. _____

2. Evaluate: $\sqrt{\dfrac{25}{144}}$

   2. _____

3. Write as an absolute value: $\sqrt{4x^2 - 12x + 9}$

   3. _____

4. Write in radical form: $\left(7b^2c\right)^{3/5}$

   4. _____

5. Write in exponential form: $\sqrt[5]{x^3 y^2}$

   5. _____

For questions 6 – 17, assume all variables represent positive real numbers.

6. Simplify: $\sqrt[4]{x^{12}}$

   6. _____

7. Simplify: $\sqrt{75}$

   7. _____

8. Simplify: $-\sqrt{20x^6 y^7 z^{12}}$

   8. _____

9. Simplify: $\sqrt[4]{\dfrac{20x^4}{81x^{-8}}}$

   9. _____

10. Simplify: $\sqrt{8} - \sqrt{12}$

   10. _____

11. Simplify: $\left(\sqrt{x} + y\right)\left(\sqrt{x} - y\right)$

   11. _____

12. Simplify: $\sqrt[4]{3x^9 y^{12}} \; \sqrt[4]{54x^4 y^7}$

   12. _____

13. Simplify: $\dfrac{x}{\sqrt{13}}$

   13. _____

14. Rationalize the denominator: $\dfrac{5}{\sqrt{2} + 1}$

   14. _____

15. Rationalize the denominator: $\dfrac{\sqrt{c} - \sqrt{2d}}{\sqrt{c} - \sqrt{d}}$

   15. _____

Chapter 11 Test Form A *(cont.)*

16. Simplify: $\dfrac{2}{\sqrt{50}} - 3\sqrt{50} - \dfrac{1}{\sqrt{8}}$

16. _____

17. Divide: $\dfrac{12 - \sqrt{-12}}{\sqrt{3} + \sqrt{-5}}$

17. _____

18. Solve and check solution(s): $\sqrt{z^2 + 3} = z + 1$

18. _____

19. Solve and check solution(s): $\sqrt{x + 1} = 2 - \sqrt{x}$

19. _____

20. Write the complex number $21 - \sqrt{-36}$ in the form $a + bi$.

20. _____

Chapter 11 Test Form B

1. Evaluate : $\sqrt[3]{\dfrac{1}{27}}$

   1. _____

2. Use absolute value to evaluate $\sqrt{(5x-8)^2}$.

   2. _____

3. Write in radical form: $\left(7x^2+2y^3\right)^{-1/6}$

   3. _____

4. Simplify: $\sqrt{y^6}$

   4. _____

5. Simplify and write the answer in exponential form without negative exponents: $\left(\dfrac{81z^{1/4}y^3}{9z^{1/4}}\right)^{1/2}$

   5. _____

6. Multiply: $-9z^{3/2}\left(z^{3/2}-z^{-3/2}\right)$

   6. _____

7. Simplify: $\sqrt[4]{80}$

   7. _____

For questions 8 – 9, assume all variables are positive. Simplify the given expressions.

8. $\sqrt[4]{48x^{11}y^{21}}$

   8. _____

9. $\dfrac{\sqrt{150a^{10}b^{11}}}{\sqrt{2ab^2}}$

   9. _____

Simplify each expression in questions 10 – 12.

10. $3\sqrt{5}+\sqrt{500}-\sqrt{80}$

    10. _____

11. $\left(3\sqrt{a}-7\sqrt{b}\right)\left(3\sqrt{a}+7\sqrt{b}\right)$

    11. _____

12. $\sqrt{3}\left(\sqrt{75}+\sqrt{15}\right)$

    12. _____

13. Simplify: $\sqrt{\dfrac{20y^4z^3}{3xy^{-2}}}$

    13. _____

Chapter 11 Test Form B *(cont.)*

Rationalize the denominator in problems 14 – 15.

14. $\dfrac{5}{\sqrt{6} + \sqrt{5}}$

14. _____

15. $\dfrac{2}{\sqrt{x + 2} - 3}$

15. _____

16. Simplify: $2\sqrt{\dfrac{8}{3}} - 4\sqrt{\dfrac{100}{6}}$

16. _____

Solve and check your solution(s) in questions 17 – 18.

17. $\sqrt{x} + 2x = 1$

17. _____

18. $\sqrt{y + 1} = \sqrt{y + 5} - 2$

18. _____

19. Add: $\left(\sqrt{20} - \sqrt{-12}\right) + \left(2\sqrt{5} + \sqrt{-75}\right)$

19. _____

20. Divide: $\dfrac{\sqrt{10} + \sqrt{-3}}{5 - \sqrt{-20}}$

20. _____

Chapter 11 Test Form C

Name:_____

Date:_____

1. Use absolute value to evaluate $\sqrt{(-6)^2}$.

2. Write as an absolute value. $\sqrt{x^2 - 10x + 25}$

3. Simplify $\left(\dfrac{27x^3}{-y^9}\right)^{1/3}$

4. Graph $f(x) = \sqrt{x-2}$

5. Write in exponential form: $\sqrt[3]{x^2 y}$

Simplify. Assume that all variables represent positive real numbers.

6. $\sqrt{24a^9 b^6}$

7. $\sqrt{14x}\sqrt{7xy^2}$

8. $\dfrac{\sqrt{60x^4}}{\sqrt{12x}}$

9. $\sqrt{\dfrac{7}{3x}}$

10. $\dfrac{-2}{1-\sqrt{2}}$

11. $\sqrt{8} + 2\sqrt{32}$

12. $7\sqrt{64x} - 2\sqrt{25x} - 4\sqrt{36x}$

13. $\left(6+\sqrt{2}\right)\left(2-\sqrt{2}\right)$

14. $\sqrt[3]{\sqrt{xy^2}}$

15. $\dfrac{\sqrt[3]{x^4}}{\sqrt{x}}$

1. _____

2. _____

3. _____

4.

5. _____

6. _____

7. _____

8. _____

9. _____

10. _____

11. _____

12. _____

13. _____

14. _____

15. _____

Solve.

16. $\sqrt[3]{n-2} = 4$

16. _____

17. $\sqrt{13-x} + 1 = x$

17. _____

18. Multiply $(2+5i)(4-2i)$

18. _____

19. Divide $\dfrac{2+3i}{1-2i}$

19. _____

20. Evaluate $x^2 - x + 1$ for $x = 3 - 2i$.

20. _____

Chapter 11 Test Form D

Name:_____

Date:_____

1. Use absolute value to evaluate $\sqrt{(-9)^2}$.

2. Write as an absolute value. $\sqrt{4x^2 - 4x + 1}$

3. Simplify $x^{1/2} \cdot x^{1/4}$.

4. Graph $f(x) = \sqrt{x} + 2$

5. Write in exponential form: $\sqrt[6]{xy^5}$

Simplify. Assume that all variables represent positive real numbers.

6. $\sqrt{45x^2y^3z^5}$

7. $\sqrt{2x}\sqrt{18y^2}$

8. $\sqrt{\dfrac{42a^3b^5}{14a^2b}}$

9. $\sqrt{\dfrac{6}{5}}$

10. $\dfrac{3}{2-\sqrt{3}}$

11. $\sqrt[3]{81} + 6\sqrt[3]{3} - \sqrt[3]{24}$

12. $2a\sqrt{27ab^5} + 3b^2\sqrt{3a^3b}$

13. $\left(\sqrt{5} - 6\right)\left(\sqrt{5} - 3\right)$

14. $\sqrt{72x^6y^4z}$

1. _____

2. _____

3. _____

4.

5. _____

6. _____

7. _____

8. _____

9. _____

10. _____

11. _____

12. _____

13. _____

14. _____

15. $\dfrac{\sqrt[3]{x^5}}{\sqrt[4]{x^3}}$

15. _____

Solve.

16. $\sqrt{7x-3} = 2$

16. _____

17. $\sqrt{2x-1} = x-2$

17. _____

18. Multiply $(4+3i)(6+i)$

18. _____

19. Divide $\dfrac{-4-7i}{6i}$

19. _____

20. Evaluate $x^2 + 5x - 2$ for $x = 2 - 3i$.

20. _____

Chapter 11 Test Form E

1. Use absolute value to evaluate $\sqrt{\left(-\dfrac{1}{8}\right)^2}$.

   1. _____

2. Write as an absolute value. $\sqrt{x^2 + 2x + 1}$

   2. _____

3. Simplify $\left(\dfrac{x^{-1/3}}{y^2}\right)^3$

   3. _____

4. Graph $f(x) = -\sqrt{x}$

   4.

5. Write in exponential form: $\sqrt[10]{x^9 y^7 z^3}$

   5. _____

Simplify. Assume that all variables represent positive real numbers.

6. $\sqrt[3]{-8x^5 y^9}$

   6. _____

7. $\sqrt{27x^3}\,\sqrt{3xy^2}$

   7. _____

8. $\sqrt{\dfrac{75a^4 b^7}{3ab^3}}$

   8. _____

9. $\sqrt[3]{\dfrac{7}{9}}$

   9. _____

10. $\dfrac{4}{3+\sqrt{2}}$

    10. _____

11. $\dfrac{3\sqrt{2}}{2} + \dfrac{1}{\sqrt{2}}$

    11. _____

12. $3\sqrt{8x^2} + 2\sqrt{18x^2} - 4x\sqrt{2}$

    12. _____

13. $\left(\sqrt{7} - 2\right)\left(\sqrt{7} - 8\right)$

    13. _____

14. $\sqrt{\sqrt[3]{x^2 y^4}}$

    14. _____

15. $\dfrac{\sqrt{x}}{\sqrt[4]{x}}$

    15. _____

Chapter 11 Test Form E *(cont.)*

Name:_____

Solve.

16. $\sqrt[3]{4x-3}=5$

16. _____

17. $\sqrt{2x+32}-4=x$

17. _____

18. Multiply $(2+3i)(4+5i)$

18. _____

19. Divide $\dfrac{4-5i}{2i}$

19. _____

20. Evaluate x^2-3 for $x=2-5i$.

20. _____

Chapter 11 Test Form F

1. Use absolute value to evaluate $\sqrt{\left(-\dfrac{4}{9}\right)^2}$.

    1. _____

2. Write as an absolute value. $\sqrt{x^2 - 20x + 100}$

    2. _____

3. Simplify $x^{2/5} \cdot x^2$

    3. _____

4. Graph $f(x) = \sqrt{x} + 2$

    4.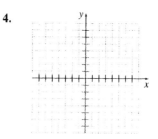

5. Write in radical form: $\left(x^3 y^2\right)^{1/5}$

    5. _____

Simplify. Assume that all variables represent positive real numbers.

6. $\sqrt{50x^2 y^6}$

    6. _____

7. $\sqrt{3}\left(\sqrt{27} - \sqrt{3}\right)$

    7. _____

8. $\sqrt[3]{\dfrac{48xy^7}{6y^4}}$

    8. _____

9. $\dfrac{5}{\sqrt[3]{a}}$

    9. _____

10. $\dfrac{2}{\sqrt{5} + 2}$

    10. _____

11. $\sqrt{300} - \sqrt{12} + \sqrt{3}$

    11. _____

12. $5\sqrt{9x} - 3\sqrt{4x} + 6\sqrt{16x}$

    12. _____

13. $\left(\sqrt{6} + 4\right)\left(\sqrt{6} - 4\right)$

    13. _____

14. $\sqrt[3]{\sqrt{x^4 y^3}}$

    14. _____

15. $\dfrac{\sqrt{x}}{\sqrt[3]{x}}$

    15. _____

Chapter 11 Test Form F *(cont.)*

Solve.

16. $\sqrt[3]{3x+2} = 2$

16. _____

17. $\sqrt{2x+14} = x+3$

17. _____

18. Multiply $(3+2i)(5+4i)$

18. _____

19. Divide $\dfrac{-5i}{2-4i}$

19. _____

20. Evaluate $x^2 + x + 1$ for $x = 2-i$.

20. _____

Chapter 11 Test Form G

Name:_____

Date:_____

1. Use absolute value to evaluate. $\sqrt{(-16)^2}$

   (a) -16　　　　　　(b) 16　　　　　　(c) -4　　　　　　(d) 4

2. Write as an absolute value. $\sqrt{x^2 - 8x + 16}$

   (a) $|x + 4|$　　　　(b) $|x - 4|$　　　　(c) $|x^2 - 8x + 16|$　　　　(d) $|x^2 + 8x + 16|$

3. Simplify $\left(\dfrac{x^{-1/3}}{y^{1/3}} \right)^3$.

   (a) $\dfrac{y^3}{x^3}$　　　　(b) $\dfrac{1}{x^3 y^3}$　　　　(c) $\dfrac{1}{xy}$　　　　(d) $\dfrac{y}{x}$

4. Write $x^{6/5}$ as a simplified radical.

   (a) $\sqrt[6]{x^5}$　　　　(b) $\sqrt[5]{x^6}$　　　　(c) $x\sqrt[6]{x}$　　　　(d) $x\sqrt[5]{x}$

5. Graph $f(x) = \sqrt{x} - 2$

   (a) 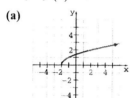　　　(b) 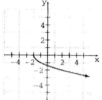　　　(c) 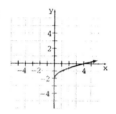　　　(d)

Simplify. Assume that all variables represent positive real numbers.

6. $\sqrt{8x^3 y^5}$

   (a) $4xy^2\sqrt{2xy}$　　(b) $2xy^2\sqrt{2xy}$　　(c) $4x^2 y^4\sqrt{2xy}$　　(d) $2xy^3\sqrt{4xy^2}$

7. $\sqrt[3]{4a^2}\,\sqrt[3]{8a}$

   (a) $2\sqrt[3]{4a}$　　　(b) $4\sqrt[3]{2a}$　　　(c) $4a\sqrt[3]{2}$　　　(d) $2a\sqrt[3]{4}$

8. $\dfrac{\sqrt{40xy^3}}{\sqrt{8x}}$

   (a) $y\sqrt{5y}$　　　(b) $5y$　　　(c) $5\sqrt{y}$　　　(d) $\sqrt{5y}$

9. $\sqrt{\dfrac{7a^2}{b}}$

   (a) $a\sqrt{7b}$　　　(b) $\dfrac{a\sqrt{7b}}{b}$　　　(c) $\dfrac{ab}{\sqrt{7}}$　　　(d) $\dfrac{ab\sqrt{7}}{b}$

10. $\dfrac{7}{2 - \sqrt{3}}$

    (a) $14 - 7\sqrt{3}$　　(b) $14 + 7\sqrt{3}$　　(c) $-14 - 7\sqrt{3}$　　(d) $-14 + 7\sqrt{3}$

11. $\sqrt[3]{16} - 2\sqrt[3]{2}$

    (a) 0   (b) $2\sqrt[3]{2}$ (c) $4\sqrt[3]{2}$ (d) $8\sqrt[3]{2}$

12. $4\sqrt{20x} + 5\sqrt{45x} - 10\sqrt{80x}$

    (a) $33\sqrt{5x}$ (b) $101\sqrt{5x}$ (c) $-17\sqrt{5x}$ (d) $51\sqrt{5x}$

13. $\left(\sqrt{3}+1\right)^2$

    (a) 4   (b) $4 + 2\sqrt{3}$ (c) $4 + \sqrt{3}$ (d) $4 + \sqrt{6}$

14. $\dfrac{8}{2\sqrt[3]{4}}$

    (a) $2\sqrt[3]{4}$ (b) $4\sqrt[3]{4}$ (c) $\sqrt[3]{2}$ (d) $4\sqrt[3]{2}$

15. $\dfrac{\sqrt[5]{x^4}}{\sqrt[3]{x}}$

    (a) $\sqrt[5]{x^3}$ (b) $\sqrt[3]{x}$ (c) $\sqrt[15]{x^7}$ (d) $\sqrt[15]{x^3}$

16. Solve. $\sqrt{x-2} - 6 = 0$

    (a) 4   (b) 8   (c) 38   (d) no real solution

17. Solve. $\sqrt{x+7} = x - 5$

    (a) 2, 9   (b) 2   (c) 9   (d) no real solution

18. Multiply $(2+3i)(4-3i)$

    (a) $8 - 9i$ (b) -1 (c) $-1 + 6i$ (d) $17 + 6i$

19. Divide $\dfrac{4}{5-2i}$

    (a) $\dfrac{20-8i}{3}$ (b) $\dfrac{20-8i}{29}$ (c) $\dfrac{20+8i}{3}$ (d) $\dfrac{20+8i}{29}$

20. Evaluate $x^2 - 2x + 3$ for $x = 1 - i$.

    (a) 3   (b) 0   (c) 1   (d) -1

Chapter 11 Test Form H

Name:_____

Date:_____

1. Use absolute value to evaluate $\sqrt{(-36)^2}$

   (a) -6 (b) 6 (c) -36 (d) 36

2. Write as an absolute value $\sqrt{x^2 - 8x + 16}$

   (a) $|x - 4|$ (b) $|x + 4|$ (c) $|x^2 - 8x + 16|$ (d) $|x^2 + 8x + 16|$

3. Simplify $x^{1/5} \cdot x^{1/2}$.

   (a) $x^{7/10}$ (b) $x^{1/10}$ (c) $x^{2/5}$ (d) $x^{5/2}$

4. Write $x^{7/5}$ in simplified radical form.

   (a) $x\sqrt[7]{x^2}$ (b) $x\sqrt[5]{x^2}$ (c) $\sqrt[7]{x^5}$ (d) $\sqrt[5]{x^7}$

5. Graph $f(x) = \sqrt{x + 1}$

   (a) 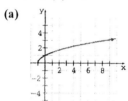   (b) 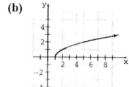   (c) 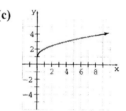   (d)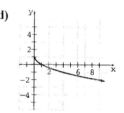

Simplify. Assume that all variables represent positive real numbers.

6. $\sqrt{32x^2 y^3}$

   (a) $8xy^2\sqrt{xy}$ (b) $2xy\sqrt{8y^2}$ (c) $8xy\sqrt{4xy}$ (d) $4xy\sqrt{2y}$

7. $\sqrt[3]{x^2}\sqrt[3]{16x^4}$

   (a) $2x\sqrt[3]{2x}$ (b) $2x^2\sqrt[3]{2}$ (c) $4x\sqrt[3]{2x}$ (d) $2x\sqrt[3]{2x^2}$

8. $\sqrt{\dfrac{50x^3 y^2}{2xy}}$

   (a) $5\sqrt{xy}$ (b) $5y\sqrt{x}$ (c) $5x\sqrt{y}$ (d) $\sqrt{5xy}$

9. $\sqrt[3]{\dfrac{3}{5}}$

   (a) $\dfrac{\sqrt[3]{45}}{5}$ (b) $\dfrac{\sqrt[3]{15}}{5}$ (c) $\dfrac{\sqrt[3]{75}}{5}$ (d) $\dfrac{3\sqrt[3]{5}}{5}$

10. $\dfrac{-1}{\sqrt{3} - 2}$

   (a) $\sqrt{3} + 2$ (b) $\sqrt{3} - 2$ (c) $-\sqrt{3} + 2$ (d) $-\sqrt{3} - 2$

11. $3\sqrt{27} + 5\sqrt{12}$

 (a) $16\sqrt{6}$ (b) $47\sqrt{3}$ (c) $19\sqrt{2}$ (d) $19\sqrt{3}$

12. $5\sqrt{27n} - \sqrt{12n} - 6\sqrt{3n}$

 (a) $35\sqrt{3n}$ (b) $7\sqrt{3n}$ (c) $-2\sqrt{3n}$ (d) $-11\sqrt{3n}$

13. $\left(\sqrt{3}+1\right)\left(\sqrt{3}-5\right)$

 (a) $-2-4\sqrt{3}$ (b) $-2+4\sqrt{3}$ (c) $2-4\sqrt{3}$ (d) $2+4\sqrt{3}$

14. $\sqrt{\sqrt{x^2 y}}$

 (a) $\sqrt[3]{x^2 y}$ (b) $x\sqrt[4]{y}$ (c) $\sqrt[4]{x^2 y}$ (d) $x\sqrt{y}$

15. $\dfrac{\sqrt[4]{x^3}}{\sqrt[3]{x^2}}$

 (a) $\sqrt[3]{x^4}$ (b) $\sqrt[12]{x}$ (c) $\sqrt[12]{x^5}$ (d) $\sqrt[4]{x^5}$

16. Solve. $\sqrt{3x-5} = 8$

 (a) 21 (b) $\dfrac{59}{3}$ (c) $\dfrac{1}{3}$ (d) no real solution

17. Solve. $\sqrt{x-2} = x-4$

 (a) $6, 3$ (b) 6 (c) 3 (d) no real solution

18. Multiply $\left(5-\sqrt{-2}\right)\left(3+\sqrt{-2}\right)$

 (a) $17+2i\sqrt{2}$ (b) $13+2i\sqrt{2}$ (c) $17-2i\sqrt{2}$ (d) $13-2i\sqrt{2}$

19. Divide $\dfrac{4}{5i}$

 (a) $\dfrac{i}{5}$ (b) $\dfrac{5i}{4}$ (c) $-\dfrac{4i}{5}$ (d) $\dfrac{4i}{5}$

20. Evaluate $x^2 - 4$ for $x = 2-3i$.

 (a) 9 (b) $-9-12i$ (c) -9 (d) $9-12i$

Chapter 12 Pretest Form A

Name:_____

Date:_____

1. Solve by completing the square: $x^2 + 2x - 8 = 0$

   1. _____

2. Solve by the quadratic formula: $5x^2 + 5x + 1 = 0$

   2. _____

3. Solve: $-3x = \dfrac{x^2}{2} + 2$

   3. _____

Determine whether each equation has two distinct real solutions, a single real solution, or no real solutions.

4. $4x^2 - 4x + 1 = 0$

   4. _____

5. $6x^2 - 5x - 6 = 0$

   5. _____

6. $x(x - 3) = -10$

   6. _____

7. Solve the formula $d = \sqrt{l^2 + w^2 + h^2}$ for $h > 0$.

   7. _____

8. Write a quadratic equation that has a solution set of $\{3, 5\}$.

   8. _____

9. Graph the function $f(x) = 2x^2 - 4x - 1$.

   9.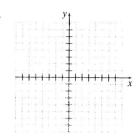

For problems 10–14, consider the quadratic equation $f(x) = x^2 - x - 6$.

10. Determine whether the parabola opens upward or downward.

    10. _____

11. Find the *axis of symmetry*.

    11. _____

12. Find the vertex.

    12. _____

13. Find the *x*-intercepts if they exist.

    13. _____

14. Draw the graph of $2y = x^2 - 2$.

14.

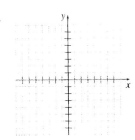

15. Graph the inequality $y = x^2 + 1$.

15.

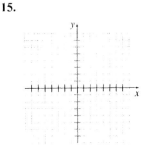

16. Solve the inequality and write the answer in interval notation.

$$\frac{x^2 - 10x + 25}{x + 5} < 0$$

16. _____

For problems 17 and 18, use the following information. A ball is thrown straight up with a velocity of 128 feet per second. The function $s = h(t) = -16t^2 + 128t$ gives the relation between s (the number of feet the ball is above the ground) and t (the time measured in seconds.)

17. How high will the ball go?

17. _____

18. How long does it take the ball to hit the ground?

18. _____

19. Solve: $\dfrac{2x}{x-3} \le \dfrac{x}{3-x}$

19. _____

20. Solve: $\dfrac{-5}{x+2} \ge \dfrac{4}{2-x}$

20. _____

Chapter 12 Pretest Form B

1. Solve by completing the square: $x^2 + 12x - 4 = 0$

    1. _____

Solve by the quadratic formula.

2. $x^2 + 8x = 20$

    2. _____

3. $x^2 - 4x + 7 = 0$

    3. _____

Determine whether each equation has two distinct real solutions, a single real solution, or no real solutions.

4. $x^2 - 20x + 100 = 0$

    4. _____

5. $2x^2 + 3x = 35$

    5. _____

6. $3x^2 + 8 = 5x$

    6. _____

7. Solve the formula $P = \sqrt{a^2 + b}$ for $a > 0$.

    7. _____

8. Write a function that has x-intercepts -2 and $\dfrac{3}{5}$.

    8. _____

9. Graph the function $f(x) = (x+2)^2 - 5$.

    9.

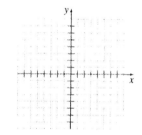

For problems 10–14, consider the quadratic equation $y = -x^2 + 4x$.

10. Determine whether the parabola opens upward or downward.

    10. _____

11. Find the *axis of symmetry*.

    11. _____

12. Find the vertex.

    12. _____

13. Find the x-intercepts if they exist.

    13. _____

14. Draw the graph.

    14.

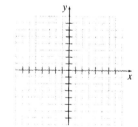

15. Solve the inequality and graph the solution on the number line.
$x^2 - x < 30$

    15. _____

16. Solve the inequality and write the answer in interval notation.
$2x^2 - 3x \le 2$

    16. _____

For problems 17 and 18, use the following information.

The cost, C, and revenue, R, equations for a company are given below. The x represents the number of items produced and sold. Profit is revenue minus cost.

$$C(x) = 7000 + 16x$$

$$R(x) = 400x - x^2$$

17. Determine the number of items that must be sold to maximize profit.

17. _____

18. Determine the maximum profit of the company.

18. _____

19. Solve the inequality and write the answer in interval notation.

19. _____

$$\frac{x+3}{2} \le \frac{3}{x-2}$$

20. Solve the inequality and write the answer in interval notation.

20. _____

$$\frac{-3}{x+6} \ge \frac{2}{5-x}$$

Mini-Lecture 12.1
Solving Quadratic Equations by Completing the Square

Learning Objectives:

1. Use the square root property to solve equations.
2. Understand perfect square trinomials.
3. Solve quadratic equations by completing the square.
4. Key vocabulary: *square root property, perfect square trinomial, completing the square.*

Examples:

1. Solve the following equations:

   a) $x^2 - 2 = 7$ b) $x^2 - 14 = 0$ c) $x^2 + 7 = 9$ d) $x^2 + 7 = 4$ e) $(x+1)^2 - 9 = 3$

2. Each of the following is a perfect square trinomial; find the missing term.

   a) $x^2 + 3x + \underline{\quad}$ b) $x^2 + \underline{\quad} + 9$

3. Solve each equation by completing the square:.

   a) $x^2 - 7x + 10 = 0$ b) $x^2 + 3x - 10 = 0$ c) $-3x^2 + 6x - 4 = 13$

   d) $2x^2 + 6x = 8$ e) $2x^2 - 12x + 23 = 5$ f) $x^2 - 6x + 8 = 5$

4. \$10,000. is invested in a savings account that compounds interest quarterly. After

   5 years the account has \$12,663.02. What is the annual interest rate? $A = P\left(1 + \frac{r}{n}\right)^{nt}$.

Teaching Notes:

- Students should understand that if p is a positive number, then $x^2 = p$ has two solutions but $x = \sqrt{p}$ has only one.
- Remind students that the first step in completing the square is always to get a lead coefficient of 1.
- Point out that solving a quadratic equation by factoring is not always (practically) possible, but the method of completing the square always gives a definitive answer.
- Mention that the process of completing the square has no natural extension to equations of higher degree.
- Be sure students know how to find the square roots of a negative number.

Answers: 1a) ± 3; 1b) $\pm\sqrt{14}$; 1c) $\pm\sqrt{2}$; 1d) $\pm i\sqrt{3}$; 1e) $-1 \pm 2\sqrt{3}$; 2a) $\frac{9}{4}$;

2b) $6x$; 3a) $5, 2$; 3b) $-5, 2$; 3c) $1 \pm \frac{i\sqrt{42}}{3}$; 3d) $-4, 1$; 3e) 3; 3f) $3 \pm \sqrt{6}$; 4) 4.75%

Mini-Lecture 12.2
Solving Quadratic Equations by the Quadratic Formula

Learning Objectives:

1. Derive the quadratic formula
2. Use the quadratic formula to solve equations.
3. Determine a quadratic equation given its solutions.
4. Use the discriminant to determine the number of real solutions to a quadratic equation.
5. Study applications that use quadratic equations.
6. Key vocabulary: *quadratic formula, discriminant.*

Examples:

1. Use the quadratic formula to find the solutions of the following equations:

   a) $2x^2 - 5x + 3 = 1$ b) $x^2 + 7x + 10 = 3$ c) $x^2 - 4x + 7 = 0$

2. Find the quadratic equation (with lead coefficient 1) whose solutions are

   a) $-3, 7$ b) $1 + \sqrt{3}, 1 - \sqrt{3}$ c) $-i\sqrt{3}, i\sqrt{3}$

3. Use the discriminant to determine the number and type of solutions the equation has

   a) $4x^2 - 12x + 9 = 4$ b) $x^2 - 6x + 11 = 2$ c) $2x^2 + 5x + 3 = -2$

4. The equation for the height of a ball thrown into the air is $h = -16t^2 + 40t + 50$

   (t is time in seconds). How long after a ball is thrown will it be 30 ft above the ground?

Teaching Notes:

- Students should memorize the quadratic formula.
- Emphasize that the equation must be in the proper form ($ax^2 + bx + c = 0$) before determining $a, b,$ and c for the quadratic formula.
- Make sure students understand how the discriminant of a quadratic equation determines the number and type of solutions.
- Explain that in section 8.5 we will see that the graphs of quadratic functions (parabolas) cross the x-axis at 0, 1 or 2 points, depending on the sign of the discriminant.

Answers: *1a)* $2, \frac{1}{2}$; *1b)* $-\frac{7}{2} \pm \frac{\sqrt{21}}{2}$; *1c)* $2 \pm i\sqrt{3}$; *2a)* $x^2 - 4x - 21 = 0$; *2b)* $x^2 - 2x - 2 = 0$; *2c)* $x^2 + 3 = 0$; *3a)* 2 *real;* *3b)* 1 *real;* *3c)* 2 *complex;* *4)* $\approx 2.93 \, sec$

Mini-Lecture 12.3
Quadratic Equations: Applications and Problem Solving

Learning Objectives:

1. Solve additional applications of quadratic equations.
2. Solve for a variable in a formula.

Examples:

1. A company's profit (in thousands of dollars) can be approximated over the next

   15 years by the function $p(n) = 1.6n^2 + 5n - 31$ (n = years from now).

   a) Estimate the profit 7 years from now.

   b) Estimate the time needed for the company to break even.

2. The function $N(t) = 0.0054t^2 - 0.46t + 95.11$ can be used to estimate the average age at death of a person who is currently t years old ($30 \le t \le 100$).

   a) If a person is currently 90 years old, how long can he expect to live?

   b) Joe is over 40 and can expect to live to age 86; how old is he?

3. a) The length of a rectangular garden is 2 feet less than 4 times its width; if the area is 3192 sq. ft., find its dimensions.

   b) Tom and Bob can paint a room together in 3 hours; working alone, it takes Bob 1.5 hours longer than Tom. How long does it take each one individually?

4. Solve each of the following for the variable w:

   a) $d = \sqrt{l^2 + w^2 + h^2}$ b) $M = N\sqrt{\left(a + \dfrac{w^2}{b^2}\right)}$ c) $u^2 + v^2 + w^2 = uv + uw + vw$

   d) $a = b\sqrt{1 - \dfrac{c^2}{w^2}}$ e) $\dfrac{w^2 - a}{w + b} = 1$

Teaching Notes:

- Remind students they must determine, from the context of the problem, whether numeric solutions make sense..
- Have students note that, in real-life situations, answers are not always integral ("nice").
- Emphasize that, in solving a literal equation for a particular value, at some point all terms involving the particular value must be isolated on one side of the equation..

Answers: *1a) $82,000.; 1b) ≈ 3.11 years; 2a) to about 97.45; 2b) ≈ 53.87; 3a) 112, 28.5;*

3b) Tom 5.34 hrs, Bob 6.84 hrs.; 4a) $w = \sqrt{d^2 - l^2 - h^2}$; 4b) $w = \dfrac{b}{N}\sqrt{M^2 - aN^2}$;

4c) $w = \dfrac{1}{2}\left[(u + v) \pm \sqrt{(u + v)^2 - 4(u^2 + v^2 - uv)}\right]$; 4d) $w = bc\sqrt{\dfrac{1}{b^2 - a^2}}$; 4e) $w = \dfrac{1}{2}\left[1 \pm \sqrt{1 + 4(a + b)}\right]$

Mini-Lecture 12.4
Factoring Expressions and Solving Equations That Are Quadratic in Form

Learning Objectives:

1. Factor trinomials that are quadratic in form.
2. Solve equations that are quadratic in form.
3. Solve equations with rational exponents.
4. Vocabulary: *expressions that are in quadratic form, rational exponent.*

Examples:

1. Factor each trinomial completely.

   a) $x^4 - 2x^2 - 3$

   b) $(x+3)^2 - 3(x+3) - 10$

   c) $2x^5 + 4x^3 - 16x$

2. Solve each of the following equations for x:

   a) $x^4 - 5x^2 + 6 = 0$

   b) $x^4 - x^2 - 20 = 0$

   c) $2(x+1)^2 - 3(x+1) = 2$

   d) $x^6 - 7x^3 - 8 = 0$

   e) $(x^2 - 3)^2 + (x^2 - 3) = 2$

   f) $4x^{-4} + 1 = 5x^{-2}$

3. Solve each of the following equations for x:

   a) $5x^{\frac{2}{3}} - 3x^{\frac{1}{3}} = 2$

   b) $6p + 6 = 13\sqrt{p}$

   c) $x - \sqrt{x} - 6 = 0$

Teaching Notes:

- Stress that the final answer(s) must be in terms of the *original* variable, and checked as such.

- Rule: whenever you raise both sides of an equation to a power, you must check all apparent solutions in the *original* equation to make sure that none is extraneous.

Answers: 4a) $(x^2 + 1)(x^2 - 3)$; 4b) $(x + 5)(x - 2)$; 4c) $2x(x^2 + 4)(x^2 - 2)$; 2a) $\pm\sqrt{2}, \pm\sqrt{3}$; 2b) $\pm\sqrt{5}, \pm 2i$; 2c) $-\dfrac{3}{2}, 1$; 2d) $2, -1$; e) $\pm 1, \pm 2$; 2f) $\pm 1, \pm 2$; 3a) $-\dfrac{8}{125}, 1$; 3b) $\dfrac{4}{9}, \dfrac{9}{4}$; 3c) 9

Mini-Lecture 12.5
Graphing Quadratic Functions

Learning Objectives:

1. Identify some key characteristics of graphs of polynomial functions.
2. Find the axis of symmetry, vertex, and *x*-intercepts of a parabola.
3. Graph quadratic functions using the axis of symmetry, vertex, and intercepts.
4. Solve maximum and minimum problems.
5. Understand translation of parabolas.
6. Write functions in the form $f(x) = a(x-h)^2 + k$
7. Vocabulary: *parabola, vertex, axis of symmetry, maximum (minimum) value, translation*

Examples:

1. For each parabola, find the *axis of symmetry, vertex,* and the *x-intercepts;* determine whether the vertex is a *max* or *min*, and *graph* the function:

   a) $f(x) = -x^2 + 2x + 8$ b) $f(x) = 4x^2 - 12x + 9$

   c) $f(x) = x^2 + 2x + 2$ d) $f(x) = \frac{1}{6}x^2 + x$

2. Write each of the following in the form $f(x) = a(x-h)^2 + k$:

   a) $y = 2x^2 - 6x + 5$ b) $y = x^2 + 6x + 9$ c) $y = -3x^2 + 12x + 1$

Teaching Notes:

- Stress that parabolas look like the (elongated) letter "U", not "V".
- Point out that the student need only memorize the formula for the *x*-coordinate of the vertex; the *y*-coordinate is obtained by substitution into the function.
- Point out that the graph of $y = ax^2$ gets narrower as $|a|$ increases.
- Explain that for $f(x) = a(x-h)^2 + k$, *h* determines the horizontal shift and *k* determines the vertical shift.
- When discussing functions of the form $f(x) = a(x-h)^2 + k$, emphasize that the amount added inside the parentheses to make a perfect square trinomial, must also be multiplied by $-a$ and added to the function.

Answers: *1)*

	axis	vertex	x – intercepts	max/min	graph
a	$x = 1$	$(1,9)$	$x = -2, 4$	*max*	*
b	$x = \frac{3}{2}$	$(\frac{3}{2}, 0)$	$x = \frac{3}{2}$	*min*	*
c	$x = -1$	$(-1, 1)$	*none*	*min*	*
d	$x = -3$	$(-3, -\frac{3}{2})$	$x = 0, -6$	*min*	*

* *see graphing solutions*

2a) $y = 2(x - \frac{3}{2})^2 + \frac{1}{2}$;   *2b)* $y = (x - (-3))^2 + 0$;   *2c)* $y = -3(x-2)^2 + 13$

Mini-Lecture 12.6
Quadratic and Other Inequalities in One Variable

Learning Objectives:

1. Solve quadratic inequalities.
2. Solve other polynomial inequalities.
3. Solve rational inequalities.
4. Vocabulary: *quadratic inequality, sign graph, boundary value, test value, polynomial inequality, rational inequality.*

Examples:

Solve each inequality and write the solution in interval notation:

1. a) $2x^2 + 5x - 3 \leq 0$

   b) $x^2 + x > 7(x+1)$

   c) $4x^2 - 4x + 7 < 6$

   d) $x^4 - 5x^2 + 4 \geq 0$

2. a) $(2x-1)(x+1)(x+3)(3x-7)(x-4) > 0$

   b) $x^3 - x^2 - 6x < 0$

   c) $-2x^3 - 7x^2 + 4x \geq 0$

   d) $x^4 - 3x^3 \geq 10x^2$

   e) $x^4 + x^2 \leq 2x^3$

3. a) $\dfrac{x+1}{x+2} < 3$

   b) $\dfrac{x+12}{x+2} \geq x$

   c) $\dfrac{x+1}{x+3} > \dfrac{2x-1}{x+1}$

Teaching Notes:

- Point out that the boundary points on the number line are the *x*-intercepts of the parabola on a coordinate graph.
- Introduce rational inequalities by first having the student graph an example and guess at the solution; then solve algebraically.

Answers: 1a) $\left[-3, \frac{1}{2}\right]$; 1b) $(-\infty, -1) \cup (7, \infty)$; 1c) $\varnothing$; 1d) $(-\infty, -2] \cup [-1, 1] \cup [2, \infty)$

2a) $(-3, -1) \cup \left(\frac{1}{2}, \frac{7}{3}\right) \cup (4, \infty)$; 2b) $(-\infty, -2) \cup (0, 3)$; 2c) $(-\infty, -4] \cup \left[0, \frac{1}{2}\right]$;

2d) $(-\infty, -2] \cup [0] \cup [5, \infty)$; 2e) $[0] \cup [1] = \{0, 1\}$

3a) $\left(-\infty, -\frac{5}{2}\right) \cup (-2, \infty)$; 3b) $(-\infty, -4] \cup (-2, 3]$; 3c) $(-4, -3) \cup (-1, 1)$

447

Additional Exercises 12.1

1. What number must be added to $x^2 - 5x$ in order to produce a trinomial that is the square of a binomial?

    1. _____

2. What number must be added to $x^2 + x$ in order to produce a trinomial that is the square of a binomial?

    2. _____

3. Find the missing term: $(x+9)^2 = x^2 + 18x +$ _____

    3. _____

4. If $x^2 +$ ___ $+ 49$ is a perfect square trinomial, fill in the blank.

    4. _____

5. Solve by completing the square: $-6x = 3x^2 - 2$

    5. _____

6. Solve by completing the square: $5x^2 + 30x = -70$

    6. _____

7. Solve by completing the square $2x^2 + 6x + 2 = 2$

    7. _____

8. Solve by completing the square: $8x = 4x^2 - 1$

    8. _____

9. What number must be added to $x^2 + 3x$ in order to produce a trinomial that is the square of a binomial?

    9. _____

10. What number must be added to $x^2 - 7x$ in order to produce a trinomial that is the square of a binomial?

    10. _____

11. Find the missing terms: $(x+3)^2 = x^2 +$ ___ $+$ _____

    11. _____

12. Find the missing term: $(x+8)^2 = x^2 + 16x +$ _____

    12. _____

13. Solve by completing the square: $-7x = 3x^2 - 1$

    13. _____

14. Solve by completing the square: $2x^2 + 8x = -14$

    14. _____

15. Solve by completing the square: $2x^2 - 2x - 6 = 0$

    15. _____

Additional Exercises 12.1 *(cont.)*

16. Solve by completing the square: $-8x = 4x^2 - 1$

16. _____

17. Solve by completing the square: $2x^2 - x + 5 = 0$

17. _____

18. Solve by completing the square: $x^2 - 3x + 2 = 0$

18. _____

19. Solve by completing the square: $x^2 + x + 1 = 0$

19. _____

20. A man puts $1000. in a savings account where interest compounded monthly. After 3 years, the account contains $1233. What is the annual interest rate?

20. _____

Additional Exercises 12.2

1. Solve for x: $px^2 + qx + r = 0$

    1. _____

2. Solve for x: $ax^2 + bx + c = 0$

    2. _____

3. Solve by the quadratic formula: $x^2 = x + 1$

    3. _____

4. Solve by the quadratic formula: $x^2 + 46 = -14x$

    4. _____

5. Find the real roots of the equation: $3x^2 - 1 = 5x$

    5. _____

6. Solve using the quadratic formula: $7x^2 + 5x = 5$

    6. _____

7. Find an equation with roots -4 and $\dfrac{5}{4}$.

    7. _____

8. Write a quadratic equation with integer coefficients that has solutions $\dfrac{2}{3}, -\dfrac{3}{2}$

    8. _____

9. Find a quadratic equation with solutions -4 and $-\dfrac{2}{7}$.

    9. _____

10. Find a quadratic equation with solutions -3 and $\dfrac{5}{3}$.

    10. _____

11. Determine whether the following equation has two distinct real solutions, a single unique solution, or no real solution. $3x^2 + 2x + 4 = 0$

    11. _____

12. Determine whether the following equation has two distinct real solutions, a single unique solution, or no real solution. $4x^2 - 4x + 5 = 4$

    12. _____

13. Determine the character of the roots of the equation: $2x^2 - 5x - 2 = 0$

    13. _____

14. Determine the character of the roots of the equation: $4x^2 + 4x + 3 = 0$

    14. _____

15. Solve for x: $gx^2 + hx + k = 0$

    15. _____

16. Solve by the quadratic formula: $10x^2 - 3x = 1$

    16. _____

17. Solve by the quadratic formula: $x^2 = 5x - 3$

    17. _____

18. Solve by the quadratic formula: $x^2 + 79 = -18x$

    18. _____

19. Find the real roots of the equation: $3x^2 + 1 = 6x$

    19. _____

20. An internet company has a special rate for quantity buying. Its gadgets ordinarily sell for $25. each, but for every gadget over 50 the price per unit is reduced by $0.10. If the company has a limit of 150 gadgets per order and John spent $2160, how many gadgets did he buy?

    20. _____

Additional Exercises 12.3

1. Solve: $(4x-5)^2 = 16$

2. Solve: $(9x-3)^2 = 30$

3. Solve for x: $x^2 = 49$

4. Solve for x: $(2x-1)^2 = 4x+6$

5. Solve $Z = \frac{1}{3}sb^2$ for b.

6. Solve $c = 3d + 7f^2$ for f.

7. Solve: $(x+4)^2 = 4$

8. Solve: $(7x-3)^2 = 15$

9. Solve for x: $(x-3)^2 = 9$

10. Solve for x: $(2x-1)^2 - (x+1)^2 + 2 = 0$

11. Solve $A = \frac{1}{6}gf^2$ for f.

12. Solve $j = 8k + 5m^2$ for m.

13. The Changs wish to plant a uniform strip of grass around their swimming pool. If the pool measures 18 feet by 25 feet and there is only enough seed to cover 408 square feet, what will be the width of the uniform strip?

14. The length of a rectangle is 4 feet greater than three times its width. Find the length and width of the rectangle if its area is 39 square feet.

15. The distance d (in meters, m) traveled by an object thrown downward with an initial velocity of v_0 after t seconds is given by the formula $d = 5t^2 + v_0 t$. Find the number of seconds it takes an object to hit the ground if the object is dropped from a height of 45 m.

16. The sum of two numbers is 20 and their product is 80. Find the two numbers.

17. The value, V, of a corn crop per acre, in dollars, d days after planting is given by the formula $V = 12d - 0.05d^2$, $20 < d < 80$. Find the value of an acre of corn after it has been planted 40 days.

1. _____

2. _____

3. _____

4. _____

5. _____

6. _____

7. _____

8. _____

9. _____

10. _____

11. _____

12. _____

13. _____

14. _____

15. _____

16. _____

17. _____

18. In a total of 2 hours, a tugboat traveled upriver 5 miles and returned. If the river's current is 4 miles per hour, find the speed of the tugboat in still water. Round your answer to the nearest 0.1 mi/hr if necessary.

18. _____

19. If the revenue is given by $R = 120x - 0.04x^2$, find the value of x that yields the maximum revenue.

19. _____

20. Kerry throws a ball upward from the top of a building. The distance, d, in feet, of the ball from the ground at any time t can be found by the formula $d = -16t^2 + 128t + 82$.
(a) Find the time the object reaches its maximum height.
(b) Find the maximum height.

20. **(a)** _____

(b) _____

Additional Exercises 12.4

Name:_____

Date:_____

1. Solve for x: $x^{-2} + 9x^{-1} + 8 = 0$

2. Solve for x: $x^{-2} + 13x^{-1} + 40 = 0$

3. Solve for x: $(x^2 + 2)^2 - 12(x^2 + 2) + 11 = 0$

4. Solve for x: $x^4 - 10x^2 + 9 = 0$

5. Solve for x: $x^{\frac{1}{2}} - 6x^{\frac{1}{4}} + 5 = 0$

6. Solve for x: $x^{2/3} - 2x^{1/3} = 3$

7. Solve for x: $x - 13\sqrt{x} + 42 = 0$

8. Solve for x: $x - 17\sqrt{x} + 70 = 0$

9. Solve for x: $x^{-2} + 4x^{-1} + 3 = 0$

10. Solve for x: $x^{-2} - 3x^{-1} - 10 = 0$

11. Solve for x: $x^4 - 11x^2 + 10 = 0$

12. Solve for x: $x^4 - 16x^2 + 15 = 0$

13. Solve for x: $x^{\frac{1}{2}} - 6x^{\frac{1}{4}} + 8 = 0$

14. Solve for x: $x^{-4} - 7x^{-2} + 10 = 0$

15. Solve for x: $x - 13\sqrt{x} + 30 = 0$

16. Solve for x: $x - 14\sqrt{x} + 33 = 0$

17. Solve for x: $x^4 + 2x^2 - 3 = 0$

18. Solve for x: $x + 6\sqrt{x} + 8 = 0$

19. Solve for x: $x - 6\sqrt{x} + 8 = 0$

20. Solve for x: $x^6 - 2x^3 - 3 = 0$

1. _____

2. _____

3. _____

4. _____

5. _____

6. _____

7. _____

8. _____

9. _____

10. _____

11. _____

12. _____

13. _____

14. _____

15. _____

16. _____

17. _____

18. _____

19. _____

20. _____

Additional Exercises 12.5

In 1-3, find the *axis of symmetry, the vertex, and the x-intercepts* of the parabola:

1. $f(x) = x^2 + 4x + 1$

2. $f(x) = x^2 - 4x - 1$

3. $f(x) = 2x^2 - 4x$

4. Graph the following equation, and determine the *x*-intercepts, if they exist. $y = -x^2 + 4x - 3$

5. Graph: $f(x) = x^2 - 4x + 1$

6. Graph: $f(x) = x^2 + 2x - 2$

7. Graph: $y = -4x^2 + 12x$

1. _____

2. _____

3. _____

4. _____

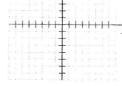

5.

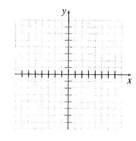

6.

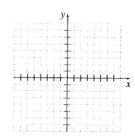

7.

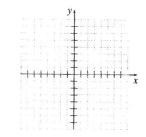

Additional Exercises 12.5 *(cont.)*

8. Graph the following equation, and determine the *x*-intercepts, if they exist. $y = x^2 - 4x + 3$.

8. _____

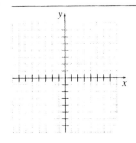

Graph.

9. $y = x^2 - 6x + 5$

9.

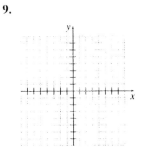

10. $y = x^2 + 6x + 4$

10.

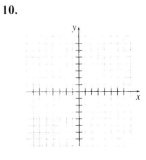

11. $y = -x^2 + 2x + 5$

11.

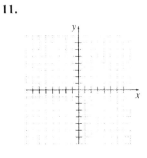

12. $y = -x^2 - 6x - 6$

12.

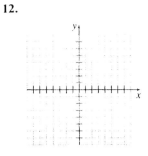

13. $y = x^2 - 4x$

13.

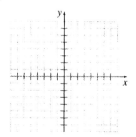

14. $y = -x^2 - 4x$

14.

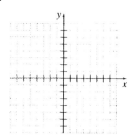

15. $y = x^2 - 2x - 5$

15.

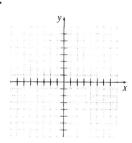

16. $y = x^2 - 3$

16.

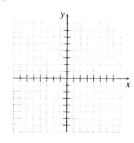

17. $y = -x^2 + 5$

17.

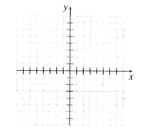

Name:_____

Write each of the following in the form $y = a(x - h)^2 + k$

18. $y = -2x^2 - 4x - 1$

19. $y = 2x^2 - 6x + 7$

20. Of all rectangles that have a perimeter of 144 inches, find the dimensions of the one with greatest area.

18. _____

19. _____

20. _____

Additional Exercises 12.6

Name:_____

Date:_____

1. Solve for x: $(x-4)(3x+4) \geq 0$

    1. _____

2. Solve the inequality and graph the solution on the number line.
   $x^2 + x \geq 6$

    2. ++++++++++++++++++++

3. Solve for x: $-x^2 - 15x - 54 > 0$

    3. _____

4. Solve for x: $2x^2 \geq 3x + 5$

    4. _____

5. Solve the inequality: $(x-1)(x+3)(x+8) > 0$

    5. _____

6. Solve the inequality: $(x-2)(x+3)(x+5) > 0$

    6. _____

7. Solve the inequality: $\dfrac{x+7}{x-3} \leq 0$

    7. _____

8. Graph the solution on the number line.
   $\dfrac{(x-1)(x-6)}{(x+5)} \geq 0$

    8. ++++++++++++++++++++

9. The graph of $y = \dfrac{x^2 + 2x - 3}{x+2}$ is graphed below. Determine
   the solutions to the following inequalities.

    (a) $\dfrac{x^2 + 2x - 3}{x+2} < 0$

    (b) $\dfrac{x^2 + 2x - 3}{x+2} > 0$

    9. (a) _____

       (b) _____

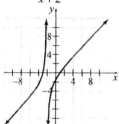

10. Solve for x: $x^3 < x^2 + 6x$

    10. _____

11. Solve for x: $(x-8)(5x+3) \leq 0$

    11. _____

12. Solve the inequality and graph the solution on the number line.
    $x^2 - x \geq 42$

    12. ++++++++++++++++++++

13. Solve for x: $-x^2 - 7x - 10 > 0$

    13. _____

14. Solve for x: $x^4 - 5x^2 + 4 < 0$

    14. _____

15. Solve the inequality: $(x-4)(x+2)(x+9) > 0$

    15. _____

16. Solve the inequality: $(x-3)(x+5)(x+8) > 0$

    16. _____

Additional Exercises 12.6 (cont.)

Name:_____

17. Solve the inequality: $\dfrac{x+2}{x-7} \le 0$

17. _____

18. Graph the solution on the number line.

$\dfrac{(x-3)(x-6)}{(x+3)} \ge 0$

18.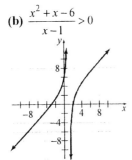

19. The graph of $y = \dfrac{x^2+x-6}{x-1}$ is graphed below. Determine the solutions to the following inequalities.

(a) $\dfrac{x^2+x-6}{x-1} < 0$

(b) $\dfrac{x^2+x-6}{x-1} > 0$

19. (a) _____

(b) _____

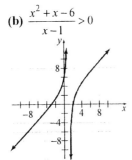

20. Solve the inequality: $\dfrac{x-3}{x+5} > x$

20. _____

Chapter 12 Test Form A

Name:_____

Date:_____

Solve each equation by completing the square.

1. $x^2 - 6x + 5 = 0$

2. $x^2 - 9x + 18 = 0$

Solve each equation using the quadratic formula.

3. $a^2 + 6a + 8 = 0$

4. $2x^2 + 5x - 3 = 0$

5. Determine whether the following equation has two distinct real solutions, a single unique solution, or no real solution: $2x^2 = 16x - 32$

6. Write a function that has the given solutions: $\{\bar{3}, \bar{5}\}$

7. Solve the formula $F = G\left(\dfrac{m_1 m_2}{r^2}\right)$ for r (Newton's Law of Gravity).

Solve each of the following equations.

8. $9d^4 - 10d^2 + 1 = 0$

9. $2b + 7\sqrt{b} = 22$

10. $\left(x^2 - 1\right)^2 + 3\left(x^2 - 1\right) + 2 = 0$

11. Find all x intercepts of the function $g(x) = x - 13\sqrt{x} + 36$.

12. Write an equation of the form $ax^4 + bx^2 + c = 0$ that has solutions $\pm\sqrt{3}$ and $\pm\sqrt{2}\,i$.

For questions 13–17, consider the function $n(x) = -x^2 - 2x + 24$.

13. Determine whether the parabola opens upward or downward.

14. Find the axis of symmetry.

15. Find the vertex.

16. Find the x-intercepts, if any.

1. _____

2. _____

3. _____

4. _____

5. _____

6. _____

7. _____

8. _____

9. _____

10. _____

11. _____

12. _____

13. _____

14. _____

15. _____

16. _____

Name:_____

17. Draw the graph of $g(x) = -2(x-3)^2 + 1$.

17.

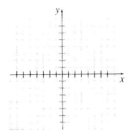

18. Find the equation of the parabola with vertex at $(2,-3)$ and containing the point $(0,7)$.

18. _____

19. Solve the inequality and give the solution in set builder notation:

$$\frac{x-4}{x+6} > 0$$

19. _____

20. Solve the inequality and give the solution in interval notation:

$$\frac{(x+1)(x-5)}{x+3} \le 0$$

20. _____

Chapter 12 Test Form B

Name:_____

Date:_____

Solve each equation by completing the square.

1. $-x^2 + 3x + 4 = 0$

1. _____

2. $2x^2 = 8x + 90$

2. _____

Solve each equation using the quadratic formula.

3. $c^2 - 3c = 0$

3. _____

4. $r^2 - 4r + 8 = 0$

4. _____

5. Determine whether the following equation has two distinct real solutions, a single unique solution, or no real solution:
$b^2 = -2b - \dfrac{9}{4}$

5. _____

6. Write a function that has the given solutions: $\left\{\sqrt{5}, -\sqrt{5}\right\}$

6. _____

7. Solve the formula $a^2 + b^2 = c^2$ for b, with $b \geq 0$.

7. _____

Solve each of the following equations.

8. $a^4 - a^2 = 30$

8. _____

9. $\sqrt{x} = 2x - 6$

9. _____

10. $8x + 2\sqrt{x} = 3$

10. _____

11. Find all x intercepts of the function $g(x) = 4x^{-2} + 12x^{-1} + 9$.

11. _____

12. Solve the equation: $1 = \dfrac{2}{x} - \dfrac{2}{x^2}$.

12. _____

For questions 13–17, consider the function $m(x) = 3x^2 + 4x + 3$.

13. Determine whether the parabola opens upward or downward.

13. _____

14. Find the y-intercept.

14. _____

15. Find the vertex.

15. _____

16. Find the x-intercepts, if any.

16. _____

17. Find the equation of a parabola whose axis of symmetry is $x = 2$, y-intercept is $(0, 5)$ and has an x-intercept of $(5, 0)$.

17. _____

Chapter 12 Test Form B *(cont.)*

18. Graph the function $f(x) = x^2 + 6x + 10$.

18.

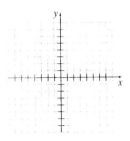

19. Solve the inequality and give the solution in set builder notation:

$$\frac{3y+6}{y+6} \le 0$$

19. _____

20. Solve the inequality and give the solution in interval notation:

$$\frac{2r+6}{r-3} \le r$$

20. _____

Chapter 12 Test Form C

Name:_____

Date:_____

Solve each equation by completing the square.

1. $x^2 + 2x - 80 = 0$

2. $16x^2 = 8x + 15$

Solve each equation using the quadratic formula.

3. $15x^2 - x - 2 = 0$

4. $2x^2 = 4x - 7$

5. Determine whether the following equation has two distinct real solutions, a single unique solution, or no real solution: $x^2 + 7x + 5 = 0$

6. Write a function that has x-intercepts $\dfrac{1}{2}$ and $\dfrac{1}{3}$.

7. Solve the formula $d = \sqrt{l^2 + w^2 + h^2}$ for l.

Solve each equation.

8. $x^4 - x^2 - 12 = 0$

9. $x + 3 = 4\sqrt{x}$

10. $\left(x^2 - 2\right)^2 - \left(x^2 - 2\right) - 6 = 0$

11. Find all x intercepts of the function $f(x) = x^{\frac{2}{3}} - 16$.

12. Write an equation that is quadratic in form and has solutions $\pm i$ and $\pm\sqrt{3}$.

For questions 13–17, consider the function $f(x) = x^2 - 4x + 3$.

13. Determine whether the parabola opens upward or downward.

14. Find the *axis of symmetry*.

15. Find the vertex.

16. Find the x-intercepts, if any.

17. Draw the graph.

1. _____

2. _____

3. _____

4. _____

5. _____

6. _____

7. _____

8. _____

9. _____

10. _____

11. _____

12. _____

13. _____

14. _____

15. _____

16. _____

17.

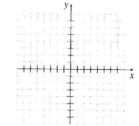

Chapter 12 Test Form C *(cont.)*

Name:_____

18. Graph the function $f(x) = -(x+1)^2$.

18.

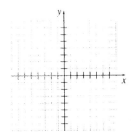

Graph the solution to the following inequalities on the number line.

19. $x^2 - 4 \geq 0$

19. ←++-+-+-++++++++++++++++++→

20. $x^2 + x - 30 < 0$

20. ←++-+-+-+++++++++++++++++++→

For questions 21 and 22, solve the inequality $2x^2 - 7x + 5 \geq 0$. Write the answer in...

21. interval notation.

21. _____

22. set notation.

22. _____

23. The product of two integers is 187, and one is 6 more than the other. Find the pair(s) of integers.

23. _____

24. Todd is constructing a tree house for his children. The flooring of the tree house is a rectangular piece of plywood. Find the dimensions of the tree house floor if the length is 2 feet less than twice its width, and the area is 24 square feet.

24. _____

25. Solve for x and write your answer in interval notation:
$$\frac{(x+13)}{(x-2)} \leq (x+1)$$

25. _____

Chapter 12 Test Form D

Solve each equation by completing the square.

1. $x^2 - 4x - 96 = 0$

2. $9x^2 - 54x + 77 = 0$

Solve each equation using the quadratic formula.

3. $x^2 - x - 30 = 0$

4. $2x^2 = 7x - 5$

5. Determine whether the following equation has two distinct real solutions, a single unique solution, or no real solution: $2x^2 + 5x + 3 = 0$

6. Write a function that has x-intercepts 7 and –4.

7. Solve the formula $A = \pi r^2$ for r.

Solve each equation.

8. $x^4 - 3x^2 - 4 = 0$

9. $2\sqrt{x} + 35 = x$

10. $\left(x^2 + 4\right)^2 - 8\left(x^2 + 4\right) + 15 = 0$

11. Find all x intercepts of the function $f(x) = x^{\frac{2}{3}} - 4x^{\frac{1}{3}} - 5$.

12. Write an equation that is quadratic in form and has solutions ± 1 and $\pm i\sqrt{2}$.

For questions 13–17, consider the function $f(x) = -x^2 + 4x - 5$.

13. Determine whether the parabola opens upward or downward.

14. Find the *axis of symmetry*.

15. Find the vertex.

16. Find the x-intercepts, if any.

17. Draw the graph.

1. _____

2. _____

3. _____

4. _____

5. _____

6. _____

7. _____

8. _____

9. _____

10. _____

11. _____

12. _____

13. _____

14. _____

15. _____

16. _____

17.

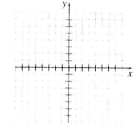

18. Find the equation of a parabola with vertex $(-2,-2)$ and y-intercept $(0,2)$.

18. _____

Graph the solution to the following inequalities on the number line.

19. $x^2 + 8x + 12 < 0$

19. ←+++++++++++++++++++++++→

20. $x^2 - 2x - 15 \geq 0$

20. ←+++++++++++++++++++++++→

For questions 21 and 22, solve the inequality and write the answer in interval notation.

21. $\dfrac{x-2}{x+1} \geq 0$

21. _____

22. $\dfrac{x+5}{x+1} \leq x - 1$

22. _____

23. The product of two consecutive odd integers is 35. Find the pair(s) of odd integers.

23. _____

24. Tom initially invested \$300 in a savings account whose interest is compounded annually. If after 2 years the amount in the account is \$318.27, find the annual interest rate.

24. _____

25. The Garcias wish to plant a uniform strip of grass around their swimming pool. If the pool measures 58 feet by 44 feet and there is only enough seed to cover 1120 square feet, what will be the width of the uniform strip?

25. _____

Chapter 12 Test Form E

Solve each equation by completing the square.

1. $x^2 + 8x - 105 = 0$

2. $25x^2 + 300x = -864$

Solve each equation using the quadratic formula.

3. $3x^2 - 5x + 2 = 0$

4. $x^2 - 2x + \dfrac{5}{4} = 0$

5. Determine whether the following equation has two distinct real solutions, a single unique solution, or no real solution: $5x^2 + 10x + 5 = 0$

6. Write a function that has x-intercepts -3 and 8.

7. Solve the formula $f_x^2 + f_y^2 = f^2$ for f_y.

Solve each equation.

8. $2x^4 + 10x^2 - 72 = 0$

9. $3\sqrt{x} = x - 4$

10. $\left(x^2 + 6\right)^2 - 10\left(x^2 + 6\right) + 24 = 0$

11. Find all x intercepts of the function $f(x) = x^{\frac{2}{3}} + 4x^{\frac{1}{3}} - 12$.

12. Write an equation that is quadratic in form and has solutions ± 2 and $\pm i$.

For questions 13–17, consider the function $f(x) = x^2 + 2x + 2$.

13. Determine whether the parabola opens upward or downward.

14. Find the *axis of symmetry*.

15. Find the vertex.

16. Find the x-intercepts, if any.

17. Draw the graph.

1. _____

2. _____

3. _____

4. _____

5. _____

6. _____

7. _____

8. _____

9. _____

10. _____

11. _____

12. _____

13. _____

14. _____

15. _____

16. _____

17.

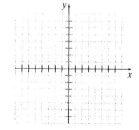

Name:_____

18. Writer the equation of a parabola whose vertex is $(0,1)$ and which contains the point $(\sqrt{3},10)$

18. _____

Graph the solution to the following inequalities on the number line.

19. $x^2 + x - 20 > 0$

19. ←+++-+++++-+++++++-+++++→

20. $x^2 - 25 \leq 0$

20. ←++-+++-+++-++++-++++-+++→

For questions 21 and 22, solve the inequality $2x^2 + 7x - 4 < 0$. Write the answer in...

21. interval notation.

21. _____

22. set notation.

22. _____

23. The product of 2 consecutive even integers is 168. Find the pair(s) of even integers.

23. _____

24. Solve for x. Write your answer in interval notation.
$$\frac{2x^2 - 5x - 1}{x+3} \geq x - 2.$$

24. _____

25. Kerry throws a ball upward from the top of a building. The distance, d, in feet, of the ball from the ground at any time t, in seconds, can be found by the formula $d = -16t^2 + 160t + 81$. Find the time the object reaches its maximum height.

25. _____

Chapter 12 Test Form F

Name:_____

Date:_____

Solve each equation by completing the square.

1. $x^2 + 14x + 45 = 0$

    1. _____

2. $9x^2 - 18x - 16 = 0$

    2. _____

Solve each equation using the quadratic formula.

3. $12x^2 - 5x - 2 = 0$

    3. _____

4. $2x^2 + 5 = -2x$

    4. _____

5. Determine the number of real solutions: $\left(\dfrac{3}{2}x\right)^2 + 3x + 1 = 0$

    5. _____

6. Write a function that has x-intercepts -2 and 8.

    6. _____

7. Solve the formula $S = 2\pi rh + 2\pi r^2$ for $r \geq 0$.

    7. _____

Solve each equation.

8. $3x^4 - 3x^2 - 6 = 0$

    8. _____

9. $5\sqrt{x} = x - 14$

    9. _____

10. $\left(x^2 - 2\right)^2 - 1 = 0$

    10. _____

11. Find all x-intercepts of the function $f(x) = 2x^{\frac{2}{3}} + 3x^{\frac{1}{3}} - 2$.

    11. _____

12. Write an equation that is quadratic in form and has solutions ± 5 and $\pm i\sqrt{7}$.

    12. _____

For questions 13–17, consider the function $f(x)$ represented by the graph below.

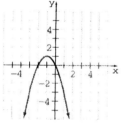

13. Determine the equation of the graph.

    13. _____

14. Find the y-intercept.

    14. _____

15. Find the vertex.

    15. _____

16. Find the x-intercepts, if any.

    16. _____

17. Determine the *axis of symmetry*.

    17. _____

18. Graph the function $f(x) = (x+3)^2 - 1$.

18.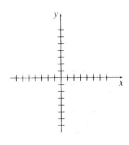

19. Graph the inequality $x^2 - x - 6 < 0$ on the number line

19.

20. Graph the inequality $x^2 + x - 12 \geq 0$ on the number line

20.

For questions 21 and 22, solve the inequality $4x^2 + 10x - 6 \leq 0$.

21. Write the answer in interval notation.

21. _____

22. Write the answer in set notation.

22. _____

23. The product of 2 positive numbers is 36 and the larger is one less than twice the smaller. Find the two numbers.

23. _____

24. Solve the inequality $\dfrac{x^2}{2x-3} \leq x + 2$ and write your answer in interval notation.

24. _____

25. The distance d (in meters, m) traveled by an object thrown downward with an initial velocity of v_o after t seconds is given by the formula $d = 5t^2 + v_0 t$. Find the number of seconds it takes an object to hit the ground if the object is dropped from a height of 20 m.

25. _____

Chapter 12 Test Form G

Name:_____

Date:_____

Solve each equation by completing the square.

1. $x^2 + 10x - 11 = 0$

    (a) $x = 1$ or 11 **(b)** $x = -1$ or 11 **(c)** $x = 1$ or -11 **(d)** $x = -1$ or -11

2. $4x^2 = 56x - 195$

    (a) $x = 3$ or 4 **(b)** $x = -3$ or 4 **(c)** $x = \dfrac{13}{2}$ or $\dfrac{15}{2}$ **(d)** $x = \dfrac{-13}{2}$ or $\dfrac{15}{2}$

Solve each equation using the quadratic formula.

3. $5x^2 + 3x - 1 = 0$

    (a) $x = -\dfrac{3}{2} \pm \sqrt{29}$ **(b)** $x = \dfrac{3}{2} \pm \dfrac{\sqrt{29}}{2}$ **(c)** $x = -\dfrac{3}{10} \pm \dfrac{\sqrt{29}}{10}$ **(d)** $x = \dfrac{3}{10} \pm \dfrac{\sqrt{29}}{10}$

4. $x^4 + 5x^2 = 0$

    (a) $x = \pm\sqrt{5}, 0$ **(b)** $x = \pm i\sqrt{5}, 0$ **(c)** $x = \pm 5$ **(d)** $x = \pm\sqrt{5}\, i$

5. Determine the number of real solutions the following equation has: $5x^2 - 4x + 1 = 0$

    (a) 0 **(b)** 1 **(c)** 2 **(d)** 4

6. Write a function that has x-intercepts -1 and 7.

    (a) $x^2 + 6x + 7$ **(b)** $x^2 + 6x - 7$ **(c)** $x^2 - 6x + 7$ **(d)** $x^2 - 6x - 7$

7. Solve the formula $c = \sqrt{b^2 + a^2}$ for $a > 0$.

    (a) $\sqrt{c^2 + b^2}$ **(b)** $\sqrt{b^2 - c^2}$ **(c)** $\sqrt{c^2 - b^2}$ **(d)** $\sqrt{c - b^2}$

Solve each equation.

8. $x^4 + 3x^2 - 10 = 0$

    (a) $x = \pm\sqrt{2}, \pm i\sqrt{5}$ **(b)** $x = 2, -5$ **(c)** $x = \pm\sqrt{2}, \pm\sqrt{5}$ **(d)** $x = -2, 5$

9. $x + 3\sqrt{x} - 28 = 0$

    (a) $x = 49, 16$ **(b)** $x = -49, 16$ **(c)** $x = 49$ **(d)** $x = 16$

10. $\left(x^2 - 5\right)^2 + 5\left(x^2 - 5\right) + 4 = 0$

    (a) $x = \pm 1, \pm 4$ **(b)** $x = \pm 1, \pm 2$ **(c)** $x = \pm 4, \pm 9$ **(d)** $x = \pm 2i, \pm 3$

11. Find all x-intercepts of the function $f(x) = x^{\frac{2}{3}} + 2x^{\frac{1}{3}} + 1$.

    (a) $x = \pm 1$ **(b)** $x = 1$ **(c)** $x = -1$ **(d)** no solution

12. Write an equation that is quadratic in form and has solutions $\pm i, \pm \sqrt{7}$.

(a) $x^4 + 6x^2 + 7 = 0$ (b) $x^4 + 6x^2 - 7 = 0$ (c) $x^4 - 6x^2 + 7 = 0$ (d) $x^4 - 6x^2 - 7 = 0$

For questions 13 – 17, consider the function $f(x)$ represented by the graph below.

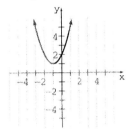

13. Determine the equation of the graph.

(a) $y = x^2 + 2x + 2$ (b) $y = x^2 + 2x + 1$ (c) $y = x^2 + x + 2$ (d) $y = x^2 + x + 1$

14. Find the y-intercept.

(a) $(0, 1)$ (b) $(0, 2)$ (c) $(1, 0)$ (d) none

15. Find the vertex.

(a) $(0, 0)$ (b) $(1, 1)$ (c) $(-1, 1)$ (d) $(-1, -1)$

16. Find the x-intercepts, if any.

(a) $(0, 0)$ (b) $(0, 1)$ (c) $(1, 0)$ (d) none

17. Determine the axis of symmetry.

(a) $x = 1$ (b) $x = -1$ (c) $y = 1$ (d) $y = -1$

18. Graph the function $f(x) = -(x-1)^2 - 1$.

(a) 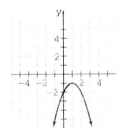 (b) 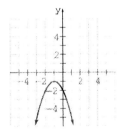 (c) 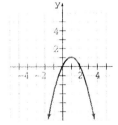 (d)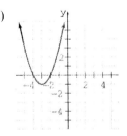

19. Which inequality represents the following number line?

(a) $x^2 + 9 > 0$ (b) $x^2 + 9 < 0$ (c) $x^2 - 9 > 0$ (d) $x^2 - 9 < 0$

20. Which inequality represents the following number line?

(a) $x^2 + x - 2 \geq 0$

(b) $x^2 + x - 2 \leq 0$

(c) $x^2 - x - 2 \geq 0$

(d) $x^2 - x - 2 \leq 0$

For questions 21 and 22, solve the inequality $3x^2 + 10x - 8 > 0$.

21. Write the answer in interval notation.

(a) $\left(-4, \dfrac{2}{3} \right)$

(b) $\left(-\dfrac{2}{3}, 4 \right)$

(c) $\left(-\infty, -4 \right) \cup \left(\dfrac{2}{3}, \infty \right)$

(d) $\left(-\infty, -\dfrac{2}{3} \right) \cup \left(4, \infty \right)$

22. Write the answer in set notation.

(a) $\left\{ x \middle| -\dfrac{2}{3} < x < 4 \right\}$

(b) $\left\{ x \middle| -4 < x < \dfrac{2}{3} \right\}$

(c) $\left\{ x \middle| x < -\dfrac{2}{3} \text{ or } x > 4 \right\}$

(d) $\left\{ x \middle| x < -4 \text{ or } x > \dfrac{2}{3} \right\}$

23. The product of 2 positive, consecutive even integers is 48. Find the larger of these 2 even integers.

(a) 4

(b) 6

(c) 8

(d) 10

24. Solve the inequality $\dfrac{2x}{x+2} \leq x - 1$ and write your answer in interval notation.

(a) $(-2, 2]$

(b) $(-2, -1] \cup [2, \infty)$

(c) $(-\infty, -2) \cup [2, \infty)$

(d) $(-2, \infty)$

25. The value, V, of a barley crop per acre, in dollars, d days after planting is given by the formula $V = 14d - 0.06d^2$, $20 < d < 80$. Find the value of an acre of barley after it has been planted 55 days.

(a) $256

(b) $588.50

(c) $736

(d) $766.70

Chapter 12 Test Form H

Solve each equation by completing the square.

1. $x^2 + 6x - 135 = 0$

    (a) $x = 9$ or 15 **(b)** $x = -9$ or 15 **(c)** $x = 9$ or -15 **(d)** $x = -9$ or -15

2. $9x^2 + 18x + 8 = 0$

    (a) $x = \dfrac{2}{3}$ or $\dfrac{4}{3}$ **(b)** $x = -\dfrac{2}{3}$ or $\dfrac{4}{3}$ **(c)** $x = \dfrac{2}{3}$ or $-\dfrac{4}{3}$ **(d)** $x = -\dfrac{2}{3}$ or $-\dfrac{4}{3}$

Solve each equation using the quadratic formula.

3. $30x^2 - 7x - 2 = 0$

    (a) $x = \dfrac{2}{5}$ or $-\dfrac{1}{6}$ **(b)** $x = -\dfrac{2}{5}$ or $\dfrac{1}{6}$ **(c)** $x = \dfrac{5}{2}$ or $-\dfrac{1}{6}$ **(d)** $x = -\dfrac{5}{2}$ or $\dfrac{1}{6}$

4. $x^2 + 3 = 0$

    (a) $x = \pm\sqrt{3}$ **(b)** $x = \pm 3$ **(c)** $x = \pm i\sqrt{3}$ **(d)** $x = \pm 3i$

5. Determine the number of real solutions the following equation has: $2x^2 + 7x + 5 = 0$

    (a) 0 **(b)** 1 **(c)** 2 **(d)** 4

6. Write the equation of a parabola whose vertex is $(2, -1)$ and whose y-intercept is 7.

    (a) $2x^2 + 8x + 7$ **(b)** $2x^2 - 8x + 7$ **(c)** $2x^2 - 8x - 7$ **(d)** $2x^2 + 8x - 7$

7. Solve the formula $x^2 + y^2 = r^2$ for y.

    (a) $\pm\sqrt{x^2 - r^2}$ **(b)** $\pm\sqrt{x^2 + r^2}$ **(c)** $\pm\sqrt{r - x}$ **(d)** $\pm\sqrt{r^2 - x^2}$

Solve each equation.

8. $x^4 + 5x^2 + 4 = 0$

    (a) $x = -1, -4$ **(b)** $x = \pm 1, \pm 2$ **(c)** $x = \pm i, \pm 2i$ **(d)** no solution

9. $7\sqrt{x} - 10 = x$

    (a) $x = 4, 25$ **(b)** $x = 25$ **(c)** $x = -2, 5$ **(d)** no solution

10. $\left(x^2 - 6\right)^2 + \left(x^2 - 6\right) - 2 = 0$

    (a) $x = \pm 2, \pm\sqrt{7}$ **(b)** $x = 4, 7$ **(c)** $x = \pm\sqrt{7}$ **(d)** no solution

11. Find all x-intercepts of the function $f(x) = x^{\frac{2}{3}} - 4x^{\frac{1}{3}} + 4$.

    (a) $x = \pm 2$ **(b)** $x = 2$ **(c)** $x = \pm 8$ **(d)** $x = 8$

Chapter 12 Test Form H *(cont.)*   Name:_____

12. Write an equation that is quadratic in form and has solutions $\pm 2\sqrt{3}, \pm 2i$.

    (a) $x^4 + 8x^2 + 48 = 0$ **(b)** $x^4 - 8x^2 - 48 = 0$ **(c)** $x^4 + 8x^2 - 48 = 0$ **(d)** $x^2 - 8x - 48 = 0$

For questions 13 – 17, consider the function $f(x)$ represented by the graph below.

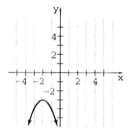

13. Determine the equation of the graph.

    (a) $y = -(x-2)^2 - 3$ **(b)** $y = -(x-2)^2 + 3$ **(c)** $y = -(x+2)^2 - 3$ **(d)** $y = -(x+2)^2 + 3$

14. Find the y-intercept.

    (a) $(0, 3)$ **(b)** $(0, -3)$ **(c)** $(0, 7)$ **(d)** $(0, -7)$

15. Find the vertex.

    (a) $(-2, 3)$ **(b)** $(2, 3)$ **(c)** $(-2, -3)$ **(d)** $(2, -3)$

16. Find the x-intercepts, if any.

    (a) $(2, 0)$ **(b)** $(-2, 0)$ **(c)** $(0, -7)$ **(d)** none

17. Determine the axis of symmetry.

    (a) $x = 2$ **(b)** $x = -2$ **(c)** $y = 2$ **(d)** $y = -2$

18. Graph the function $f(x) = (x+2)^2 - 3$.

    (a) 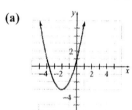   **(b)** 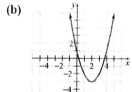   **(c)** 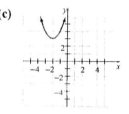   **(d)**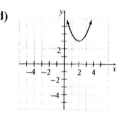

19. Which inequality represents the following number line?

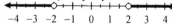

    (a) $x^2 - 4 > 0$ **(b)** $x^2 - 4 < 0$ **(c)** $x^2 + 4 > 0$ **(d)** $x^2 + 4 < 0$

20. Which inequality represents the following number line?

    (a) $x^2 - 2x - 15 \geq 0$ **(b)** $x^2 - 2x - 15 \leq 0$ **(c)** $x^2 + 2x - 15 \geq 0$ **(d)** $x^2 + 2x - 15 \leq 0$

Chapter 12 Test Form H *(cont.)*

Name:_____

For questions 21 and 22, solve the inequality $5x^2 - 29x - 6 < 0$.

21. Write the answer in interval notation.

    (a) $\left(-\dfrac{1}{5}, 6\right)$ (b) $\left(-6, \dfrac{1}{5}\right)$ (c) $\left(-\infty, -\dfrac{1}{5}\right) \cup (6, \infty)$ (d) $(-\infty, -6) \cup \left(\dfrac{1}{5}, \infty\right)$

22. Write the answer in set notation.

    (a) $\left\{x \middle| x < -6 \text{ or } x > \dfrac{1}{5}\right\}$ (b) $\left\{x \middle| x < -\dfrac{1}{5} \text{ or } x > 6\right\}$ (c) $\left\{x \middle| -6 < x < \dfrac{1}{5}\right\}$ (d) $\left\{x \middle| -\dfrac{1}{5} < x < 6\right\}$

23. The product of two positive integers is 78 and the larger is one more than twice the smaller. Find the smaller of these integers.

    (a) 3   (b) 2   (c) 13   (d) 6

24. Solve the inequality $\dfrac{-2x}{x+2} \le x + 3$ and write your answer in interval notation.

    (a) $[-6, -2) \cup [-1, \infty)$ (b) $(-2, -1]$ (c) $(-\infty, -2) \cup (-2, \infty)$ (d) $(-\infty, -6] \cup [-1, \infty)$

25. If the revenue is given by $R = 300x - 0.06x^2$, find the value of x that yields the maximum revenue.

    (a) 5000   (b) 375,000   (c) 2500   (d) 10,000

Cumulative Review Test 1–12 Form A

1. Evaluate: $\dfrac{-3|3-45| \div 6 + 2}{\sqrt{4} + 80 \div 4^2}$

   1. _____

2. The circle graph shows the leading cotton producing states by percent of U.S. cotton produced in 1996. If the U.S. produced 1.84×10^7 bushels of cotton in 1996, how many bushels were produced in Georgia?

   2. _____

U.S. Cotton Production

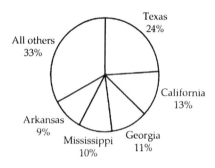

3. Solve for x: $\dfrac{x-4}{8} = \dfrac{x+2}{7}$

   3. _____

4. Find the solution set to the inequality $|2x-8| + 8 > 18$.

   4. _____

5. Solve for x: $|x-3| = |x-9|$

   5. _____

6. Is the relation $\{(6, 2), (-4, 2), (-5, 2)\}$ a function?

   6. _____

7. Find the domain and range for the relation graphed below.

   7. _____

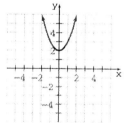

8. Graph $x = 3$.

   8.

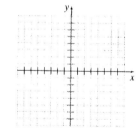

Cumulative Review Test 1–12
Form A *(cont.)*

9. Use the *x*- and *y*-intercepts to graph the linear equation $-y - 2x = 2$.

9.

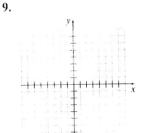

10. Determine the equation of a line perpendicular to the graph of $2y = -7x + 3$ that passes through $(3, -1)$. Write the equation in point-slope form.

10. _____

11. Solve the system using the addition method:
$3x + 2y = 7$
$4x - 3y = -2$

11. _____

12. Evaluate: $\begin{vmatrix} 3 & 1 & 5 \\ 4 & 5 & 3 \\ 1 & 3 & 1 \end{vmatrix}$

12. _____

13. Factor completely: $x^7 y - xy^7$

13. _____

For 14 and 15, let $f(x) = 4 - x^2$, $g(x) = 2 - x$.

14. Find $(f + g)(x)$.

14. _____

15. Find $(f \cdot g)(x)$.

15. _____

16. A rock is thrown from the top of a tall building. The distance, in feet, between the rock and the ground *t* seconds after it is thrown is given by $d = -16t^2 - 2t + 532$. How long after the rock is thrown is it 427 feet from the ground?

16. _____

17. Solve for *x*: $2x^2 + x < 6$

17. _____

18. The intensity, *I*, of light received at a source varies inversely as the square of the distance, *d*, from the source. If the light intensity is 30 foot-candles at 14 feet, find the light intensity at 17 feet. Round your answer to the nearest hundredth if necessary.

18. _____

19. Simplify: $\dfrac{3 + 4i}{8 + 5i}$

19. _____

20. Solve for *x*: $\dfrac{x - 2}{2x + 1} = \dfrac{x - 1}{x + 8}$

20. _____

Cumulative Review Test 1–12 Form B

Name:_____

Date:_____

1. Evaluate: $\left[\dfrac{9+(-4)}{-9-3}\right]\left[\dfrac{72+(-24)}{2-4}\right]$

   (a) 0   (b) 10   (c) 40   (d) −10

2. The circle graph shows the leading cotton producing states by percent of U.S. cotton produced in 1996. If the U.S produced 1.84×10^7 bushels of cotton in 1996, how many bushels were produced in California?

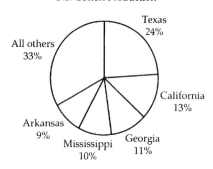

U.S. Cotton Production

   (a) about 2.024×10^7 bushels   (b) about 2.024×10^6 bushels
   (c) about 2.392×10^7 bushels   (d) about 2.392×10^6 bushels

3. Solve for x: $-4(x+5) = 2\left[7-(x-3)\right] - 5x$

   (a) $x = -1$ (b) $x = 0$ (c) $x = \dfrac{40}{3}$ (d) $x = \dfrac{40}{8}$

4. Find the solution set to the inequality $|x-2| - 3 < 0$.

   (a) $\{x | -3 \le x \le 3\}$ (b) $\{x | -1 < x < 5\}$ (c) $\{x | x \le -1 \text{ or } x \ge 5\}$ (d) $\{x | -5 < x < 1\}$

5. Solve for x: $|x-6| = |3-2x|$

   (a) $x = -3, x = 3$ (b) $x = 3$ (c) $x = 6, x = \dfrac{3}{2}$ (d) $x = -3$

6. Which of the answers below is a function?

   (a) $\{(6,-4),(-6,-4),(4,1)\}$ (b) $\{(x,y) | x^2 + y^2 = 36\}$
   (c) $\{(-4,4),(-4,1),(1,6)\}$ (d) $\{(x,y) | x = y^2 - 4\}$

Cumulative Review Test 1–12
Form B *(cont.)*

Name:_____

7. Find the domain and range for the relation graphed below.

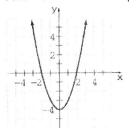

(a) $D = \{x \mid x > -4\}$

  $R = \{y \mid y - 4\}$

(b) $D = \{x \mid x \text{ is a real number}\}$

  $R = \{y \mid y \text{ is a real number}\}$

(c) $D = \{x \mid x \text{ is a real number}\}$

  $R = \{y \mid y \geq -4\}$

(d) $D = \{x \mid x \leq -4\}$

  $R\{y \mid y \text{ is a real number}\}$

8. Which equation matches the graph?

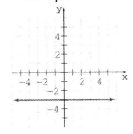

(a) $x = -3$

(b) $y = -3$

(c) $x = 3$

(d) $y = 3$

9. Use the *x*- and *y*-intercepts to decide which equation matches the graph.

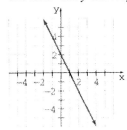

(a) $2x - y = -2$

(b) $2x - y = 2$

(c) $2x + y = -2$

(d) $2x + y = 2$

10. Determine the equation of a line perpendicular to the graph of $3y = -5x + 4$ that passes through $(-5, 1)$.

(a) $y - 1 = \frac{3}{5}(x + 5)$

(b) $y + 5 = \frac{3}{5}(x - 1)$

(c) $y - 1 = -\frac{3}{5}(x + 5)$

(d) $y + 5 = -\frac{3}{5}(x - 1)$

11. Solve the system using substitution:

$-x - 3y = -\dfrac{2}{3}$

$4x + 9y = 1$

(a) $\left(-1, -\dfrac{9}{5}\right)$

(b) $(5, 3)$

(c) $\left(-1, \dfrac{5}{9}\right)$

(d) $\left(1, \dfrac{16}{15}\right)$

Cumulative Review Test 1–12
Form B *(cont.)*

12. Evaluate: $\begin{vmatrix} -4 & -3 & 0 \\ 2 & -2 & 2 \\ -4 & 3 & -2 \end{vmatrix}$

(a) 20   (b) 4   (c) –4   (d) –20

13. Factor completely: $5x^3y^3 + 625x^3$

(a) $5x^3(y+5)(y^2-5y+25)$ (b) $x^3(5y+5)(y+25)^2$

(c) $x^3(y+25)(5y^2+25)$ (d) $5x^3(y+5)(y^2+10y+25)$

14. Let $f(x)=16-x^2$, $g(x)=4-x$. Find $\dfrac{f}{g}(x)$.

(a) $-x^2+x+12$ (b) $x+4$ (c) $x^3-4x^2-16x+64$ (d) $-x^2-x+20$

15. Let $f(x)=16-x^2$, $g(x)=4-x$. Find $(f\cdot g)(x)$.

(a) $64-4x^2+16x-x^3$ (b) $64+4x^2-16x-x^3$ (c) $x^3+4x^2+16x-64$ (d) $x^3-4x^2-16x+64$

16. A rock is thrown from the top of a tall building. The distance, in feet, between the rock and the ground t seconds after it is thrown is given by $d=-16t^2-4t+372$. How long after the rock is thrown is it 370 feet from the ground?

(a) $\dfrac{3}{2}$ sec   (b) $\dfrac{3}{4}$ sec   (c) $\dfrac{1}{4}$ sec   (d) $\dfrac{1}{2}$ sec

17. Solve for x: $\dfrac{x-6}{x+3}=\dfrac{x+1}{x-2}$

(a) $x=-\dfrac{5}{4}$ (b) $x=\dfrac{3}{4}$ (c) $x=-\dfrac{15}{8}$ (d) $x=\dfrac{1}{4}$

18. The wattage rating of an appliance, W, varies jointly as the square of the current, I, and the resistance, R. If the wattage is 10 watts when the current is 0.2 ampere and the resistance is 250 ohms, find the wattage when the current is 0.1 ampere and the resistance is 100 ohms.

(a) 10 watts   (b) 20 watts   (c) 1 watt   (d) 1000 watts

19. Rationalize the denominator: $\dfrac{6+i}{9+4i}$

(a) $\dfrac{58}{97}-\dfrac{15i}{97}$ (b) $-\dfrac{58}{97}+\dfrac{15i}{97}$ (c) $-\dfrac{58}{97}-\dfrac{15i}{97}$ (d) $\dfrac{58}{97}+\dfrac{15i}{97}$

20. Solve for x: $3x^2+3=2x$

(a) $\dfrac{-1\pm\sqrt{2}\,i}{3}$ (b) $\dfrac{-1\pm 2\sqrt{2}\,i}{3}$ (c) $\dfrac{1\pm\sqrt{2}\,i}{3}$ (d) $\dfrac{1\pm 2\sqrt{2}\,i}{3}$

Chapter 13 Pretest Form A

Name:_____
Date:_____

For questions 1–4, use the following function: $\{(1, 1), (2, 4), (3, 9), (4, 16)\}$

1. Is the function a one-to-one function?

1. _____

2. Name the domain and range of the function.

2. _____

3. List the set of ordered pairs in the inverse function.

3. _____

4. Name the domain and range of the inverse function.

4. _____

For questions 5 – 7, use the following functions: $f(x) = 2x + 1$ and $g(x) = x^2 - 1$

5. Find: $(f \circ g)(2)$

5. _____

6. Find: $(g \circ f)(-3)$

6. _____

7. Find: $(f \circ g)(x)$

7. _____

8. Find $f^{-1}(x)$ for $f(x) = \dfrac{x-4}{x+5}$.

8. _____

9. Graph $y = 5^x$.

9.

10. Graph $y = \log_3 x$.

10.

11. Write $3^x = 17$ in logarithmic form.

11. _____

Write in exponential form and find the missing values.

12. $\log_7 x = 2$

12. _____

13. $y = \log_{1/2} 8$

13. _____

14. $\log_{2\sqrt{2}} x = 2$

14. _____

15. Write as a logarithm of a single expression: $\log_b (x+1) - 3 \cdot \log_b x$

15. _____

16. Evaluate: $\log_{27} 9$

16. _____

Solve for x:

17. $\log x + \log(x-3) = 1$

17. _____

18. $3^{x^2+4x} = \dfrac{1}{27}$

18. _____

19. $\ln x = 2 \cdot \ln(x-1)$

19. _____

20. Use the change of base formula to evaluate $\log_3 7$.

20. _____

Chapter 13 Pretest Form B

Name:_____

Date:_____

For problems 1–4, use the following function.

$\{(-6, 5), (-5, 4), (2, -3), (1, -1)\}$

1. Is the function a one-to-one function?

2. Name the domain and range of the function.

3. List the set of ordered pairs in the inverse function.

4. Name the domain and range of the inverse function.

1. _____

2. _____

3. _____

4. _____

For problems 5–7, use the following functions.

$f(x) = x^2 - 6$ and $g(x) = \sqrt{x + 3}, x \geq -3$

5. Find $(f \circ g)(x)$

6. Find $(g \circ f)(x)$

7. Find $(g \circ f)(2)$

8. Find $f^{-1}(x)$ for $y = f(x) = \dfrac{9 - 4x}{x}$

5. _____

6. _____

7. _____

8. _____

9. Graph $y = 5^x$.

9.

10. Graph $y = \log_5 x$.

10.

11. Write $5^{2.3} = x$ in logarithmic form.

11. _____

Write in exponential form and find the missing values.

12. $4 = \log_{10} x$

13. $y = \log_{49} 7$

14. Expand $\log_6 \dfrac{x^3(x+1)}{y}$.

15. Write as a logarithm of a single expression: $4\log_2(x + 7) - \dfrac{1}{5}\log_2 x$

12. _____

13. _____

14. _____

15. _____

16. Evaluate $2\log_{\sqrt{8}}\sqrt[3]{2}$.

16. _____

Solve for *x*.

17. $\log 6x = \log(x+5) + \log 4$

17. _____

18. $10^x = 1{,}000{,}000$

18. _____

Use change of base to find x :

19. $\log_3 7 = x$

19. _____

20. $\log_x 7 = 3$

20. _____

Mini-Lecture 13.1
Composite and Inverse Functions

Learning Objectives:

1. Find composite functions.
2. Understand one-to-one functions.
3. Find inverse functions.
4. Find the composition of a function and its inverse.
5. Key vocabulary: *composition of f with g, one-to-one function, horizontal line test, inverse function.*

Examples:

1. For each pair of functions, find $(f \circ g)(x)$ and $(g \circ f)(4)$:

   a) $f(x) = x^2 - x - 3, g(x) = x + 5$ b) $f(x) = \sqrt{x} + 3, g(x) = (x+1)^2$

   c) $f(x) = \dfrac{3x+2}{2x-1}, g(x) = x^2 - 2x - 1$ d) $f(x) = x^2 + 1, \; g(x) = \sqrt{x^2 + 1}$

2. If $f(x) = \{(-2,5), (-1,3), (0,0), (2,-2), (5,-5)\}$, find $f^{-1}(x)$.

3. Determine if $f(x)$ is one-to-one, and, if so, find $f^{-1}(x)$:

   a) $f(x) = \dfrac{3}{x} + 1$ b) $f(x) = \sqrt{|x-2|}$ c) $f(x) = \sqrt[5]{x^3}$ d) $f(x) = \dfrac{1}{\frac{1}{x} + 2}$ e) $f(x) = \dfrac{x+3}{x+2}$

Teaching Notes:

- Stress that, in $(f \circ g)(x)$, x is the input for g and $g(x)$ is the input for f.
- Make sure students understand the difference between $(f \circ g)(x)$ and $(f \cdot g)(x)$.
- Point out that not all functions are one-to-one.
- Note that only one-to-one functions have inverses.
- Note that the range of the function is the domain of the inverse, and vice versa.
- Point out the symmetry of the graphs of f and f^{-1} about the line $y = x$.

Answers:

1. a) $(f \circ g)(x) = x^2 + 9x + 17$; $(g \circ f)(4) = 14$

   b) $(f \circ g)(x) = |x+1| + 3$; $(g \circ f)(4) = 36$

   c) $(f \circ g)(x) = \dfrac{3x^2 - 6x - 1}{2x^2 - 4x - 3}$; $(g \circ f)(4) = -1$

   d) $(f \circ g)(x) = x^2 + 2$; $(g \circ f)(4) = \sqrt{290}$

2. $f^{-1}(x) = \{(-5,5), (-2,2), (0,0), (3,-1), (5,-2)\}$

3. a) yes; $f^{-1}(x) = \dfrac{3}{x-1}$

   b) not one-to-one

   c) yes; $f^{-1}(x) = x\sqrt[3]{x^2}$

   d) yes; $f^{-1}(x) = \dfrac{x}{1-2x}$

   e) yes; $f^{-1}(x) = \dfrac{3-2x}{x-1}$

Mini-Lecture 13.2
Exponential Functions

Learning Objectives:

1. Graph exponential functions.
2. Solve applications of exponential functions.
3. Key vocabulary: *exponential function*

Examples:

1. Graph the following:

   a) $y = 4^x$ b) $y = \left(\dfrac{3}{7}\right)^x$ c) $y = 3 \cdot (2^x)$ d) $y = 5^x + 2$

2. The formula used for carbon dating is: $A(t) = A_0 \cdot 2^{-t/5600}$. If an organism contained 230 grams of Carbon 14 when it died, how many grams will be found in the fossil 1500 years later?

3. If Joe invests $5000 in an account paying 4.75% interest, compounded quarterly, how much will he have after 5 years?

4. Logan bought a dishwasher for $399. If it depreciates 12% per year, what is it worth after 5 years?

Teaching Notes:

- Be sure students learn the basic properties of $y = a^x$:

    domain is $(-\infty, \infty)$, range is $(0, \infty)$

    graph contains the points $\left(-1, \dfrac{1}{a}\right)$, (0.1), and $(1, a)$

    graph approaches the x-axis

- Stress that the variable in an exponential function is the *exponent*.

- Point out that functions like $f(x) = ka^x$ and $f(x) = a^x + k$ are also exponential..

Answers:

1. See graphing answers. 2. 191.03 grams 3. $6331.51 4. $210.57

Mini-Lecture 13.3
Logarithmic Functions

Learning Objectives:

1. Convert from exponential form to logarithmic form.
2. Graph logarithmic functions.
3. Compare the graphs of exponential and logarithmic functions.
4. Solve applications of logarithmic functions.
5. Key vocabulary: *logarithms, logarithmic function*

Examples:

1. Graph the following:

   a) $y = \log_4 x$ b) $y = \log_{1/7} x$ c) $y = 3 \cdot \log_2 x$ d) $y = (\log_5 x) + 2$

2. Evaluate the following without using a calculator:

   a) $\log_2 64$ b) $\log_{0.1} 0.001$ c) $\log_3 \dfrac{1}{27}$ d) $\log_{1/3} 27$ e) $\log_{11} 1$

3. The magnitude, R, of an earthquake on the Richter Scale is given by $R = \log_{10} I$, where I represents the number of times more intense the earthquake is than the smallest activity that can be measured on a seismograph. How much more intense is a 6.2 earthquake than a 3.7 earthquake?

Teaching Notes:

- Be sure students understand that $v = \log_a u$ and $u = a^v$ are equivalent statements; further, $y = \log_a x$ and $y = a^x$ are inverse functions..

- Be sure students learn the basic properties of $y = \log_a x$:

  domain is $(0, \infty)$, range is $(-\infty, \infty)$, inverse function of $y = a^x$

  graph contains the points $\left(\dfrac{1}{a}, -1 \right)$, $(1, 0)$, and $(a, 1)$

  graph approaches the y-axis

- Point out that $y = \log_a x$ has no y-intercept and $y = a^x$ has no x-intercept.

Answers:

1. See graphing answers. *2. a) 6, b) 3, c) -3, d) -3, e) 0*

3. 316.23 times as intense.

Mini-Lecture 13.4
Properties of Logarithms

Learning Objectives:

1. Use the product rule for logarithms.
2. Use the quotient rule for logarithms.
3. Use the power rule for logarithms.
4. Use additional properties of logarithms.
5. Key vocabulary: *argument of a logarithm*

Examples:

1. Expand each of the following:

   a) $\log_3\left(\dfrac{a+b}{c}\right)$ b) $\log_7\left(\dfrac{6x}{uv}\right)$ c) $\log_5\left(\dfrac{x^2 y^3}{z^4}\right)$ d) $\log_4\left(\dfrac{\sqrt[3]{x}}{\sqrt[5]{yz}}\right)$

2. Simplify:

   a) $\log_5 x + \log_5 y - \log_5 17$ b) $3\log_4 x - 2\log_4\left(\dfrac{1}{x}\right)$ c) $\dfrac{1}{2}\left[\log_8(x+y) - \log_8 z\right]$

   d) $\log_3 7 - \log_3 4$ e) $\log_3(2x+1) + \log_2(3x+1)$ f) $\log_3\left(4^{\log_4 3}\right) - \log_3\left(\dfrac{1}{27}\right)$

3. Use the fact that $\log_5 3 = 0.68261$ and $\log_5 7 = 1.20906$ to evaluate the following:

   a) $\log_5 \dfrac{3}{7}$ b) $\log_5 \sqrt[3]{7}$ c) $\log_5 \dfrac{1}{21}$ d) $\log_5 63$

Teaching Notes:

- Students should memorize the basic properties of logarithms.
- Point out that $\log_a(u+v)$ has no simplifying formula in general.
- Derive Property 3A: $\log_a \dfrac{1}{x} = -\log_a x$.

Answers:

1. a) $\log_3(a+b) - \log_3 c$;   b) $\log_7 6 + \log_7 x - \log_7 u - \log_7 v$;

   c) $2\log_5 x + 3\log_5 y - 4\log_5 z$;   d) $\dfrac{1}{3}\log_4 x - \dfrac{1}{5}\log_4 y - \dfrac{1}{5}\log_4 z$.   2. a) $\log_5\left(\dfrac{xy}{17}\right)$;

   b) $\log_4 x^5$;   c) $\log_8 \sqrt{\dfrac{x+y}{z}}$;   d) $\log_3\left(\dfrac{7}{4}\right)$;   e) *cannot be simplified*   f) 4 ;

   3. a) -0.52645 ;   b) 0.40302 ;   c) -1.89167 ;   d) 2.57428

Mini-Lecture 13.5
Common Logarithms

Learning Objectives:

1. Find common logarithms of powers of 10.

2. Find common logarithms.

3. Find antilogarithms.

4. Key vocabulary: *common logarithms, antilogarithm* (or *inverse logarithm*), *significant digits.*

Examples:

1. Evaluate the following:

    a) $\log 27.36$ b) $\log 0.5291$ c) $\log 1000$ d) antilog 2.38 e) antilog -1.76 f) antilog 0

2. The seismic energy (E) released by an earthquake is sometimes given by $\log E = 11.8 + 1.5 m_s$, where m_s =surface wave magnitude.

    a) Find E when $m_s = 3.7$ b) Find m_s when $E = 6.3 \times 10^{17}$

3. Sound pressure level (s_p) is related to sound pressure (p_r) in dynes $/ cm^2$ by $s_p = 20 \cdot \log\left(\dfrac{p_r}{0.0002}\right)$.

    a) Find s_p when $p_r = 0.0041$ dynes/cm^2 b) Find p_r when $s_p = 17.8$

4. The acidity (*pH*) of a solution is defined to be $pH = -\log\left[H_3O^+\right]$, where H_3O^+ represents the hydronium ion concentration of the solution. Find the *pH* of a solution when $H_3O^+ = 5.7 \times 10^{-4}$.

Teaching Notes:

- Point out that, when $\log x$ is written without a base, *common* log (base 10) is assumed.

- Mention that, although equality is stated, most of the time $\log_a x$ is an approximation.

- Emphasize that $L = \log N$ and $N = $ antilog L are equivalent statements; $\log N$ is an *exponent*, antilog L is the *number* 10^L.

- Point out that $f(x) = \log x$ is an increasing function so that $10^a < 10^b \Leftrightarrow a < b$.

Answers: *1. a) 1.4371; b) -0.2765; c) 3; d) 239.88; e) 0.01738; f) 1.*

2. a) $\approx 2.24 \times 10^{17}$; b) ≈ 4.00.   3. a) 26.235; b) 0.00155 dynes/cm^2.

4. 3.244

Mini-Lecture 13.6
Exponential and Logarithmic Equations

Learning Objectives:

1. Solve exponential and logarithmic equations.
2. Solve applications.

Examples:

1. Solve each equation for x; use a calculator if necessary:

   a) $\log_2(x+2) - \log_2(x+1) = 3$ b) $2^{3x+1} = 4^{x+5}$ c) $\log_3(x-2) + \log_3(x+1) = 1$

   d) $\log_3(x) + \log_3(x+2) = 1$ e) $\log(x+5) + \log 2 = 2 \cdot \log(x+1)$ f) $6^{x+1} = 5^{2x-1}$

2. Sam invests \$6000 in a savings account that pays 4.8% compounded monthly. How long will it take him to double his money?

3. The power gain (P) of an amplifier is defined by $P = 10 \cdot \log\left(\dfrac{P_{out}}{P_{in}}\right)$ where P_{out} = power output (watts) and P_{in} = power input (watts). If an amp has a power input of 0.328 watts and a power gain of 17.1 watts, what is its output power?

4. The scrap value of an item, t years after purchase, is given by $S = c(1-r)^t$, where c is the original cost and r is the annual rate of depreciation (as a decimal). If a refrigerator depreciates 14% per year and is worth \$259 after $5\frac{1}{2}$ years, what did it cost originally?

Teaching Notes:

- Reiterate the one-to-oneness of $y = \log_a x$ and $y = a^x$ so that
  $$\log_a u = \log_a v \Leftrightarrow u = v \quad \text{and} \quad a^u = a^v \Leftrightarrow u = v.$$

- Emphasize that logarithmic equations must be checked for extraneous solutions.

Answers: 1. a) $-\dfrac{6}{7}$*; b) 9; c) 2.791; d) 1; e) 3; f) 2.383*

    2. ≈ 174 *months (14.5 yrs.); 3. 16.822 watts; 4. \$593.69*

Mini-Lecture 13.7
Natural Exponential and Natural Logarithmic Functions

Learning Objectives:

1. Identify the natural exponential function.
2. Identify the natural logarithmic function.
3. Find values on a calculator.
4. Find logarithms using the change of base formula.
5. Solve natural logarithmic and natural exponential equations.
6. Solve applications.
7. Key vocabulary: *natural exponential function, natural logarithmic function, e, change of base formula, exponential growth (decay)*.

Examples:

1. Solve each equation for x; use a calculator if necessary:

   a) $x = \ln 3.82$ b) $x + 1 = e^{5.39}$ c) $\ln(x-2) = -1.76$ d) $e^{2x+1} = 17.88$ e) $x = \log_5 12.7$

   f) $\log_{9.2} x = -0.8622$ g) $\log_x 17 = 3.1$ h) $\ln x - \ln y = 2y$

2. Josh invests $5000 in a savings account that compounds interest continuously $\left(P = P_0 e^{rt} \right)$. What interest rate must he get to have $7500 after 4 years?

3. The *Jenss Model* $\left(h(x) = 79.041 + 6.39x - e^{(3.261 - 0.993x)} \right)$ predicts the height, h *(in cm)*, of preschoolers based on their age, x, in years $\left(\frac{1}{4} \le x \le 6 \right)$. What is the average height of a 3-yr. old?

4. Half life for radioactive substances is given by $0.5 = e^{-\lambda \cdot t}$ where t is the half-life and λ is a decay constant $\left(in\ \dfrac{1}{time\ units} \right)$ peculiar to the substance. If Carbon 10 has a half-life of 19.255 seconds, find its decay constant.

Teaching Notes:

- Point out the importance of e both in nature and in higher mathematics.

- Stress that $\ln x$ is short for $\log_e x$.

- Emphasize that the *change of base* relationship enables us to solve <u>any</u> logarithmic or exponential equation using a <u>single</u> table of logarithms from a random base.

- Point out that $P = P_0 e^{k \cdot t}$ represents exponential "growth" or "decay" as k is positive or negative.

Answers: 1. a) 1.3403; b) 218.203; c) 2.172; d) 0.942; e) 1.579; f) 0.1476; g) 2.494; h) ye^{2y}

2. 10.14 %.   3. 96.885 cm or 38.14 inches.   4. 0.036

Additional Exercises 13.1

Name:_____

Date:_____

1. Given $f(x) = x^3$ and $g(x) = 4 + 3x$, find $(g \circ f)(x)$.

2. Given $f(x) = x^2 - 4$ and $g(x) = \sqrt{x-2}$, find $(f \circ g)(x)$.

3. Given $f(x) = \dfrac{x+3}{x}$ and $g(x) = x^2 + 8$, find $(g \circ f)(1)$.

4. If $f(x) = x^3$ and $g(x) = -6 - 3x$, find $f \circ g(-1)$.

1. _____

2. _____

3. _____

4. _____

Determine whether each function is a one-to-one function.

5.

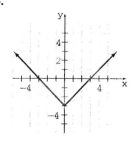

6. $\{(-1, 4), (0, -3), (1, -5), (-3, 0)\}$

7. $f(x) = 4x + 5$.

5. _____

6. _____

7. _____

For each function, **(a)** determine if it is one-to-one; **(b)** if it is one-to-one find its inverse function.

8. $f(x) = \{(7, -6), (-6, 7), (8, 1), (10, -6)\}$

9. $f(x) = 5x^3 - 3$

10. $f(x) = \sqrt{x-4}$

8. (a) _____
    (b) _____

9. (a) _____
    (b) _____

10. (a) _____
    (b) _____

For the one-to-one functions given, find $f^{-1}(x)$ and graph $f(x)$ and $f^{-1}(x)$ on the same axes.

11. $f(x) = 2x - 3$.

11.

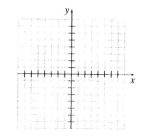

494

12. $f(x) = 3x + 1$

12.

13. $f(x) = \sqrt[3]{x-1}$

13.

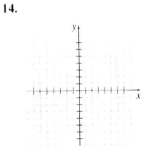

14. $f(x) = \sqrt{x+4}$, $x \ge -4$

14.

For $f(x) = 4x + 3$:

  15. find $f^{-1}(x)$

  15. _____

  16. show that $\left(f^{-1} \circ f\right)(x) = x$

  16. _____

For $f(x) = \sqrt[3]{x+5}$:

  17. find $f^{-1}(x)$

  17. _____

  18. show that $\left(f^{-1} \circ f\right)(x) = x$

  18. _____

Additional Exercises 13.1 *(cont.)* Name:_____

Determine if the pairs of functions shown in each graph are inverses.

19.

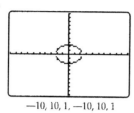

—10, 10, 1, —10, 10, 1

19. _____

20.

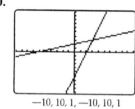

—10, 10, 1, —10, 10, 1

20. _____

Additional Exercises 13.2

Graph the exponential function.

1. $y = 6^x$

1.

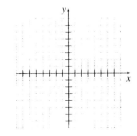

2. $y = \left(\dfrac{1}{4}\right)^x$

2.

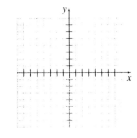

3. $y = 5^{-x}$

3.

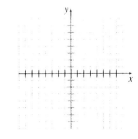

4. $y = 2^{x+3}$

4.

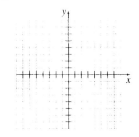

5. $y = 3^{x-2}$

5.

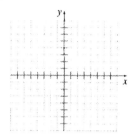

6. $y = 2^x - 5$

6.

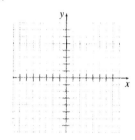

7. $y = \left(\dfrac{1}{2}\right)^x - 2$

7.

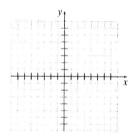

8. $y = 2^{3x} - 4$

8.

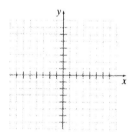

The amount of money, A, accrued at the end of n years when a certain amount, P, is invested at a compound annual rate, r, is given by $A = P(1+r)^n$.

9. If a person invests \$290 at 8% interest compounded annually, find the approximate amount obtained at the end of 15 years.

9. _____

10. If a person invests \$340 at 5% interest compounded annually, find the approximate amount obtained at the end of 10 years.

10. _____

11. Find the amount of money, P, originally invested, if Say-Chun has \$3436.57 after 5 years in a savings account yielding 12% interest per year.

11. _____

Additional Exercises 13.2 *(cont.)* Name:_____

The amount of money, A, accrued at the end of Z years when a certain amount, P, is invested at a rate, r, compounded n times in a year, is given by $A = P\left(1 + \dfrac{r}{n}\right)^{nt}$.

12. If a person invests \$450 at 6% interest compounded quarterly, find the amount accrued at the end of 10 years.

12. _____

13. If a person invests \$760 at 11% interest compounded quarterly, find the amount accrued at the end of 15 years.

13. _____

14. Find the amount of money, P, originally invested if Manvella has \$8447.64 after 8 years in a savings account yielding 12% interest compounded monthly.

14. _____

For 15–18, use $A = A_0 \cdot 2^{-t/5600}$ where A_0 is the amount of carbon 14 in a sample and A is the amount of carbon 14 in the sample after t years.

15. If 45 grams of carbon 14 are originally present in a certain fossil found at an archeological site, how much will remain after 4000 years?

15. _____

16. If 45 grams of carbon 14 are originally present in a certain fossil found at an archeological site, how much will remain after 6000 years?

16. _____

17. If 40 grams of carbon 14 are present in an animal bone at the end of 5000 years, how much was originally present?

17. _____

18. If 30 grams of carbon 14 are originally present in an animal bone, how much will remain at the end of 8000 years?

18. _____

19. The expected future population of a town which presently has 5000 residents can be approximated by the formula $y = 5000(1.2)^{0.1x}$, where x is the number of years in the future. Find the expected population of the town in 20 years.

19. _____

20. A car originally cost \$16,500. and depreciates at a rate of 18% per year (i.e., $V = 16500(0.82)^t$, after t years). What will the car be worth after 7 years?

20. _____

Additional Exercises 13.3

Graph the logarithmic function.

1. $y = \log_{1/4} x$

1.

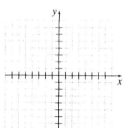

2. $y = \log_6 x$

2.

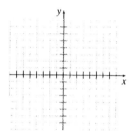

3. $y = \log_4 x$

3.

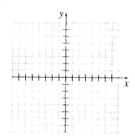

Graph each pair of functions on the same axes.

4. $y = \log_3 x, \ y = 3^x$

4.

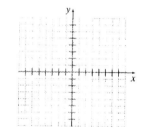

Additional Exercises 13.3 *(cont.)*

Name:_____

5. $y = \log_{1/3} x$, $y = \left(\dfrac{1}{3}\right)^x$

5.

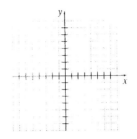

6. $y = \log_3 x$, $y = \left(\dfrac{1}{3}\right)^x$

6.

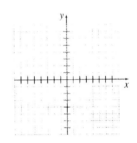

Write each expression in logarithmic form.

7. $3^4 = 81$

8. $5^d = f$

9. $x = \left(\dfrac{9}{8}\right)^y$

10. $x^y = \dfrac{2}{7}$

11. $4^k = m$

12. $x = \left(\dfrac{7}{8}\right)^y$

13. $z = x^y$

Write each expression in exponential form.

14. $\log_{1/3} \dfrac{1}{9} = 2$

15. $\log_{10} 10{,}000 = 4$

7. _____

8. _____

9. _____

10. _____

11. _____

12. _____

13. _____

14. _____

15. _____

Additional Exercises 13.3 *(cont.)*

Name:_____

Evaluate each of the following.

16. $\log_5 1$

17. $\log_{10} 0.001$

18. $\log_2 \dfrac{1}{32}$

19. Graph: $\log_{25} 5$

20. Graph: $y = \log_2 (x + 3)$

16. _____

17. _____

18. _____

19. _____

20.

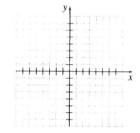

Additional Exercises 13.4

Name:_____

Date:_____

Use properties 1–3 to expand.

1. $\log_p \dfrac{Q}{5}$

2. $\log_a \dfrac{7xy^3}{z^5}$

3. $\log_a \dfrac{5xy^3}{z^4}$

4. $\log_5 \left[\dfrac{x(x+6)}{x^2} \right]$

5. $\log_3 \left[\dfrac{x(x+5)}{x^6} \right]$

6. $\log_b \sqrt[7]{\dfrac{x^7 y^2}{z^6}}$

7. $\log_b \sqrt{\sqrt[3]{x} \sqrt[4]{y}}$

8. $\log_8 x^5 (x-3)^4$

Write as a logarithm of a single expression.

9. $\left[3\log_2 (x+2) + 7\log_2 (x+4) \right] - \dfrac{1}{2}\log_2 x$

10. $2\log_b x - 7\log_b y$

11. $\log_5 \sqrt{x} + \log_5 \sqrt[3]{x^2}$

12. $\dfrac{1}{2}\log_9 (x+2) + \log_9 x$

13. $\log_a 4 + \log_a x + 2\log_a y - 5\log_a z$

14. $4\log_5 (x+6) + 3\log_5 (x+4) - \dfrac{1}{2}\log_5 x$

15. $\log_2 x + \log_4 y$

16. Given $\log_{10} 2 = 0.301$ and $\log_{10} 11 = 1.041$, find $\log_{10} 22$.

17. Find the value of $\log_{10} 4$ using the value $\log_{10} 2 = 0.3010$.

1. _____

2. _____

3. _____

4. _____

5. _____

6. _____

7. _____

8. _____

9. _____

10. _____

11. _____

12. _____

13. _____

14. _____

15. _____

16. _____

17. _____

Additional Exercises 13.4 *(cont.)* Name:_____

Evaluate.

18. $\log_6 36 + \log_2 8$

18. _____

19. $7^{\log_7 12}$

19. _____

20. $\dfrac{1}{2}\log_4 \sqrt[3]{4}$

20. _____

Additional Exercises 13.5

Name:_____

Date:_____

Find the common logarithm of the number. Round to four decimal places.

1. $\log 0.0000940$

2. $\log 3926$

Find the antilog of the logarithm. Round to three significant digits.

3. 2.8987

4. 3.9965

5. 1.4624

6. -1.0845

7. 0.0000

8. -0.0025

Find the number N. Round N to three significant digits.

9. $\log N = 2.0350$

10. $\log N = -3.1072$

To what exponent must the base 10 be raised to obtain each of the following values? Round to four decimal places.

11. 720

12. 29

13. 0.65

Find the value obtained when 10 is raised to the following exponents. Round to three significant digits.

14. 4.8315

15. -2.5017

By changing the logarithm to exponential form, evaluate the common logarithm without the use of a calculator.

16. $\log 0.0001$

17. $\log 100,000$

Use the properties $\log 10^x = x$ and $10^{\log x} = x \, (x > 0)$ to evaluate each of the following.

18. $\log 10^{8.7}$

19. $2\left(10^{\log 1.9}\right)$

20. $7\log 10^{4.5} - 2\log 10^{2.5}$

1. _____

2. _____

3. _____

4. _____

5. _____

6. _____

7. _____

8. _____

9. _____

10. _____

11. _____

12. _____

13. _____

14. _____

15. _____

16. _____

17. _____

18. _____

19. _____

20. _____

Additional Exercises 13.6

Name:_____

Date:_____

Solve the exponential equation without using a calculator.

1. $3^x = 729$

2. $5^x = \sqrt{125}$

3. $49^x = 7^{3x-5}$

4. $\left(\dfrac{1}{4}\right)^{2x-3} = 8$

Use a calculator to solve. Round to three significant digits.

5. $5.50^x = 43$

6. $5.15^x = 32$

7. $3^x = 133$

Solve the logarithmic equation. Use a calculator where appropriate. If the answer is irrational, round to the nearest hundredth.

8. $\log_2(3x+4) = 6$

9. $\log x + \log 3 = 0.7292$

10. $\log 11x = \log(x+21) + \log 4$

11. $\log x + \log 3 = 0.6243$

12. $\log 8 + \log x = \log(x+18) + \log 2$

13. If there are initially 2000 bacteria in a culture, and the number of bacteria double each hour, the number of bacteria after t hours can be found using the formula $N = 2000(2^t)$. How long will it take for the culture to grow to 65,000 bacteria?

14. The amount, A, of 159 grams of a certain radioactive material remaining after t years can be found by the equation $A = 159(0.800)^t$. After how long will there be 53 grams of the material remaining? Round your answer to two decimal places.

15. The compound amount, A, accrued with quarterly compounding is given by $A = P\left(1 + \dfrac{r}{4}\right)^{4t}$, where P is the principal, r is the annual interest rate and t is the time in years. If the rate is 5.5%, find how long it takes for the money to double, that is for $A = 2P$.

16. If there are initially 1000 bacteria in a culture, and the number of bacteria double each hour, the number of bacteria after t hours can be found using the formula $N = 1000(2^t)$. How long will it take for the culture to grow to 45,000 bacteria?

1. _____

2. _____

3. _____

4. _____

5. _____

6. _____

7. _____

8. _____

9. _____

10. _____

11. _____

12. _____

13. _____

14. _____

15. _____

16. _____

Additional Exercises 13.6 *(cont.)*

17. The amount, A, of 108 grams of a certain radioactive material remaining after t years can be found by the equation
$A = 108(0.850)^t$. After how long will there be 27 grams of the material remaining? Round your answer to two decimal places.

17. _____

18. The compound amount, A, accrued with continuous compounding is given by $A = P \cdot 10^{0.4343rt}$, where P is the principal, r is the rate, and t is the time in years. If the rate is 11.5%, find how long it takes for the money to double, that is for $A = 2P$. ($\log 2 = 0.30103$)

18. _____

19. If \$100,000 is invested at 6% compounded monthly for 7 years, the compounded amount is given by $A = 100,000(1.005)^{84}$.
Given that $\log 1.005 = 0.00217$, find $\log A$.
(Note that $100,000 = 10^5$.)

19. _____

20. If an item has a new cost of C and an annual depreciation rate of r (as a decimal), its value, V, after t years is given by
$V = C(1-r)^t$. Monica bought a microwave oven 4 years ago. It has depreciated 13% annually, and currently has a value of \$246.32. What did she pay for it new?

20. _____

Additional Exercises 13.7

Find the following values. Round values to four decimal places.

1. $\ln 626$

2. $\ln 0.630$

Find the value of N. Round values to three significant digits.

3. $\ln N = -1.614$

4. $\ln N = 2.016$

Use the change of base formula to find the value of the following logarithms.

5. $\log_8 33$

6. $\log_8 90$

Solve for x:

7. $\log_7 24.3 = x + 2$

8. $\log_{x+1} 17 = 3.5$

Solve the following logarithmic equations.

9. $\ln x + \ln(x-1) = \ln 20$

10. $\ln(x^2 - 3) - \ln(x-1) = \ln 1$

Solve each equation.

11. $P = 400e^{1.6(2.4)}$

12. $e^{8.3x} = e$

13. $100 = 20e^{8.6t}$

14. $\ln x - \ln(y+3) = t$ for x

15. $\ln x + \ln(y-5) = t$ for x

16. $3\ln x - 2\ln(y-1) = t$ for x

Use a calculator to solve.

17. If \$6500 is invested at a rate of 12% compounded continuously, find the balance in the account after 5 years. Use the formula $P = P_0 e^{rt}$.

1. _____

2. _____

3. _____

4. _____

5. _____

6. _____

7. _____

8. _____

9. _____

10. _____

11. _____

12. _____

13. _____

14. _____

15. _____

16. _____

17. _____

Additional Exercises 13.7 *(cont.)*

18. A radioactive material decays exponentially at a rate of 3.9%
    per year. The amount of the material left after t years can
    be found by the formula $P = P_0 e^{-0.039t}$. Assume there are
    8000 grams of the material.
    (a) Find the number of grams of the material remaining after 50 years.
    (b) Find the half-life of the material. Round your answers to the
        nearest tenth.

18. (a) _____
    (b) _____

19. The number of a certain product that will be sold t years after the
    product is introduced is given by $S = 4100\ln(6t + 3)$. How many
    of the product will be sold 6 years after the product is introduced?
    Round the answer to the nearest whole number.

19. _____

20. The number of a certain product that will be sold t years after the
    product is introduced is given by $S = 5200\ln(2t + 7)$. How
    long will it take to sell 16,000 of the product?

20. _____

Chapter 13 Test Form A

1. Determine whether the function is a one-to-one function
   $f(x) = x^2 - 4, x \leq 0$

   1. _____

For questions 2 and 3, use $f(x) = x + 2, and\ g(x) = x^2 + 4x - 2.$

2. Find: $(f \circ g)(x)$

   2. _____

3. Find: $(g \circ f)(x)$

   3. _____

For questions 4 and 5, use $f(x) = -3x + 6.$

4. Find: $f^{-1}(x)$

   4. _____

5. Graph: $f(x)$ and $f^{-1}(x)$ on the same axes

   5.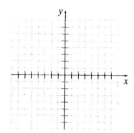

6. Graph: $y = 5^x$

   6.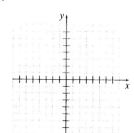

7. Graph: $y = \log_4 x$.

   7.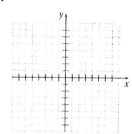

8. Write $81^{1/2} = 9$ in logarithmic form.

   8. _____

9. Write $\log_7 \dfrac{1}{49} = y$ in exponential form, then find the unknown value.

   9. _____

510

10. Use the properties of logarithms to expand $\log_8 \dfrac{\sqrt{x}}{13}$.

10. _____

For 11 and 12, write as the logarithm of a single expression.

11. $5\log_7(a+3) + 2\log_7(a-1) - \dfrac{1}{2}\log_7 a$

11. _____

12. $5\log_6(x+3) - \left[2\log_6(x-4) + 3\log_6 2 \right]$

12. _____

For problems 13 and 14, evaluate the given expressions..

13. $2\log_9 \sqrt{3}$

13. _____

14. $\dfrac{1}{2}\log_6 \sqrt[3]{6}$

14. _____

15. If $\log N = -1.1469$, find N. Round to 3 significant digits.

15. _____

16. Solve the following exponential equation without using a calculator: $4^{x+1} = 8$

16. _____

17. Solve for x: $\log(x+3) + \log x = \log 4$

17. _____

18. Solve for x: $\ln(x+3) + \ln(x-2) = \ln 14$

18. _____

19. If $\ln N = 0.543$, find N. Round to three significant digits.

19. _____

20. Use the change of base formula to find the value of $2\log_{15} 7$.

20. _____

Chapter 13 Test Form B

1. Determine whether the function is a one-to-one function:
   $f(x) = |x|$

1. _____

For questions 2 and 3, use $f(x) = x^2 + 1$, and $g(x) = x + 5$.

2. Find: $(f \circ g)(x)$

2. _____

3. Find: $(g \circ f)(4)$

3. _____

For questions 4 and 5, use $f(x) = \sqrt{x+3}$, $x \geq -3$.

4. Find $f^{-1}(x)$.

4. _____

5. Graph $f(x)$ and $f^{-1}(x)$ on the same axes.

5.

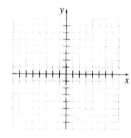

6. Graph: $y = \left(\dfrac{1}{3}\right)^{x+1}$

6.

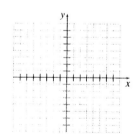

7. Graph: $\log_{1/3} x$

7.

8. Write $a^n = b$ in logarithmic form.

8. _____

9. Write $\log_{1/3} x = 4$ in exponential form, then find the unknown value.

9. _____

10. Use the properties of logarithms to expand $\log_3 \dfrac{d^6}{(a-5)^4}$.

10. _____

For questions 11 and 12, write as the logarithm of a single expression.

11. $\dfrac{1}{2}\left[\log_5(x-4) - \log_5 x\right]$

11. _____

12. $4\log_6 3 - \left[2\log_6(x+3) + 4\log_6 x\right]$

12. _____

For questions 13 and 14, evaluate the given expressions.

13. $5^{\log_5 10}$

13. _____

14. $\left(2^3\right)^{\log_2 5}$

14. _____

15. If $\log N = 1.9036$, find $\log 2N$. Round to 4 decimal places.

15. _____

16. Solve the following exponential equation without using a calculator: $\left(\sqrt{27}\right)^x = 3^{2x+3}$

16. _____

17. Solve for x: $\log(x+7) - \log(x+3) = \log(x-2)$

17. _____

18. Solve for x: $\ln x = \dfrac{5}{2}\ln 9$

18. _____

19. If $\ln N = 4.1$, find N. Round to three significant digits.

19. _____

20. Use the change of base formula to find the value of $\log_5 719$.

20. _____

Chapter 13 Test Form C

Name:_____

Date:_____

1. Determine whether the function is a one-to-one function.
   $f(x) = \sqrt[3]{x+4}$

   1. _____

For questions 2 and 3, use $f(x) = x^2 - 2$ and $g(x) = \sqrt{x+3}$.

2. Find $(f \circ g)(x)$.

   2. _____

3. Find $(g \circ f)(x)$.

   3. _____

For questions 4 and 5, use $f(x) = 4x + 2$.

4. Find $f^{-1}(x)$.

   4. _____

5. Graph $f(x)$ and $f^{-1}(x)$ on the same axes.

   5.
   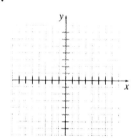

6. Graph $y = \left(\dfrac{1}{3}\right)^x$.

   6.
   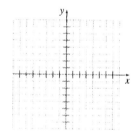

7. Graph $y = \log_{1/3} x$.

   7.
   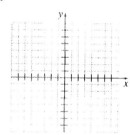

8. Write $5^x = 74$ in logarithmic form.

   8. _____

9. Write $\log_2 x = 4$ in exponential form, then find the missing value.

   9. _____

10. Expand and simplify $\log_2 \dfrac{\sqrt{x}\sqrt[3]{x}}{(x-1)}$.

10. _____

For 11 and 12, write as the logarithm of a single expression.

11. $\log_6(x+1) + 5\log_6(x-1) - 7\log_6(x-2)$

11. _____

12. $3\log_{12} x - \left[4\log_{12}(x-3) + 5\log_{12}(x-8)\right]$

12. _____

Evaluate.

13. $4\log_5 \sqrt{5}$

13. _____

14. $10^{\log 4.7}$

14. _____

15. If $\log N = 2.1453$, find N. Round to 3 significant digits.

15. _____

16. If $\ln N = -0.8265$, find $\sqrt{N}$. Round to 3 significant digits.

16. _____

Solve for x.

17. $\left(\dfrac{1}{4}\right)^x = 32$

17. _____

18. $\log_8(x-3)^2 = 2$

18. _____

19. $\log_3(x+5) - \log_3 2 = \log_3(x^2+1)$

19. _____

For 20 and 21, evaluate using the change of base formula.

20. $\log_{12} 0.864$

20. _____

21. $\log_2 13.5$

21. _____

Solve.

22. $100 = P_0 e^{kt}$ for k.

22. _____

23. $\ln(x-1) - \ln(y+1) = \ln 3$ for y.

23. _____

24. If $7000 is invested at 7% compounded continuously, determine the balance in the account after 5 years.

24. _____

25. Suppose that the population of a city is estimated at 3.21 million people, and that the population will grow exponentially at a rate of 3.5% per year. In how many years will this city's population double?

25. _____

Chapter 13 Test Form D

1. Determine whether the function is a one-to-one function.
   $f(x) = x^2 - 12$

   1. _____

For questions 2 and 3, use $f(x) = \sqrt{x+5}$ and $g(x) = x^2 - 7$.

2. Find $(f \circ g)(x)$.

   2. _____

3. Find $(g \circ f)(x)$.

   3. _____

For questions 4 and 5, use $f(x) = 2x - 4$.

4. Find $f^{-1}(x)$.

   4. _____

5. Graph $f(x)$ and $f^{-1}(x)$ on the same axes.

   5.

   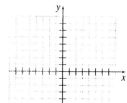

6. Graph $y = \left(\dfrac{1}{2}\right)^x$.

   6.

   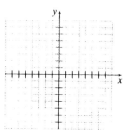

7. Graph $y = \log_{1/2} x$.

   7.

8. Write $b^L = N$ in logarithmic form.

   8. _____

9. Write $y = \log_{64}\left(\dfrac{1}{8}\right)$ in exponential form, then find the missing value.

   9. _____

516

10. Expand and simplify $\log_9 \dfrac{\sqrt{x}\,(x-2)^3}{(x+1)^5}$.

10. _____

For 11 and 12, write as the logarithm of a single expression.

11. $3\log_{12}(x-7) + 2\log_{12}5 - 7\log_{12}x$

11. _____

12. $\log_2(x+11) - \left[5\log_2(x-8) + 5\log_2(x+8)\right]$

12. _____

Evaluate.

13. $\dfrac{1}{3}\log_{14}(14)^3$

13. _____

14. $10^{\log 13.2}$

14. _____

15. If $\log N = 0.7312$, find N^3. Round to 3 significant digits.

15. _____

16. If $\ln N = -0.2318$, find N. Round to 3 significant digits.

16. _____

Solve for x.

17. $256^x = 4$

17. _____

18. $\log_7(x-1)^2 = 2$

18. _____

19. $\log_9(x+5) - \log_9 2 = \log_9 1$

19. _____

For 20 and 21, evaluate using the change of base formula.

20. $\log_5 25.5$

20. _____

21. $\log_{13}1.78$

21. _____

Solve.

22. $300 = 400e^{-0.02t}$ for t.

22. _____

23. $\ln y = \ln(x-2) + \ln(x+3)$ for y.

23. _____

24. If $10,000 is invested at 4% compounded quarterly, how long would it take the value of the account to double?

24. _____

25. Suppose that the population of a city is estimated at 8.99 million people in 2007, and it is expected to grow exponentially at a rate of 5% per year. Find the expected population of this city in the year 2020.

25. _____

Chapter 13 Test Form E

1. Determine whether the function is a one-to-one function.
$f(x) = 5x - 8$

1. _____

For questions 2 and 3, use $f(x) = x^2 + 9$ and $g(x) = x - 8$.

2. Find $(f \circ g)(x)$.

2. _____

3. Find $(g \circ f)(x)$.

3. _____

For questions 4 and 5, use $f(x) = 2x^3 - 4$.

4. Find $f^{-1}(x)$.

4. _____

5. Graph $f(x)$ and $f^{-1}(x)$ on the same axes.

5.

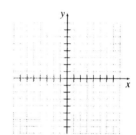

6. Graph $y = 5^x$.

6.

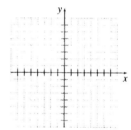

7. Graph $y = \log_5 x$.

7.

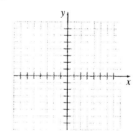

8. Write $10^4 = 10,000$ in logarithmic form.

8. _____

9. Write $\log_9 \dfrac{1}{3} = x$ in exponential form, then find the missing value.

9. _____

10. Expand and simplify $\log_7\left(\dfrac{x}{5}\right)^3$.

10. _____

For 11 and 12, write as the logarithm of a single expression.

11. $8\log_8(x-4)+8\log_8 x-3\log_8(x-1)$

11. _____

12. $4\log_{1/2}(x+1)-\left[3\log_{1/2}5+5\log_{1/2}(x-2)\right]$

12. _____

Evaluate.

13. $3\left(10^{\log 6.1}\right)$

13. _____

14. $6\log_7\sqrt[3]{7}$

14. _____

15. If $\log N = 0.1789$, find $2\sqrt{N}$. Round to 3 significant digits.

15. _____

16. If $\ln N = -0.3342$, find N. Round to 3 significant digits.

16. _____

Solve for x.

17. $\dfrac{1}{216}=36^x$

17. _____

18. $\log_4(x+12)^5=5$

18. _____

19. $\log_3(x-4)-\log_3 6=\log_3\dfrac{1}{3}$

19. _____

For 20 and 21, evaluate using the change of base formula.

20. $\log_9 45.3$

20. _____

21. $\log_3 0.479$

21. _____

Solve.

22. $100=500e^{-0.023t}$ for t.

22. _____

23. $\ln y - \ln(x+5)=3$ for y.

23. _____

24. If \$11,999.27 has accrued over 3 years at 9% compounded continuously, how much was originally invested?.

24. _____

25. The radioactive element, carbon 14, decays at a rate of 0.01205% per year. The amount of carbon 14 in an object after t years is given by the function $f(t)=200e^{-0.0001205t}$, when there were initially 200 grams present. Find the amount of carbon 14 remaining after 100 years.

25. _____

Chapter 13 Test Form F

1. Is the following function one-to-one?

    $\{(-2,3),(-1,4),(0,2),(1,5)\}$

1. _____

For questions 2 and 3, use $f(x)=x^2-7$ and $g(x)=\sqrt{x+3}$.

2. Find $(f\circ g)(x)$.

2. _____

3. Find $(g\circ f)(x)$.

3. _____

For questions 4 and 5, use $f(x)=\sqrt[3]{x-4}$.

4. Find $f^{-1}(x)$.

4. _____

5. Find the domain and range of $f^{-1}(x)$.

5. _____

In problems 6 and 7, write an exponential or logarithmic equation for the given graph using base $\frac{1}{5}$.

6. Write an equation for the following graph.

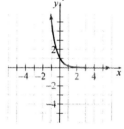

6. _____

7. Write an equation for the following graph.

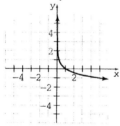

7. _____

8. Write $x^z=y+1$ in logarithmic form.

8. _____

9. Write $\log_x 5=\frac{1}{3}$ in exponential form, then find the missing value.

9. _____

10. Expand $\log_{1/3}\dfrac{(x+4)(x-2)^3}{16x^8}$.

10. _____

For 11 and 12, write as the logarithm of a single expression.

11. $5\log_2 x+3\log_2(x+1)-8\log_2(x-1)$

11. _____

12. $2\log_{1/2}(x+1)-\left[\log_{1/2}x+3\log_{1/2}(x-1)\right]$

12. _____

Evaluate.

13. $4^{\log_4 8.2}$

13. _____

14. $\dfrac{1}{5}\log_8\left(8^5\right)$

14. _____

15. If $\log N = 0.5432$, find N. Round to 3 significant digits.

15. _____

16. If $\ln N = 1.4358$, find N. Round to 3 significant digits.

16. _____

Solve for x.

17. $100^x = \dfrac{1}{10}$

17. _____

18. $\log_2(x-5)^3 = 9$

18. _____

19. $\log_7(x+9) - \log_7 4 = \log_7 3$

19. _____

For 20 and 21, evaluate using the change of base formula.

20. $\log_{14} 0.892$

20. _____

21. $\log_3 29.4$

21. _____

Solve.

22. $30 = 50e^{-0.05t}$ for t

22. _____

23. $\ln y - \ln(x + 4) = \ln 8.1$ for y

23. _____

24. If \$4,200 is invested at 6% compounded monthly, how long would it take the value of the account to double?

24. _____

25. The percentage of doctors who accept and prescribe a new drug is given by the function $P(t) = 1 - e^{-0.22t}$, where t is the time in months since the drug was placed on the market. What percentage of doctors accepted the new drug 2 months after it is placed on the market?

25. _____

Chapter 13 Test Form G

1. Determine which function is a one-to-one function.

   (a) $\{(4, 3), (5, 1), (-6, 2), (7, 3\}$

   (b) $\{(4, 3), (5, 3), (-6, 2), (7, 1)\}$

   (c) $\{(4, 3), (5, 1), (-6, 2), ((7, 4)\}$

   (d) $\{(4, 4), (5, 1), (-6, 2), (7, 4)\}$

For questions 2 and 3, use $f(x) = x^2 - 10$ and $g(x) = x + 6$.

2. Find $(f \circ g)(x)$.

   (a) $x^3 + 6x^2 - 10x - 60$ **(b)** $x^2 + 12x + 26$ **(c)** $x^2 + 12x - 26$ **(d)** $x^2 - 4$

3. Find $(g \circ f)(x)$.

   (a) $x^3 + 6x^2 - 10x - 60$ **(b)** $x^2 + 12x + 26$ **(c)** $x^2 + 12x - 26$ **(d)** $x^2 - 4$

For questions 4 and 5, use $f(x) = \sqrt[3]{x + 3}$.

4. Find $f^{-1}(x)$.

   (a) $x^3 - 3$ **(b)** $x^3 + 3$ **(c)** $x + 3$ **(d)** $\sqrt[3]{x - 3}$

5. Find the domain and range of $f^{-1}(x)$.

   (a) $D: \mathbb{R}$, $R: \mathbb{R}$ **(b)** $D: \mathbb{R}$, $R: x \neq 3$ **(c)** $D: x \geq 3$, $R: y \geq 0$ **(d)** $D: x \geq -3$, $R: \mathbb{R}$

6. Which of the following is the equation of the following graph?

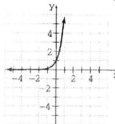

   (a) $y = 6^x$ **(b)** $y = \left(\dfrac{1}{6}\right)^x$ **(c)** $y = \log_6 x$ **(d)** $y = \log_{1/6} x$

7. Which of the following is the equation of the following graph?

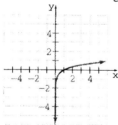

   (a) $y = 6^x$ **(b)** $y = \left(\dfrac{1}{6}\right)^x$ **(c)** $y = \log_6 x$ **(d)** $y = \log_{1/6} x$

8. Write $a^5 = b$ in logarithmic form.

(a) $\log_b 5 = a$ (b) $\log_a b = 5$ (c) $\log_5 b = a$ (d) $\log_5 a = b$

9. Write $\log_{1/3} x = -2$ in exponential form, then find the missing value.

(a) $x = \dfrac{1}{9}$ (b) $x = -\dfrac{1}{8}$ (c) $x = -8$ (d) $x = 9$

10. Expand $\log_7 \dfrac{\sqrt{x}\sqrt[3]{x+1}}{8\sqrt[4]{x-1}}$.

(a) $\dfrac{1}{2}\log_7 x + \dfrac{1}{3}\log_7(x+1) - \left[\dfrac{1}{4}\log_7(x-1) + 3\log_7 2\right]$ (b) $2\log_7 x + 3\left[\log_7(x+1) + \log_7 2\right] - 4\log_7(x-1)$

(c) $7\log\sqrt{x} + 7\log\sqrt[3]{x+1} - 7\log\sqrt[4]{x-1} - 7\log 8$ (d) $\dfrac{1}{2}\log x + \dfrac{1}{3}\log(x+1) - \dfrac{1}{4}\log(x-1) - \dfrac{1}{3}\log 8$

For 11 and 12, write as the logarithm of a single expression.

11. $2\log_5(x-4) + 3\log_5 x - 7\log_5(x+3)$

(a) $\log_5 \dfrac{6x(x-4)}{7(x+3)}$ (b) $\log_5 \dfrac{x(x-4)}{x+3}$ (c) $\log_5 \dfrac{x^3(x-4)^2}{(x+3)^7}$ (d) $\log_5 \dfrac{5x-8}{7x+21}$

12. $\log_{10}(x+3) - \left[5\log_{10} x + 2\log_{10}(x-8)\right]$

(a) $\log_{10} \dfrac{1}{2x+13}$ (b) $\log_{10} \dfrac{x+3}{x^5(x-8)^2}$ (c) $\log_{10} \dfrac{x^5(x-8)^2}{x+3}$ (d) $\dfrac{\log_{10}(x+3)(x-8)^2}{x^5}$

Evaluate.

13. $\left(3^2\right)^{\log_9 4}$

(a) 3   (b) 4   (c) 9   (d) 27

14. $5\log 10^3$

(a) 3   (b) 5   (c) 10   (d) 15

15. If $\log N = -0.7114$, find $\dfrac{1}{N}$. Round to 3 significant digits.

(a) 5.15   (b) 0.039   (c) 25.4   (d) 1.94

16. If $\ln N = 0.3211$, find N. Round to 3 significant digits.

(a) −1.14   (b) 1.38   (c) 2.09   (d) 3.21

Solve for x.

17. $8^{x+1} = \dfrac{1}{32}$

   (a) $x = -\dfrac{8}{3}$ (b) $x = -\dfrac{3}{8}$ (c) $x = \dfrac{3}{8}$ (d) $x = \dfrac{8}{3}$

18. $\log_2 (x+8)^5 = 5$

   (a) $x = -10$ (b) $x = -6$ (c) $x = -3$ (d) $x = 2$

19. $\log_3 (x+12) - \log_3 2 = \log_3 (2x-3)$

   (a) $x = 6.5$ (b) $x = 2$ (c) $x = 6$ (d) $x = 4$

For 20 and 21, evaluate using the change of base formula.

20. $\log_8 17.4$

   (a) 0.7280 (b) 0.9031 (c) 1.2405 (d) 1.3737

21. $\log_4 29.9$

   (a) 0.4080 (b) 0.6021 (c) 1.4757 (d) 2.4510

Solve.

22. $120 = R_0 e^{kt}$ for t

   (a) $\dfrac{1}{k} \ln\left(\dfrac{120}{R_0}\right)$ (b) $k \ln\left(\dfrac{120}{R_0}\right)$ (c) $\dfrac{1}{k} e^{120}$ (d) $\dfrac{1}{k} e R_0$

23. $\ln y - \ln x = 4.7$ for y

   (a) $\ln x + 4.7$ (b) $x e^{4.7}$ (c) $x + e^{4.7}$ (d) x

24. If \$4,900 is invested at 8.9% compounded monthly, determine the value of the account after 4 years.

   (a) \$5065.50   (b) \$6290.00   (c) \$6986.09   (d) \$17,435.

25. At what rate, compounded continuously, must a sum of money be invested if it is to double in 5 years?

   (a) 5%   (b) 10%   (c) 11.2%   (d) 13.9%

Chapter 13 Test Form H

Name:_____

Date:_____

1. Determine which function is a one-to-one function.

   (a) $\{(4, 11), (-3, 2), (1, 5), (-4, 2)\}$
   (b) $\{(4, 11), (-3, 4), (1, 5), (-4, 4)\}$
   (c) $\{(4, 11), (-3, 5), (1, 5), (-4, 2)\}$
   (d) $\{(4, 11), (-3, 4), (1, 5), (-4, 1)\}$

For questions 2 and 3, use $f(x) = x^2 - 8$ and $g(x) = \sqrt{x + 8}$.

2. Find $(f \circ g)(x)$.

   (a) x^2 (b) $x + 8$ (c) $x - 8$ (d) x

3. Find $(g \circ f)(x)$.

   (a) x^2 (b) $|x| + 8$ (c) $\sqrt{x^2 + 16}$ (d) $|x|$

For questions 4 and 5, use $f(x) = x^3 - 8$.

4. Find $f^{-1}(x)$.

   (a) $x - 2$ (b) $\sqrt[3]{x - 8}$ (c) $\sqrt[3]{x + 8}$ (d) $x^3 + 8$

5. Find the domain and range of $f^{-1}(x)$.

   (a) $D: \mathbb{R}$, $R: \mathbb{R}$ (b) $D: \mathbb{R}$, $R: y \geq 0$ (c) $D: x \geq -8$, $R: y \geq 0$ (d) $D: x \geq 8$, $R: \mathbb{R}$

6. Which of the following is the equation of the following graph?

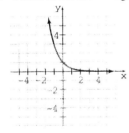

   (a) $y = 3^x$ (b) $y = \left(\dfrac{1}{3}\right)^x$ (c) $y = \log_3 x$ (d) $y = \log_{1/3} x$

7. Which of the following is the equation of the following graph?

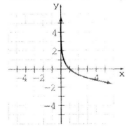

   (a) $y = 3^x$ (b) $y = \left(\dfrac{1}{3}\right)^x$ (c) $y = \log_3 x$ (d) $y = \log_{1/3} x$

8. Write $\left(\dfrac{1}{2}\right)^{-3} = 8$ in logarithmic form.

   (a) $\log_8\left(\dfrac{1}{2}\right) = -3$ (b) $\log_{1/2} 8 = -3$ (c) $\log_8(-3) = \dfrac{1}{2}$ (d) $\log_{1/2}(-3) = 8$

9. Write $y = \log_{36}\left(\dfrac{1}{6}\right)$ in exponential form, then find the missing value.

   (a) $y = 2$ (b) $y = \dfrac{1}{2}$ (c) $y = -\dfrac{1}{2}$ (d) $y = -2$

10. Expand $\log_3\left[\left(\dfrac{x-4}{5}\right)^5\left(\dfrac{1}{x}\right)\right]$.

    (a) $5\log_3(x-4) + \log_3 5x$

    (c) $5\log_3(x-4) - \log_3 5x$

    (b) $5\log_3(x-4) - 5\log_3 x - \log_3 5$

    (d) $5\log_3(x-4) - \left[5\log_3 5 + \log_3 x\right]$

For 11 and 12, write as the logarithm of a single expression.

11. $8\log_{20}(x-12) + 6\log_{20} 2 - 3\log_{20}(x+10)$

    (a) $\log_{20}\dfrac{14x-10}{3x+30}$ (b) $\log_{20}\dfrac{64(x-12)^8}{(x+10)^3}$ (c) $\log_{20}\dfrac{36(x-12)^8}{(x+10)^3}$ (d) $\log_{20}\dfrac{2(x-12)}{x+10}$

12. $4\log_a(3+x) - \left[\log_a x + 5\log_a(x-1)\right]$

    (a) $\log_a\dfrac{(x+3)^4}{x(x-1)^5}$ (b) $\log_a\dfrac{(3+x)}{x(x-1)}$ (c) $\log_a\dfrac{(x+3)^4(x-1)^5}{x}$ (d) $\log_a\dfrac{x+3}{2x-1}$

Evaluate.

13. $\dfrac{1}{2}\log_{12}\left(12^6\right)$

    (a) $\dfrac{1}{2}$ (b) 2   (c) 3   (d) 6

14. $1.5\left(10^{\log 3}\right)$

    (a) 1.5   (b) 3   (c) 4.5   (d) 7

15. If $\log N = 0.9255$, find $3N$. Round to 3 significant digits.

    (a) −2.92   (b) 2.53   (c) 25.3   (d) 29.2

16. If $\ln N = -0.6211$, find N. Round to 3 significant digits.

    (a) −0.621   (b) 0.239   (c) 0.537   (d) undefined

Chapter 13 Test Form H *(cont.)*   Name:_____

Solve for x.

17. $\left(\dfrac{1}{9}\right)^{x} = 27$

    (a) $x = -\dfrac{1}{3}$ **(b)** $x = \dfrac{3}{2}$ **(c)** $x = -\dfrac{3}{2}$ **(d)** $x = -3$

18. $\log_{13}(x-8)^{3} = 3$

    (a) $x = 6$ **(b)** $x = 8$ **(c)** $x = 13$ **(d)** $x = 21$

19. $\log_{8}(x-3) - \log_{8}4 = \log_{8}\left(\dfrac{1}{2}x - 1\right)$

    (a) $x = 1$ **(b)** $x = 4$ **(c)** $x = 3$ **(d)** $x = 2$

For 20 and 21, evaluate using the change of base formula.

20. $\log_{2}8.31$

    (a) 0.3010 **(b)** 0.3273 **(c)** 0.9196 **(d)** 3.0548

21. $\log_{12}0.794$

    (a) 1.0792 **(b)** -0.0928 **(c)** -0.1002 **(d)** -10.7725

Solve.

22. $P = 100e^{2t}$ for t

    (a) $\ln\left(\dfrac{P}{10}\right)$ **(b)** $\ln\left(\dfrac{P}{100}\right)$ **(c)** $\dfrac{1}{2}\ln\left(\dfrac{P}{100}\right)$ **(d)** $\ln\left(\dfrac{P}{50}\right)$

23. $\ln(x+1) - \ln(y-1) = \ln 3$ for y

    (a) $\dfrac{x+4}{3}$ **(b)** $\dfrac{x+1}{3}$ **(c)** $\dfrac{x-2}{3}$ **(d)** $\dfrac{x-3}{3}$

24. If \$1200 is invested at 4.5% compounded continuously, how long would it take the value of the account to double?

    (a) 6.9 yrs.   **(b)** 15.4 yrs.   **(c)** 31.7 yrs.   **(d)** 45 years

25. At what rate, compounded continuously, must a sum of money be invested if it is to triple in 5 years?

    (a) 16%   **(b)** 18%   **(c)** 20%   **(d)** 22%

Chapter 14 Pretest Form A

1. Find the distance between the points $P(-2, 3)$ and $Q(4, -5)$.

   1. _____

2. Find the midpoint of the line segment whose endpoints are $P(-2, 3)$ and $Q(3, 5)$.

   2. _____

3. Given the line segment with endpoints $P(-4, 8)$ and $Q(6, -2)$, find the points on $\overline{PQ}$ that break it into four equal parts.

   3. _____

4. Find equations for two different parabolas with vertex $(2, 3)$ and containing the point $(5, 1)$.

   4. _____

For problems 5 and 6, use the equation $x = -2y^2 + 4y + 2$.

5. Write the equation in the form $x = a(y - k)^2 + h$.

   5. _____

6. Graph the equation.

   6.

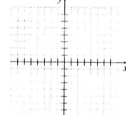

7. Write the equation of a circle with center at $(5, -4)$ and radius 6.

   7. _____

8. Draw the graph of the circle described in the previous problem.

   8.

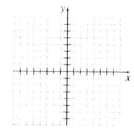

9. Graph $4x^2 + 25y^2 = 100$.

   9.

10. Determine the center of the ellipse given by $25(x + 1)^2 + 9y^2 = 225$.

   10. _____

Chapter 14 Pretest Form A *(cont.)* Name:_____

For problems 11 and 12, use the equation $(x+1)^2 - (y+2)^2 = 4$.

11. What are the equations of the asymptotes of the graph of the equation?

11. _____

12. Graph the equation and its asymptotes.

12.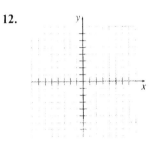

Determine whether the graph of the equation is a parabola, circle, ellipse, or hyperbola.

13. $x^2 + 9y^2 = 36$

13. _____

14. $y^2 - x + 2y = 9$

14. _____

15. $y^2 = x^2 + 16$

15. _____

16. $3y^2 = 7 - 3x^2$

16. _____

17. Solve: $3x - y = -2$
    $2x^2 - y = 0$

17. _____

18. Solve: $x^2 - 2y^2 = 2$
    $xy = 2$

18. _____

19. Find two numbers such that their sum is 3 and their product is 1.

19. _____

20. Solve: $x^2 + y^2 = 4$
    $3y^2 - 4x^2 = 5$

20. _____

Chapter 14 Pretest Form B

Name:_____

Date:_____

1. Find the length of the line segment whose endpoints are (−5, 3) and (−7, −4).

    1. _____

2. Find the midpoint of the line segment whose endpoints are (−5, 3) and (−7, −4).

    2. _____

3. The midpoint of a line segment is $(3, 7)$ and one endpoint is $(−2, −2)$. Find the other endpoint.

    3. _____

4. Find equations for two different parabolas with vertex $(1, −1)$ and containing the point $(0, 3)$.

    4. _____

For problems 5 and 6, use the equation $x = 2y^2 + 4y − 3$.

5. Write the equation in the form $x = a(y − k)^2 + h$.

    5. _____

6. Graph the equation.

    6.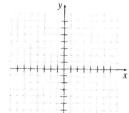

7. Write the equation of a circle with center at $(1, −1)$ and radius 5.

    7. _____

8. Draw the graph of the circle described in the previous problem.

    8.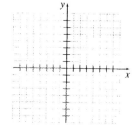

9. Graph $25x^2 + 16y^2 = 400$.

    9.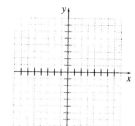

10. Determine the center of the ellipse given by $9(x − 8)^2 + 25(y + 7)^2 = 225$.

    10. _____

For problems 11 and 12, use the equation $\dfrac{y^2}{4} − \dfrac{x^2}{25} = 1$.

11. What are the equations of the asymptotes of the graph of the equation?

    11. _____

Chapter 14 Pretest Form B *(cont.)*

12. Graph the equation and its asymptotes.

12.

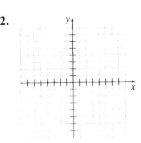

Determine whether the graph of the equation is a parabola, circle, ellipse, or hyperbola.

13. $3y^2 = 100 - 3x^2$

13. _____

14. $4x^2 - 100y^2 = 400$

14. _____

15. $4x - 12 = x^2 - y$

15. _____

16. Solve the system of equations.

$x^2 + y^2 = 100$

$y = 3x - 10$

16. _____

17. Solve the system of equations.

$2x^2 + y^2 = 16$

$3x^2 - 2y^2 = -4$

17. _____

18. Solve the system of equations.

$x^2 + y^2 = 16$

$x^2 + y^2 - 2x = 3$

18. _____

19. Solve the system of equations.

$x^2 + y^2 - 6x - 4y = 12$

$3y + 4x = -7$

19. _____

20. A rectangular field has a perimeter of 60 ft. and an area of 221 ft^2. Find its dimensions.

20. _____

Mini-Lecture 14.1
The Parabola and the Circle

Learning Objectives:

1. Identify and describe the conic sections.
2. Review parabolas.
3. Graph parabolas of the form $x = a(y-k)^2 + h$.
4. Learn the distance and midpoint formulas.
5. Graph circles with centers at the origin.
6. Graph circles with centers at (h,k).
7. Key vocabulary: *distance between two points, midpoint of a line segment, circle and its center.*

Examples:

1. Given the parabola $y = 2x^2 - 4x - 16$,

    a) Write it in the form $y = a(x-h)^2 + k$ b) find the vertex   c) find the *x*-intercepts

2. Given the parabola $x = -y^2 + 8y - 15$,

    a) Write it in the form $x = a(y-k)^2 + h$ b) which way does it open?   c) find the vertex

    d) find all the intercepts   e) sketch the graph

3. Find the *length* and *midpoint* of the line segment whose endpoints are $(1,7)$ and $(-3,-2)$

4. Graph: a) $x^2 + y^2 = 49$ and   b) $y = -\sqrt{49 - x^2}$

5. Write the equation of the circle whose center is $(-2,5)$ and whose radius is $\sqrt{11}$.

6. Find the center and radius of the circle $x^2 + y^2 - 2x + 6y - 5 = 0$

Teaching Notes:

- Point out how conic sections are related to a (two-nappe) cone
- Point out that, in parabolic equations, only *one* variable gets squared.
- Show the connection between the distance formula and the length of a hypotenuse.
- Mention that the center of a circle is the midpoint of any diameter.
- Note that the range of the function is the domain of the inverse, and vice versa.
- Point out that a circle is not a function.

Answers: *1a)* $y = 2(x-1)^2 - 18$;   *b)* $(1,-18)$;   *c)* $x = -2, 4$.

   2a) $x = -(y-4)^2 + 1$;   *b) left (negative x-direction);*   *c)* $(1,4)$;   *d)* $x = -15$ *and*

   $y = 5, 3$;   *e) see graphing answers.*

   3. $l = \sqrt{97}$, $m = \left(-1, \frac{5}{2}\right)$.   *4. see graphing answers.*   *5.* $(x+2)^2 + (y-5)^2 = 11$

   6. $C = (1,-3)$, $r = \sqrt{15}$

Mini-Lecture 14.2
The Ellipse

Learning Objectives:

1. Graph ellipses
2. Graph ellipses with center at (h, k).
3. Key vocabulary: *ellipse, major and minor axes, center, vertices, foci.*

Examples:

For each of the following ellipses, determine

a) the center, b) the major axis, c) the vertices, d) the area, e) the graph

1. $\dfrac{x^2}{18} + \dfrac{y^2}{50} = 2$

2. $\dfrac{(x-1)^2}{9} + \dfrac{(y+2)^2}{4} = 1$

3. $9x^2 + 25y^2 - 18x = 216$

4. $2x^2 + 3y^2 + 12x - 6y + 9 = 0$

5. Write the equation of the ellipse whose vertices are $(2,12), (2,-2)$ and whose area is 35π

Teaching Notes:

- Point out that the major axis of the ellipse $\dfrac{(x-h)^2}{a^2} + \dfrac{(y-k)^2}{b^2} = 1$ is parallel to the coordinate axis corresponding to the largest denominator.
- Point out that the area of an ellipse $(ab\pi)$ depends only on its shape, not its location or orientation. .
- Mention some concrete examples of an ellipse, such as planetary orbits, lithotripters, whispering galleries, luminary reflectors, etc.

Answers:

	Center	Maj.Axis	Vertices	Area	Graph
1.	$(0,0)$	$x=0$	$(0,-10),(0,10)$	60π	see
2.	$(1,-2)$	$y=-2$	$(-2,-2),(4,-2)$	6π	graphing
3.	$(1,0)$	$y=0$	$(-4,0),(6,0)$	15π	answers
4.	$(-3,1)$	$y=1$	$(-3-\sqrt{6},1),(-3+\sqrt{6},1)$	$2\sqrt{6}\,\pi$	

5. $\dfrac{(x-2)^2}{25} + \dfrac{(y-5)^2}{49} = 1$

Mini-Lecture 14.3
The Hyperbola

Learning Objectives:

1. Graph hyperbolas.
2. Review conic sections.
3. Key vocabulary: *hyperbola, transverse axis, center, vertices, asymptotes.*

Examples:

For each of the following hyperbolas, determine

a) the vertices, b) the transverse axis, c) the asymptotes, d) the graph

1. $\dfrac{x^2}{9} - \dfrac{y^2}{4} = 1$

2. $\dfrac{y^2}{4} - \dfrac{x^2}{9} = 1$

3. $4x^2 - y^2 = 4$

4. $2x^2 + 20 = 5y^2$

5. Write the equation of an hyperbola that has $(0,3)$ as a vertex and $y = -0.6x$ as an asymptote.

6. By writing the following in standard form, determine whether the graph is a *parabola, circle, ellipse* or *hyperbola*:

   a) $x^2 + 2y^2 + 2x - 8y + 5 = 0$ b) $4x^2 + 100 = y^2$

   c) $x + 2y^2 = 8y - 5$ d) $2x^2 + 2y^2 + 4x - 8y + 3 = 0$

Teaching Notes:

- Point out that asymptotes are dotted lines because they have no points in common (do not intersect) with the graph of the hyperbola.
- Point out that , for these hyperbolas, the transverse axis is the axis of the variable with the *positive* coefficient, no matter which denominator is larger.

Answers:

	Vertices	Transv. Axis	Asymptotes	Graph
1.	$(3,0),(-3,0)$	$x-axis$	$y = \pm \dfrac{2}{3}x$	see
2.	$(0,2),(0,-2)$	$y-axis$	$y = \pm \dfrac{2}{3}x$	graphing
3.	$(1,0),(-1,0)$	$x-axis$	$y = \pm 2x$	
4.	$(0,2),(0,-2)$	$y-axis$	$y = \pm \dfrac{\sqrt{10}}{5}x$	answers

5. $\dfrac{y^2}{9} - \dfrac{x^2}{25} = 1$

6. a) *ellipse*
   b) *hyperbola*
   c) *parabola*
   d) *circle*

Mini-Lecture 14.4
Nonlinear Systems of Equations and Their Applications

Learning Objectives:

1. Solve nonlinear systems using substitution.
2. Solve nonlinear systems using addition.
3. Solve applications.
4. Key vocabulary: *non-linear system of equations.*

Examples:

1. Solve each of the following systems by either the *substitution* or the *addition* method:

   a) $x^2 - y = 5$
   $x^2 + y^2 = 7$

   b) $2x^2 - 3y^2 = 15$
   $x^2 + y^2 = 10$

   c) $\sqrt{3}\, y - x = 2$
   $x^2 + y^2 = 4$

   d) $4x^2 - 9y^2 = 36$
   $5x - 6y = 0$

   e) $x^2 + y^2 - 2x + 4y = 0$
   $x - y = 2$

2. Given two whole numbers, x and y: The product of the two numbers, added to the smallest, is 70; three times the smallest plus twice the largest is 41 Find the two numbers.

3. A right triangle has a perimeter of 30 in. and an area of 30 in.^{2}. Find its dimensions.

Teaching Notes:

- Point out to students that they may want to review *substitution* and *addition* methods in Chapter 4.
- Remind students that answers to these systems (if there are any) are in the form of ordered pairs.
- Remind students to check for extraneous solutions.

Answers:

1. a) $(\sqrt{3}, -2), (-\sqrt{3}, -2), (\sqrt{6}, 1), (-\sqrt{6}, 1)$; b) $(-3, -1), (-3, 1), (3, -1), (3, 1)$;
   c) $(-2, 0), (1, \sqrt{3})$; d) *no solution;* e) $(-1, -3), (2, 0)$

2. The numbers are 5 and 13.

3. The sides are 5, 12, and 13 inches.

Additional Exercises 14.1

Sketch the graph of each equation.

1. $y = (x+2)^2 - 1$

1.

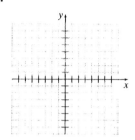

2. $x = 2(y-4)^2 - 2$

2.

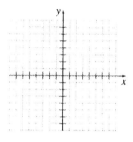

3. $x = 2(y+3)^2 + 4$

3.

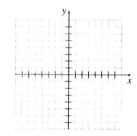

Write each equation in the form $y = a(x-h)^2 + h$ or $x = a(y-h)^2 + h$, and then sketch the graph of the equation.

4. $x = y^2 + 8y + 15$

4. _____

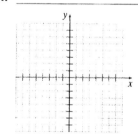

Additional Exercises 14.1 *(cont.)*

5. $y = x^2 - 4x - 2$

5. _____

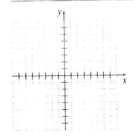

6. $y = 3x^2 - 6x + 4$

6. _____

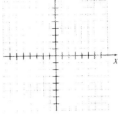

Determine the distance between the points. Use a calculator where appropriate and round your answer to the nearest hundredth.

7. $(3, 3)$ and $(8, 3)$

7. _____

8. $(-4, -4)$ and $(9, 2)$

8. _____

9. Find the perimeter of the triangle whose vertices are $(2,1), (5,3),$ and $(7,-1)$

9. _____

Determine the midpoint of the line segment between the points.

10. $(-6, 3)$ and $(0, 5)$

10. _____

11. $(-3, 4)$ and $(-4, 1)$

11. _____

12. If $P(2,5)$ and $Q(14,21)$ are endpoints, find the points on $\overline{PQ}$ that divide it into four equal parts.

12. _____

Write the equation of the circle

13. with center $(0, 0)$, containing the point $(4, 2\sqrt{5})$

13. _____

14. with center $(-3, -1)$ and radius 2

14. _____

15. Find the equation of the parabola that:
  has vertex at $(5, -2)$,
  opens to the left,
  and contains the point $(2, -5)$.

15. _____

Additional Exercises 14.1 *(cont.)*

16. Write the equation of the circle below. Assume the radius is a whole number.

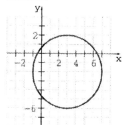

16. _____

Sketch the graph of each equation.

17. $x^2 + y^2 = 5$

17.

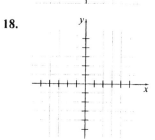

18. $(x-2)^2 + (y+3)^2 = 16$

18.

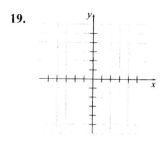

19. $y = -\sqrt{25-x^2}$

19.

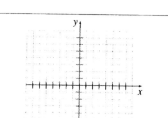

20. Write the equation in standard form, then sketch the graph of the equation.

$x^2 + y^2 + 6x + 10y + 25 = 0$

20. _____

Additional Exercises 14.2

Name:_____

Date:_____

1. Graph: $\dfrac{x^2}{16} + \dfrac{y^2}{64} = 1$

1.

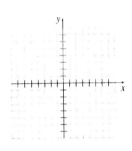

2. Graph: $\dfrac{x^2}{4} + \dfrac{y^2}{9} = 1$

2.

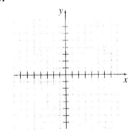

For 3 to 5, use the equation $16x^2 + y^2 = 16$.

3. Write the equation in standard form.

3. _____

4. Give the coordinates of the center of the ellipse.

4. _____

5. Graph the equation.

5.

For 6 to 9, use the equation $4(x+1)^2 + (y-2)^2 = 20$.

6. Write the equation in standard form.

6. _____

7. Give the coordinates of the center of the ellipse.

7. _____

8. Give the x-intercepts of the graph.

8. _____

9. Give the y-intercepts of the graph.

9. _____

10. Graph the equation $36x^2 + 64y^2 = 2304$.

10.

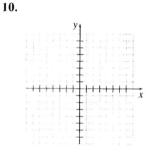

For 11 and 12, use the equation $\dfrac{(x+2)^2}{9} + \dfrac{(y-3)^2}{16} = 1$.

11. Give the coordinates of the center of the ellipse.

11. _____

12. Graph the equation.

12.

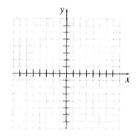

For 13 to 15, use the equation $(x+4)^2 + 4(y+3)^2 = 16$.

13. Write the equation in standard form.

13. _____

14. Give the coordinates of the center of the ellipse.

14. _____

15. Graph the equation.

15.

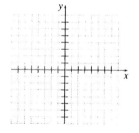

For 16 to 20, use the equation $9(x-1)^2 + 16(y+2)^2 = 144$.

16. Write the equation in standard form.

16. _____

17. Give the coordinates of the center of the ellipse.

17. _____

18. If the ellipse contains $(3, y)$ and $y > 0$, find y.

18. _____

19. What is the length of the major axis of the ellipse?

19. _____

20. What is the area of the ellipse?

20. _____

Additional Exercises 14.3

Name:_____

Date:_____

For 1 and 2, use the equation $\dfrac{x^2}{16} - \dfrac{y^2}{36} = 1$.

   1. Find the equations of the asymptotes.

   2. Graph the equation and asymptotes.

1. _____

2.

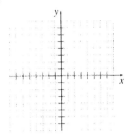

For 3 and 4, use the equation $\dfrac{y^2}{9} - \dfrac{x^2}{9} = 1$.

   3. Find the equations of the asymptotes.

   4. Graph the equation and asymptotes.

3. _____

4.

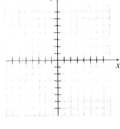

For 5 and 6, use the equation $\dfrac{y^2}{9} - \dfrac{x^2}{4} = 1$.

   5. Find the equations of the asymptotes.

   6. Graph the equation and asymptotes.

5. _____

6.

For 7 to 10, use the equation $9x^2 - 16y^2 = 144$.

   7. Write the equation in standard form.

   8. Find the equations of the asymptotes.

   9. Find the x- or y-intercepts of the graph.

7. _____

8. _____

9. _____

Additional Exercises 14.3 *(cont.)* Name:_____

10. Graph the equation and asymptotes.

10.
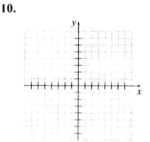

For 11 to 14, use the equation $4y^2 - x^2 = 20$.

11. Write the equation in standard form.

11. _____

12. Find the equations of the asymptotes.

12. _____

13. Find the *x*- or *y*-intercepts of the graph.

13. _____

14. If the hyperbola contains the point $(\sqrt{5}, y)$ and $y > 0$ find y.

14. _____

Indicate whether the graph of the equation is a parabola, circle, ellipse, or hyperbola.

15. $81x^2 = 9 + 81y^2$

15. _____

16. $144x^2 = 36 - 64y^2$

16. _____

17. $16x^2 = 64 + 121y^2$

17. _____

18. $7x^2 + 7y^2 - 11x + 2y + 3 = 0$

18. _____

19. $25x^2 = 121 + 4y^2$

19. _____

20. $30x = 3x^2 - y + 77$

20. _____

Additional Exercises 14.4

Find all real solutions to the system using the substitution method.

1. $x^2 + y^2 = 49$
   $x + y = 7$

2. $x^2 + y^2 = 41$
   $x - 3y = -11$

3. $16x^2 + 25y^2 = 400$
   $5y - 4x = -20$

4. $y = x^2 - 7$
   $x^2 + y^2 = 49$

5. $x^2 + y^2 = 25$
   $x + y = 5$

6. $x^2 + y^2 = 17$
   $x + 2y = 6$

7. $4x^2 + 9y^2 = 36$
   $3y + 2x = 6$

8. $x + 3y = 10$
   $x^2 + y^2 - 2x + 4y = 20$

Find all real solutions to the system of equations using the addition method.

9. $x^2 + y^2 = 144$
   $x^2 - 4y^2 = 64$

10. $x^2 + y^2 = 2$
    $2x^2 - y^2 = 1$

11. $x^2 + y^2 = 64$
    $64x^2 + 36y^2 = 2304$

12. $x^2 + y^2 = 20$
    $x^2 - 2y^2 = 8$

13. $x^2 + y^2 = 25$
    $3x^2 - y^2 = 39$

14. $x^2 + y^2 = 63$
    $x^2 - 3y^2 = 27$

1. _____

2. _____

3. _____

4. _____

5. _____

6. _____

7. _____

8. _____

9. _____

10. _____

11. _____

12. _____

13. _____

14. _____

15. $5x^2 - 2y^2 = 1$
    $x^2 + y^2 - 2x = 1$

    15. _____

16. The area of a rectangle is 85 square feet and its perimeter is 37 feet. Find the dimensions of the rectangle.

    16. _____

17. The area of a rectangle is 621 square feet and its length is 4 feet longer than its width. Find the length of the rectangle.

    17. _____

18. A right triangle has a perimeter of 24 inches and a hypotenuse of 10 inches. Find the length of the legs of the triangle.

    18. _____

19. The cost C of manufacturing and selling x units of a product is $C = 22x - 7$, and the corresponding revenue R is $R = x^2 - 55$. Find the break-even value of x.

    19. _____

20. The product of two positive numbers is 12 and the sum of their squares is 26. Find the two numbers.

    20. _____

Chapter 14 Test Form A

1. Graph the equation $y = -(x + 4)^2$.

1.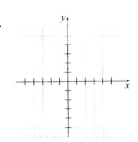

For problems 2 and 3, write the equation in the form $y = a(x - h)^2 + k$ or $x = a(y - k)^2 + h$.

2. $y = x^2 + 2x - 7$

2. _____

3. $x = 3y^2 - 12y - 36$

3. _____

4. Determine the distance between $(-3, -5)$ and $(3, 3)$.

4. _____

5. If the midpoint of a line segment is $(3, 7)$ and one endpoint is $(-1, 2)$, find the other endpoint.

5. _____

6. Write the equation of the circle with center at $(-5, 2)$ and radius 1.

6. _____

7. Use the method of completing the square to write the following equation in standard form: $x^2 + y^2 + 2x - 4y - 4 = 0$.

7. _____

8. Graph the equation $x^2 + 16y^2 = 16$.

8.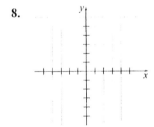

9. Find the equation of the ellipse if $(-7, 4)$ and $(3, 4)$ are the endpoints of the major axis and $(-2, 6)$ and $(-2, 2)$ are the endpoints of the minor axis.

9. _____

10. Find the area of the ellipse in Problem 9.

.

10. _____

11. Write the equation $25y^2 - x^2 = 25$ in standard form and determine the equations of the asymptotes.

11. _____

For problems 12 – 15, indicate whether the given equation represents a parabola, a circle, an ellipse, or a hyperbola.

12. $x = 5y^2 + 5y + 1$

12. _____

13. $9x^2 = -18y^2 + 36$

13. _____

14. $9x^2 = -9y^2 + 54$

14. _____

15. $3x^2 - 6x = 3y^2 + 6y$

15. _____

16. Solve using the substitution method: $x^2 + y^2 = 4$
$$x - 2y = 4$$

16. _____

17. Find the perimeter of the triangle whose vertices are $(1,1)$, $(5,4)$, and $(7,-2)$.

17. _____

18. Solve using the method of addition: $x^2 + y^2 = 16$
$$2x^2 - 5y^2 = 25$$

18. _____

19. A rectangle has perimeter 68 and area 273. Find its dimensions.

19. _____

20. The sum of the squares of two numbers is 109, and the product of the two numbers is 30. Find the two numbers.

20. _____

Chapter 14 Test Form B

Name:_____

Date:_____

1. Graph the equation $y = -2\left(x + \dfrac{1}{2}\right)^2 + 2$.

1.

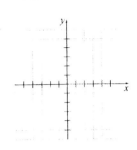

For problems 2 and 3, write the equation in the form $y = a(x - h)^2 + k$ or $x = a(y - k)^2 + h$.

2. $x = -y^2 - 5y - 4$

2. _____

3. $y = x^2 + 7x + 10$

3. _____

4. Determine the distance between $\left(\dfrac{1}{4}, 2\right)$ and $\left(-\dfrac{1}{2}, 6\right)$.

4. _____

5. Determine the midpoint of the line segment between $\left(\dfrac{5}{2}, 3\right)$ and $\left(2, \dfrac{9}{2}\right)$.

5. _____

6. Write the equation of the circle with center at $(-6, -1)$ and containing the point $(-5, 1)$.

6. _____

7. Use the method of completing the square to write the following equation in standard form: $x^2 + y^2 - x + 3y - \dfrac{3}{2} = 0$

7. _____

8. Graph the equation $25x^2 + 4y^2 = 100$.

8.

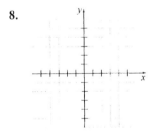

9. Find the equation of the ellipse if $(4, 4)$ and $(4, -6)$ are the endpoints of the major axis and $(0, -1)$ and $(8, -1)$ are the endpoints of the minor axis.

9. _____

10. Find the area of the ellipse in Problem 9

10. _____

Chapter 14 Test Form B *(cont.)*   Name:_____

11. Write the equation $64y^2 - 25x^2 = 1600$ in standard form   11. _____
and determine the equations of the asymptotes.

For problems 12 – 15, indicate whether the given equation represents a parabola, a circle, an ellipse, or a hyperbola.

12. $6x^2 + 6y^2 = 36$ 12. _____

13. $9x^2 = -18y^2 + 36$ 13. _____

14. $3x^2 = 12y^2 + 48$ 14. _____

15. $x^2 + 2x = y^2 - 4y$ 15. _____

16. Solve using the substitution method: $x^2 + y = 4$ 16. _____
$$y = x^2 + 2$$

17. Find the area of the right triangle whose vertices are   17. _____
$(-14,7), (1,1),$ and $(3,6)$.

18. Solve using the addition method: $x^2 + y^2 = 13$ 18. _____
$$2x^2 + 3y^2 = 30$$

19. Solve the system of equations using any method: 19. _____
$$x^2 + y^2 = 3$$
$$6x^2 + 5y^2 = 30$$

20. The length of the hypotenuse of a right triangle is 10 cm. If the   20. _____
area of the triangle is $24\,\text{cm}^2$, find the length of the two legs
of the triangle.

Chapter 14 Test Form C

1. Find the length of the line segment whose endpoints are $(6, 3)$ and $(8, 2)$.

1. _____

2. Find the midpoint of the line segment whose endpoints are $(6, -1)$ and $(-10, 15)$.

2. _____

For 3 and 4, refer to the equation $y = 2(x - 3)^2 - 4$.

3. Indicate the vertex of the graph of the equation.

3. _____

4. Draw the graph of the equation.

4.

For 5 and 6, refer to the equation $x = 2y^2 + 8y + 7$.

5. Write the equation in the form $x = a(y - k)^2 + h$.

5. _____

6. Draw the graph of the equation.

6.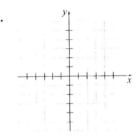

Problems 7 through 9 refer to the equation: $2x^2 + 3y^2 - 12x + 12y + 24 = 0$

7. Write the equation in standard form.

7. _____

8. Find the center of the ellipse.

8. _____

9. Find the area of the ellipse.

9. _____

10. Find the equation of the circle with center $(-1, 5)$ and containing the point $(3, 3)$

10. _____

For 11 to 14, refer to the equation $25x^2 + 64y^2 = 1600$.

11. Write the equation in standard form.

11. _____

12. Give the x-intercepts of the graph of the equation.

12. _____

13. Give the y-intercepts of the graph of the equation.

13. _____

14. Draw the graph of the equation.

14.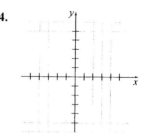

For 15 to 17, refer to the equation $36y^2 - 9x^2 = 324$.

15. Write the equation in standard form.

15. _____

16. Find the equations of the asymptotes

16. _____

17. Draw the graph of the equation and the asymptotes.

17.

18. Solve: $y = x^2 - 7$
$$x^2 + y^2 = 27$$

18. _____

19. A right triangle has a hypotenuse of 25 and an area of 84. Find its other two sides.

19. _____

20. The cost C of manufacturing and selling x units of a product is $C = 18x + 53$, and the corresponding revenue R is $R = x^2 - 35$. Find the break-even value of x.

20. _____

Chapter 14 Test Form D

1. Find the length of the line segment whose endpoints are (8, 2) and (−5, 1).

1. _____

2. Find the midpoint of the line segment whose endpoints are (1, −3) and (4, 3).

2. _____

3. Find the equation of the perpendicular bisector of the segment whose endpoints are $(-2,5)$ and $(6,-3)$.

3. _____

4. Find the perimeter of the square having a diagonal with endpoints $(1,-1)$ and $(7,7)$.

4. _____

For 5 and 6, refer to the equation $x = y^2 - 4y + 2$.

5. Write the equation in the form $x = a(y - k)^2 + h$.

5. _____

6. Draw the graph of the equation.

6.

7. Write the equation of an ellipse with center at (1, 2) and containing the points $(1,5)$ and $(2,4)$.

7. _____

8. Find the area of the ellipse in problem 7.

8. _____

For 9 and 10, refer to the equation $x^2 + y^2 - 6x + 14y + 33 = 0$.

9. Write the equation in standard form.

9. _____

10. Find the center and radius of the circle.

10. _____

For 11 to 14, refer to the equation $25x^2 + 16y^2 = 400$.

11. Write the equation in standard form.

11. _____

12. Give the x-intercepts of the graph of the equation.

12. _____

13. Give the y-intercepts of the graph of the equation.

13. _____

14. Draw the graph of the equation.

14.

For 15 to 17, refer to the equation $4y^2 - 49x^2 = 196$.

15. Write the equation in standard form.

15. _____

16. Find the equations of the asymptotes

16. _____

17. Draw the graph of the equation and the asymptotes.

17.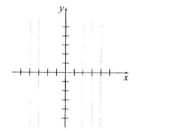

18. Solve : $\dfrac{x^2}{9} + \dfrac{y^2}{2} = 1$

$x^2 + y^2 = 2$

18. _____

19. Solve: $y = x^2 - 8$

$x^2 + y^2 = 64$

19. _____

20. A rectangular rug has an area of 48 square feet. A diagonal stripe in a contrasting color is 10 feet long. Find the dimensions of the rug.

20. _____

Chapter 14 Test Form E

1. Find the length of the line segment whose endpoints are (–6, 2) and (–9, 6).

2. Given $P(3,8)$ and $Q(7,-4)$, find the points that divide $\overline{PQ}$ into four equal segments.

For 3 and 4, refer to the equation $x = 3(y+1)^2 - 5$.

3. Indicate the vertex of the graph of the equation.

4. Draw the graph of the equation.

For 5 and 6, refer to the equation $x = 2y^2 - 8y + 5$.

5. Write the equation in the form $x = a(y-k)^2 + h$.

6. Draw the graph of the equation.

7. Write the equation of a circle with center at (–1, 3) and radius 8.

8. Graph the circle.

For 9 and 10, refer to the equation $x^2 + 4y^2 + 24y + 24 = 4x$.

9. Find the center of the ellipse.

10. Find the area of the ellipse.

1. _____

2. _____

3. _____

4.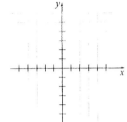

5. _____

6.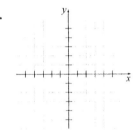

7. _____

8.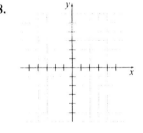

9. _____

10. _____

Chapter 14 Test Form E *(cont.)*

For 11 to 14, refer to the equation $9x^2 + 25y^2 = 225$.

11. Write the equation in standard form.

11. _____

12. Give the *x*-intercepts of the graph of the equation.

12. _____

13. Give the *y*-intercepts of the graph of the equation.

13. _____

14. Draw the graph of the equation.

14.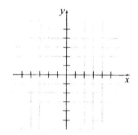

For 15 to 17, refer to the equation $4x^2 - 25y^2 = 100$.

15. Write the equation in standard form.

15. _____

16. Find the equations of the asymptotes

16. _____

17. Draw the graph of the equation and the asymptotes.

17.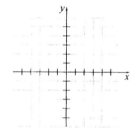

18. Solve: $x^2 + y^2 = 100$
    $y = x - 2$

18. _____

19. Solve: $x^2 + y^2 = 12$
    $y = x^2$

19. _____

20. A farmer has 32 meters of fencing to enclose a rectangular pen with an area of 63 square meters. Find the dimensions she should make the pen.

20. _____

Chapter 14 Test Form F

1. Find the distance between the points (5, 6) and (–2, –2).

2. Determine the midpoint of the line segment between points (2, 4) and (0, –7).

3. Indicate the vertex of the graph of the equation $y = 3(x-8)^2 + 7$.

4. Which equation is shown in the graph?

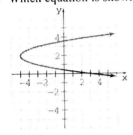

5. Write the equation $y = x^2 + 6x + 5$ in the form $y = a(x-h)^2 + k$.

6. Indicate whether the following equation when graphed will be a parabola, circle, ellipse, or hyperbola: $2x^2 + 3 = 2y^2 - 1$

7. Write the equation of a circle with center (9, –7) and radius 12.

8. Which equation is shown in the graph?

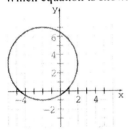

9. Write the equation $x^2 + y^2 - 10x - 4y + 25 = 0$ in standard form.

10. Find the center and radius of the circle with equation $x^2 + y^2 - 10x - 4y + 25 = 0$.

Problems 11 and 12 refer o the equation $x^2 + 4y^2 + 10x + 16y + 33 = 0$

11. Find the center of the ellipse.

12. Find the area of the ellipse.

13. Give the y-intercepts of the graph of the equation $4x^2 + 25y^2 = 100$.

1. _____

2. _____

3. _____

4. _____

5. _____

6. _____

7. _____

8. _____

9. _____

10. _____

11. _____

12. _____

13. _____

14. Draw the graph of the equation
$$4x^2 + 25y^2 = 100$$

14.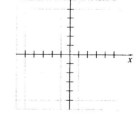

15. Write the equation $49x^2 - 4y^2 = 196$ in standard form.

15. _____.

16. Find the equations of the asymptotes of the hyperbola $49x^2 - 4y^2 = 196$.

16. _____.

17. Draw the graph of the equation
$$49x^2 - 4y^2 = 196.$$

17.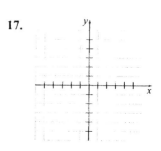

18. Solve the system of equations.
$$x^2 + y^2 = 25$$
$$y = x^2 - 5$$

18. _____.

19. Solve the system of equations.
$$x^2 + y^2 = 63$$
$$x^2 - 3y^2 = 27$$

19. _____.

20. The cost C of manufacturing and selling x units of a product is $C = 16x + 55$, and the corresponding revenue R is $R = x^2 - 50$. Find the break-even value of x.

20. _____.

Chapter 14 Test Form G

Name:_____

Date:_____

Choose the correct answer to each problem.

1. Find the distance between the points $(-4, 1)$ and $(-2, -6)$.

   (a) $\sqrt{53}$ units   (b) 53 units   (c) $\sqrt{61}$ units   (d) 61 units

2. Determine the midpoint of the line segment joining the points $P_1 = (-1, 4)$ and $P_2 = (-3, 2)$.

   (a) $(-1.9, 2.9)$ (b) $(-1.7, 3.3)$ (c) $(-2.0, 3.0)$ (d) $(1.0, 1.0)$

3. Indicate the vertex of the graph of the equation $x = 7(y + 9)^2 + 6$.

   (a) $(6, 9)$ (b) $(9, 6)$ (c) $(-9, 6)$ (d) $(6, -9)$

4. Which equation is shown in the graph?

   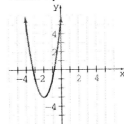

   (a) $y = 2(x - 2)^2 - 3$ (b) $y = 2(x + 2)^2 - 3$ (c) $x = 2(y - 2)^2 - 3$ (d) $x = 2(y + 2)^2 - 3$

5. Write the equation $y = 5y^2 - 40y + 87$ in the form $y = a(x - h)^2 + k$.

   (a) $x = 5(y + 3)^2 + 42$ (b) $x = 5(y - 3)^2 + 42$ (c) $y = 5(y + 4)^2 + 7$ (d) $x = 5(y - 4)^2 + 7$

6. Which equation describes an ellipse?

   (a) $3x^2 + 10x + 6y^2 - y = -7$ (b) $-3y^2 + 5x - 2y = -5$

   (c) $5y^2 - 7y - 12x^2 + 4x + 3 = 0$ (d) $4x^2 - 7x + 4y^2 + 5y - 1 = 0$

7. Write the equation of a circle with center $(10, -3)$ and radius 9.

   (a) $(x + 10)^2 + (y - 3)^2 = 81$ (b) $(x - 10)^2 + (y + 3)^2 = 81$

   (c) $(x + 10)^2 + (y - 3)^2 = 9$ (d) $(x - 10)^2 + (y + 3)^2 = 9$

8. Which equation is shown in the graph?

   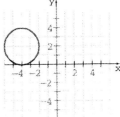

   (a) $(x + 4)^2 + (y - 2)^2 = 2$ (b) $(x - 4)^2 + (y + 2)^2 = 2$

   (c) $(x + 4)^2 + (y - 2)^2 = 4$ (d) $(x - 4)^2 + (y + 2)^2 = 4$

9. Find the center and radius of the circle with equation $x^2 + y^2 + 12x + 10y + 52 = 0$.

    (a) center $(-6, -5)$; $r = 9$ **(b)** center $(-6, -5)$; $r = 3$ **(c)** center $(6, 5)$; $r = 3$ **(d)** center $(6, 5)$; $r = 9$

10. Find the area of the ellipse $\dfrac{x^2}{12} + \dfrac{y^2}{3} = 1$

    (a) 36   **(b)** 6π **(c)** 36π **(d)** 6

11. Write the equation $9x^2 + 16y^2 = 144$ in standard form.

    (a) $\dfrac{x^2}{16} + \dfrac{y^2}{9} = 1$ **(b)** $\dfrac{x^2}{9} + \dfrac{y^2}{16} = 1$ **(c)** $\dfrac{x^2}{16} - \dfrac{y^2}{9} = 1$ **(d)** $\dfrac{x^2}{9} - \dfrac{y^2}{16} = 1$

12. Give the x-intercepts of the graph of the equation in number eleven.

    (a) $(3, 0)$ and $(-3, 0)$ **(b)** $(4, 0)$ and $(-4, 0)$ **(c)** $(0, 3)$ and $(0, -3)$ **(d)** $(0, 4)$ and $(0, -4)$

13. Write an equation of the ellipse with center at $(0,1)$ and containing the points $(4,1)$ and $(2\sqrt{3},0)$.

    (a) $2x^2 + (y-1)^2 = 32$ **(b)** $x^2 - 4(y-1)^2 = 16$ **(c)** $2x^2 + 8(y+1)^2 = 32$ **(d)** $x^2 + 4(y-1)^2 = 16$

14. Which of the following is the graph of the equation $9x^2 + 16y^2 = 144$?

    (a) 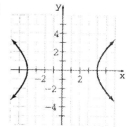   **(b)**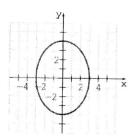

    (c) 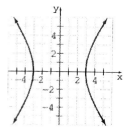   **(d)**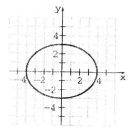

15. Write the equation $4x^2 - 25y^2 = 100$ in standard form.

    (a) $\dfrac{x^2}{25} + \dfrac{y^2}{4} = 1$ **(b)** $\dfrac{x^2}{4} + \dfrac{y^2}{25} = 1$ **(c)** $\dfrac{x^2}{25} - \dfrac{y^2}{4} = 1$ **(d)** $\dfrac{x^2}{4} - \dfrac{y^2}{25} = 1$

16. Find the equations of the asymptotes of the hyperbola $4x^2 - 25y^2 = 100$.

    (a) $y = \dfrac{2}{5}x$ and $y = -\dfrac{2}{5}x$ **(b)** $y = \dfrac{5}{2}x$ and $y = -\dfrac{5}{2}x$

    (c) $y = \dfrac{4}{25}x$ and $y = -\dfrac{4}{25}x$ **(d)** $y = \dfrac{25}{4}x$ and $y = -\dfrac{25}{4}x$

17. Solve the system of equations: $y = (x+2)^2 - 4$

$$y = \frac{7}{5}x - \frac{6}{5}$$

(a) $(-2,-4)$ (b) $(-2,-4); (2,4)$ (c) $(-2,-4); (-0.6,-2.04)$ (d) $(-0.6,-2.04)$

18. Solve the system of equations.

$9x^2 + 25y^2 = 225$

$3x - 5y = 15$

(a) $(-5, 0), (0, 3)$ (b) $(5, 0), (0, -3)$
(c) $(5, 0), (-5, 0), (0, 3), (0, -3)$ (d) $(0, 5), (0, -5), (3, 0), (-3, 0)$

19. Solve the system of equations.

$x^2 + y^2 = 144$

$x^2 - 4y^2 = 64$

(a) $\left(8\sqrt{2}, 4\right), \left(8\sqrt{2}, -4\right), \left(-8\sqrt{2}, 4\right), \left(-8\sqrt{2}, -4\right)$ (b) $\left(1, \sqrt{143}\right), \left(1, -\sqrt{143}\right), \left(-1, \sqrt{143}\right), \left(-1, -\sqrt{143}\right)$
(c) $\left(-8\sqrt{2}, 4\right), \left(-8\sqrt{2}, -4\right)$ (d) $\left(8\sqrt{2}, 4\right), \left(8\sqrt{2}, -4\right)$

20. A rectangle has perimeter 20 and a diagonal of $2\sqrt{13}$. Find its dimensions.

(a) 5 and 5   (b) 4.5 and 5.5   (c) 4 and 6   (d) 2 and 3

Chapter 14 Test Form H

Name:_____

Date:_____

Choose the correct answer to each problem.

1. Find the length of the straight line between the points $P(-5, -8)$ and $Q(-5, -3)$.

    (a) 21   **(b)** 11   **(c)** 10   **(d)** 5

2. Determine the midpoint of the line segment joining the points $P_1 = (-4, -2)$ and $P_2 = (-2, 1)$.

    (a) $(-3.0, -0.5)$ **(b)** $(-2.9, -0.6)$ **(c)** $(-1.0, -1.5)$ **(d)** $(-3.3, -1.0)$

3. Indicate the vertex of the graph of the equation $y = 6(x-8)^2 - 11$.

    (a) $(-11, 8)$ **(b)** $(-11, -8)$ **(c)** $(8, -11)$ **(d)** $(-8, -11)$

4. Which equation is shown in the graph?

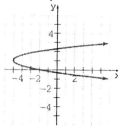

    (a) $x = 3(y+1)^2 - 5$ **(b)** $x = 3(y-1)^2 - 5$ **(c)** $x = 3(x+1)^2 - 5$ **(d)** $y = 3(x-1)^2 - 5$

5. Write the equation $x = 2y^2 - 28y + 101$ in the form $x = a(y-k)^2 + h$.

    (a) $x = 2(y-7)^2 + 3$ **(b)** $x = 2(y+7)^2 + 3$ **(c)** $x = 2(y-6)^2 + 29$ **(d)** $x = 2(y+6)^2 + 29$

6. Identify the following curve: $49x^2 = 25 - 49y^2$

    (a) hyperbola   **(b)** ellipse   **(c)** circle   **(d)** parabola

7. Write the equation of a circle with center $(12, -5)$ and radius 11.

    (a) $(x+12)^2 + (y-5)^2 = 11$ **(b)** $(x-12)^2 + (y+5)^2 = 11$
    (c) $(x+12)^2 + (y-5)^2 = 121$ **(d)** $(x-12)^2 + (y+5)^2 = 121$

8. Which equation is shown in the graph?

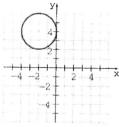

    (a) $(x-2)^2 + (y+4)^2 = 4$ **(b)** $(x+2)^2 + (y-4)^2 = 4$
    (c) $(x+4)^2 + (y-2)^2 = 4$ **(d)** $(x-4)^2 + (y+2)^2 = 4$

9. Write the equation $x^2 + y^2 - 6x - 10y + 9 = 0$ in standard form.

    (a) $(x+3)^2 + (y+5)^2 = 9$ **(b)** $(x-3)^2 + (y-5)^2 = 9$

    (c) $(x+3)^2 + (y+5)^2 = 25$ **(d)** $(x-3)^2 + (y-5)^2 = 25$

10. Write the equation $5y^2 + 8 = x + 10y$ in standard form.

    (a) $5x = (y-1)^2 + 3$ **(b)** $x = 5(y-1)^2 + 8$ **(c)** $x = 5(y-1)^2 + 3$ **(d)** $x = 3(y-1)^2 + 5$

11. Write the equation $4x^2 + 49y^2 = 196$ in standard form.

    (a) $\dfrac{x^2}{49} + \dfrac{y^2}{4} = 1$ **(b)** $\dfrac{x^2}{4} + \dfrac{y^2}{49} = 1$ **(c)** $\dfrac{x^2}{49} - \dfrac{y^2}{4} = 1$ **(d)** $\dfrac{x^2}{4} - \dfrac{y^2}{49} = 1$

12. Give the *x*-intercepts of the graph of the equation in number eleven.

    (a) $(2, 0)$ and $(-2, 0)$ **(b)** $(7, 0)$ and $(-7, 0)$ **(c)** $(0, 2)$ and $(0, -2)$ **(d)** $(0, 7)$ and $(0, -7)$

13. Give the *y*-intercepts of the graph of the equation in number eleven.

    (a) $(7, 0)$ and $(-7, 0)$ **(b)** $(0, 7)$ and $(0, -7)$ **(c)** $(2, 0)$ and $(-2, 0)$ **(d)** $(0, 2)$ and $(0, -2)$

14. Find the area of the graph in number eleven.

    (a) 14π **(b)** 196   **(c)** 14   **(d)** 196π

15. Which of the following is the graph of the equation $4x^2 + 49y^2 = 196$?

    (a) 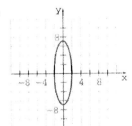   **(b)**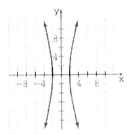

    (c) 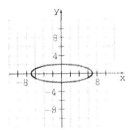   **(d)**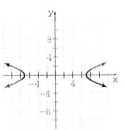

16. Find the equations of the asymptotes of the hyperbola $16x^2 - 9y^2 = 144$.

    (a) $y = \dfrac{4}{3}x$ and $y = -\dfrac{4}{3}x$ **(b)** $y = \dfrac{3}{4}x$ and $y = -\dfrac{3}{4}x$

    (c) $y = \dfrac{16}{9}x$ and $y = -\dfrac{16}{9}x$ **(d)** $y = \dfrac{9}{16}x$ and $y = -\dfrac{9}{16}x$

17. Which of the following is the graph of the equation $16x^2 - 9y^2 = 144$?

(a)
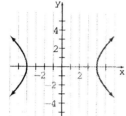

(b)
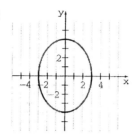

(c)
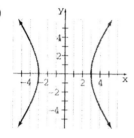

(d)
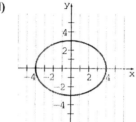

18. Solve the system of equations.

$$y^2 - 16x^2 = 9$$
$$5x - y = 0$$

(a) $(1, 5), (-1, 5), (1, -5), (-1, -5)$
(c) $(1, 5), (-1, -5)$

(b) $(5, 1), (-5, 1), (5, -1), (-5, -1)$
(d) $(5, 1), (-5, -1)$

19. Solve the system of equations.

$$x^2 + y^2 = 2x + 2y + 7$$
$$y^2 + 5 = x + 2y$$

(a) $\left(-3, 1+\sqrt{5}\right); \left(-3, 1-\sqrt{5}\right)$
(c) $(4, 1); (-4, -1)$

(b) $\left(-3, 1+\sqrt{5}\right)$
(d) $(4, 1)$

20. Kara and Karl are making a rectangular quilt. They have enough fabric to make it 3 square yards in area. They also have 7 yards of bias tape for the hem. What dimensions should they make the quilt?

(a) 3 yd by 1 yd   (b) 2 yd by 1.5 yd   (c) 4 yd by 0.75 yd   (d) 2.5 yd by 1.2 yd

Cumulative Review Test 1–14 Form A

1. Simplify: $\sqrt{18} - \sqrt{8} + \sqrt{32}$

1. _____

2. Solve the equation $2(x-4)+5 = x-(4x-3)$.

2. _____

3. Simplify and write without negative exponents: $\dfrac{6x^2y^4z}{2x^5yz^2}$

3. _____

4. Factor completely: $x^3 - 3x^2 - 4x + 12$

4. _____

5. Solve for x: $|2x-3| = 8-x$

5. _____

6. Simplify: $\dfrac{1+i}{1-i}$

6. _____

7. Find the domain and range for the relation graphed below.

7. _____

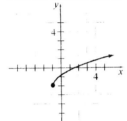

8. Graph $3x - 4y = 6$ using the x- and y-intercepts.

8.

9. Subtract $\dfrac{x+2}{x^2-5x+6} - \dfrac{1}{x-2}$

9. _____

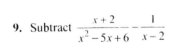

10. Determine the equation of a line parallel to the graph of $y = \dfrac{3}{2}x - 5$ that passes through $(0,6)$. Write the equation in slope-intercept form.

10. _____

11. Solve the system using substitution or addition:
    $4x - 5y = 8$
    $x + y = 3$

11. _____

12. Solve $\sqrt{x+3} - 2 = 5$

12. _____

13. Solve for x: $9^{2x+1} = 27^{x-2}$

13. _____

14. Let $f(x) = 3x - 7$, $g(x) = x^2 + x - 4$. Find $(f \circ g)(x)$.

14. _____

Cumulative Review Test 1–14
Form A *(cont.)*

15 Solve $\log_2(x-3) + \log_2(2x) = 3$

15. _____

16. Pebbles, Inc. estimates that the average cost, in dollars, per wicket that it produces is given by $C(x) = .2x^2 - .9x + 3.2$ where x is the number of wickets produced. What is the minimum average cost?

16. _____

17. Solve for x: $\dfrac{1}{x} + \dfrac{5}{x+2} = 1$

17. _____

18. Write the augmented matrix for the system.
$2x + y = 4$
$x - 3y + z = 3$
$5x + 2y - z = 7$

18. _____

19. Given $P(-1,8), Q(5,6)$, find the equation of the perpendicular bisector of $\overline{PQ}$

19. _____

20. Find the center and the area of the ellipse
$2x^2 + 3y^2 - 12x + 12y + 24 = 0$.

20. _____

Cumulative Review Test 1–14 Form B

1. Evaluate: $\dfrac{3+\sqrt{9}}{5-2^2}+\dfrac{|4-7|}{6-3}$

   (a) 0   (b) 3   (c) 5   (d) 7

2. Write 0.000002398 in scientific notation.

   (a) 2.398×10^6 (b) 2.398×10^7 (c) 2.398×10^{-6} (d) 2.398×10^{-7}

3. Solve for x: $9x-(10x-1)=3(1-x)-2$

   (a) $x=-\dfrac{1}{16}$ (b) $x=0$ (c) $x=\dfrac{1}{2}$ (d) $x=\dfrac{40}{3}$

4. The manager of a gourmet popcorn shop has learned that her store will sell 93 popcorn tins each day if they charge $1.50 for each tin. Raising the price to $2.50 will cause sales to fall to 49 tins per day. Let y be the number of popcorn tins the store sells at x dollars each. Write a linear equation that models the number of tins sold per day when the price is x dollars each.

   (a) $y=-44x-159$ (b) $y=-\dfrac{1}{44}x+\dfrac{81}{88}$ (c) $y=44x+27$ (d) $y=-44x+159$

5. Solve for x: $\left|\dfrac{5}{4}x-20\right|+15=25$

   (a) $x=8$ (b) $x=10,\ x=15$ (c) $x=24$ (d) $x=8,\ x=24$

6. Solve using substitution or the addition method.
   $2x+14y=-2$
   $-2x+4y=2$

   (a) $(0,-1)$ (b) $(1,-1)$ (c) $(-1,0)$ (d) No solution

7. Payton sells wood carvings at craft shows. She spends $26.50 each month on general supplies and $3.49 for each piece of wood that she carves. Last year her monthly costs ranged from $47.44 to $99.79. Over what range did the number of carvings made each month vary?

   (a) from 5 to 20 inclusive   (b) from 6 to 21 inclusive
   (c) from 7 to 22 inclusive   (d) from 5 to 22 inclusive

8. Which system of inequalities matches the graph?

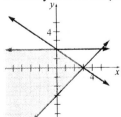

   (a) $2x+3y\ge6$ (b) $2x+3y\le6$ (c) $2x+3y\ge6$ (d) $2x+3y\ge6$
       $x-y\ge3$ $x-y\le3$ $x-y\le3$ $x-y\ge3$
       $y\le2$ $y\le2$ $y\le2$ $y\ge2$

Cumulative Review Test 1–14
Form B *(cont.)*

9. Use the *x*- and *y*-intercepts to decide which equation matches the graph.

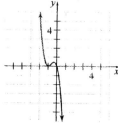

(a) $y = x^2 + 2x + 1$ **(b)** $y = -3x^3 - 6x^2 - 3x$ **(c)** $y = \left| -3x(x+1)^2 \right|$ **(d)** $y = 3x(x-1)^2$

10. Subtract: $\left(6x^5 - 14x^2y + 8y^2\right) - \left(2x^5 - 10x^2y - 17y^2\right)$

(a) $4x^5 - 4x^2y + 25y^2$ **(b)** $4x^5 - 12x^2y - 9y^2$ **(c)** $23x^8y^3$ **(d)** $4x^5 - 4x^2y - 9y^2$

11. Multiply: $(p + 10q)(p - 10q)$

(a) $p^2 - 100q^2$ **(b)** $p^2 - 2pq - 100q^2$ **(c)** $p^2 + 10pq - 100q^2$ **(d)** $p^2 - 5q^2$

12. Divide: $\dfrac{2x^3 + 15x^2 - 22x + 45}{x + 9}$

(a) $x^2 + 3x + 2$ **(b)** $x^2 + 4x + 5$ **(c)** $2x^2 + 3x + 5$ **(d)** $2x^2 - 3x + 5$

13. Factor completely: $15x^2 + 14x - 8$

(a) $(15x + 4)(x - 2)$ **(b)** $(5x - 2)(3x + 4)$
(c) $(5x + 2)(3x - 4)$ **(d)** does not factor

14. Solve: $2x^2 + 10x + 4 = 0$

(a) $x = \dfrac{-5 \pm \sqrt{17}}{2}$ **(b)** $x = \dfrac{-5 \pm \sqrt{33}}{2}$ **(c)** $x = \dfrac{-10 \pm \sqrt{17}}{2}$ **(d)** $x = \dfrac{-5 \pm \sqrt{17}}{4}$

15. Find the length and the midpoint of the line segment whose endpoints are $(-3, 1)$ and $(5, 7)$.

(a) $10; (4, 3)$ **(b)** $10; (1, 4)$ **(c)** $\sqrt{68}; (4, 3)$ **(d)** $\sqrt{68}; (1, 4)$

16. Simplify: $\dfrac{1 - i}{1 + i}$

(a) $\dfrac{1}{2} - \dfrac{1}{2}i$ **(b)** $\dfrac{1}{2} + \dfrac{1}{2}i$ **(c)** $-i$ **(d)** -1

17. Solve: $7 + \log_3(x+1) = 5$

(a) $x = -\dfrac{8}{9}$ (b) $x = -\dfrac{7}{5}$ (c) $x = 0$ (d) $x = -5$

18. Add: $\dfrac{3}{(x-2)(x-1)} + \dfrac{7}{x^2-1}$

(a) $\dfrac{10x-11}{(x^2-1)(x-2)}$ (b) $\dfrac{11x-10}{(x^2-1)(x-2)}$ (c) $\dfrac{42x-11}{(x^2-1)(x-2)}$ (d) $\dfrac{10x-11}{(x+1)(x-2)}$

19. Find the center of the ellipse: $4x^2 + 5y^2 + 8x + 4 = 20y$

(a) $(-1, 2)$ (b) $(4, -10)$ (c) $(-4, 10)$ (d) $(1, -2)$

20. If $f(x) = 3x - 2$ and $g(x) = 5x + 1$, find $(f \circ g)(x)$.

(a) $8x - 1$ (b) $-2x - 3$ (c) $15x + 1$ (d) $15x^2 - 7x - 2$

Chapter 15 Pretest Form A

Are the following sequences arithmetic, geometric, or neither?

1. 5, 9, 13, 17, 21, 25, …

2. 5, 15, 45, 135, 405, …

3. 2, 3, 5, 8, 13, …

In 4–6, write the first four terms of each sequence.

4. $a_1 = 3, r = 2$

5. $a_n = 2n + 1$

6. $a_6 = 9, d = 1.5$

7. Write the general term for the following arithmetic sequence.
5, 9, 13, 17, 21, …

8. Write the general term for the following geometric sequence.
$2, -1, \dfrac{1}{2}, -\dfrac{1}{4}, …$

9. Find a_8 when $a_3 = 2$, and $d = 6$.

10. Find s_{10} for the arithmetic sequence
$^-9, ^-5, ^-1, 3, …$

11. Find a_4 when $a_1 = -3$ and $r = \dfrac{1}{3}$.

12. Find s_6 when $a_1 = 250$ and $r = \dfrac{1}{5}$.

13. Write 0.2525… as a ratio of integers.

14. Write 0.82525… as a ratio of integers.

For problems 15 and 16, use the following series.

$$\sum_{n=1}^{4} \left(n^2 + 2 \right)$$

15. Write out the series.

16. Find the sum of the series.

17. Find the sum of the following infinite geometric series.
$3 + 2 + \dfrac{4}{3} + …$

18. Find the sum of the following infinite geometric series.
$3 - 2 + \dfrac{4}{3} - …$

19. Use the binomial theorem to expand $(x + 2a)^3$.

20. Use the binomial theorem to expand $(2x + 3y)^4$.

1. _____

2. _____

3. _____

4. _____

5. _____

6. _____

7. _____

8. _____

9. _____

10. _____

11. _____

12. _____

13. _____

14. _____

15. _____

16. _____

17. _____

18. _____

19. _____

20. _____

Chapter 15 Pretest Form B

Name:_____

Date:_____

Are the following sequences arithmetic, geometric, or neither?

1. 5, 10, 20, 40, 80, …

2. 2, 3, 5, 8, 13, …

3. 1, 2.9, 4.8, 6.7, …

In 4–6, write the first four terms of each sequence.

4. $a_1 = 15, \ d = -4$

5. $a_n = \dfrac{n+6}{n}$

6. $a_6 = \dfrac{3}{2}, r = \dfrac{1}{2}$

7. Write the general term for the following arithmetic sequence.

$1, \ 1\dfrac{1}{2}, \ 2, \ 2\dfrac{1}{2}, \ \ldots$

8. Write the general term for the following geometric sequence.
4, 8, 16, 32, …

9. Find a_{10} when $a_3 = 100$ and $d = -4$.

10. Find s_{10} for the arithmetic sequence
$-3, -9, -15, \ldots$

11. Find a_5 when $a_1 = 36$ and $r = \dfrac{1}{9}$.

12. Find s_6 when $a_1 = \dfrac{5}{6}$ and $r = 6$.

13. Write 0.7272… as a ratio of integers.

14. Write 0.37272… as a ratio of integers.

For problems 15 and 16, use the following series.

$$\sum_{n=1}^{4} (5n^2 - 6)$$

15. Write out the series.

16. Find the sum of the series.

17. Find the sum of the following infinite geometric series.

$30 + 12 + \dfrac{24}{5} + \dfrac{48}{25} + \ldots$

18. Find the sum of the following infinite geometric series.

$30 - 12 + \dfrac{24}{5} - \dfrac{48}{25} + \ldots$

19. Use the binomial theorem to expand $(x-3)^4$.

20. Use the binomial theorem to expand $(3x+2y)^4$.

1. _____

2. _____

3. _____

4. _____

5. _____

6. _____

7. _____

8. _____

9. _____

10. _____

11. _____

12. _____

13. _____

14. _____

15. _____

16. _____

17. _____

18. _____

19. _____

20. _____

Mini–Lecture 15.1
Sequences and Series

Learning Objectives:

1. Find the terms of a sequence.
2. Write a series.
3. Find partial sums.
4. Use summation notation, Σ.
5. Key vocabulary: *finite/infinite sequence, general term (a_n) of a sequence, increasing/decreasing sequence, alternating sequence, finite/infinite series, partial sum, sigma (Σ), index, upper/lower limit of summation, summation notation.*

Examples:

In problems 1 and 2, the nth tern of a sequence is defined. Do the following: a) write the first 5 terms of the sequence, b) find a_{13}, and c) write s_5 in sigma notation and evaluate it.

1. $a_n = \dfrac{5n-2}{2n+1}$

2. $a_n = (-1)^{n+1}(2n+3)$

3. Find the next 3 terms of the sequence:

   a) $2, \dfrac{3}{2}, \dfrac{4}{3}, \dfrac{5}{4}, \dfrac{6}{5}, \ldots$

   b) $2, -4, 8, -16, 32, \ldots$

4. Write out the series and evaluate it:

   a) $\displaystyle\sum_{i=1}^{5}\left(\dfrac{2i-1}{i}\right)$

   b) $\displaystyle\sum_{i=3}^{7}\left(\dfrac{i+1}{i-1}\right)$

Teaching Notes:

- Point out how to read the notation "a_n".

- Make sure students understand the relation between the factor $(-1)^n$ and an alternating sequence.

- Point out that an nth partial sum is the sum of (the first) n elements of a sequence.

Answers: 1a) $1, \dfrac{8}{5}, \dfrac{13}{7}, \dfrac{18}{9}, \dfrac{23}{11}$; b) $\dfrac{63}{27}$; c) $s_5 = \displaystyle\sum_{i=1}^{5}\left(\dfrac{5i-2}{2i+1}\right) = \dfrac{29,619}{3,465}$.

2a) $5, -7, 9, -11, 13$; b) 29; c) $s_5 = \displaystyle\sum_{i=1}^{5}(-1)^{i+1}(2i+3) = 9$.

3a). $\dfrac{7}{6}, \dfrac{8}{7}, \dfrac{9}{8}$; b) $-64, 128, -256$

4a). $1 + \dfrac{3}{2} + \dfrac{5}{3} + \dfrac{7}{4} + \dfrac{9}{5} = \dfrac{463}{60}$ b) $2 + \dfrac{5}{3} + \dfrac{6}{4} + \dfrac{7}{5} + \dfrac{8}{6} = \dfrac{474}{60}$

Mini–Lecture 15.2
Arithmetic Sequences and Series

Learning Objectives:

1. Find the common difference in an arithmetic sequence.
2. Write the nth term of an arithmetic sequence.
3. Find the nth partial sum of an arithmetic sequence.
4. Key vocabulary: *arithmetic sequence, common difference, arithmetic series.*

Examples:

All sequences are assumed to be arithmetic.

1. If $a_1 = 15$ and $d = -2$, find

   a) a_8 b) s_{12}

2. If $a_7 = 4$ and $a_{15} = 24$, find a_1 and d.

3. If $s_7 = 105$ and $s_{13} = 286$, find a_1 and d.

4. Find the sum of the natural numbers from 112 to 349 (inclusive).

Teaching Notes:

- Stress that the common difference is negative when the sequence is decreasing.
- Point out that s_n is also given by

$$s_n = n\left(a_1 + \frac{d}{2}(n-1) \right)$$

which does not require the calculation of a_n.

Answers: *1a)* $a_8 = 1$;   *b)* $s_{12} = 48$

           2. $a_1 = -11, \; d = 2.5$

           3. $a_1 = 8, \; d = \dfrac{7}{3}$

           4.   54,859

Mini–Lecture 15.3
Geometric Sequences and Series

Learning Objectives:

1. Find the common ratio in a geometric sequence.
2. Write the nth term of a geometric sequence.
3. Find the nth partial sum of a geometric sequence.
4. Identify infinite geometric series.
5. Find the sum of an infinite geometric series.
6. Study applications of geometric series.
7. Key vocabulary: *geometric sequence, common ratio, geometric series.*

Examples:

All sequences are assumed to be geometric.

1. If $a_1 = 7$ and $r = 0.95$, find

    a) a_8 b) s_{12}

2. If $a_5 = 23.2$ and $a_{10} = 742.4$, find a_1 and r.

3. If $a_1 = 100$ and $r = 0.95$, find s_∞.

4. If you invest \$10,000. at an interest rate of 4.5% compounded annually, how much will you have after 12 years?

5. Write the repeating decimal $0.17569569569\ldots = 0.17\overline{569}$ as a fraction.

Teaching Notes:

- Point out that a negative common ratio results in an alternating sequence.
- Point out that "series" refers to the addition of the terms in the sequence, whereas "the sum of the series" is the answer.
- Note that an infinite geometric series has a sum if and only if $|r| < 1$.

Answers: *1a)* $a_8 = 5.1456$;   *b)* $s_{12} = 64.35$

            2. $a_1 = 1.45$, $r = 2$

            3. $s_\infty = 2000$

            4. $a_{12} = 16,958.81$

            5. $0.17\overline{569} = \dfrac{17552}{99900} = \dfrac{4388}{24975}$

Mini–Lecture 15.4
The Binomial Theorem

Learning Objectives:

1. Evaluate factorials.
2. Use Pascal's triangle.
3. Use the binomial theorem.
2. Key vocabulary: *factorial, Pascal's triangle, binomial theorem.*

Examples:

1. Evaluate: a) $9!$ b) $0!$ c) $\binom{15}{3}$.   d) $\binom{15}{12}$

2. Expand: a) $(a+b)^7$ b) $(2x-3y)^5$

3. Show that $\sum_{r=0}^{n} \binom{n}{r} = 2^n$ by expanding $(1+1)^n$.

4. Prove: For every n and $0 \le r \le n-1$, $\binom{n}{r} + \binom{n}{r+1} = \binom{n+1}{r+1}$, which is the basis of Pascal's triangle.

Teaching Notes:

- Emphasize that $0! = 1$ (not 0), and is defined this way so certain formulas make sense.
- Point out that the symbol for the binomial coefficient, $\binom{n}{r}$, does *not* contain a fraction line.
- Emphasize that there are $n + 1$ terms in the expansion of $(a+b)^n$ (r ranges from 0 to n)
- Point out the symmetric pattern in the terms of $(a+b)^n$; make sure students know the relationships

$$\binom{n}{0} = \binom{n}{n} = 1, \quad \binom{n}{1} = \binom{n}{n-1} = n, \text{ and } \binom{n}{r} = \binom{n}{n-r}$$ which explain this symmetric pattern.

- Show the notation $(a+b)^n = \sum_{r=0}^{n} \binom{n}{r} a^r b^{n-r}$ for the binomial theorem.

Answers: 1a). 362,880;   b) 1;   c) 455;   d) 455

2.a) $a^7 + 7a^6b^1 + 21a^5b^2 + 35a^4b^3 + 35a^3b^4 + 21a^2b^5 + 7a^1b^6 + b^7$

b) $32x^5 - 240x^4y^1 + 720x^3y^2 - 1080x^2y^3 + 810x^1y^4 - 243y^5$

3. $2^n = (1+1)^n = \sum_{r=0}^{n} \binom{n}{r} 1^r 1^{n-r} = \sum_{r=0}^{n} \binom{n}{r}$

4. $\binom{n}{r} + \binom{n}{r+1} = \dfrac{n!}{r!(n-r)!} + \dfrac{n!}{(r+1)!(n-r-1)!} = \dfrac{n!(r+1)}{(r+1)!(n-r)!} + \dfrac{n!(n-r)}{(r+1)!(n-r)!} =$

$\dfrac{n!r + n! + n!n - n!r}{(r+1)!(n-r)!} = \dfrac{n!(n+1)}{(r+1)!(n-r)!} = \dfrac{(n+1)!}{(r+1)!(n-r)!} = \binom{n+1}{r+1}$

Additional Exercises 15.1

Name:_____

Date:_____

Write the first five terms of the sequence whose nth term is shown.

1. $a_n = n^2 - 1$

2. $a_n = \dfrac{n-3}{n+2}$

Find the indicated term of the sequence whose nth term is shown.

3. $a_n = 6n - 2$, twentieth term

4. $a_n = \dfrac{n(n+1)}{4}$, eighth term

5. $a_n = n(5n+3)$, sixth term

6. $a_n = (-1)^n \dfrac{n^2}{3n-4}$, seventh term

7. Given the sequence $a_n = (-2)^n - 1$, find the first and third partial sum.

8. Find the first and third partial sums of $a_n = \dfrac{3n+4}{n}$.

Write the next three terms of each sequence.

9. $1, 2, 3, 5, 8, 13, \ldots$

10. $1, \dfrac{1}{4}, \dfrac{1}{9}, \dfrac{1}{16}, \ldots$

Write out the series and evaluate it.

11. $\displaystyle\sum_{i=1}^{3}\left(-2i^2 - 3\right)$

12. $\displaystyle\sum_{i=1}^{3}\left(2i^2 - 3\right)$

13. $\displaystyle\sum_{j=3}^{6}\left(\dfrac{1}{2}\right)^j$

14. $\displaystyle\sum_{j=1}^{4}\left(\dfrac{1}{3}\right)^j$

1. _____

2. _____

3. _____

4. _____

5. _____

6. _____

7. _____

8. _____

9. _____

10. _____

11. _____

12. _____

13. _____

14. _____

Additional Exercises 15.1 *(cont.)*

For the general term a_n write an expression using $\sum$ to represent the indicated partial sum.

15. $a_n = n^2 - 5$, fifth partial sum

15. _____

16. $a_n = \dfrac{n-1}{n^2 + 2}$, fourth partial sum

16. _____

For the set of values $x_1 = 4$, $x_2 = -1$, $x_3 = -5$, $x_4 = 3$, and $x_5 = 2$, find each of the following.

17. $\displaystyle\sum_{i=1}^{5}(2 - x_i)$

17. _____

18. $\displaystyle\sum_{i=1}^{5} x_i^2$

18. _____

Find the arithmetic mean, $\bar{x}$, of the following sets of data.

19. 12, 8, 15, 19, 21, 7, 2

19. _____

20. 85, 72, 93, 88, 97

20. _____

Additional Exercises 15.2

Write the first five terms of the arithmetic sequence with the given first term and common difference. Write the expression for the general (or nth) term, a_n, of the arithmetic sequence.

1. $a_1 = -6, d = 2$

    1. _____

2. $a_1 = \dfrac{7}{2}, d = -\dfrac{1}{2}$

    2. _____

3. In an arithmetic sequence $a_1 = 10$ and $d = -3$. Find a_{21}.

    3. _____

4. In an arithmetic sequence $a_5 = 4$ and $d = -\dfrac{1}{2}$, Find a_{17}.

    4. _____

5. The first term in an arithmetic sequence is 5 and the nth term is 32. If the number of terms is 10, find the common difference.

    5. _____

6. If $a_4 = 7$ and $a_{13} = 29$ in an arithmetic sequence, find d.

    6. _____

7. Find the common difference for the arithmetic sequence: 32, 49, 66, …

    7. _____

8. Find the common difference for the arithmetic sequence: 34, 27, 20 …

    8. _____

9. Find s_8 when $a_1 = 6$, and $d = 1$.

    9. _____

10. Find the sum of the first 14 terms of the arithmetic sequence 7, 15, 23, 31, …

    10. _____

11. Find the sum of the first 11 terms of the sequence: –5, 1, 7, 13, …

    11. _____

12. Find the sum of the first 22 terms of the arithmetic sequence: 10, 7, 4, 1, …

    12. _____

Find s_{10} for each of the following.

13. $a_1 = 6, d = \dfrac{1}{3}$

    13. _____

14. $a_1 = 30, d = -8$

    14. _____

15. $a_1 = \dfrac{4}{9}, d = \dfrac{1}{9}$

    15. _____

For 16 and 17, use the sequence 13, 25, 37, 49, …, 1177.

16. Find the number of terms (n) in the arithmetic sequence.

    16. _____

17. Find s_n.

    17. _____

For 18 and 19, use the sequence 8, 18, 28, 38, …, 288.

18. Find the number of terms (n) in the arithmetic sequence.

    18. _____

19. Find s_n.

    19. _____

20. A 50 row theater has 40 seats in the front row. The second row has 41 seats. Each row has one more than the row in front of it, how many seats are there in the theater?

    20. _____

Additional Exercises 15.3

Name:_____

Date:_____

Determine the first five terms of the geometric sequence.

1. $a_1 = 5, r = -2$

2. $a_1 = 64, r = \dfrac{1}{2}$

Find the indicated term of the geometric sequence.

3. $a_1 = 10, r = 2$; find a_{10} .

4. $a_3 = 4, r = \dfrac{5}{2}$; find a_6 .

5. $a_1 = -\dfrac{2}{3}, r = \dfrac{3}{4}$; find a_6 .

6. $a_1 = \dfrac{1}{2}, r = 3$; find a_7 .

Find the sum.

7. $a_1 = -8, r = 3$; find s_8 .

8. $a_1 = 2, r = \dfrac{2}{3}$; find s_7 .

9. $a_4 = 8, r = -2$; find s_9 .

For 10 and 11, use the geometric sequence $7, \dfrac{21}{4}, \dfrac{63}{16}, \dfrac{189}{64}, \ldots$

10. Find the common ratio, r, for the geometric sequence.

11. Write an expression for the general (or nth) term, a_n , for the geometric sequence.

For 12 and 13, use the geometric sequence $-\dfrac{2}{3}, -\dfrac{5}{3}, -\dfrac{25}{6}, \ldots$

12. Find the common ratio, r, for the geometric sequence.

13. Write an expression for the general (or nth) term, a_n , for the geometric sequence.

Find the sum of each infinite series.

14. $3, 2, \dfrac{4}{3}, \dfrac{8}{9}, \ldots$

15. $0.9 + 0.09 + 0.009 \ldots$

16. $0.2 + 0.02 + 0.002 \ldots$

17. $-\dfrac{1}{2} + \dfrac{1}{4} - \dfrac{1}{8} + \ldots$

1. _____

2. _____

3. _____

4. _____

5. _____

6. _____

7. _____

8. _____

9. _____

10. _____

11. _____

12. _____

13. _____

14. _____

15. _____

16. _____

17. _____

Additional Exercises 15.3 *(cont.)*

Write the repeating decimal as a ratio of integers.

18. 0.57272…

19. 0.6969…

20. A woman made $40,000 during the first year of her new job at city hall. Each year she received a 10% raise. Find her total earnings during her first eight years on the job.

18. _____

19. _____

20. _____

Additional Exercises 15.4

Evaluate each of the following combinations.

1. $\begin{pmatrix} 8 \\ 3 \end{pmatrix}$

2. $\begin{pmatrix} 7 \\ 3 \end{pmatrix}$

3. $\begin{pmatrix} 10 \\ 4 \end{pmatrix}$

4. $\begin{pmatrix} 8 \\ 0 \end{pmatrix}$

5. $\begin{pmatrix} 11 \\ 3 \end{pmatrix}$

6. $\begin{pmatrix} 11 \\ 8 \end{pmatrix}$

Use the binomial theorem to expand each expression.

7. $(a-3b)^4$

8. $(x+4y)^4$

9. $(s-w)^5$

10. $(5x-y)^4$

11. $(3a-b)^4$

12. $(x-2y)^4$

13. $(2x+3y)^4$

14. $(6x-y)^4$

Write the first four terms of each expression.

15. $(f+g)^5$

16. $(b+c)^6$

17. $(2x+y)^{10}$

18. $(p+2q)^{11}$

19. $(x-3y)^8$

20. $\left(\dfrac{x}{3}+\dfrac{y}{2}\right)^7$

1. _____

2. _____

3. _____

4. _____

5. _____

6. _____

7. _____

8. _____

9. _____

10. _____

11. _____

12. _____

13. _____

14. _____

15. _____

16. _____

17. _____

18. _____

19. _____

20. _____

Chapter 15 Test Form A

1. Write the first five terms of the sequence whose nth term is shown: $a_n = (-2)^{n+1}$

   1. _____

2. Find the ninth term of the sequence $a_n = n(n+2)$.

   2. _____

3. Find the first and third partial sums, s_1 and s_3, for the sequence: $a_n = \dfrac{n^2}{n+4}$

   3. _____

4. Write out the series $\displaystyle\sum_{i=1}^{4} \dfrac{i^2}{2}$ and evaluate it.

   4. _____

5. Write out the series $\displaystyle\sum_{i=3}^{7} \dfrac{i+1}{i-1}$ and evaluate it.

   5. _____

6. Find a_{13} for the arithmetic sequence with $a_1 = -2, d = \dfrac{5}{3}$.

   6. _____

7. Find a_{13} for the arithmetic sequence with $a_5 = 3$, $a_9 = 4$.

   7. _____

8. Find the sum, s_n, and common difference, d, for the sequence with $a_1 = \dfrac{5}{3}, a_8 = 2, n = 8$.

   8. _____

9. Find a_{10} for the geometric sequence with $a_1 = 50, r = \dfrac{1}{3}$.

   9. _____

10. Find s_{12} for the geometric sequence with $a_1 = 1, r = -2$.

    10. _____

11. Find the sum of the infinite geometric series:
    $2 - 1 + \dfrac{1}{2} - \dfrac{1}{4} + \dfrac{1}{8} - \cdots$

    11. _____

12. Find the sum of the infinite geometric series:
    $2 + 1 + \dfrac{1}{2} + \dfrac{1}{4} + \dfrac{1}{8} + \cdots$

    12. _____

13. Use the binomial theorem to expand the expression $\left(2x + \dfrac{1}{2}\right)^3$.

    13. _____

14. Use the binomial theorem to expand the expression $(x-2)^5$.

    14. _____

15. Write out the first four terms of the expansion of $(3x+2)^7$.

    15. _____

16. Write the next three terms of the sequence: $7, 1, 9, 17, \ldots$

    16. _____

17. Write the first four terms of the sequence with $a_1 = \dfrac{9}{5}, d = \dfrac{3}{5}$; also find a_{10} and s_{10}.

    17. _____

18. Find the number of terms and s_n for the sequence $9, 12, 15, 18, \ldots, 93$.

    18. _____

19. Write $0.375375\ldots$ as a ratio of integers.

    19. _____

20. Write $0.5484848\ldots$ as a ratio of integers.

    20. _____

Chapter 15 Test Form B

1. Write the first five terms of the sequence whose nth term is shown: $a_n = 3^{n-1}$

    1. _____

2. Find the ninth term of the sequence $a_n = \dfrac{n(n+1)}{n^2}$.

    2. _____

3. Find the first and third partial sums, s_1 and s_3 , for the sequence $a_n = \dfrac{n^2}{2}$.

    3. _____

4. Write out the series $\displaystyle\sum_{k=1}^{5}\left(2^k + k\right)$ and evaluate it.

    4. _____

5. Write out the series $\displaystyle\sum_{j=2}^{4}\dfrac{j^2 + j}{j+1}$ and evaluate it.

    5. _____

6. Find n for the arithmetic sequence with $a_1 = 4, a_n = 28, d = 3$.

    6. _____

7. For an arithmetic sequence with $a_5 = 4$ and $a_9 = 3$, find a_{13} .

    7. _____

8. Find the sum, s_n, and common difference, d, for the sequence with $a_1 = 7, a_{11} = 67, n = 11$.

    8. _____

9. Find a_{10} for the geometric sequence with $a_1 = -10, r = -2$.

    9. _____

10. Find s_7 for a geometric sequence with $a_1 = \dfrac{3}{5}, r = 3$.

    10. _____

11. Find the sum of the infinite geometric series: $4 - \dfrac{8}{3} + \dfrac{16}{9} - \dfrac{32}{27} + \cdots$

    11. _____

12. Find the sum of the infinite geometric series: $5 + 3 + \dfrac{9}{5} + \dfrac{27}{25} + \cdots$

    12. _____

13. Use the binomial theorem to expand the expression $(2x - 3)^3$.

    13. _____

14. Use the binomial theorem to expand the expression $(2 - x)^5$.

    14. _____

15. Write out the first four terms of the expansion of $\left(2x + \dfrac{y}{5}\right)^9$.

    15. _____

16. Write the next three terms of the sequence: $-\dfrac{2}{3}, \dfrac{1}{3}, -\dfrac{1}{6}, \dfrac{1}{12}, \ldots$

    16. _____

17. Write the first four terms of the sequence with $a_1 = 35, d = 3$; also find a_{10} and s_{10}.

    17. _____

18. Find the number of terms and s_n for the sequence $-12, 16, 20, \ldots, 52$.

    18. _____

19. Write 0.742742… as a ratio of integers.

    19. _____

20. Write 0.1742742… as a ratio of integers.

    20. _____

Chapter 15 Test Form C

1. Is the sequence $1, 2, 4, 5, 10, \ldots$ arithmetic, geometric, or neither.

1. _____

For questions 2 and 3, write the first four terms of the sequence with:

2. $a_n = \dfrac{n-4}{n^2}$

2. _____

3. First term $a_1 = \frac{1}{4}$ and common ratio $r = -\frac{1}{3}$.

3. _____

4. Find the first and third partial sums of $a_n = 2^n - 2$.

4. _____

5. Write out the series $\sum_{n=1}^{4} \dfrac{n}{2}$ and evaluate the sum.

5. _____

6. Write out the series $\sum_{n=3}^{7} \left(2^n - n\right)$ and evaluate the sum.

6. _____

7. Find d if $a_4 = 1$ and $a_{12} = 17$.

7. _____

8. Find the sum, s_n, and common difference, d, if $a_1 = 3$, $a_n = 0$, and $n = 4$.

8. _____

9. Find the number of terms in the arithmetic sequence $12, 5, -2, \ldots, -37$.

9. _____

10. Find the term a_5 of the geometric sequence with $a_1 = 4$ and $r = \frac{3}{2}$.

10. _____

11. Find a_1 and r if $a_6 = 72$ and $a_9 = 9$.

11. _____

12. Find the sum s_3 of the geometric sequence with $a_3 = \frac{4}{3}$ and $r = \frac{2}{3}$.

12. _____

13. Find the sum of the infinite geometric series:
$$-8 - 2 - \frac{1}{2} - \frac{1}{8} - \ldots$$

13. _____

14. Find the sum of the infinite geometric series:
$$-8 + 2 - \frac{1}{2} + \frac{1}{8} - \ldots$$

14. _____

15. Write the repeating decimal, $0.262626\ldots$ as a ratio of two integers.

15. _____

16. Write the repeating decimal, $0.4262626\ldots$ as a ratio of two integers.

16. _____

17. Find the common ratio for the sequence $-\dfrac{5}{6}, -\dfrac{5}{4}, -\dfrac{15}{8}, -\dfrac{45}{16}, \ldots$

17. _____

18. Use the binomial theorem to expand the expression $(x - y)^5$.

18. _____

19. Use the binomial theorem to expand the expression $(x + 2)^5$.

19. _____

20. Use the binomial theorem to expand the expression $(2x - 3y)^4$.

20. _____

Chapter 15 Test Form D

Name:_____

Date:_____

1. Is the sequence $\frac{1}{2}, \frac{1}{3}, \frac{1}{4}, \frac{1}{5}, \frac{1}{6}, \ldots$ arithmetic, geometric, or neither.

1. _____

For questions 2 – 4, write the first four terms of the sequence with:

2. First term $a_1 = 5$ and common difference $d = 6$.

2. _____

3. First term $a_1 = 7$ and common ratio $r = -\frac{1}{5}$.

3. _____

4. Arithmetic sequence with $a_6 = 19$ and $a_{10} = 13$.

4. _____

5. Find the first and third partial sums of $a_n = \frac{2}{n^2}$.

5. _____

6. Write out the series $\sum_{k=1}^{3} \frac{k^2 - 1}{k^2 + 1}$ and evaluate the sum.

6. _____

7. Write out the series $\sum_{k=3}^{6} \left(3^k - 2^k\right)$ and evaluate the sum.

7. _____

8. Find d if $a_3 = -8$ and $a_{10} = -22$.

8. _____

9. Find the sum, s_n, and common difference, d, if $a_1 = 2$, $a_n = \frac{1}{2}$, and $n = 4$.

9. _____

10. Find the number of terms in the arithmetic sequence $-17, -5, 7, \ldots, 55$.

10. _____

11. Find the term a_4 of the geometric sequence with $a_1 = 64$ and $r = \frac{7}{8}$.

11. _____

12. Find the sum s_3 of the geometric sequence with $a_1 = 5$ and $r = -\frac{1}{2}$.

12. _____

13. Find the sum of the infinite geometric series:
$$-3 + 2 - \frac{4}{3} + \frac{8}{9} - \frac{16}{27} + \ldots$$

13. _____

14. Find the sum of the infinite geometric series:
$$3 + 2 + \frac{4}{3} + \frac{8}{9} + \frac{16}{27} + \ldots$$

14. _____

15. Write the repeating decimal, 0.353535… as a ratio of two integers.

15. _____

16. Write the repeating decimal, 0.10353535… as a ratio of two integers.

16. _____

17. Find the common ratio for the sequence 8, –20, 50, –125, …

17. _____

18. Use the binomial theorem to expand the expression $(x - 3)^4$.

18. _____

19. Use the binomial theorem to expand the expression $\left(2x + \frac{1}{2}\right)^4$.

19. _____

20. Use the binomial theorem to expand the expression $(2x + 3)^4$.

20. _____

Chapter 15 Test Form E

1. Is the sequence $\frac{3}{4}, -\frac{3}{2}, 3, -6, 12, \ldots$ arithmetic, geometric, or neither.

1. _____

For questions 2 – 4, write the first four terms of the sequence with:

2. $a_n = \frac{n^2 - 2}{n}$

2. _____

3. Ninth term $a_9 = 113$ and common difference $d = -4$.

3. _____

4. First term $a_1 = -\frac{3}{4}$ and common ratio $r = 3$.

4. _____

5. Find the first and third partial sums of $a_n = \frac{2n}{n+3}$.

5. _____

6. Write out the series $\sum_{n=1}^{3} \frac{n^2 + 1}{n}$ and evaluate the sum.

6. _____

7. Write out the series $\sum_{n=3}^{6} \left(\frac{n^2}{2^n} \right)$ and evaluate the sum.

7. _____

8. Find d if $a_4 = 3$ and $a_{10} = 12$.

8. _____

9. Find the sum, s_n, and common difference, d, if $a_1 = 4$, $a_n = 0$, and $n = 3$.

9. _____

10. Find the number of terms in the arithmetic sequence $-11, -3, 5, \ldots, 69$.

10. _____

11. Find the term a_4 of the geometric sequence with $a_1 = 6$ and $r = -\frac{1}{3}$.

11. _____

12. Find the term a_7 of the geometric sequence with $a_{10} = 2$ and $r = -\frac{1}{3}$.

12. _____

13. Find the sum s_4 of the geometric sequence with $a_1 = 6$ and $r = -\frac{1}{3}$.

13. _____

14. Find the sum of the infinite geometric series:
$7 + \frac{7}{6} + \frac{7}{36} + \frac{7}{216} + \ldots$

14. _____

15. Find the sum of the infinite geometric series:
$\frac{5}{6} + \frac{5}{36} + \frac{5}{216} + \frac{5}{1296} + \ldots$

15. _____

16. Write the repeating decimal, 0.5474747… as a ratio of two integers.   **16.** _____

17. Write the repeating decimal, 0.0212121… as a ratio of two integers.   **17.** _____

18. Find the common ratio for the sequence $-\dfrac{3}{8}, \dfrac{3}{4}, -\dfrac{3}{2}, 3, -6, \ldots$ **18.** _____

19. Use the binomial theorem to expand the expression $(x-2)^4$.   **19.** _____

20. Use the binomial theorem to expand the expression $(3a+2b)^3$.   **20.** _____

Chapter 15 Test Form F

1. Is the sequence: $-4, -1, 2, 5, 8, \ldots$ arithmetic, geometric, or neither?

    1. _____

2. Is the sequence: $1, \frac{1}{2}, 2, \frac{1}{3}, 3, \frac{1}{4}, \ldots$ arithmetic, geometric, or neither?

    2. _____

For questions 3 – 5, write the first four terms of the sequence with:

3. $a_n = \dfrac{3n}{n^2 + 1}$

    3. _____

4. First term $a_1 = -2$ and common ratio $r = -\dfrac{3}{2}$.

    4. _____

5. Common difference $d = -1.5$ and $a_5 = 5.5$.

    5. _____

6. Find the first and third partial sums of $a_n = 4n - 1$.

    6. _____

7. Write out the series $\displaystyle\sum_{k=1}^{4} -k^2 + 1$ and evaluate the sum.

    7. _____

8. Write out the series $\displaystyle\sum_{k=5}^{8} \dfrac{2k}{k+1}$ and evaluate the sum.

    8. _____

9. Find d if $a_1 = -5$ and $a_4 = 1$.

    9. _____

10. Find r if $a_5 = 8$ and $a_8 = -27$.

    10. _____

11. Find the sum, s_n, and common difference, d, if $a_1 = -1$, $a_n = 8$, and $n = 4$.

    11. _____

12. Find a_1 and d if $a_4 = 1$ and $a_8 = -5$.

    12. _____

13. Find the term a_3 of the geometric sequence with $a_1 = \dfrac{2}{3}$ and $r = \dfrac{3}{4}$.

    13. _____

14. Find the sum s_3 of the geometric sequence with $a_1 = \dfrac{2}{3}$ and $r = \dfrac{3}{4}$.

    14. _____

15. Find the sum of the infinite geometric series: $6 + \dfrac{6}{7} + \dfrac{6}{49} + \dfrac{6}{343} + \ldots$

    15. _____

16. Find the sum of the infinite geometric series: $5 - \dfrac{5}{3} + \dfrac{5}{9} - \dfrac{5}{27} + \ldots$

    16. _____

17. Write the repeating decimal $0.189189\ldots$ as a ratio of two integers.

    17. _____

18. Write the repeating decimal $0.5181818\ldots$ as a ratio of two integers.

    18. _____

For questions 19–20, use the binomial theorem to expand each expression.

19. $\left(2x + y^2\right)^3$

    19. _____

20. $(x - 4)^4$

    20. _____

Chapter 15 Test Form G

Name:_____

Date:_____

1. What type of sequence is: $\frac{1}{2}, -\frac{1}{6}, \frac{1}{9}, -\frac{1}{27}, \frac{1}{81}, \ldots$?

    (a) arithmetic $\qquad$ (b) infinite sum $\qquad$ (c) geometric $\qquad$ (d) none of these

For questions 2–4, write the first four terms of the sequence with:

2. $a_n = \frac{1}{2}n + 1$

    (a) $1, \frac{3}{2}, 2, \frac{5}{2}$ $\qquad$ (b) $\frac{3}{2}, 2, \frac{5}{2}, 3$ $\qquad$ (c) $\frac{3}{2}, \frac{5}{2}, \frac{7}{2}, \frac{9}{2}$ $\qquad$ (d) $1, \frac{3}{2}, \frac{5}{2}, \frac{7}{2}$

3. First term $a_1 = -5$ and common ratio $r = -\frac{2}{5}$.

    (a) $-5, -\frac{27}{5}, -\frac{29}{5}, -\frac{31}{5}$ $\quad$ (b) $2, -\frac{4}{5}, \frac{8}{25}, -\frac{16}{125}$ $\quad$ (c) $-5, \frac{2}{5}, -\frac{4}{25}, \frac{8}{125}$ $\quad$ (d) $-5, 2, -\frac{4}{5}, \frac{8}{25}$

4. Fifth term $a_5 = -7$ and common difference $d = 3$.

    (a) $-7, -4, -1, 2$ $\qquad$ (b) $-19, -16, -13, -10$ $\qquad$ (c) $-16, -13, -10, -7$ $\qquad$ (d) $-4, -1, 2, 5$

5. Find the first and third partial sums of $a_n = \frac{n+1}{n+2}$.

    (a) $\frac{2}{3}; \frac{133}{60}$ $\qquad$ (b) $\frac{2}{3}; \frac{9}{60}$ $\qquad$ (c) $\frac{1}{2}; \frac{23}{12}$ $\qquad$ (d) $\frac{1}{2}; \frac{21}{12}$

6. Write out the series $\sum_{n=1}^{3}(2n+2)$ and evaluate the sum.

    (a) $2 + 4 + 6; 12$ $\qquad$ (b) $2 + 4 + 6; 14$ $\qquad$ (c) $4 + 6 + 8; 18$ $\qquad$ (d) $4 + 6 + 8; 16$

7. Write out the series $\sum_{k=3}^{5}\left(\frac{k-1}{k+1}\right)$ and evaluate the sum.

    (a) $\frac{1}{2} + \frac{3}{5} + \frac{2}{3} = \frac{53}{30}$ $\qquad$ (b) $0 + \frac{1}{3} + \frac{1}{2} = \frac{5}{6}$ $\qquad$ (c) $\frac{2}{4} + \frac{3}{5} + \frac{4}{6} = \frac{9}{15}$ $\qquad$ (d) $\frac{1}{2} + \frac{3}{5} + \frac{5}{7} = \frac{127}{70}$

8. Find d if $a_4 = -3$ and $a_8 = 5$.

    (a) $d = \frac{1}{2}$ $\qquad$ (b) $d = 1$ $\qquad$ (c) $d = 2$ $\qquad$ (d) $d = \frac{3}{2}$

9. Find the sum, s_n, and common difference, d, if $a_1 = -2$, $a_n = 2$, and $n = 3$.

    (a) $s_3 = 4; d = 2$ $\qquad$ (b) $s_3 = 0; d = 2$ $\qquad$ (c) $s_3 = 4; d = 3$ $\qquad$ (d) $s_3 = 3; d = 3$

10. Find the number of terms in the arithmetic sequence $-8, 3, 14, \ldots, 69$.

    (a) 5 $\qquad$ (b) 6 $\qquad$ (c) 7 $\qquad$ (d) 8

11. Find the term a_4 of the geometric sequence with $a_1 = 10$ and $r = -\dfrac{3}{5}$.

(a) $\dfrac{18}{5}$ 　　　　(b) $-\dfrac{54}{25}$ 　　　　(c) $\dfrac{54}{25}$ 　　　　(d) $-\dfrac{162}{125}$

12. Find the first term a_1 of the geometric sequence with $a_4 = -81$ and $r = -\dfrac{3}{2}$.

(a) 54 　　　　(b) -54 　　　　(c) -36 　　　　(d) 24

13. Find the sum s_3 of the geometric sequence with $a_1 = 10$ and $r = -\dfrac{3}{5}$.

(a) $\dfrac{18}{5}$ 　　　　(b) $\dfrac{22}{5}$ 　　　　(c) $\dfrac{38}{5}$ 　　　　(d) 8

14. Find the sum of the infinite geometric series: $\dfrac{1}{2} + \dfrac{1}{6} + \dfrac{1}{18} + \dfrac{1}{54} + \dfrac{1}{162} + \ldots$

(a) $\dfrac{3}{8}$ 　　　　(b) $\dfrac{1}{2}$ 　　　　(c) $\dfrac{3}{4}$ 　　　　(d) $\dfrac{3}{2}$

15. Find the sum of the infinite geometric series: $\dfrac{1}{2} - \dfrac{1}{6} + \dfrac{1}{18} - \dfrac{1}{54} + \dfrac{1}{162} - \ldots$

(a) $\dfrac{3}{8}$ 　　　　(b) $\dfrac{1}{2}$ 　　　　(c) $\dfrac{3}{4}$ 　　　　(d) $\dfrac{3}{2}$

16. Write the repeating decimal $0.121212\ldots$ as a ratio of two integers.

(a) $\dfrac{3}{25}$ 　　　　(b) $\dfrac{6}{25}$ 　　　　(c) $\dfrac{4}{33}$ 　　　　(d) $\dfrac{4}{99}$

17. Write the repeating decimal $0.8121212\ldots$ as a ratio of two integers.

(a) $\dfrac{812}{999}$ 　　　　(b) $\dfrac{12}{99}$ 　　　　(c) $\dfrac{12}{33}$ 　　　　(d) $\dfrac{134}{165}$

18. Find the common ratio for the sequence $\dfrac{7}{8}, -\dfrac{7}{4}, \dfrac{7}{2}, -7, 14, \ldots$

(a) $\dfrac{1}{2}$ 　　　　(b) $-\dfrac{1}{2}$ 　　　　(c) 2 　　　　(d) -2

For questions 19–20, use the binomial theorem to expand each expression.

19. $(x - 2y)^3$

(a) $x^3 - 6x^2 y + 12xy^2 - 8y^3$ 　　　　(b) $x^3 - 6x^2 y - 12xy^2 + 8y^3$
(c) $x^3 - 3x^2 y + 3xy^2 - 8y^3$ 　　　　(d) $x^3 - 3x^2 y - 3xy^2 + 8y^3$

20. $(3a + b)^4$

(a) $81a^4 - 108a^3 b + 54a^2 b^2 - 12ab^3 + b^4$ 　　　　(b) $81a^4 - 4a^3 b + 6a^2 b^2 - 4ab^3 + b^4$
(c) $81a^4 + 108a^3 b + 54a^2 b^2 + 12ab^3 + b^4$ 　　　　(d) $81a^4 + 4a^3 b + 6a^2 b^2 + 4ab^3 + b^4$

Chapter 15 Test Form H

1. What type of sequence is: $\dfrac{3}{4}, \dfrac{4}{5}, \dfrac{5}{6}, \dfrac{6}{7}, \dfrac{7}{8}, \ldots$?

   (a) arithmetic   (b) infinite sum   (c) geometric   (d) none of these

For questions 2–4, write the first four terms of the sequence with:

2. $a_n = -n^2 - 3$

   (a) $-4, -7, -12, -19$ (b) $-4, -7, -10, -13$ (c) $-2, -1, -6, -13$ (d) $-2, -5, -8, -11$

3. First term $a_1 = \dfrac{1}{3}$ and common ratio $r = \dfrac{3}{2}$.

   (a) $\dfrac{1}{3}, \dfrac{11}{6}, \dfrac{10}{3}, \dfrac{9}{2}$ (b) $\dfrac{1}{3}, \dfrac{1}{2}, \dfrac{3}{4}, \dfrac{9}{8}$ (c) $\dfrac{1}{2}, \dfrac{3}{4}, \dfrac{9}{8}, \dfrac{27}{16}$ (d) $\dfrac{1}{3}, \dfrac{2}{9}, \dfrac{4}{27}, \dfrac{8}{81}$

4. Fifth term $a_5 = 12$ and common difference $d = -2$.

   (a) $10, 8, 6, 4$ (b) $20, 18, 16, 14$ (c) $18, 16, 14, 12$ (d) $16, 14, 12, 10$

5. Find the second and third partial sums of $a_n = \dfrac{2n}{n+1}$.

   (a) $\dfrac{7}{3}; \dfrac{3}{2}$ (b) $\dfrac{7}{3}; \dfrac{23}{6}$ (c) $1; \dfrac{5}{3}$ (d) $\dfrac{4}{3}; \dfrac{7}{3}$ s

6. Write out the series $\displaystyle\sum_{k=1}^{4} \left(k^2 - 2\right)$ and evaluate the sum.

   (a) $-1 + 2 + 14 + 23; 38$ (b) $-1 + 2 + 7 + 14; 22$ (c) $2 + 7 + 14 + 23; 46$ (d) $-1 + 7 + 14 + 23; 43$

7. Write out the series $\displaystyle\sum_{k=3}^{5} \left(\dfrac{k+2}{2k-1}\right)$ and evaluate the sum.

   (a) $1 + \dfrac{6}{7} + \dfrac{7}{9} = \dfrac{166}{63}$ (b) $3 + \dfrac{4}{3} + 1 = \dfrac{16}{3}$ (c) $3 + \dfrac{4}{3} + 1 + \dfrac{6}{7} + \dfrac{7}{9} = \dfrac{439}{63}$ (d) $\dfrac{6}{7} + \dfrac{7}{9} = \dfrac{103}{63}$

8. Find d if $a_3 = -13$ and $a_8 = 12$.

   (a) $d = \dfrac{15}{4}$ (b) $d = 3$ (c) $d = 4$ (d) $d = 5$

9. Find the sum, s_n, and common difference, d, if $a_1 = 4$, $a_n = 2$, and $n = 3$.

   (a) $s_3 = 9; d = -1$ (b) $s_3 = 6; d = -1$ (c) $s_3 = 9; d = 1$ (d) $s_3 = 6; d = 1$

10. Find the number of terms in the arithmetic sequence $-2, 6, 14, \ldots, 62$.

    (a) 7 (b) 8 (c) 9 (d) 10

11. Find the term a_5 of the geometric sequence with $a_1 = 4$ and $r = \dfrac{3}{2}$.

(a) $\dfrac{81}{4}$ (b) $\dfrac{27}{2}$ (c) $\dfrac{23}{2}$ (d) 30

12. Find the term a_1 of the geometric sequence with $a_4 = 24$ and $r = \dfrac{2}{3}$.

(a) 54   (b) 27   (c) 16   (d) 81

13. Find the sum s_4 of the geometric sequence with $a_1 = 4$ and $r = \dfrac{3}{2}$.

(a) $\dfrac{23}{2}$ (b) 19   (c) $\dfrac{65}{2}$ (d) $\dfrac{211}{4}$

14. Find the sum of the infinite geometric series: $6 + 4 + \dfrac{8}{3} + \dfrac{16}{9} + \dfrac{32}{27} + \ldots$

(a) 30   (b) $\dfrac{18}{5}$ (c) 18   (d) $\dfrac{422}{27}$

15. Find the sum of the infinite geometric series: $6 - 4 + \dfrac{8}{3} - \dfrac{16}{9} + \dfrac{32}{27} - \ldots$

(a) 4   (b) 9   (c) 18   (d) $\dfrac{18}{5}$

16. Write the repeating decimal $0.454545\ldots$ as a ratio of two integers.

(a) $\dfrac{4}{5}$ (b) $\dfrac{5}{11}$ (c) $\dfrac{9}{20}$ (d) $\dfrac{5}{4}$

17. Write the repeating decimal $0.7454545\ldots$ as a ratio of two integers.

(a) $\dfrac{39}{50}$ (b) $\dfrac{33}{40}$ (c) $\dfrac{149}{200}$ (d) $\dfrac{41}{55}$

18. Find the common ratio of the sequence $16, -24, 36, -54, \ldots$

(a) $\dfrac{3}{4}$ (b) $-\dfrac{3}{2}$ (c) $-\dfrac{3}{4}$ (d) $\dfrac{3}{2}$

For questions 14–15, use the binomial theorem to expand each expression.

19. $(a - 3b)^3$

(a) $a^3 - 3a^2b + 3ab^2 - 27b^3$ (b) $a^3 - 3a^2b - 3ab^2 - 27b^3$

(c) $a^3 - 9a^2b + 27ab^2 - 27b^3$ (d) $a^3 - 9a^2b - 27ab^2 - 27b^3$

20. $(x + y)^4$

(a) $x^4 + 2x^3y + 4x^2y^2 + 2xy^3 + y^4$ (b) $x^4 - 2x^3y + 4x^2y^2 - 2xy^3 + y^4$

(c) $x^4 + 4x^3y + 6x^2y^2 + 4xy^3 + y^4$ (d) $x^4 - 4x^3y + 6x^2y^2 - 4xy^3 + y^4$

Final Exam Form A

Name:_____

Date:_____

For questions 1–2, use the set $\left\{-3, -\dfrac{1}{2}, -1, 0, 1, 2, \dfrac{5}{3}, \sqrt{7}, 3.25, 6, 9\right\}$.

1. List the elements of the set that are natural numbers.

2. List the elements of the set that are whole numbers.

3. Let $A = \{2, 4, 6\}$, $B = \{2, 4, 6, 8, 11\}$. Find $A \cup B$.

1._____

2._____

3._____

Solve the following equations.

4. $\dfrac{x-4}{4} = -3$

5. $\sqrt{x-3} - 7 = 0$

6. $2x^2 + 3 = 7$

7. $\dfrac{x}{x-3} = \dfrac{x+2}{x}$

8. $\dfrac{2x+1}{x-3} = \dfrac{x+8}{x-4}$

9. $x + y = 3$
   $x + y + z = 0$
   $x - y - z = 2$

10. Write in standard notation: 4.86×10^8

11. Solve: $x^3 - 4x = 0$

12. Simplify: $\dfrac{w^2 - 3w + 6}{w - 5} + \dfrac{9 - w^2}{w - 5}$

13. Multiply: $\dfrac{a^2 + a - 2}{a^3 - 1} \cdot \dfrac{a^2 + a + 1}{a^2 - 4}$

14. Simplify $\left(\dfrac{4b}{3}\right)^{-2}$ and write the answer without negative exponents.

15. Find the determinant: $\begin{vmatrix} 1 & 3 \\ -2 & 5 \end{vmatrix}$

16. Find the determinant: $\begin{vmatrix} 2 & 1 & 3 \\ 4 & 5 & 2 \\ -6 & 1 & -3 \end{vmatrix}$

4._____

5._____

6._____

7._____

8._____

9._____

10._____

11._____

12._____

13._____

14._____

15._____

16._____

For numbers 17–18, $A = (7, 4)$ and $B = (3, 8)$

17. Find the length of $\overline{AB}$

17._____

591

Final Exam Form A *(continued)*

Name:_____

18. Find the points that break $\overline{AB}$ into four equal parts.

18. _____

Factor.

19. $9x^2 - 24x + 16$

19. _____

20. $-a^3 + 25a$

20. _____

21. Divide: $\left(x^3 + 6x^2 - 3x - 5\right) \div \left(x - 3\right)$

21. _____

22. Multiply: $\left(3w - 2x\right)^2$

22. _____

23. Simplify: $\dfrac{2 + i}{2 - i}$

23. _____

24. Solve: $\left|\dfrac{5}{2} - x\right| = \left|2 - \dfrac{x}{2}\right|$

24. _____

25. Solve using the quadratic formula: $4x^2 - 4x + 1 = 0$

25. _____

26. If $f(x) = x^2 + 4x - 2$ and $g(x) = 2x - 1$, find $(f \circ g)(x)$.

26. _____

27. Find the solution set to the inequality: $|5 - 3x| \le 6$

27. _____

28. Solve the inequality $2x^2 - 7x - 15 \ge 0$. Write the answer in interval notation.

28. _____

29. Find the equation of the line through $\left(2, \dfrac{1}{2}\right)$, with slope $m = \dfrac{1}{4}$.

29. _____

30. Graph: $\dfrac{x^2}{25} + \dfrac{(y + 3)^2}{4} = 1$

30.

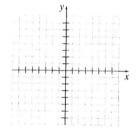

31. Graph: $x^2 - 4x + y^2 + 4y = 1$

31.

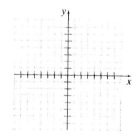

Final Exam Form A *(continued)*

Name:_____

Problems 32–33 refer to the ellipse: $x^2 + 4y^2 + 16y + 9 = 2x$

32. Find the center of the ellipse.

32. _____

33. Find the area of the ellipse.

33. _____

34. Evaluate the sum of the series $\sum_{k=1}^{4} (3k - 4)$.

34. _____

35. Solve for x: $4^{2x+1} = 8^{x+3}$

35. _____

36. Write as a logarithm of a single expression:
$\log_b x + \log_b (x+2) - \log_b 8$

36. _____

37. Find the sum of the infinite geometric sequence $125, 25, 5 \cdots$

37. _____

38. Find the sum of the infinite geometric sequence $125, -25, 5, -1, \cdots$

38. _____

39. In an arithmetic sequence, $a_3 = -2$ and $a_7 = 8$. Find a_1 and d.

39. _____

40. Write out the first four terms of the expansion of $(2a + 5b)^7$.

40. _____

Final Exam Form B

Name:_____

Date:_____

For questions 1–2, use the set $\left\{-3, -\dfrac{1}{2}, -1, 0, 1, 2, \dfrac{5}{3}, \sqrt{7}, 3.25, 6, 9\right\}$.

1. List the elements of the set that are real numbers.

2. List the elements of the set that are irrational numbers.

3. Let $A = \{2,4,6\}$, $B = \{3,4,6,8,11\}$. Find $A \cap B$.

Solve the following equations or inequalities.

4. $|x - 5| = 2$

5. $\dfrac{x}{x+2} > -1$

6. $4x^2 - 12x + 9 = 0$

7. $\sqrt{a-1} - 6 = 0$

8. $8x + 3y = -1$
   $2x + y = 1$

9. Evaluate the determinant:
$$\begin{vmatrix} 2 & -1 & 3 \\ 1 & 0 & 2 \\ 0 & 2 & -3 \end{vmatrix}$$

Simplify the following:

10. $\dfrac{1 + \sqrt{2}}{\sqrt{3} - 1}$

11. $\dfrac{2 - 3i}{3 + 4i}$

12. $\dfrac{a + \dfrac{3}{b}}{\dfrac{b}{a} + \dfrac{1}{b}}$

For problems 13 and 14, use $f(x) = 2x - 3$, $g(x) = x^2 + 3x$.

13. Find: $(f \circ g)(1)$

14. Find: $f^{-1}(x)$

15. Simplify: $\left(1 + 2^{-1}\right)^{-2}$

16. Simplify using positive exponents only: $\left(\dfrac{2r^{-3}t^{-1}}{10r^5t^2t^{-3}}\right)$

1. _____

2. _____

3. _____

4. _____

5. _____

6. _____

7. _____

8. _____

9. _____

10. _____

11. _____

12. _____

13. _____

14. _____

15. _____

16. _____

Final Exam Form B *(continued)*

17. Express 0.0015 in scientific notation.

17. _____

18. Simplify: $\left(\dfrac{7.2 \times 10^6}{1.2 \times 10^8} \right)$

18. _____

19. Solve the equation: $\dfrac{3x}{2} - 6 = 9$

19. _____

20. Solve: $x^2 + 6x - 16 = 0$

20. _____

21. Solve: $(2a - 1)^2 + 2(2a - 1) - 8 = 0$

21. _____

22. Solve: $x^4 - 10x^2 + 9 = 0$

22. _____

23. Solve the equation $\dfrac{2x + 1}{x - 1} = \dfrac{x - 2}{x + 2}$

23. _____

24. Solve using the quadratic formula: $4 - 10x = -5x^2$

24. _____

25. Solve the inequality. Write the solution in interval notation.
$-6 < 3(x + 2) < 9$

25. _____

26. Find the solution: $|3x + 24| = 0$

26. _____

27. Solve the inequality. Write the answer in interval notation.
$7 - x \le 3x - 1$

27. _____

28. Write the equation of the line, in slope–intercept form, that passes
through (1, 1) and is perpendicular to the graph $y = \dfrac{1}{2}x - 3$.

28. _____

29. Find the maximum area of a rectangular field that has a
perimeter of 200 feet.

29. _____

Graph problems 30 and 31.

30. $x^2 + 4y^2 - 2x - 16y = -13$

30.

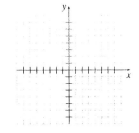

595

31. $y \le -x + 4$

31.

For problems 32–33, $A = (13,12)$ and $B = (-3,6)$.

32. Find the length of $\overline{AB}$

32. _____

33. Find the midpoint of $\overline{AB}$

33. _____

34. In an arithmetic sequence, $a_5 = 2$ and $a_{11} = -7$. Find a_1 and d.

34. _____

35. Evaluate the sum of the series $\sum_{k=2}^{6}(k^2 + 1)$.

35. _____

36. Write the repeating decimal $0.327272727\ldots$ as a ratio of two integers.

36. _____

37. Write as a logarithm of a single expression:
$-2\log_b x - 3\log_b y + \log_b z$

37. _____

38. Solve for x: $\log(3x + 1) = \log(x - 2) + 1$

38. _____

39. Find the sum of the infinite geometric series
$1 - \dfrac{1}{2} + \dfrac{1}{4} - \dfrac{1}{8} + \cdots$

39. _____

40. Use the binomial theorem to expand the expression $(2x + 3)^5$.

40.. _____

Final Exam Form C

For questions 1–2, consider the set $\left\{-2.87, -1, 0, \dfrac{7}{32}, \sqrt{72}, \sqrt{-8}, 3i\right\}$.

1. List the elements of the set that are irrational numbers.

2. List the elements of the set that are imaginary numbers.

3. Find $A \cap B$ for $A = \left\{0, \dfrac{1}{2}, 1, \dfrac{3}{2}, 2\right\}$ and $B = \left\{-\dfrac{1}{2}, \dfrac{3}{2}, \dfrac{5}{2}, \dfrac{7}{2}\right\}$.

4. Solve the equation: $7(x - 5) = 3(2 - x) - 1$

5. Solve $\sqrt{3 - x} = 2x$.

6. Solve $5x^4 + 13x^2 = 6$.

7. Solve: $\dfrac{3x + 1}{x + 3} = \dfrac{x + 2}{x + 6}$.

8. Solve the equation $x^2 - 12x + 5 = 0$ using the quadratic formula.

9. Solve the inequality $\dfrac{6 - 3x}{7} \geq 6$. Write the solution in interval notation.

10. Find the solution set to the inequality $|5x - 2| < 8$.

11. Solve the inequality $2x^2 + 2x < 12$. Write the answer in interval notation.

12. Find the equation of the line through $(-1, 5)$ that is parallel to the graph of $y = -3x + 7$, written in slope–intercept form.

13. Solve the system of equations using any method.
$$\begin{cases} 2x - y = 2 \\ 3x - 2y = 2 \end{cases}$$

14. Solve the system of equations using any method.
$$\begin{cases} x + y + 2z = 0 \\ 2x + y + z = 2 \\ x + 2y + 2z = 1 \end{cases}$$

15. Evaluate the determinant.
$$\begin{vmatrix} 6 & -3 \\ 4 & -2 \end{vmatrix}$$

16. Evaluate the determinant.
$$\begin{vmatrix} 2 & 1 & -1 \\ 3 & 2 & 5 \\ 7 & -3 & -1 \end{vmatrix}$$

17. Multiply $(3x + 4)(2x - 5)$.

18. Multiply $\left(7 + \sqrt{-4}\right)\left(3 - \sqrt{-9}\right)$.

1. _____

2. _____

3. _____

4. _____

5. _____

6. _____

7. _____

8. _____

9. _____

10. _____

11. _____

12. _____

13. _____

14. _____

15. _____

16. _____

17. _____

18. _____

Final Exam Form C *(continued)*

19. Factor $x^4 - 81$.

19. _____

For questions 20–22, perform the indicated operations.

20. $\dfrac{x+5}{3x-1} \cdot \dfrac{9x^2-1}{x^3+125}$

20. _____

21. $\dfrac{a^3-64b^3}{2a+b} \div \dfrac{a^2+4ab+16b^2}{4a^2-b^2}$

21. _____

22. $(2+3i) \div (1+i)$

22. _____

23. Use long division to find $(8x^2 - 6x + 7) \div (2x + 1)$.

23. _____

24. Simplify $\sqrt[5]{\dfrac{64a^7}{b^5c^9}}$.

24. _____

25. Simplify $2\sqrt{20} - 6\sqrt{5} + 3\sqrt{45}$.

25. _____

26. Solve for x: $4^{2x-1} = \dfrac{1}{8^{x+1}}$.

26. _____

27. Write $\log_4 x = 4$ in exponential form, then solve for x.

27. _____

28. Write as the logarithm of a single expression:
$5\log_3(x+1) - 3\log_3 x + 6\log_3(x-4)$

28. _____

29. Solve for x: $\log_7(3x-2) - \log_7 8 = \log_7 5$

29. _____

30. Solve for x: $\log_b(3x-2) - \log_b 2 = \log_b(x+1)$

30. _____

31. If $\log_{13} 5 = 0.62747356$ and $\log_{13} 17 = 1.1045884$, evaluate $\log_5 17$.

31. _____

32. Evaluate the sum of the series $\displaystyle\sum_{n=1}^{4}(3n+2)$.

32. _____

33. Find the sum of the infinite geometric series
$18 + 12 + 8 + \dfrac{16}{3} + \dfrac{32}{9} + \cdots$

33. _____

34. Find the sum of the infinite geometric series
$8 - \dfrac{16}{3} + \dfrac{32}{9} - \dfrac{64}{27} + \cdots$

34. _____

35. Graph: $y \le -\dfrac{1}{2}x + 2$

35.

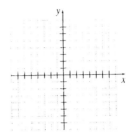

36. Graph: $\dfrac{x^2}{9} - \dfrac{y^2}{25} = 1$

36.

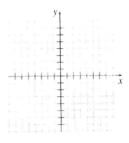

37. If the largest angle of a triangle is eight times the measure of the smallest angle, and the middle–sized angle is 30° greater than the measure of the smallest angle, find the measures of the three angles.

37. _____

38. The compound interest formula is $A = P\left(1 + \dfrac{r}{n}\right)^{nt}$, where A is the amount, P is the principal invested, r is the annual rate of interest, n is the number of times the interest is compounded annually, and t is the time in years. Find the value of A if you invest $3000 at an annual interest rate of 4% compounded quarterly for 3 years.

38. _____

39. Write out the first four terms of the expansion of $(2x + y)^9$.

39. _____

40. Write the repeating decimal 0.243243243… as the quotient of two integers.

40. _____

Final Exam Form D

Name:_____

Date:_____

For questions 1–2, consider the set $\left\{-4, -2.3\overline{7}, \sqrt{5}, i\sqrt{2}, -\sqrt{36}, \frac{2}{7}\right\}$.

1. List the elements of the set that are rational numbers.

1. _____

2. List the elements of the set that are irrational numbers.

2. _____

3. Find $A \cap B$ for $A = \{-7, -5, -3, -1\}$ and $B = \{1, 2, 5, 8\}$.

3. _____

4. Solve the equation: $7(x + 4) = 3(5 - x) + 3$

4. _____

5. Solve $\sqrt{5x-3} = \sqrt{2}x$.

5. _____

6. Solve $x^4 - 3x^2 - 4 = 0$.

6. _____

7. Solve the equation $\dfrac{x^2}{a^2} + \dfrac{y^2}{b^2} = 1$ for the variable x.

7. _____

8. Solve the equation $x^2 = 8x - 13$ using the quadratic formula.

8. _____

9. Solve the equation: $\dfrac{3x - 2}{x + 2} = \dfrac{2x + 1}{x - 1}$

9. _____

10. Find the solution set to the inequality $|2x + 7| < 3$.

10. _____

11. Solve the inequality $2x^2 - 7x - 15 \geq 0$. Write the answer in interval notation.

11. _____

12. Find the equation of the line through $(-3, 5)$ that is perpendicular to the graph of $y = -\dfrac{1}{3}x + 7$, written in slope–intercept form.

12. _____

13. Solve the system of equations using any method.
$$\begin{cases} 2x + y = 3 \\ x - y = 6 \end{cases}$$

13. _____

14. Solve the system of equations using any method.
$$\begin{cases} 3x + 4y + z = 4 \\ x - 2y + z = 6 \\ 2x + 2y + z = 4 \end{cases}$$

14. _____

15. Evaluate the determinant.
$$\begin{vmatrix} 4 & -3 \\ 3 & -2 \end{vmatrix}$$

15. _____

16. Evaluate the determinant.
$$\begin{vmatrix} 2 & 1 & 0 \\ 3 & -2 & 1 \\ 4 & 7 & 2 \end{vmatrix}$$

16. _____

17. Multiply $(x + 2y)(x^2 - 2xy + 4y^2)$.

17. _____

18. Multiply $\left(7 + \sqrt{-16}\right)\left(3 - \sqrt{-81}\right)$.

18. _____

Final Exam Form D *(continued)*

19. Factor $27a^3 + 1000b^3$

19. _____

For questions 20–22, perform the indicated operations.

20. $\dfrac{5x+1}{x-7} \cdot \dfrac{x^2-49}{125x^3+1}$

20. _____

21. $\dfrac{a^3-64b^3}{4a^2-9b^2} \div \dfrac{a^2+4ab+16b^2}{2a+3b}$

21. _____

22. $\dfrac{x+4}{x^2+x-6} - \dfrac{3x}{x+3}$

22. _____

23. Use synthetic division to find $(5x^2 - 3x - 10) \div (x+2)$.

23. _____

24. Simplify $\sqrt[5]{\dfrac{-32x^{10}}{5y^{12}z^{20}}}$.

24. _____

25. Simplify $2\sqrt{24} - 5\sqrt{6} + \sqrt{600}$.

25. _____

26. Simplify: $\dfrac{3-i}{2+i}$

26. _____

27. Write as the logarithm of a single expression:
$3\log_7(x+3) - 5\log_7(x-1) - \log_7 x$

27. _____

28. Solve for x : $\left(\dfrac{1}{3}\right)^{x+1} = 9^{2x-3}$

28. _____

29. Solve for x: $\log_2(5x-4) - \log_2 7 = \log_2(x-2)$

29. _____

30. If $\log_8 11 = 1.153144$ and $\log_8 5 = 0.773976$, calculate $\log_5 11$.

30. _____

31. Evaluate the sum of the series $\sum\limits_{n=1}^{3}(2n^2 - 1)$.

31. _____

32. Find the sum of the infinite geometric series:
$12 + 6 + 3 + \dfrac{3}{2} + \dfrac{3}{4} + \cdots$

32. _____

33. If $f(x) = \sqrt{x^2+1}$ and $g(x) = 2x+5$, find $(f \circ g)(x)$

33. _____

601

Final Exam Form D *(continued)*

Name:_____

For questions 34–35, graph.

34. $y \leq \dfrac{1}{5}x - 3$

34.

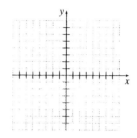

35. $x^2 + 6x + y^2 - 4y = -4$

35.

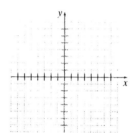

36. If the largest angle of a triangle is 10° more than three times the measure of the smallest angle, and the middle–sized angle is 20° greater than the measure of the smallest angle, find the measures of the three angles.

36. _____

37. If $3200 is invested at 6% compounded continuously, how long would it take the value of the account to reach $6400?

37. _____

38. In an arithmetic sequence, $a_8 = 16$ and $a_{18} = 46$; find a_1 and d.

38. _____

39. Write out the first four terms of the expansion of $(x - 4y)^{10}$.

39. _____

40. Write the infinite repeating decimal 0.12818181... as the quotient of two integers.

40. _____

Final Exam Form E

Name:_____

Date:_____

For question 1–2, consider the set $\left\{-3, -\sqrt{3}, \sqrt{-3}, i, \sqrt{36}, \sqrt{-4}\right\}$.

1. List the elements of the set that are imaginary numbers.

2. List the elements of the set that are irrational numbers.

3. Find $A \cap B$ for $A = \{-2, -1, 0, 1, 2\}$ and $\{-1.5, -0.5, 0.5, 1.5\}$.

4. Solve the equation: $2(x - 4) = 3(7 - x) - 14$

5. Solve $\sqrt{4x + 7} = x\sqrt{3}$.

6. Solve $x^4 - 6x^2 = 27$.

7. Solve: $\dfrac{2x+1}{x-2} = \dfrac{x+4}{2x+7}$

8. Solve the equation $x^2 - 8x + 3 = 0$ using the quadratic formula.

9. Solve the inequality $\dfrac{x+7}{4} \le 9$. Write the solution in interval notation.

10. Find the solution set to the inequality $|2x + 5| < 3$.

11. Solve the inequality $x^2 - 5x > 24$, and write the answer in interval notation.

12. Find the equation of the line through $(-1, 4)$ that is parallel to the graph of $y = 2x - 7$, written in slope–intercept form.

13. Solve the system of equations using any method.
$$\begin{cases} x - 2y = -4 \\ 2x + 5y = 1 \end{cases}$$

14. Solve the system of equations using any method.
$$\begin{cases} x + y + z = 4 \\ 2x - y + z = 3 \\ x + 2y - z = 1 \end{cases}$$

15. Evaluate the determinant.
$$\begin{vmatrix} -5 & 4 \\ -6 & 5 \end{vmatrix}$$

16. Evaluate the determinant.
$$\begin{vmatrix} 0 & -2 & 5 \\ 3 & -1 & 3 \\ 2 & 5 & 2 \end{vmatrix}$$

17. Multiply $(2x - 3y)(4x^2 + 6xy + 9y^2)$.

18. Multiply $\left(8 + \sqrt{-4}\right)\left(2 - \sqrt{-64}\right)$.

19. Factor $125x^3 + 64y^3$.

1. _____

2. _____

3. _____

4. _____

5. _____

6. _____

7. _____

8. _____

9. _____

10. _____

11. _____

12. _____

13. _____

14. _____

15. _____

16. _____

17. _____

18. _____

19. _____

Final Exam Form E *(continued)*

Name:_____

For questions 20–21, perform the indicated operations.

20. $\dfrac{x+7}{5x-1} \cdot \dfrac{25x^2-1}{x^3+343}$

20. _____

21. $\dfrac{27a^3-b^3}{2a+5b} \div \dfrac{9a^2+3ab+b^2}{4a^2-25b^2}$

21. _____

22. Find the length and midpoint of the line segment whose endpoints are $(2,7)$ and $(-4,-5)$

22. _____

23. Use long division to find $(9x^2+3x-2) \div (3x-1)$.

23. _____

24. Simplify $\sqrt[4]{\dfrac{81x^5}{2y^{28}}}$.

24. _____

25. Simplify $3\sqrt{20}-6\sqrt{5}+2\sqrt{45}$.

25. _____

26. Simplify: $\dfrac{-13i}{2-3i}$.

26. _____

27. Write as the logarithm of a single expression:
$-3\log_8(x+2)+5\log_8(2x-1)-2\log_8 x$

27. _____

28. If $\log_{13}5 = 0.62747356$ and $\log_{13}17 = 1.1045884$, evaluate $\log_5 17$.

28. _____

29. Solve for x: $4^{2x+1}=32^{x-2}$

29. _____

30. In an arithmetic sequence, $a_5 = 3$ and $a_{12} = -11$. Find a_1 and d.

30. _____

31. Evaluate the sum of the series $\displaystyle\sum_{n=1}^{3}(3n^2+1)$.

31. _____

32. Find the sum of the infinite geometric series:
$10+6+\dfrac{18}{5}+\dfrac{54}{25}+\dfrac{162}{125}+\cdots$

32. _____

33. Find the sum of the infinite geometric series:
$-6+\dfrac{18}{5}-\dfrac{54}{25}+\dfrac{162}{125}-\cdots$

33. _____

34. Write the repeating decimal $0.6151515\ldots$ as the ratio of two integers.

34. _____

Final Exam Form E *(continued)*

For questions 35–37, graph.

35. $f(x) = -(x+2)^2 + 4$

35.

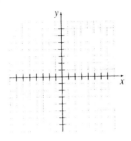

36. $y > \dfrac{2}{3}x + 2$

36.

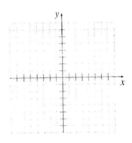

37. $x^2 + 2x + y^2 + 2y = 14$

37.

38. If the largest angle of a triangle is $10°$ more than twice the measure of the smallest angle, and the middle-sized angle is $10°$ greater than the measure of the smallest angle, find the measures of the three angles.

38. _____

39. If $8000 is invested at 7.5% compounded continuously, how long would it take the value of the account to triple?

39. _____

40. Write out the first four terms of the expansion of $(2x - y)^{10}$.

40. _____

Final Exam Form F

Name:_____

Date:_____

For questions 1–2, consider the set $\left\{ \sqrt{-4}, -2.3\overline{7}, \sqrt{8}, i\sqrt{2}, -\sqrt{36}, \dfrac{\pi}{7} \right\}$.

1. List the elements of the set that are rational numbers.

   1. _____

2. List the elements of the set that are irrational numbers.

   2. _____

3. Find $A \cap B$ for $A = \{..., -3, -2, -1, 0, 1, 2, 3, ...\}$ and $B = \{\text{all negative, real numbers}\}$.

   3. _____

4. Solve the equation: $3(2 - x) = 5(2x - 1) + 37$

   4. _____

5. Solve $\sqrt{5x - 2} = \sqrt{2}x$.

   5. _____

6. Solve $2x^4 + 5x^2 - 3 = 0$.

   6. _____

7. Solve the equation $\dfrac{2x - 5}{x + 8} = \dfrac{4 - x}{x - 2}$.

   7. _____

8. Solve the equation $x^2 - 12x + 4 = 0$ using the quadratic formula.

   8. _____

9. Solve the inequality $\dfrac{5x + 8}{3} \leq 11$. Write the solution in interval notation.

   9. _____

10. Find the solution set to the inequality $|3x + 10| < 2$.

    10. _____

11. Solve the inequality $x^2 - 5x \leq 14$. Write the answer in interval notation.

    11. _____

12. Find the equation of the line through $(2, -3)$ that is parallel to the graph of $y = -3x - 1$.

    12. _____

13. Solve the system of equations using any method.
$$\begin{cases} 2x - y = -5 \\ x + 3y = 1 \end{cases}$$

    13. _____

14. Solve the system of equations using any method.
$$\begin{cases} x + y + z = 3 \\ x - y + z = -1 \\ 2x + y + z = 2 \end{cases}$$

    14. _____

15. Solve the system of equations using any method.
$$\begin{cases} 3x^2 - 2y = -2 \\ 5x + 3y = 11 \end{cases}$$

    15. _____

16. Evaluate the determinant.
$$\begin{vmatrix} 2 & 1 & 3 \\ 1 & 3 & 2 \\ -1 & 2 & -3 \end{vmatrix}$$

    16. _____

17. Multiply $(x + 5)(3x - 2)$.

    17. _____

18. Multiply $\left(-3 - \sqrt{-9}\right)\left(2 + \sqrt{-4}\right)$.

    18. _____

19. Factor $x^4 - 256$.

    19. _____

Final Exam Form F *(continued)*

For questions 20–22, perform the indicated operations.

20. $\dfrac{x+3}{5x-1} \cdot \dfrac{25x^2-1}{x^3+27}$

20. _____

21. $\dfrac{8a^3-64b^3}{a+7b} \div \dfrac{2a-4b}{a^2-49b^2}$

21. _____

22. $(21+i) \div (3+2i)$

22. _____

23. Use synthetic division to find $(5x^2-3x+8) \div (x+1)$.

23. _____

24. Simplify $\sqrt[3]{\dfrac{-125x^5}{8y^9z^{15}}}$.

24. _____

25. Simplify $3\sqrt{8} - \sqrt{200} + 3\sqrt{18}$.

25. _____

26. Write $3^4 = 81$ in logarithmic form.

26. _____

27. Write as the logarithm of a single expression:
$3\log_8 x - 2\log_8(x+4) - 7\log_8(2x-1)$

27. _____

28. Solve for x: $\log_8(6x+5) - \log_8(2x-1) = \log_8 4$

28. _____

29. If $\log_8 11 = 1.153144$ and $\log_8 5 = 0.773976$,
calculate $\log_5 11$.

29. _____

30. Find the length and midpoint of the line segment with endpoints $(6, -3)$ and $(-2, -9)$.

30. _____

31. Evaluate the sum of the series $\displaystyle\sum_{n=1}^{3}(4n^2-1)$.

31. _____

32. Find the sum of the infinite geometric series: $5 + 2 + \dfrac{4}{5} + \dfrac{8}{25} + \dfrac{16}{125} + \cdots$

32. _____

33. Find the sum of the infinite geometric series: $2 - \dfrac{4}{5} + \dfrac{8}{25} - \dfrac{16}{125} + \cdots$

33. _____

34. Graph: $y = -(x+2)^2 + 3$

34.

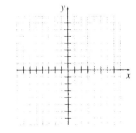

35. Graph: $y > \dfrac{3}{2}x - 1$

35.

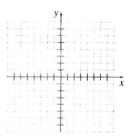

36. Graph: $(x+2)^2 + (y-2)^2 = 4$

36.

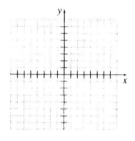

37. If the largest angle of a triangle is three times the measure of the smallest angle, and the middle–sized angle is $30°$ greater than the measure of the smallest angle, find the measures of the three angles.

37. _____

38. If $7000 is invested at 2.5% compounded continuously, how long would it take the value of the account to double?

38. _____

39. Find the fourth term in the expansion of $(3a - 2b)^8$.

39. _____

40. In an arithmetic sequence, $a_4 = 2$ and $a_9 = -23$; find a_1 and d.

40. _____

Final Exam Form G

Name:_____

Date:_____

Choose the correct answer to each problem.

1. Which of the following sets consists of only integers?

   (a) $\left\{-3, -\dfrac{8}{6}, \sqrt{4}, 6\right\}$ (b) $\left\{-3, -\dfrac{8}{6}, \sqrt{3}, 5\right\}$ (c) $\left\{-3, -\dfrac{8}{4}, \sqrt{9}, 7\right\}$ (d) $\left\{-3, -\dfrac{8}{4}, \sqrt{3}, 5\right\}$

2. Which of the following sets consists of only irrational numbers?

   (a) $\left\{-\sqrt{2}, \sqrt{3}, \sqrt{12}\right\}$ (b) $\left\{\sqrt{-2}, \sqrt{3}, \sqrt{12}\right\}$ (c) $\left\{-\sqrt{2}, \sqrt{-3}, \sqrt{12}\right\}$ (d) $\left\{\sqrt{-2}, \sqrt{-3}, \sqrt{12}\right\}$

3. Find $A \cap B$ for $A = \{-3, -1, 0, 1, 7, 13\}$ and $B = \{-7, -1, 0, 2, 3, 4, 6, 11\}$.

   (a) $\{-3, -1, 0, 4\}$ (b) $\{-1, 0\}$ (c) $\{0, 4\}$ (d) $\{-1, 0, 4\}$

4. Solve the equation: $3(x + 2) = 4(x - 5) + 23$

   (a) $x = 3$ (b) $x = 2$ (c) $x = -3$ (d) $x = -2$

5. Solve $\sqrt{7x + 8} = x$.

   (a) $x = 8$ (b) $x = -1$ (c) $x = 1, -8$ (d) $x = -1, 8$

6. Solve $x^4 + x^2 = 20$.

   (a) $x = \pm 4, \pm 5$ (b) $x = \pm 4, \pm i\sqrt{5}$ (c) $x = \pm 2, i\sqrt{5}$ (d) $x = \pm 2, \pm i\sqrt{5}$

7. Solve the equation : $\dfrac{x + 2}{x + 3} = \dfrac{x - 2}{2x + 5}$.

   (a) 4 (b) $-4, 4$ (c) -4 (d) $4, -2$

8. Solve the equation $2x^2 - 10x + 3 = 0$ using the quadratic formula.

   (a) $x = 5 \pm \sqrt{19}$ (b) $x = \dfrac{5 \pm \sqrt{19}}{2}$ (c) $x = \dfrac{5 \pm \sqrt{76}}{2}$ (d) $x = \dfrac{10 \pm \sqrt{76}}{2}$

9. Solve the inequality $\dfrac{3x - 5}{8} > 2$. Write the solution in interval notation.

   (a) $(-\infty, 7)$ (b) $(7, \infty)$ (c) $(-\infty, 7]$ (d) $[7, \infty)$

10. Find the solution set to the inequality $|7x - 1| \ge 20$.

    (a) $\left\{x \,\Big|\, -\dfrac{19}{4} < x < 3\right\}$ (b) $\left\{x \,\Big|\, -\dfrac{19}{4} \le x \le 3\right\}$ (c) $\left\{x \,\Big|\, x < -\dfrac{19}{7} \text{ or } x > 3\right\}$ (d) $\left\{x \,\Big|\, x \le -\dfrac{19}{7} \text{ or } x \ge 3\right\}$

11. Solve the inequality $x^2 - x - 2 > 0$. Write the answer in interval notation.

    (a) $(-1, 2)$ (b) $(-\infty, -1) \cup (2, \infty)$ (c) $(-\infty, -2) \cup (1, \infty)$ (d) $(-2, 1)$

12. Find the equation of the line through (4, 2) that is perpendicular to the graph of $y = 2x - 7$.

(a) $y = -\dfrac{1}{2}x + 4$ 　　　(b) $y = -\dfrac{1}{2}x - 4$ 　　　(c) $y = 2x - 6$ 　　　(d) $y = 2x + 6$

13. Solve the system of equations using any method.
$$\begin{cases} x + y = 1 \\ 2x + y = -2 \end{cases}$$

(a) $(-4, 5)$ 　　　(b) $(3, 4)$ 　　　(c) $(-3, 4)$ 　　　(d) $(3, -4)$

14. Solve the system of equations using any method.
$$\begin{cases} x + y + z = 1 \\ 2x - y + z = -2 \\ x + 2y + z = 2 \end{cases}$$

(a) $(-2, 2, 1)$ 　　　(b) $(2, -2, 1)$ 　　　(c) $(1, -1, 1)$ 　　　(d) $(-1, 1, 1)$

15. Solve the system of equations using any method. Which of the following is a solution to the system?
$$\begin{cases} 2x + 5y^2 = -1 \\ 3x + y = -8 \end{cases}$$

(a) $(-3, 1)$ 　　　(b) $(3, 1)$ 　　　(c) $(3, -1)$ 　　　(d) $(-3, -1)$

16. Evaluate the determinant.
$$\begin{vmatrix} 0 & 2 & 1 \\ 3 & -7 & 6 \\ 2 & -6 & 4 \end{vmatrix}$$

(a) 4 　　　(b) -4 　　　(c) 32 　　　(d) -32

17. Multiply $(5x + 3)(x - 7)$.

(a) $5x^2 - 21$ 　　　(b) $5x^2 - 32x - 21$ 　　　(c) $6x - 4$ 　　　(d) $6x - 21$

18. Multiply $\left(2 + \sqrt{-25}\right)\left(-3 - \sqrt{-100}\right)$.

(a) $-6 - 5i$ 　　　(b) $-6 - 35i$ 　　　(c) $50 - 35i$ 　　　(d) $44 - 35i$

19. Factor $x^4 - 1296$.

(a) $(x + 36)(x - 36)$ 　　　(b) $(x^2 + 6)(x^2 - 6)$
(c) $(x^2 + 36)(x + 6)(x - 6)$ 　　　(d) $(x^2 + 216)(x^2 - 216)$

For questions 20–22, perform the indicated operations.

20. $\dfrac{3x - 4}{x + 7} \cdot \dfrac{x^3 + 343}{9x^2 - 16}$

(a) $\dfrac{x^2 - 7x + 49}{3x + 4}$ 　　(b) $\dfrac{x^2 + 7x + 49}{3x + 4}$ 　　(c) $\dfrac{3x + 4}{x^2 - 7x + 49}$ 　　(d) $\dfrac{3x + 4}{x^2 + 7x + 49}$

21. $\dfrac{a-4b}{a^3+27b^3} \div \dfrac{a^2-16b^2}{a^2-3ab+9b^2}$

    (a) $\dfrac{1}{(a+3b)(a+4b)}$ **(b)** $(a+3b)(a+4b)$ **(c)** $\dfrac{1}{(a-3b)(a+4b)}$ **(d)** $(a-3b)(a+4b)$

22. $(8+i) \div (2-i)$

    (a) $4-i$ **(b)** 4 **(c)** $2-3i$ **(d)** $3+2i$

23. Use long division to find $(12x^2+11x+3) \div (4x+1)$.

    (a) $3x+2+\dfrac{1}{4x+1}$ **(b)** $3x+2+\dfrac{5}{4x+1}$ **(c)** $3x+2+1$ **(d)** $3x+\dfrac{7}{2}$

24. Simplify $\sqrt[4]{\dfrac{162x^{12}}{y^9z^{20}}}$.

    (a) $\dfrac{40.5x^8}{y^5z^{16}}$ **(b)** $\dfrac{9x^8\sqrt[4]{2}}{y^5z^{16}}$ **(c)** $\dfrac{3x^3\sqrt[4]{2y^3}}{y^3z^5}$ **(d)** $\dfrac{3x^3}{y^2z^5}$

25. Simplify $2\sqrt{75} - \sqrt{48} + \sqrt{12}$.

    (a) $6\sqrt{3}$ **(b)** $8\sqrt{3}$ **(c)** $12\sqrt{3}$ **(d)** $2\sqrt{39}$

26. Write $\log_5 x = -2$ in exponential form, then solve for x.

    (a) $-2^5 = x; x = -32$ **(b)** $(-2)^5 = x; x = 32$ **(c)** $5^{-2} = x; x = -25$ **(d)** $5^{-2} = x; x = \dfrac{1}{25}$

27. Write as the logarithm of a single expression: $-3\log_7(x-1) + 5\log_7 x + 7\log_7(x+1)$

    (a) $\log_7 \dfrac{(x-1)^3}{x^5(x+1)^7}$ **(b)** $\log_7(9x+10)$

    (c) $\log_7 \dfrac{x^5(x+1)^7}{(x-1)^3}$ **(d)** $\log_7[(x-1)^{-3} + x^5 + (x+1)^7]$

28. Solve for x: $\log_4(7x+1) - \log_4 5 = \log_4(2x-1)$.

    (a) $x = -2$ **(b)** $x = 2$ **(c)** $x = -1$ **(d)** $x = 1$

29. Evaluate the sum of the series $\displaystyle\sum_{n=1}^{3}(n^2-4)$.

    (a) 2 **(b)** 3 **(c)** 5 **(d)** 8

30. Find the sum of the infinite geometric series: $6 - 4 + \dfrac{8}{3} - \dfrac{16}{9} + \dfrac{32}{27} - \cdots$

    (a) 18 **(b)** 2 **(c)** $\dfrac{18}{5}$ **(d)** 10

Final Exam Form G *(continued)*

Name:_____

For questions 31–34, find the equation or inequality that best matches the given graph.

31.

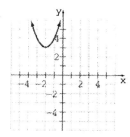

(a) $y = -(x-2)^2 + 3$ **(b)** $y = -(x+2)^2 + 3$ **(c)** $y = (x-2)^2 + 3$ **(d)** $y = (x+2)^2 + 3$

32.

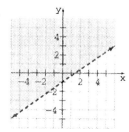

(a) $y \geq \dfrac{2}{3}x - 1$ **(b)** $y > \dfrac{2}{3}x - 1$ **(c)** $y \geq \dfrac{3}{2}x - 1$ **(d)** $y > \dfrac{3}{2}x - 1$

33.

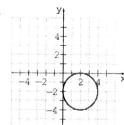

(a) $(x-2)^2 + (y+2)^2 = 4$ **(b)** $(x-2)^2 + (y+2)^2 = 2$ **(c)** $(x+2)^2 + (y-2)^2 = 4$ **(d)** $(x+2)^2 + (y-2)^2 = 2$

34.

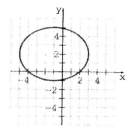

(a) $\dfrac{(x-1)^2}{16} + \dfrac{(y+2)^2}{9} = 1$ **(b)** $\dfrac{(x+1)^2}{16} + \dfrac{(y-2)^2}{9} = 1$

(c) $\dfrac{(x-1)^2}{16} - \dfrac{(y+2)^2}{9} = 1$ **(d)** $\dfrac{(x+2)^2}{16} - \dfrac{(y-2)^2}{9} = 1$

35. Find the center of the ellipse $x^2 + 4y^2 + 16y + 9 = 2x$.

(a) $(-1, 2)$ **(b)** $(1, 2)$ **(c)** $(1, -2)$ **(d)** $(-1, -2)$

36. Find the area of the ellipse in problem 35.

(a) 8π (b) 16π (c) 4π (d) 2π

37. If the middle–sized angle of a triangle is 30° greater than the measure of the smallest angle, and the largest angle is 10° less than six times the measure of the smallest angle, find the measures of the three angles.

(a) $20°, 50°, 110°$ (b) $30°, 60°, 90°$ (c) $35°, 65°, 80°$ (d) $40°, 50°, 90°$

38. Find the length and the midpoint of the line segment whose endpoints are $(6, -1)$ and $(-6, 5)$.

(a) $6\sqrt{5}; (0,2)$ (b) $2; (6,3)$ (c) $\sqrt{18}; (0,3)$ (d) $4; (0,2)$

39. Find the fourth term in the expansion of $(3a - 4b)^8$.

(a) $-672a^5b^3$ (b) $-870,912a^5b^3$ (c) $-56a^5b^3$ (d) $-54,432a^5b^3$

40. In an arithmetic sequence, $a_4 = 2$ and $a_9 = -23$; find a_1 and d.

(a) $a_1 = 17$ and $d = 5$ (b) $a_1 = -17$ and $d = -5$ (c) $a_1 = 17$ and $d = -5$ (d) $a_1 = -17$ and $d = 5$

Final Exam Form H

Name:_____

Date:_____

Choose the correct answer to each problem.

1. Which of the following sets consists of only integers?

   (a) $\left\{-4, -\dfrac{12}{8}, -\sqrt{16}, 5\right\}$
   (b) $\left\{-4, -\dfrac{12}{3}, -\sqrt{16}, 5\right\}$
   (c) $\left\{-4, -\dfrac{12}{8}, \sqrt{-16}, 5\right\}$
   (d) $\left\{-4, -\dfrac{12}{13}, \sqrt{-16}, 5\right\}$

2. Which of the following sets consists of only imaginary numbers?

   (a) $\left\{-3i, -\sqrt{5}, 2i\right\}$
   (b) $\left\{-3i, -\sqrt{5}, 2i-8\right\}$
   (c) $\left\{-3i, \sqrt{-5}, 2i\right\}$
   (d) $\left\{-3i, \sqrt{-5}, -\sqrt{8}\right\}$

3. Find $A \cap B$ for A = {integers} and B = {rational numbers}.

   (a) {real numbers}
   (b) {integers}
   (c) {rational numbers}
   (d) { }

4. Solve the equation: $4(7 - x) = 2(3x + 1) + 36$

   (a) $x = -1$
   (b) $x = 1$
   (c) $x = -6$
   (d) $x = 6$

5. Solve $\sqrt{4x+9} = \sqrt{5}x$.

   (a) $x = \dfrac{9}{5}$
   (b) $x = -1$
   (c) $x = \dfrac{9}{5}, -1$
   (d) $x = \dfrac{5}{9}, -1$

6. Solve $x^4 - 24x^2 = 25$.

   (a) $x = 25, -1$
   (b) $x = \pm 5, i$
   (c) $x = \pm 5, \pm i$
   (d) $x = -25, 1$

7. Solve the equation $V = 2\pi r^2 h$ for the variable r.

   (a) $r = \dfrac{V}{2\pi h}$
   (b) $r = \left(\dfrac{V}{2\pi h}\right)^2$
   (c) $r = 2\pi hV$
   (d) $r = \pm\dfrac{\sqrt{2\pi hV}}{2\pi h}$

8. Solve the equation $x^2 - 13x + 36 = 0$ using the quadratic formula.

   (a) $x = \dfrac{13 \pm \sqrt{133}}{2}$
   (b) $x = 4, 9$
   (c) $x = 6, 9$
   (d) $x = \dfrac{-13 \pm \sqrt{133}}{2}$

9. Solve the equation: $\dfrac{x-2}{x+2} = \dfrac{2x+5}{x+3}$.

   (a) 4
   (b) -4
   (c) $-4, 4$
   (d) $-2, 4$

10. Find the solution set to the inequality $|8x + 5| < 3$.

    (a) $\left\{x \,\middle|\, -1 < x < -\dfrac{1}{4}\right\}$
    (b) $\left\{x \,\middle|\, -1 \le x \le -\dfrac{1}{4}\right\}$
    (c) $\left\{x \,\middle|\, x < -1 \text{ or } x > -\dfrac{1}{4}\right\}$
    (d) $\left\{x \,\middle|\, x \le -1 \text{ or } x \ge -\dfrac{1}{4}\right\}$

11. Solve the inequality $x^2 + 7x + 10 \ge 0$. Write the answer in interval notation.

    (a) $(2, 5)$
    (b) $(-\infty, 2] \cup [5, \infty)$
    (c) $[-5, -2]$
    (d) $(-\infty, -5] \cup [-2, \infty)$

Final Exam Form H *(continued)* · Name:_____

12. Find an equation of the line through (–2, 2) that is parallel to the graph of $y = \frac{1}{2}x - 4$.

   (a) $y = \frac{1}{2}x + 3$ **(b)** $y = \frac{1}{2}x + 1$ **(c)** $y = -2x - 2$ **(d)** $y = -2x + 6$

13. Solve the system of equations using any method.
$$\begin{cases} 2x + y = 1 \\ x - y = 5 \end{cases}$$

   (a) $(-1, 3)$ **(b)** $(2, -3)$ **(c)** $(-2, 3)$ **(d)** $(1, -3)$

14. Solve the system of equations using any method.
$$\begin{cases} x + y + z = 4 \\ 2x - y + z = 4 \\ x + 2y - z = 3 \end{cases}$$

   (a) $(1, 2, 1)$ **(b)** $(2, 1, 1)$ **(c)** $(1, 1, 2)$ **(d)** $(3, 1, 0)$

15. Solve the system of equations using any method. Which of the following is a solution to the system?
$$\begin{cases} 2x + 5y^2 = 8 \\ 3x - y = 12 \end{cases}$$

   (a) $(4, 0)$ **(b)** $(-6, 2)$ **(c)** $(0, -4)$ **(d)** $(5, 3)$

16. Evaluate the determinant.
$$\begin{vmatrix} 5 & -2 & 0 \\ -1 & 6 & 3 \\ 1 & 2 & 1 \end{vmatrix}$$

   (a) -8 **(b)** 8 **(c)** -4 **(d)** 4

17. Multiply $(x + 7)(3x - 4)$.

   (a) $3x^2 - 28$ **(b)** $3x^2 + 21x - 28$ **(c)** $3x^2 + 17x - 28$ **(d)** $4x + 3$

18. Multiply $\left(10 - \sqrt{-4}\right)\left(2 + \sqrt{-49}\right)$.

   (a) $34 + 66i$ **(b)** $20 + 66i$ **(c)** $20 - 14i$ **(d)** $12 + 5i$

19. Factor $27a^3 - 512$.

   (a) $(3a + 8)(9a^2 - 24a + 64)$ **(b)** $(3a - 8)(9a^2 + 24a + 64)$
   (c) $(3a - 8)(9a^2 - 24a + 64)$ **(d)** $(3a + 8)(9a^2 + 24a + 64)$

Final Exam Form H *(continued)*

Name:_____

For questions 20–22, perform the indicated operations.

20. $\dfrac{6x-5}{x+1} \cdot \dfrac{x^3+1}{36x^2-25}$

    (a) $\dfrac{6x+5}{x^2+x+1}$
    (b) $\dfrac{6x+5}{x^2-x+1}$
    (c) $\dfrac{x^2+x+1}{6x+5}$
    (d) $\dfrac{x^2-x+1}{6x+5}$

21. $\dfrac{a+2b}{27a^3+8b^3} \div \dfrac{a^2-4b^2}{3a+2b}$

    (a) $\dfrac{1}{(a-2b)(9a^2-6ab+4b^2)}$
    (b) $\dfrac{1}{(a-2b)(9a^2+6ab+4b^2)}$

    (c) $(a-2b)(9a^2-6ab+4b^2)$
    (d) $(a-2b)(9a^2+6ab+4b^2)$

22. $(23+i) \div (7-2i)$

    (a) $\dfrac{24}{5}i$
    (b) $\dfrac{23}{7} - \dfrac{1}{2}i$
    (c) $3+i$
    (d) $5-2i$

23. Use long division to find $(15x^2-7x+7) \div (5x+1)$.

    (a) $3x-2$
    (b) $3x-\dfrac{4}{5}$
    (c) $3x-\dfrac{4}{5}+\dfrac{39}{25x+5}$
    (d) $3x-2+\dfrac{9}{5x+1}$

24. Simplify $\sqrt[3]{\dfrac{-8x^6}{y^{12}z^{15}}}$.

    (a) $\dfrac{2x^2}{y^4z^5}$
    (b) $\dfrac{-2x^2}{y^4z^5}$
    (c) $\dfrac{2x^3}{y^9z^{12}}$
    (d) $\dfrac{-2x^3}{y^9z^{12}}$

25. Simplify $\sqrt{45} - \sqrt{80} + \sqrt{180}$.

    (a) $3\sqrt{5}$
    (b) $4\sqrt{5}$
    (c) $5\sqrt{5}$
    (d) $6\sqrt{5}$

26. Write $\log_4 x = -3$ in exponential form, then solve for x.

    (a) $4^{-3}=x;\ x=\dfrac{1}{64}$
    (b) $4^{-3}=x;\ x=-64$
    (c) $(-3)^4=x;\ x=81$
    (d) $-3^4=x;\ x=-81$

27. Write as the logarithm of a single expression: $2\log_5(x+3)+5\log_5(2x)-7\log_5(x-1)$

    (a) $\log_5[(x+3)^2+(2x)^5-(x-1)^7]$
    (b) $\log_5(5x+13)$

    (c) $\log_5 \dfrac{(x+3)^2(2x)^5}{(x-1)^7}$
    (d) $\log_5 \dfrac{(x-1)^7}{(x+3)^2(2x)^5}$

28. Solve for x: $\log_3(2x-7) - \log_3 3 = \log_3(x-4)$

    (a) 6
    (b) 2
    (c) 4
    (d) 5

Name:_____

29. Solve for x: $27^{2x+2}=81^{x+5}$

    (a) 3   **(b)** 7   **(c)** 13   **(d)** 4

30. Evaluate the sum of the series $\displaystyle\sum_{n=1}^{3}(4n+2)$.

    (a) 16   **(b)** 24   **(c)** 30   **(d)** 48

31. Find the sum of the infinite geometric series: $4+2+1+\dfrac{1}{2}+\dfrac{1}{4}+\cdots$

    (a) 2   **(b)** 4   **(c)** 8   **(d)** 16

32. Find the sum of the infinite geometric series: $4-2+1-\dfrac{1}{2}+\dfrac{1}{4}-\cdots$

    (a) 2   **(b)** $\dfrac{1}{2}$ **(c)** $\dfrac{8}{3}$ **(d)** 4

For questions 33–36, find the equation that best matches the given graph.

33.

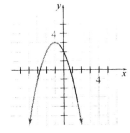

    (a) $y=(x-1)^2+3$ **(b)** $y=(x+1)^2+3$ **(c)** $y=-(x-1)^2+3$ **(d)** $y=-(x+1)^2+3$

34.

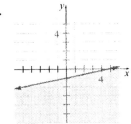

    (a) $y\le 5x-1$ **(b)** $y\le\dfrac{1}{5}x-1$ **(c)** $y\ge 5x-1$ **(d)** $y\ge\dfrac{1}{5}x-1$

35.

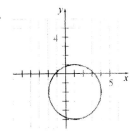

    (a) $(x-1)^2+(y+2)^2=9$ **(b)** $(x-1)^2+(y+2)^2=3$ **(c)** $(x+1)^2+(y-2)^2=9$ **(d)** $(x+1)^2+(y-2)^2=3$

36.

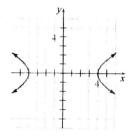

(a) $\dfrac{x^2}{4}+\dfrac{y^2}{16}=1$

(b) $\dfrac{x^2}{4}-\dfrac{y^2}{16}=1$

(c) $\dfrac{x^2}{16}+\dfrac{y^2}{4}=1$

(d) $\dfrac{x^2}{16}-\dfrac{y^2}{4}=1$

37. Find the length and the midpoint of the line segment with endpoints $(7, 3)$ and $(-5, 9)$.

(a) $18;\ (6,-3)$

(b) $6\sqrt{5};\ (1,6)$

(c) $2\sqrt{10};\ (1,6)$

(d) $6\sqrt{5};\ (6,-3)$

38. If $\$11,000$ is invested at 8.5% compounded continuously, how long would it take the value of the account to double?

(a) 8.2 years

(b) 9.1 years

(c) 10.4 years

(d) 18.3 years

39. Find the fourth term in the expansion of $(2a-5b)^8$.

(a) $-56a^5b^3$

(b) $-560a^5b^3$

(c) $-224,000a^5b^3$

(d) $-8960a^5b^3$

40. In an arithmetic sequence, $a_4=-1.5$ and $a_7=-6$; find a_1 and d.

(a) $a_1=-3$ and $d=1.5$

(b) $a_1=3$ and $d=1.5$

(c) $a_1=-3$ and $d=-1.5$

(d) $a_1=3$ and $d=-1.5$

Chapter 1 Answers

Chapter 1 Pretest Form A

1. $\left\{-12, 31, 0, \dfrac{3}{4}, -\dfrac{1}{8}, 2.25\right\}$

2. $\{-12, 31, 0\}$

3. $\left\{-\sqrt{3}, \sqrt{7}\right\}$

4. $<$

5. $>$

6. -12

7. -22

8. $1\dfrac{1}{4}$

9. $\dfrac{2}{7}$

10. 24

11. $\dfrac{5}{6}$

12. 0

13. 1000

14. $3^5 \cdot 5^2 \cdot y^3 \cdot z^2$

15. 100

16. 29

17. commutative, multiplication

18. distributive

19. $4.16

20. $73.45

Chapter 1 Pretest Form B

1. $\left\{-4, -\sqrt{25} = -5, 0, 15\right\}$

2. $\{15\}$

3. $\left\{\sqrt{6}\right\}$

4. $<$

5. $<$

6. -3

7. -12

8. $\dfrac{9}{20}$

9. $\dfrac{1}{5}$

10. -40

11. $\dfrac{11}{14}$

12. undefined

13. 64

14. $11^3 \cdot m^2 \cdot n^4$

15. 4

16. 8

17. commutative, addition

18. associative, multiplication

19. $1.72

20. $26.25

Additional Exercises 1.2

1. $135,000

2. $69°$

3. 86

4. $61.75

5. $36,250

6. $2700

7. $1865.30

8. $40

9. 24 mpg

10. $18,194.50

11. $37,200

12. $11

13. $924

14. $756

15. $546

16. $252

17. $1848

18. 78

19. $23.75

20. $12.16

Chapter 1 Answers

Additional Exercises 1.3

1. $\dfrac{2}{5}$

2. $\dfrac{2}{3}$

3. $\dfrac{2}{5}$

4. $4\dfrac{1}{6}$

5. $\dfrac{77}{9}$

6. $\dfrac{1}{10}$

7. $\dfrac{1}{21}$

8. $\dfrac{3}{143}$

9. $\dfrac{7}{81}$

10. $6\dfrac{3}{4}$

11. $4\dfrac{4}{5}$

12. $\dfrac{30}{7}$ or $4\dfrac{2}{7}$

13. $\dfrac{27}{14}$ or $1\dfrac{13}{14}$

14. $1\dfrac{1}{24}$

15. $8\dfrac{3}{8}$

16. $\dfrac{14}{15}$

17. $\dfrac{1}{3}$

18. $\dfrac{2}{7}$

19. $4\dfrac{8}{15}$

20. $2\dfrac{5}{8}$ inches

Additional Exercises 1.4

1. $13, 0, -5$

2. $0.5, 13, 0, -5, \dfrac{1}{4}$

3. $-\sqrt{13}, \sqrt{7}$

4. $40, 0$

5. 40

6. $-\sqrt{7}, \sqrt{3}, \pi$

7. $-16, 40, -\sqrt{7}, 0, \sqrt{3}, \pi, \dfrac{1}{5}, -\dfrac{1}{5}, 3.25$

8. irrational numbers, real numbers

9. true

10. false

11. false

12. true

13. false

14. true

15. $-7, 0$

16. $\dfrac{5}{6}, -0.72$

17. $0, 4, \dfrac{5}{6}$

18. 0

19. π

20. $-7, \dfrac{5}{6}, -0.72$

Additional Exercises 1.5

1. 22

2. -22

3. -14

4. 6

5. $<$

6. $>$

7. $<$

8. $>$

9. $>$

10. $>$

11. <
12. <
13. >
14. =
15. =
16. <
17. =
18. >
19. <
20. <

Additional Exercises 1.6

1. 9
2. $-\dfrac{22}{3}$
3. $2\dfrac{3}{8}$
4. 3.4
5. $-\dfrac{3}{8}$
6. 53
7. -8
8. -6
9. 2
10. -22
11. -19
12. -207
13. -312
14. $-\dfrac{1}{12}$
15. $-\dfrac{13}{24}$
16. $-\dfrac{13}{30}$
17. 4.91
18. $15° + (-23°)$; $-8°$ F
19. -218 ft $+ (-54$ ft$)$; -272 ft
20. -12 yd $+ 5$ yd; -7 yd

Additional Exercises 1.7

1. -26
2. -11
3. 20
4. -149
5. -173
6. -9
7. 11
8. $-\dfrac{1}{6}$
9. $\dfrac{5}{12}$
10. $-\dfrac{1}{36}$
11. -20
12. -19
13. 5
14. -0.8
15. -1
16. -2
17. 11
18. $28°$
19. 170 ft
20. 121 ft

Additional Exercises 1.8

1. 35
2. -24
3. -13.34
4. 24
5. -30
6. $\dfrac{16}{27}$
7. $-\dfrac{3}{22}$
8. $\dfrac{54}{11}$ or $4\dfrac{10}{11}$
9. -8
10. -6

Chapter 1 Answers

11. -0.5

12. 69

13. 74

14. $\dfrac{9}{10}$

15. $-\dfrac{64}{71}$

16. 4

17. -60

18. 6

19. undefined

20. 0

Additional Exercises 1.9

1. 32

2. 81

3. -100

4. -625

5. 9

6. -343

7. $\dfrac{8}{27}$

8. $\dfrac{25}{36}$

9. 48

10. -9

11. 412

12. 28

13. 7

14. -361

15. 659

16. 748

17. 484

18. 49

19. $\dfrac{126}{23}$ or $5\dfrac{11}{23}$

20. $\dfrac{110}{21}$ or $5\dfrac{5}{21}$

Additional Exercises 1.10

1. (a) 13 (b) $-\dfrac{1}{13}$

2. (a) $-\dfrac{7}{9}$ (b) $\dfrac{9}{7}$

3. (a) -2.5 (b) 0.4 or $\dfrac{2}{5}$

4. (a) $\dfrac{5}{16}$ (b) $-\dfrac{16}{5}$

5. commutative, addition

6. associative, addition

7. associative, multiplication

8. distributive

9. $5(-3)$

10. $3 + (-7 + 5)$

11. $8x + 8y$

12. $(-8) + 10$

13. $(-5 \cdot 2)8$

14. $4y + 4x + 12$

15. $(5 + 3) \cdot 8$

16. $-5(xy)$

17. $(x + 6)7$

18. $2x + (7 + 6)$

19. $6(8 + x)$

20. $10x + 10y + 70$

Chapter 1 Test Form A

1. $\$6.08$

2. 79

3. 78

4. 12

5. $\{-4, 0\}$

6. $<$

7. $=$

8. -11

Chapter 1 Answers

9. $-\dfrac{13}{20}$

10. -5

11. -36

12. $\dfrac{15}{16}$

13. -6

14. -9

15. undefined

16. 2

17. -8

18. 80

19. 13

20. 2

21. 32

22. -11

23. 0

24. -1

25. commutative, multiplication

Chapter 1 Test Form B

1. 35

2. $5.57

3. 42

4. 6

5. $\{-3, 0\}$

6. $<$

7. $>$

8. -4

9. 13

10. -6

11. 120

12. $-\dfrac{10}{27}$

13. $-\dfrac{20}{9}$

14. -16

15. 0

16. 8

17. -4

18. $-\dfrac{11}{24}$

19. -26

20. -9

21. -3

22. -12

23. 2

24. 7

25. distributive

Chapter 1 Test Form C

1. $297.50

2. 1230 ft

3. 60

4. 30

5. $\left\{-\dfrac{3}{8}, 0, 3.7, 1\right\}$

6. $<$

7. $<$

8. -18

9. -5

10. $-\dfrac{1}{24}$

11. -30

12. $\dfrac{6}{55}$

13. -9

14. -81

15. undefined

16. 10

17. 19

18. $-\dfrac{1}{64}$

Chapter 1 Answers

19. −15

20. −3

21. 3

22. −1

23. 11

24. 28

25. associative, addition

Chapter 1 Test Form D

1. $18,550

2. 135

3. 72

4. 156

5. $\left\{\sqrt{5}, -\sqrt{3}\right\}$

6. >

7. <

8. −8

9. $-\dfrac{1}{6}$

10. −39

11. −330

12. $-\dfrac{9}{25}$

13. −8

14. $-\dfrac{2}{7}$

15. 0

16. −54

17. −25

18. −12

19. −168

20. −6

21. −109

22. 5

23. 8

24. −2

25. associative, multiplication

Chapter 1 Test Form E

1. $264

2. 5802.5 computers per month

3. $\left\{\sqrt{4} = 2, 8\right\}$

4. <

5. =

6. −19

7. 34

8. 1

9. 42

10. $-\dfrac{2}{7}$

11. 22

12. −50

13. undefined

14. 28

15. 8

16. $\dfrac{16}{81}$

17. −114

18. −6

19. −79

20. −81

21. $2\dfrac{13}{24}$

22. −7

23. 12

24. 16

25. commutative, addition

Chapter 1 Test Form F

1. 97 miles per day

2. $1680

3. $\left\{3, -\dfrac{7}{3}, -0.5\right\}$

4. <

5. >

6. −5

624

Chapter 1 Answers

7. 15

8. −6

9. 44

10. $\dfrac{6}{35}$

11. 3

12. −16

13. undefined

14. 18

15. 6

16. $\dfrac{9}{64}$

17. −6

18. 0

19. −5

20. −144

21. $2\dfrac{11}{20}$

22. 9

23. 80

24. −60

25. distributive

Chapter 1 Test Form G

1. d
2. b
3. a
4. d
5. a
6. d
7. c
8. b
9. b
10. a
11. d
12. c
13. d
14. d
15. c

16. b
17. d
18. a
19. a
20. d
21. c
22. b
23. d
24. c
25. a .

Chapter 1 Test Form H

1. c
2. c
3. d
4. a
5. a
6. b
7. d
8. c
9. d
10. b
11. d
12. a
13. c
14. a
15. d
16. b
17. c
18. b
19. a
20. b
21. c
22. d
23. b
24. c
25. a

Chapter 2 Answers

Chapter 2 Pretest Form A

1. $56x - 32y$
2. $-3x - 5y + 9$
3. $-5x + 3$
4. $5x - 8y$
5. $-10x - 13$
6. $\dfrac{5}{3}m + \dfrac{8}{5}n - \dfrac{3}{2}$
7. $2x + 17y$
8. $x = 15$
9. $x = 54$
10. $x = 8$
11. $x = 7$
12. $x = -4$
13. $x = 6$
14. no solution
15. $m = 10$
16. $x = zs + m$
17. $x = 6$
18. $\dfrac{14}{25}$
19. $3\dfrac{3}{4}$ pounds
20. 72 parts

Chapter 2 Pretest Form B

1. $54 - 36x$
2. $-6a + 4b - 14$
3. $-7x + 10$
4. $\dfrac{1}{12}x + \dfrac{32}{7}$
5. $3 + 2x$
6. $16x + 8y - 8$
7. $-2x + 13y$
8. $m = 23$
9. $x = 125$
10. $y = 2$
11. $x = 4$

12. $x = 5$
13. $k = 3.4$
14. no solution
15. $x = -24$
16. $y = -\dfrac{2}{3}x + 4$
17. $x = 6$
18. $\dfrac{4}{3}$
19. $1\dfrac{2}{3}$ ft
20. 112 gallons

Additional Exercises 2.1

1. $4x^3, -8x^2, -3x, 1$
2. $-2w^5, 3w^4, -w^3, -1$
3. no
4. $4x + 6y$
5. $4x - 11y - 4$
6. $-8p - 8$
7. $-x - 10y - 4$
8. $8x - 2.07$
9. $-5p^2 + 5p - 7$
10. $10x + 15y$
11. $2.4 - 0.48k$
12. $-3x - 3y + 2$
13. $\dfrac{1}{2}x - 6y + 12$
14. $10x - 3y - 9.5$
15. $-6x - 8y + 2$
16. $-20x + 5y + 10$
17. $-6x - 8$
18. $4a - 5b$
19. $1.8x + 1.2$
20. $-\dfrac{1}{6}x + \dfrac{4}{3}$

Chapter 2 Answers

Additional Exercises 2.2

1. yes
2. no
3. yes
4. no
5. yes
6. no
7. $x = 2$
8. $m = 35$
9. $t = 35$
10. $x = -3$
11. $x = -42$
12. $x = 7$
13. $m = 1$
14. $c = -41$
15. $x = 0$
16. $x = 66.45$
17. $x = -33$
18. $x = -0.64$
19. $x = 2.5$
20. $x = 408$

Additional Exercises 2.3

1. $x = -48$
2. $x = 12$
3. $m = 6$
4. $x = -7$
5. $x = -56$
6. $x = 14$
7. $x = -8$
8. $x = \dfrac{1}{9}$
9. $x = 5$
10. $x = -8$
11. $x = 15$
12. $x = 45$
13. $x = 9$

14. $x = 25$
15. $p = 21$
16. $x = 12$
17. $x = 144$
18. $q = -14.3$
19. $x = 25$
20. $x = 81$

Additional Exercises 2.4

1. $x = -9$
2. $x = -3$
3. $y = 16$
4. $x = 4$
5. $x = -\dfrac{1}{6}$
6. $x = -3$
7. $x = 15$
8. $x = 63$
9. $z = -23$
10. $d = \dfrac{19}{2}$
11. $y = 18$
12. $x = 2.5$
13. $x = 12$
14. $x = -10$
15. $x = -44$
16. $x = \dfrac{9}{4}$
17. $x = -\dfrac{13}{8}$
18. $x = 4$
19. $n = \dfrac{9}{40}$
20. $x \approx -2.3$

Chapter 2 Answers

Additional Exercises 2.5

1. $x = -6$
2. $x = -1$
3. $x = -\dfrac{5}{2}$
4. $c = -5$
5. $x = -\dfrac{1}{4}$
6. $x = -4$
7. $x = \dfrac{17}{9}$
8. no solution
9. $m = 3$
10. all real numbers
11. $x = \dfrac{12}{11}$
12. no solution
13. no solution
14. $m = -\dfrac{3}{2}$
15. all real numbers
16. no solution
17. all real numbers
18. no solution
19. $x = -5$
20. $x = 1$

Additional Exercises 2.6

1. $C = 56.52$ inches
2. $A = 19.625$ square feet
3. $P = 20$
4. $P = 11.6$
5. $h = 8$
6. $h = 4$
7. $A = \dfrac{5B + 22}{2}$
8. $y = \dfrac{5}{2}x - 2$
9. $F = \dfrac{9}{5}C + 32$
10. $y = 2x + 10$

11. $t = \dfrac{I}{Pr}$
12. $t = \dfrac{A}{6s^2}$
13. $r = \dfrac{C}{2\pi}$
14. 184 square inches
15. $\dfrac{45}{4}\pi \approx 35.343$ cubic inches
16. $80\pi \approx 251.327$ square feet
17. $552.00
18. $247.50
19. $8840
20. $42.50

Additional Exercises 2.7

1. 17:32
2. 3:2
3. 5:7
4. 13:14
5. 6:1
6. 6:7
7. $j = 12$
8. $x = -24$
9. $j = 30$
10. $x = -9$
11. 67.5
12. $12\dfrac{1}{11}$
13. 7.2
14. 6.8
15. 130 parts
16. 96 gallons
17. 7:13
18. 48 oz
19. $\dfrac{3}{4}$ ft
20. $15\dfrac{1}{4}$ or 15.25 pounds

Chapter 2 Answers

Chapter 2 Test Form A

1. $24x - 12$
2. $2x + 5y - 4$
3. $-3x + 6$
4. $3x - 3$
5. $2x - 4y$
6. $-\dfrac{3}{5}x - \dfrac{2}{21}$
7. $11x - 17$
8. $x = 2$
9. $x = 3$
10. $x = 3$
11. no solution
12. $x = 1$
13. all real numbers
14. $w = \dfrac{P - 2l}{2}$
15. $y = 6x - \dfrac{21}{2}$
16. 94.2 cubic inches
17. 1:6
18. $x = 2\dfrac{2}{5}$ in.
19. 11 eggs
20. 19 pounds

Chapter 2 Test Form B

1. $15x - 80$
2. $-4x - 6y + 10$
3. $-x + 6y - 3$
4. $4x - y + 4$
5. $-7x + 5$
6. $5x + 3$
7. $x - 1$
8. $x = 2$
9. all real numbers
10. $x = -2$
11. $x = \dfrac{2}{5}$

12. $x = 9$
13. no solution
14. $h = \dfrac{3V}{\pi r^2}$
15. $y = -\dfrac{5}{4}x + \dfrac{11}{2}$
16. 12.4 yards
17. 1:4
18. 12.5 in.
19. 30 pounds
20. 8 gallons

Chapter 2 Test Form C

1. $12x - 40$
2. $-9x + 9y - 12$
3. $6x + 6$
4. $10x - 3y - 1$
5. $17x - 9y$
6. $\dfrac{1}{4}x + \dfrac{5}{6}$
7. $x - 34$
8. $x = -3$
9. $x = 2$
10. all real numbers
11. $x = \dfrac{1}{8}$
12. no solution
13. $x = 21$
14. $z = \dfrac{x - m}{s}$
15. $y = -\dfrac{1}{4}x + \dfrac{11}{16}$
16. $r = 0.045$
17. 1:4
18. 1 ft.
19. 12 gallons
20. 585 miles

Chapter 2 Answers

Chapter 2 Test Form D

1. $2x + 3$
2. $60x - 42 + 12y$
3. $-11x - 11$
4. $-9x + 6$
5. $x - y - 2$
6. $\dfrac{1}{2}x + \dfrac{3}{4}$
7. $4x + 41$
8. $x = -11$
9. all real numbers
10. $x = 2$
11. no solution
12. $x = 2$
13. $x = 2$
14. $h = \dfrac{A - 2\pi r^2}{2\pi r}$
15. $y = 0.5x - 1.4$
16. $b = 11.2$ centimeters
17. 3:8
18. $x = 3$ inches
19. 15.1 gallons
20. 6 cups

Chapter 2 Test Form E

1. $-6x - 10$
2. $-4x + 12y + 28$
3. $6x + 2$
4. $3x - 4y + 3$
5. $2x - 7y + 4$
6. $\dfrac{5}{8}x + \dfrac{5}{24}$
7. $5x - 32$
8. $x = -2$
9. $x = 4$
10. $x = \dfrac{9}{4} = 2\dfrac{1}{4}$
11. $x = 11$

12. all real numbers
13. $x = 45$
14. $r = \dfrac{C}{2\pi}$
15. $y = -\dfrac{1}{2}x - \dfrac{14}{5}$
16. $w = 3$ feet
17. 1:3
18. $x = 7\dfrac{1}{5}$ in. (7.2 in.)
19. 20 gallons
20. 6020 sq ft

Chapter 2 Test Form F

1. $2x + 3$
2. $2x + 5y - 4$
3. $-12x + 12$
4. $6x + 2$
5. $-x + 4$
6. $-\dfrac{5}{24}x + \dfrac{1}{18}$
7. $9x + 21$
8. $x = 2$
9. all real numbers
10. $x = 3$
11. $x = 1$
12. no solution
13. $x = \dfrac{5}{2}$
14. $t = \dfrac{A - P}{Pr}$
15. $y = 2x + 5$
16. $t = 6.4$ hours
17. 3:8
18. 12 inches
19. 45 minutes
20. 115 ounces

Chapter 2 Answers

Chapter 2 Test Form G

1. d
2. c
3. a
4. a
5. b
6. b
7. a
8. a
9. c
10. c
11. b
12. d
13. b
14. c
15. b
16. d
17. a
18. a
19. d
20. d

Chapter 2 Test Form H

1. b
2. c
3. a
4. c
5. d
6. a
7. c
8. d
9. c
10. a
11. b
12. d
13. c
14. a
15. b

16. a
17. d
18. c
19. d
20. b

Chapters 1–2 Cumulative Test Form A

1. $\dfrac{2}{45}$
2. $4\dfrac{1}{5}$
3. $>$
4. 1
5. -32
6. 26
7. 8
8. 784
9. Associative property of multiplication
10. $-3x + 8$
11. $5x + y + 9$
12. $x = -1$
13. $x = -4$
14. $x = -\dfrac{3}{4}$
15. $x = \dfrac{15}{7}$
16. $x = -\dfrac{10}{9}$
17. $y = \dfrac{2}{5}x - 3$
18. 2:3
19. $926.62
20. 95 oz

Chapter 2 Answers

Chapters 1–2 Cumulative Test Form B

1. b
2. a
3. c
4. a
5. d
6. c
7. c
8. d
9. a
10. d
11. a
12. d
13. d
14. a
15. a
16. d
17. b
18. c
19. d
20. a

Chapter 3 Answers

Chapter 3 Pretest Form A

1. $x + 25$
2. $x + 0.04x$, or $1.04x$
3. $12f$
4. $\frac{1}{2}h - 3$
5. $x + 8 = 21$
6. 13
7. $x + (3x + 7) = 59$
8. 13 and 46
9. $x + (x - 3) = 17$
10. Caleb = 10; Clay = 7
11. $x + 0.15x = 230$
12. 200
13. 200 miles
14. 36, 38, 40
15. $9000 at 11%; $10,000 at 13%
16. 50 months
17. $3800
18. 5 cm by 15 cm
19. 84 liters
20. 20 hours

Chapter 3 Pretest Form B

1. $x + 5$
2. $x - 0.40x$, or $0.60x$
3. $60h$
4. $\frac{1}{3}x + 2$
5. $x - 9 = 31$
6. 40
7. $x + (2x - 13) = 50$
8. 21 and 29
9. $x + (x + 4) = 14$
10. Katie = 5; Josh = 9
11. $x + 0.30x = 33,800$
12. $26,000

13. 250 miles
14. 46, 48, and 50
15. $9500 at 11% and $12,500 at 13%
16. 24 months
17. 8 nickels, 6 dimes, and 9 quarters
18. 5 inches by 12 inches
19. 14 liters
20. 3.5 hours

Additional Exercises 3.1

1. $\frac{15}{x}$
2. $x - 50$
3. $C - 0.14C$
4. $P + 0.12P$
5. $\frac{1}{2}w + 35$
6. $5x - \frac{1}{4}$
7. $5n + 10d$
8. $60m + s$
9. $2d + 120$
10. $3w - 8$
11. $4x - 13 = 11$
12. $4x + 14 = 42$
13. $x - 0.12x = 200$
14. $x - 0.15x = 180$
15. $x(6x + 9) = 81$
16. $x + 4x = 24$
17. $x + (x + 7) = 13$
18. $x + (x + 1) = 51$
19. $x + (x + 8) = 15$
20. $x + 3x = 24$

Additional Exercises 3.2

1. 48
2. -25
3. 12
4. -1

Chapter 3 Answers

5. 85

6. 18 and 48

7. 38

8. 19 and 69

9. 43 months

10. 28 months

11. 7 ten-dollar bills and 4 twenty-dollar bills

12. 16 nickels, 32 dimes, and 20 quarters

13. 12 hours

14. $625,000

15. $23,600

16. $6000

17. $35,000

18. $58,360

19. 7375 students

20. 425,000 people

Additional Exercises 3.3

1. $A = 57°$, $B = 33°$

2. $A = 69°$, $B = 21°$

3. $A = 42°$, $B = 138°$

4. $A = 65°$, $B = 115°$

5. 36°, 46°, 98°

6. smaller angles 75°, larger angles 105°

7. 50 ft by 70 ft

8. 10 in; 34 in; 30 in

9. 8 in

10. 60 ft by 40 ft

11. 50 ft by 94 ft

12. 18 m by 9 m

13. 8 in, 39 in, 24 in

14. 10 cm

15. width = 3.5 ft; height = 6 ft

16. width = 3.75 ft; height = 8 ft

17. 90 ft by 90 ft

18. 280 m

19. 7 ft by 10 ft

20. 4.5 yds

Additional Exercises 3.4

1. 15 gallons

2. 2340 parts

3. 37 miles per hour

4. 36 miles per hour

5. 17 gallons

6. 330 miles

7. 8 hours

8. 200 miles

9. 16 hours

10. 12:24 p.m.

11. commuter = 25 mph; express = 50 mph

12. $8000 at 11%; $11,000 at 12%

13. $14,000 at 6%; $10,000 at 9%

14. 80 liters

15. 8 lb

16. 3 lb

17. 180 ml of 10%; 420 ml of 40%

18. 5 private students; 8 group students

19. 7 lb

20. 8 lb

Chapter 3 Test Form A

1. $x + 35$

2. $1200 - m$

3. $(3n + 2) + n$

4. $x + 0.065x$, or $1.065x$

5. $2(b + 8) + b$

6. 38 and 83

7. 34 and 35

8. 40 and 63

9. $600

10. $42.00

11. 8950 students

12. 100 miles

13. 24 cuttings

14. 20°, 60°, and 100°

15. 12 cm by 32 cm

16. 75°, 75°, 105°, 105°

17. $6000 at 8%; $18,000 at 10%

18. 55 mph and 65 mph

19. 3 gallons

20. 2.5 liters of 12%, 1.5 liters of 20%

Chapter 3 Test Form B

1. $x + 0.55$

2. $2t - 10$

3. $\dfrac{1}{3}h$

4. $155 - p$

5. $n + \left(\dfrac{1}{4}n - 15 \right)$

6. 80 and 22

7. 42 and 44

8. 48 and 79

9. $299.80

10. $71.40

11. 1150 crimes

12. 225 miles

13. 15 cuttings

14. 15 cm, 45 cm, and 45 cm

15. 10 ft by 22 ft

16. 81°, 81°, 99°, 99°

17. $29,000 at 11%; $9000 at 9%

18. 50 mph

19. 8 liters

20. 4 liters of 12%, 12 liters of 24%

Chapter 3 Test Form C

1. $3x$

2. $2a - 7$

3. $8 - r$

4. $g + 0.12$

5. $n + (5n + 1)$

6. 41 and 90

7. 11, 12, and 13

8. 31 and 48

9. $275

10. $149.60

11. 11,750 students

12. 500 minutes

13. $1.08

14. 70°, 68°, and 42°

15. 12 inches by 28 inches

16. 25°, 25°, 155°, 155°

17. $17,500 at 9%; $13,500 at 12%

18. 45 mph and 60 mph

19. 2 liters

20. 6 gallon of 12%, 3 gallons of 18%

Chapter 3 Test Form D

1. $1869 - p$

2. $3a + 19$

3. $3a + 1$

4. $\dfrac{x}{6}$.

5. $n + \left(\dfrac{2}{3}n + 4 \right)$

6. 28 and 62

7. 35 and 37

8. 12 and 40

9. $800

10. 225 pounds

11. 48,275 people

12. 20 sodas

13. 200 miles

14. 18°, 54°, and 108°

15. 12 inches, 36 inches, and 36 inches

16. 5 inches by 33 inches

17. $4500 at 11%, $13,500 at 14%

18. 55 mph and 60 mph

19. 4 liters

20. 5.25 liters of 4%, 2.75 liters of 36%

Chapter 3 Answers

Chapter 3 Test Form E

1. $x - 23$

2. $8270 - f$

3. $1.5r$

4. $x + 0.075x$, or $1.075x$

5. $\dfrac{x}{12}$

6. 28 and 89

7. 33 and 34

8. 20 and 42

9. $300

10. $32.48

11. $25.40

12. 250 miles

13. 20 feet

14. $90°$, $45°$, and $45°$

15. 17 cm by 10 cm

16. $77°$, $77°$, $103°$, and $103°$,

17. $5500 at 10%; $7500 at 8%

18. Car A = 68 mph, Car B = 56 mph

19. 2 liters

20. 5 liters of 12%, 3 liters of 20%

Chapter 3 Test Form F

1. $x + 0.90$

2. $3a - 9$

3. $\dfrac{1}{4}h$

4. $72 - p$

5. $n + \left(\dfrac{1}{3}n - 8 \right)$

6. 12 and 41

7. 87 and 89

8. 15 and 75

9. $98.00

10. $92.80

11. 1020 crimes

12. 320 miles

13. 16 cuttings

14. $35°$, $70°$, and $75°$

15. 34 mm by 65 mm

16. $36°$, $36°$, $144°$, and $144°$

17. $3000 at 7%; $8000 at 9%

18. 62 mph

19. 2.5 liters

20. 150 milliliters of 8%, 1050 milliliters of 24%

Chapter 3 Test Form G

1. b

2. d

3. a

4. a

5. b

6. c

7. c

8. d

9. d

10. b

11. a

12. b

13. d

14. b

15. b

16. c

17. b

18. c

19. b

20. d

Chapter 3 Test Form H

1. b

2. d

3. c

4. b

5. a

6. d

7. b

8. d

9. d

10. c

11. a

12. b

13. b

14. c

15. a

16. c

17. a

18. d

19. c

20. a

Chapter 4 Answers

Chapter 4 Pretest Form A

1. II
2. III
3. yes
4. no
5. $-\dfrac{7}{8}$
6. $\dfrac{2}{3}$
7. $m = 2;\ b = -2$
8.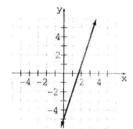
9.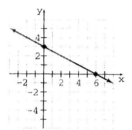
10.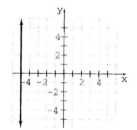
11. $y = -\dfrac{3}{2}x + 3$
12. parallel
13. neither
14. $y = 8x - 3$
15. $y = 5x + 1$
16. $y = -\dfrac{5}{3}x + 1$
17. $y = -3$
18. $y = -2x + 3$
19. $3x + 4y = -40$
20. undefined

Chapter 4 Pretest Form B

1. III
2. IV
3. yes
4. no
5. $-\dfrac{9}{4}$
6. $\dfrac{7}{3}$
7. $m = 2;\ b = 8$
8.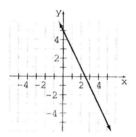
9.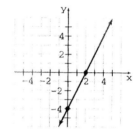
10.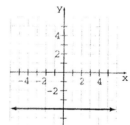
11. $y = -\dfrac{2}{3}x + 2$
12. neither
13. perpendicular
14. $y = 3x - 4$
15. $y = -4x - 5$
16. $y = -4x + 9$
17. $x = 5$
18. $y = -\dfrac{1}{2}x + \dfrac{9}{2}$
19. $2x - 5y = -15$
20. 0

Chapter 4 Answers

Mini-Lecture 4.1

1a)-1f)

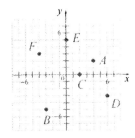

Mini-Lecture 4.2

1a)

1b)

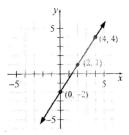

2a)

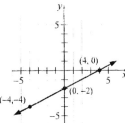

2b)

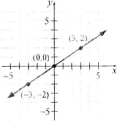

3a)

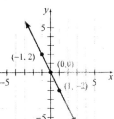

3b)

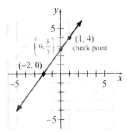

4a)

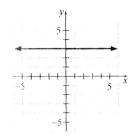

4b)

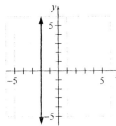

5b)

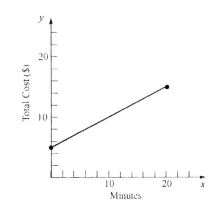

Mini-Lecture 4.3

3a)

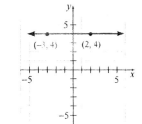

639

Chapter 4 Answers

3b)

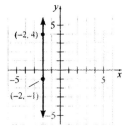

3c)

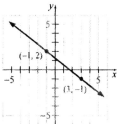

Additional Exercises 4.1

1. II

2. IV

3. I

4. III

5. (4, 3)

6. (−2, 5)

7. (−1, −6)

8. (0, −2)

9. (−4, 0)

10. (5, −3)

11–14.

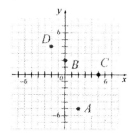

15. Yes

16. No

17. Yes

18. $u = -9$

19. No

20. No

Additional Exercises 4.2

1. 4

2. 0

3. −3

4. 0

5. 13

6. 4

7.

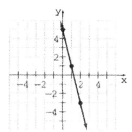

8.

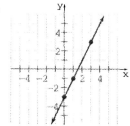

9.

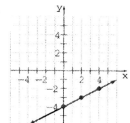

10.

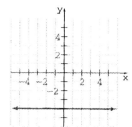

11.

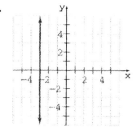

Chapter 4 Answers

12.

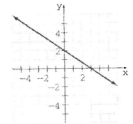

13.

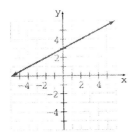

14.

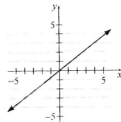

15. $x = 1$

16. $y = -5$

17. $c = 120 + 0.50m$

18.

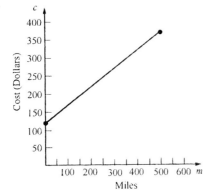

19. \$220

20. 280 miles

Additional Exercises 4.3

1. $\dfrac{2}{9}$

2. $\dfrac{1}{3}$

3. 2

4. -2

5. Undefined

6. $\dfrac{1}{5}$

7. $-\dfrac{4}{5}$

8. 0

9. 5

10. $\dfrac{4}{5}$

11. 7

12. 2

13. -3

14. -2

15. 3

16. Undefined

17. 0

18. $u = 6$

19. $u = 3$

20. $u = 3$

Additional Exercises 4.4

1. $m = -1; b = 12$

2. $m = 7; b = 0$

3. $m = 3; b = 27$

4. $m = \dfrac{5}{4}; b = -1$

5.

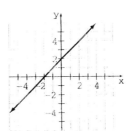

6.

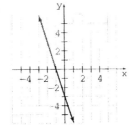

Chapter 4 Answers

7.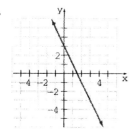

8. $y = 4x$

9. $y = -3x + 3$

10. $y = x + 2$

11. parallel

12. perpendicular

13. neither

14. perpendicular

15. $y = 7x - 2$

16. $y = 5x - 5$

17. $y = x + 1$

18. $y = 4x - 8$

19. $y = -2x + 8$

20. $y = 3x - 10$

Chapter 4 Test Form A

1. II

2. IV

3. Yes

4. 7

5.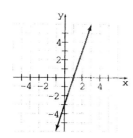

6.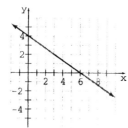

7.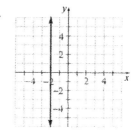

8.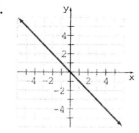

9. $m = 2$

10. $m = -1$

11. $m = \dfrac{4}{3}; b = \dfrac{-5}{3}$

12. neither

13. $y = -x + 3$

14. parallel

15. $y = -x + 7$

16. $y = \dfrac{3}{2}x - 5$

17. $x = 5$

18. $y = \dfrac{1}{3}x$

19.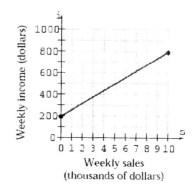

20. $430

Chapter 4 Answers

Chapter 4 Test Form B

1. IV
2. III
3. No
4. 4
5.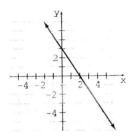
6.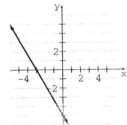
7.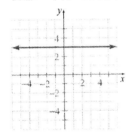
8.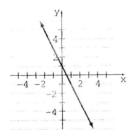
9. $m = -\dfrac{5}{7}$
10. $m = 0$
11. $m = \dfrac{7}{3}; b = 21$
12. neither
13. $y = x - 5$
14. perpendicular
15. $y = -4x + 17$
16. $y = -\dfrac{4}{3}x + 2$
17. $x = -4$
18. $y = \dfrac{3}{2}x - 8$
19.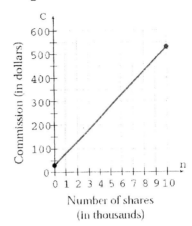

Number of shares
(in thousands)

20. 180

Chapter 4 Test Form C

1. III
2. I
3. Yes
4. 8
5.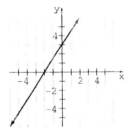
6.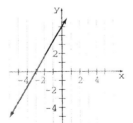
7.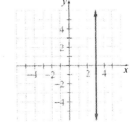

643

Chapter 4 Answers

8.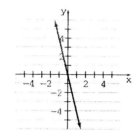

9. $m = -1$

10. slope is undefined

11. $m = 2; b = -5$

12. $\dfrac{5}{3}$

13. $2x - 5y = 10$

14. perpendicular

15. $y = -\dfrac{1}{2}x + 8$

16. $y = -\dfrac{7}{2}x + 4$

17. $y = -2$

18. $y = 5x - 5$

19.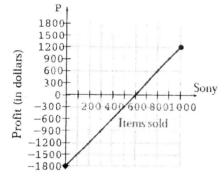

20. $0 (break even)

Chapter 4 Test Form D

1. II

2. IV

3. Yes

4. −10

5.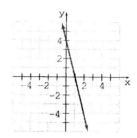

6.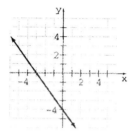

7.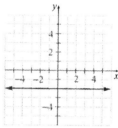

8.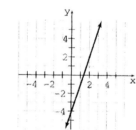

9. $m = -\dfrac{2}{3}$

10. $m = -2$

11. $m = \dfrac{7}{4}; b = -4$

12. perpendicular

13. $y = -2x + 5$

14. neither

15. $y = -\dfrac{1}{3}x$

16. $y = \dfrac{1}{4}x - 5$

17. $x = 2$

18. $y = \dfrac{2}{5}x - 1$

Chapter 4 Answers

19.

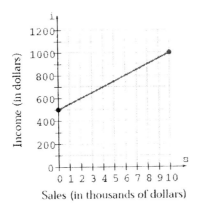

Sales (in thousands of dollars)

20. $4500

Chapter 4 Test Form E

1. IV
2. III
3. no
4. 3
5.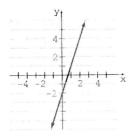
6.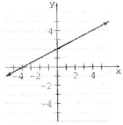
7.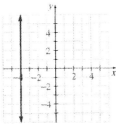
8.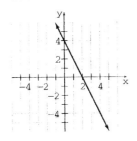

9. $m = -1$

10. $m = \dfrac{2}{3}$

11. $m = \dfrac{2}{5}; b = 2$

12. neither

13. $x - y = 4$

14. parallel

15. $y = 5x + 6$

16. $y = \dfrac{1}{2}x - 3$

17. $y = -8$

18. $y = -\dfrac{3}{4}x + 7$

13. yes

14. Function

15. -4

16. 3

19. 800

20. $6,000

Chapter 4 Test Form F

1. II
2. I
3. yes
4. I
5.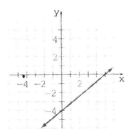
6.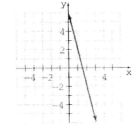

Chapter 4 Answers

7.

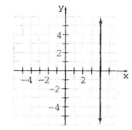

8.

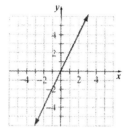

9. $m = -\dfrac{7}{4}$

10. $m = \dfrac{3}{2}$

11. $m = -\dfrac{2}{3};\ b = 6$

12. x-intercept = 4.5, y-intercept = 3

13. perpendicular

14. $y = \dfrac{3}{2}x + 3$

15. $y = -2x + 8$

16. $y = 5$

17. $y = -\dfrac{1}{3}x + 2$

18. $y = \dfrac{1}{3}x + 9$

19. $i = 10,000 + 0.02s$

20. $40,000

Chapter 4 Test Form G

1. c
2. d
3. c
4. b
5. a
6. c
7. a

8. d
9. a
10. a
11. a
12. a
13. d
14. d
15. d
16. a
17. b
18. a
19. d
20. a

Chapter 4 Test Form H

1. d
2. c
3. c
4. c
5. d
6. b
7. c
8. a
9. d
10. b
11. a
12. c
13. a
14. b
15. b
16. a
17. b
18. c
19. a
20. d

Chapter 4 Answers

Chapters 1–4 Cumulative Test Form A

1. $-\dfrac{1}{2}$

2. $x = 13$

3. $C = \dfrac{5}{9}(F - 32)$

4. $5.6, \dfrac{2}{3}, -\dfrac{5}{9}, 40$

5. 0

6. $5a^2 + 3ab + b^2$

7. $x = 3$

8. commutative property of addition

9. $y = -\dfrac{3}{7}x - \dfrac{5}{7}$

10.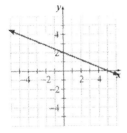

11. $m = \dfrac{2}{3}; \left(0, -\dfrac{2}{3}\right)$

12. 1

13. $2x - 3y = 3$

14. $x = \dfrac{6}{5}$

15. 22 minutes

16. -2

17. $2000 in CD; $3000 in bonds

18. 11 inches, 21 inches, 22 inches.

19. 45 liters

20. 8 hr

6. b

7. a

8. c

9. c

10. d

11. d

12. a

13. b

14. c

15. b

16. a

17. b

18. c

19. d

20. d

Chapters 1–4 Cumulative Test Form B

1. b

2. c

3. a

4. a

5. d

Chapter 5 Answers

Chapter 5 Pretest Form A

1. $6x^5$
2. $16x^8 y^4$
3. $6x^3$
4. $\dfrac{1}{x^9}$
5. 9
6. 2.117×10^6
7. 2400
8. 2.494×10^{-2}
9. binomial
10. second
11. $2x^4 - x^3 + 5x^2 - 8$
12. $-9x^5 + 7x + 14$
13. $11p^3 + 5p + 9$
14. $3x + 2$
15. $4x^2 - 2x + 6$
16. $24x^3 y^2 - 12xy^3$
17. $x^2 + 8x + 7$
18. $9p^3 + 18p^2 - p - 2$
19. $3x - 2 + \dfrac{4}{x}$
20. $2x + 3 - \dfrac{2}{2x+1}$

Chapter 5 Pretest Form B

1. $20x^6$
2. $10,000a^4 b^{12}$
3. $3x^3$
4. $\dfrac{1}{x^5}$
5. 1
6. 2.932×10^6
7. $391,000,000$
8. 1.428×10^{-1}

9. trinomial
10. fifth
11. $x^5 + 4x^3 - 7x^2 - 19$
12. $12c^5 + c - 8$
13. $4x^2 - x - 7$
14. $2x + 2$
15. $7x^2 - 3x + 8$
16. $-15x^4 y + 10x^3 y^2$
17. $x^2 + x - 30$
18. $x^4 - 8x^3 + 6x^2 + 40x + 25$
19. $2x - 5 + \dfrac{41}{6x+1}$
20. $x - 2 + \dfrac{2}{3x}$

Additional Exercises 5.1

1. 192
2. $8x^6$
3. x^8
4. $4x^5$
5. 36
6. $\dfrac{1}{x^3}$
7. 162
8. -1
9. y^{12}
10. $x^8 y^4$
11. $3x^3$
12. 2
13. $4c^7 d^{13}$
14. $16f^{10} g^{22}$
15. $x^4 y^4$
16. xy
17. $\dfrac{x^{12}}{64y^9}$
18. $2x^3 y^6$

Chapter 5 Answers

19. $a^2 + 2ab + b^2$

20. $2a^2 + ab + b^2$

Additional Exercises 5.2

1. $\dfrac{1}{y^7}$

2. $\dfrac{1}{49}$

3. x^5

4. 16

5. $\dfrac{1}{x^6}$

6. x^2

7. $\dfrac{1}{j^3}$

8. x^7

9. $4x^{13}$

10. $\dfrac{1}{f^4}$

11. $\dfrac{1}{x^{10}}$

12. $2x^{14}$

13. $\dfrac{1}{4x^4 y^6}$

14. $\dfrac{y^6}{4x^6}$

15. $\dfrac{1}{4x^3 y^7}$

16. $\dfrac{64^{12}}{y^{12}}$

17. $\dfrac{10x^6}{y^2}$

18. $\dfrac{81a^4 b^4 c^{12}}{7^4}$

19. $\dfrac{9c^{14}}{16a^8 b^2}$

20. $\dfrac{5}{4}$

Additional Exercises 5.3

1. 5.3×10^{-5}

2. 9×10^8

3. 5.5×10^6

4. 1.7×10^{-3}

5. 3.472856921×10^9

6. 1.89×10^{-6}

7. 46,700,000

8. 569,000,000

9. 31,300

10. 6.24

11. 0.000000478

12. 0.0000201

13. 13,530

14. 3,485,000

15. 21.28

16. 0.0000812

17. 0.42

18. 0.00000055

19. 2.03×10^{-4}

20. 15,000

Additional Exercises 5.4

1. not polynomial

2. binomial

3. trinomial

4. monomial

5. $5x^2 - 3x - 9$;
second

6. $-4x^3 + 6x + 2$; third

7. already in descending order; zero

8. $11a^5 + a - 6$

9. $8m^5 + 9m$

10. $12m^5 + 3m - 1$

11. $-8x^2 - 6x + 6$

12. $7x^3 - 7x - 2$

Chapter 5 Answers

13. $3x^4 - x^3 - 2x^2 + 3x + 2$

14. $2x^2 + 9x - 2$

15. $16x^2 - 8x + 5$

16. $6x^2 + 3x + 5$

17. $-9x^3 - 6x$

18. $-2x^3 + 10x^2 - 6x + 18$

19. $-x^3 + 11x^2 - 8x + 10$

20. $-2x^3 + 12x^2 - 8x + 10$

Additional Exercises 5.5

1. $-24x^5 y^5$

2. $72x^5 y^4$

3. $a^{10} b^4$

4. $18x^5 y + 10x^3 y^2$

5. $15m^5 + 25m^4 - 20m^3 + 30m^2$

6. $-4x^4 y - 28x^3 y^2 + 24x^2 y^4$

7. $x^2 + 8x + 12$

8. $x^2 + 3x - 54$

9. $12x^2 - 29x + 15$

10. $30a^2 + 17a + 2$

11. $10x^2 + 33xy + 20y^2$

12. $8x^2 - 14xy - 9y^2$

13. $36r^2 - 24r - 5$

14. $36c^2 - 49$

15. $9x^2 + 30xy + 25y^2$

16. $x^3 - 3x^2 + 5x - 6$

17. $10x^3 - 17x^2 - 6x - 35$

18. $x^5 - 1$

19. $x^4 + 10x^3 + 37x^2 + 60x + 36$

20. $6x^3 + 31x^2 + 30x + 8$; 1683 cu units

Additional Exercises 5.6

1. $3x - 2$

2. $2x - \dfrac{7}{2}$

3. $x - 3 + \dfrac{5}{x}$

4. $-x + 5 - \dfrac{2}{5x}$

5. $x^8 - \dfrac{5}{4}x^7 - 5x^2$

6. $-x^2 + x + \dfrac{7}{x^2} - \dfrac{9}{x^4}$

7. $x + 2 + \dfrac{6}{x}$

8. $-1 + \dfrac{15x - 5}{3x^2}$

9. $x^8 - \dfrac{5}{2}x^4 - 5x^3$

10. $-2x^2 + 5x + \dfrac{9}{x^2} + \dfrac{2}{x^4}$

11. $f^2 - 4f + 16$

12. $3p + 7 + \dfrac{5}{p-1}$

13. $2m - 7 + \dfrac{31}{m+5}$

14. $j^2 + 4j + 16$

15. $p^4 - 4p^2 + 16$

16. $3p - 8 + \dfrac{31}{p-2}$

17. $1 - 4x$

18. $-2x^2 + 2x - 1 - \dfrac{2}{x-1}$

19. $4x - 3$

20. $3x - 1 - \dfrac{4x+3}{2x^2+1}$

Chapter 5 Answers

Chapter 5 Test Form A

1. $-10x^5$
2. $1000x^{12}y^3$
3. $\dfrac{25}{x^6 y^4}$
4. $3x^2$
5. $\dfrac{x^6 y^6}{9}$
6. not a polynomial
7. binomial
8. 5.59×10^{10}
9. 4×10^{-10}
10. $2x^3 + x^2 - 3x + 5$, third degree
11. $2x^2 + 7x - 1$
12. $-3x^2 + 5$
13. $-4x^3 y + 4x^2 y^3$
14. $-2x^2 + 17x - 35$
15. $4x^3 - 2x - 2$
16. $x^2 + 3x - \dfrac{2}{x}$
17. $10x^2 + x + 9 + \dfrac{9}{x-1}$
18. $x - 1 + \dfrac{40}{2x+5}$
19. $5x^2 + 4x - 8$
20. $\$3.4 \times 10^{10}$

Chapter 5 Test Form B

1. $15x^7$
2. $16x^4 y^6$
3. $2x^4$
4. $\dfrac{y^3}{8x^9}$
5. $\dfrac{x^9}{27 y^9}$

6. trinomial
7. not a polynomial
8. 3.3×10^6
9. 7×10^{-10}
10. $-4x^5 + 3x^3 + x^2 + 4$; fifth degree
11. $5x^2 - x + 3$
12. $3x^2 + 5x - 6$
13. $-3x^4 y - 6x^2 y^3$
14. $-3x^2 + 12$
15. $x^3 + 5x^2 - 27x + 26$
16. $x - 2 - \dfrac{9}{x}$
17. $-3x^2 - 5x - 15 - \dfrac{30}{x-2}$
18. $4x - 11$
19. $3x^2 - 4x - 5$
20. 7×10^6 sec

Chapter 5 Test Form C

1. $32x^6$
2. $8x^6 y^3$
3. $\dfrac{x^4}{3}$
4. $\dfrac{27 y^9}{x^3}$
5. $\dfrac{y^6}{4x^6}$
6. not a polynomial
7. binomial
8. 1.441×10^6
9. 2.5×10^{-10}
10. $2x^5 - 5x^3 + 3x - 4$; fifth degree
11. $3x^2 - 2x - 10$
12. $x^2 + x$
13. $-5x^3 y^2 + 20x^2 y^3$

Chapter 5 Answers

14. $8x^2 - 26x - 7$

15. $-2x^3 + 13x^2 - 22x + 3$

16. $2x - 1 + \dfrac{1}{4x}$

17. $4x^2 + 5x + 9$

18. $x - 2 + \dfrac{4}{3x - 1}$

19. $4 + \dfrac{4x + 17}{x^2 - 4}$

20. 4.05×10^{-8} light year

Chapter 5 Test Form D

1. $12x^7$

2. $8x^6 y^9$

3. $3x^4$

4. $\dfrac{49}{x^6 y^2}$

5. $\dfrac{x^4 y^6}{64}$

6. trinomial

7. not a polynomial

8. 7.371×10^{10}

9. 4×10^7

10. $5x^3 - 3x^2 + x + 12$; third degree

11. $3x^2 + 6x - 9$

12. $-7x^2 + x - 5$

13. $-9x^3 y^2 + 27x^2 y^3$

14. $3x^2 - 9x + 6$

15. $-x^3 + 6x^2 - 6x + 1$

16. $2x + \dfrac{1}{3} - \dfrac{1}{x}$

17. $2x^2 + x - 5$

18. $2x - 1$

19. $9x^4 - 36x^3 + 35x^2 + 4x - 4$

20. 2.35×10^{20} miles

Chapter 5 Test Form E

1. $-15x^5$

2. $36x^4 y^6$

3. $2x^2$

4. $\dfrac{8y^6}{27x^6}$

5. $216 \dfrac{x^9}{y^{12}}$

6. not a polynomial

7. monomial

8. 8.211×10^9

9. 8×10^{-9}

10. $x^4 + 7x^3 - 3x + 5$; fourth degree

11. $2x^2 + 2x - 2$

12. $2x^2 + 4x + 10$

13. $-7x^4 y + 7x^2 y^4$

14. $8x^2 + 6x - 5$

15. $16x^3 - 20x^2 - 10x + 12$

16. $3x^4 - x - 2$

17. $-4x^2 + 11x - 22 + \dfrac{26}{x + 2}$

18. $4x + 3$

19. $8x^2 + 2x - 2$

20. $\$7.9115 \times 10^4$

Chapter 5 Test Form F

1. $15x^5$

2. $256x^4 y^8$

3. $5x^2$

4. $\dfrac{x^2 y^2}{9}$

5. $\dfrac{y^6}{9x^4}$

6. not a polynomial

7. trinomial

Chapter 5 Answers

8. 8.5×10^5

9. 2×10^{-6}

10. $4x^4 + 5x^2 - 7x - 2$; fourth

11. $3x^2 + 2x + 3$

12. $4x + 9$

13. $x^2 + 2x^2y - y^2 - 2xy^2$

14. $12x^2 + 5x - 3$

15. $-3x^3 + 8x^2 + 4x - 15$

16. $3x^2 - x + \dfrac{2}{x}$

17. $3x^2 - 7x + 31$

18. $x + 5$

19. $4x^2 + 16 + \dfrac{25}{x^2 - 1}$

20. 7.631×10^{19} miles

Chapter 5 Test Form G

1. d
2. a
3. b
4. c
5. b
6. d
7. b
8. c
9. a
10. b
11. d
12. d
13. a
14. c
15. a
16. c
17. b
18. c
19. b
20. d

Chapter 5 Test Form H

1. c
2. a
3. d
4. d
5. b
6. c
7. b
8. c
9. a
10. c
11. d
12. a
13. d
14. a
15. d
16. b
17. c
18. b
19. c
20. a

Chapter 6 Answers

Chapter 6 Pretest Form A

1. $4x$

2. $5x^3(2 - 5x^2)$

3. $4x^2(5x^2 - 3x + 7)$

4. $(4x + 5)(x + 3)$

5. $(x + 8)(x - 5)$

6. $(x + 5)(x - 3)$

7. $3x(x - 2)(x - 3)$

8. $(2x - 3)(x + 7)$

9. $(4x - 5)(3x + 4)$

10. $3x(x - 2)(x^2 + 2x + 4)$

11. $(2x - 5)^2$

12. $5x(x - 2)(x + 2)$

13. $\dfrac{4}{7}, -2$

14. 2, 15

15. $2, -2, -\dfrac{3}{2}$

16. 12 cm, 5 cm, 13 cm

Chapter 6 Pretest Form B

1. $2x$

2. $3x^2(3 - x)$

3. $2x(2x^2 - 5x + 9)$

4. $(5x + 2)(x - 7)$

5. $(x + 2)(x + 1)$

6. $(x - 6)(x + 8)$

7. $4x(x - 1)(x - 5)$

8. $(3x - 1)(x + 4)$

9. $(6x - 5)(2x + 3)$

10. $5(x + 1)(x^2 - x + 1)$

11. $x(5x - 6)(5x + 6)$

12. $3x(x - 4)^2$

13. $\dfrac{5}{9}, -3$

14. 0, 5, –5

15. 9

16. 8 m

Additional Exercises 6.1

1. $2^2 \cdot 3 \cdot 5^2$

2. $2^3 \cdot 3^2 \cdot 5$

3. $2 \cdot 7^2$

4. 14

5. 15

6. 1

7. x^2

8. $8x$

9. $x^3 y^3$

10. $4x$

11. x

12. 2

13. $2(4x - 3)$

14. cannot be factored

15. $6x^4(2 - 7x^2)$

16. $7x(3xy - 2x + 6)$

17. $5x(x^2 + x + 3)$

18. $3x^2(4x^3 - 3x^2 + 5x + 3)$

19. $4xy^2(6x^2 - 7x + 3y)$

20. $3x(x^2 + x + 2)$

Additional Exercises 6.2

1. $(x + 6)(x - 2)$

2. $(x + 5)(x - 4)$

3. $(x + 7)(x - 4)$

4. $(x + 3)(x + 8)$

5. $(5x + 3)(x + 1)$

6. $(3x + 5)(x + 6)$

7. $(x - 2)(x + 2)$

8. $(x + 2)(2x^2 + 1)$

Chapter 6 Answers

9. $(3x + 1)(3x + 1)$

10. $(x+1)(2x^2+1)$

11. $(4x + 3)(x + 5)$

12. $(x-2)(3x^2+1)$

13. $(3x + 4)(x + 6)$

14. $(2x + y)(x - 2y)$

15. $(3x - y)(x + 5y)$

16. $(2x + y)(x - 4y)$

17. $(y + a)(y + 2b)$

18. $(5x + 4y)(x - y)$

19. $4(x - 2)(x - 2)$

20. $5(2x + 3y)(3x + 2y)$

Additional Exercises 6.3

1. $(x + 2)(x - 8)$

2. $(x - 4y)(x - 6y)$

3. $(x + 2)(x + 3)$

4. $(x - 2)(x - 1)$

5. $(x + 5)(x - 6)$

6. $(x + 11)(x - 4)$

7. cannot be factored

8. $(x - 2y)(x - 3y)$

9. $(x + 8)(x + 2)$

10. $(x + 9)(x - 8)$

11. $(x + 1)(x - 9)$

12. $(x + 2)(x + 2)$

13. $(x - 6y)(x - 7y)$

14. $(x + 6)(x + 2)$

15. $4x(x + 1)(x - 3)$

16. $2x(x + 1)(x - 2)$

17. $2x(x + 1)(x - 3)$

18. $2x(x - 3y)(x + y)$

19. $5xy(x - 5)(x - 2)$

20. $3xy(x + 7)(x - 1)$

Additional Exercises 6.4

1. $(2x - 1)(2x - 1)$

2. $(3x + 4)(5x + 3)$

3. $(2r + 3)(3r - 2)$

4. $(5a + 4)(7a - 5)$

5. $(5x + 4)(5x + 2)$

6. $(3x + 1)(4x + 3)$

7. $(3x - 5y)(3x - 4y)$

8. $(5u + 3)(5u - 6)$

9. $(4f - 3)(3f + 2)$

10. cannot be factored

11. $(2x + 1)(2x - 1)$

12. $(5x + 2)(4x + 3)$

13. $(2x - 5y)(3x - 2y)$

14. $(7p + 4)(5p - 1)$

15. $2(5x + 3)(5x - 2)$

16. $-5(2x + 3)(3x - 2)$

17. $-2(3x - 5)(3x + 2)$

18. $x(3x - 1)(x - 7)$

19. $3(5x + 4)(x - 2)$

20. $x(3x^2 + 4x + 5)$

Additional Exercises 6.5

1. $(7x - 3y)^2$

2. $(6x + 7y)^2$

3. $(9r + 4s)(9r - 4s)$

4. $3(2x - 5y)(2x + 5y)$

5. $(8x + 3y)(8x - 3y)$

6. $(x + 8)(x - 8)$

7. $(3m + 4)(9m^2 - 12m + 16)$

8. $(4p + 5)(16p^2 - 20p + 25)$

9. $(3x + 2y)^2$

10. $(7x - 6y)^2$

11. $(9a - 4b)(9a + 4b)$

Chapter 6 Answers

12. $y^2(x-8)(x+8)$

13. $(5x+9y)(5x-9y)$

14. $(x+2)(x-2)(x+3)$

15. $(3a+2)(9a^2-6a+4)$

16. $(c+5)(c^2-5c+25)$

17. $3(5x-4)(5x+4)$

18. $(x-1)^2(x+1)(x^2+x+1)$

19. $(x+2)^2(x-3)(x+3)$

20. $5(c+2)(c^2-2c+4)$

Additional Exercises 6.6

1. $0, 6$

2. $-2, 11$

3. $-7, -12$

4. $0, 19$

5. $0, 9$

6. $3, 2$

7. $-5, \dfrac{5}{4}$

8. $-5, 6$

9. $0, 3$

10. $6, 3$

11. $0, 5, -5$

12. $3, -2$

13. $-5, \dfrac{2}{3}$

14. $\dfrac{3}{4}, -\dfrac{1}{5}$

15. $-\dfrac{7}{2}, \dfrac{4}{3}$

16. 7

17. $8, -13$

18. $-\dfrac{8}{3}, 4$

19. $-5, 9$

20. $9, -18$

Additional Exercises 6.7

1. no

2. no

3. yes

4. 12

5. 40

6. 34

7. 6

8. 4

9. 15

10. 11

11. 7

12. 6, 15

13. 17, 19

14. 6, 7, 8

15. 5, 6

16. 7,14

17. 11, 22

18. 12.5 seconds

19. 24

20. 4.5 feet

Chapter 6 Test Form A

1. $9x^2$

2. $5x^2y^2$

3. $3x^2y(2xy^2+3)$

4. $5ab^2(2a-b-3)$

5. prime

6. $(2x-1)(x+5)$

7. $(a+2b)(a-8b)$

8. $(x+2y)(3x-4y)$

9. $(2a-z)^2$

10. $3(x+4)(x+5)$

11. $(t+w)(t-w)$

12. $(x-1)(x+1)(x+2)(x^2-2x+4)$

Chapter 6 Answers

13. $(a+4)(a^2-4a+16)$

14. $-1, \dfrac{2}{7}$

15. $0, 3, -3$

16. -9

17. $-1, -2, 3$

18. $\dfrac{5}{3}, 6$

19. $14, 15$

20. 24 feet, 7 feet, 25 feet

Chapter 6 Test Form B

1. $5x^2$

2. $7x^2y$

3. $4x^2y^3(2x+3y)$

4. $3ab(a^2-2ab-1)$

5. prime

6. $(2a-1)(a+6)$

7. $(2x-y)(x+5y)$

8. $(x+3)(x+5)$

9. $(3a+1)^2$

10. $(t+z)(t-z)$

11. $2a(a+4)(2a-3)$

12. $(x-2y)(x^2+2xy+4y^2)$

13. $(x-1)^2(x+1)(x^2+x+1)$

14. $-5, \dfrac{3}{2}$

15. $0, 2$

16. 6

17. $-3, -2, 5$

18. $-5, 2$

19. $-11, -12$

20. 24 feet, 10 feet, 26 feet

Chapter 6 Test Form C

1. $6y^3$

2. $4x^2y^3$

3. $xy(13x^2y-5)$

4. $6a^2b(a-ab^2+4b)$

5. prime

6. $(3x+1)(x-2)$

7. $(x+2y)(3x+y)$

8. $(t+u)(3t-2u)$

9. $(a+2b)(a-2b)$

10. $4x^2(x-1)(x+2)$

11. $(2-a)(4+2a+a^2)$

12. $(2x-1)(2x+1)(x+1)(x^2-x+1)$

13. $(3a+2)^2$

14. $0, 2$

15. $2, 3, -5$

16. 5

17. $0, \dfrac{1}{3}, -\dfrac{1}{3}$

18. $-3, 7$

19. $17, 18$

20. $x=10$

Chapter 6 Test Form D

1. $4a^2$

2. $3x^2y^3$

3. $5xy(2x^2y+1)$

4. $2x^2y(2x-3xy^3+5)$

5. prime

6. $(2x-1)(3x-1)$

7. $(a+4)(2a+5)$

8. $(2x+3)(x-11)$

9. $(x-2y)(x+2y)$

Chapter 6 Answers

10. $(t-2u)^2$

11. $3(x+2)(x-9)$

12. $(2x+1)(4x^2-2x+1)$

13. $(x-2)(x+2)(x+3)(x^2-3x+9)$

14. $0, 11, -7$

15. $0, 4, -4$

16. $3, 10$

17. $2, -\dfrac{1}{2}, -3$

18. $\dfrac{5}{7}, 2$

19. $24, 26$

20. A = 480 sq. feet and d = 34 feet

Chapter 6 Test Form E

1. $6x^3$

2. $5ab^2$

3. $2a^2b(3a-b^2)$

4. $3x^2y(y+5x)$

5. prime

6. $(t+v)(2t-3v)$

7. $(a-2)(2a-7)$

8. $(2x-3)(3x+5)$

9. $5(x-3)(2x+3)$

10. $(x-1)(x+1)(x+2)(x^2-2x+4)$

11. $(x-2a)^2$

12. $(x+2)(x^2-2x+4)$

13. $(t-5)(t+5)$

14. $-3, 3$

15. $6, 7$

16. $0, 1$

17. $-\dfrac{2}{3}, \dfrac{5}{8}$

18. $-3, -5, 5$

19. $-16, -18$

20. 12 inches, 9 inches, and 15 inches

Chapter 6 Test Form F

1. $5x^3$

2. $6x^2y^2$

3. $3ax(4a^2x^2+7)$

4. $4x^2y^2(2x-3y+1)$

5. prime

6. $(2a+3)(a-6)$

7. $(x-3y)(x-2y)$

8. $(3a-5x)(2a+5x)$

9. $5(5a-2)(5a-3)$

10. $(2x-3y)^2$

11. $(x-2)(x+2)(x-1)(x^2+x+1)$

12. $(5t+2w)(5t-2w)$

13. $(3x-1)(9x^2+3x+1)$

14. $-\dfrac{2}{3}, 7, -1$

15. $0, 2, -2$

16. 11

17. 2

18. $2, 4$

19. $-23, -21$

20. 5

Chapter 6 Test Form G

1. a

2. b

3. d

4. b

5. d

6. d

7. a

8. a

9. c

10. b

11. d

Chapter 6 Answers

12. b

13. a

14. c

15. b

16. c

17. d

18. d

19. a

20. b

Chapter 6 Test Form H

1. d
2. b
3. d
4. c
5. a
6. c
7. b
8. d
9. d
10. b
11. a
12. c
13. b
14. a
15. a
16. a
17. c
18. b
19. d
20. d

Chapters 1–6 Cumulative Test Form A

1. -134

2. $\left\{ -17, 31, 0, \dfrac{4}{7}, -\sqrt{\dfrac{4}{9}}, 1.75 \right\}$

3. -42

4. no solution

5. $d = \dfrac{S}{9c^2}$

6. 200 gallons

7. -3.5

8. 8 inches, 26 inches, 24 inches

9. express 30 mph; commuter 20 mph

10. 1

11.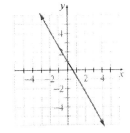

12. $\dfrac{x^{20}}{243y^{10}}$

13. $\dfrac{9x^6 y^4}{4z^8}$

14. $5x^2 - 4x - 8$

15. $-30x^5 y^6$

16. $24x^3 + 44x^2 - 4x - 20$

17. $x^3 + x - 1$

18. $(2x - 5)(4x + 7)$

19. $5m(m + 4)(m - 4)$

20. $\left\{ -4, \dfrac{2}{3} \right\}$

Chapters 1–6 Cumulative Test Form B

1. c
2. c
3. c
4. a
5. a
6. c
7. b
8. d
9. b
10. b

Chapter 6 Answers

11. a
12. d
13. a
14. d
15. b
16. c
17. b
18. c
19. d
20. d

Chapter 7 Answers

Chapter 7 Pretest Form A

1. All real numbers except $x = -\dfrac{3}{4}$

2. All real numbers except $x = 1$ and $x = -6$

3. $\dfrac{1}{p-3}$

4. $-(x+2)$

5. $\dfrac{3y^3}{8}$

6. $\dfrac{2x-y}{4x-5}$

7. $\dfrac{3}{y(y+6)}$

8. $\dfrac{3}{x}$

9. $\dfrac{6x+5}{x+7}$

10. $30x^3y^5$

11. $(x+7)(x+2)(x-2)$

12. $\dfrac{9x+8y^2}{8x^3y^4}$

13. $\dfrac{5x+27}{(x+4)^2}$

14. $\dfrac{6}{7}$

15. $\dfrac{3x^2y+12x^2}{5y}$

16. $x = 20$
17. $x = 7$
18. $x = 24$
19. $y = 2$
20. 1.2 hours, or 1 hour and 12 minutes

Chapter 7 Pretest Form B

1. $x \neq -3$
2. $x \neq -1,\ x \neq -5$

3. $\dfrac{1}{m-7}$

4. $-(x+2)$

5. $\dfrac{3xy}{5}$

6. $a + 2$

7. $\dfrac{4}{y(y+5)}$

8. $\dfrac{1}{3}$

9. $\dfrac{4x+5}{x+8}$

10. $20x^3y^5$

11. $(x+1)(x-4)(x+2)$

12. $\dfrac{5x^3-4y}{14x^5y^3}$

13. $\dfrac{9x+31}{(x+3)^2}$

14. $\dfrac{4}{5}$.

15. $\dfrac{x}{x-2}$

16. $x = 28$
17. $x = 9$
18. $x = 40$
19. $y = 2$
20. 2.4 hours, or 2 hours and 24 minutes

Additional Exercises 7.1

1. All real numbers except $x = -4,\ x = 4$

2. All real numbers except $x = \dfrac{3}{5}$

3. All real numbers except $x = -6,\ x = -1$

4. All real numbers except $x = 0,\ x = -\dfrac{5}{4}$

5. All real numbers except $x = 6,\ x = 2$

6. All real numbers

Chapter 7 Answers

7. $\dfrac{5+3b}{2}$

8. $\dfrac{5+q}{8}$

9. $\dfrac{8}{6+n}$

10. -3

11. -2

12. $-(x+7)$

13. $-(x+2)$

14. $\dfrac{x+2}{x+7}$

15. $\dfrac{1}{u+1}$

16. $\dfrac{x+3}{x+4}$

17. $\dfrac{2}{y-8}$

18. $-\dfrac{1}{x+3}$

19. $\dfrac{x+9}{x+2}$

20. $\dfrac{-(x+3)}{9+3x+x^2}$

7. $-\dfrac{y+4x}{3x+5y}$

8. $a+4$

9. $\dfrac{3x-y}{2x+1}$

10. $a+6$

11. $35pr$

12. $\dfrac{r^2t^2}{s^2}$

13. $\dfrac{40e^2f}{c}$

14. $\dfrac{u^4y^3}{x}$

15. $\dfrac{(x-3)^2}{x-2}$

16. -5

17. -5

18. $\dfrac{5}{y(y+5)}$

19. $\dfrac{3}{y(y+4)}$

20. $\dfrac{x-4}{x+2}$

Additional Exercises 7.2

1. $6xy$

2. $\dfrac{2x^4z}{5y^4}$

3. $\dfrac{4xy}{3}$

4. $\dfrac{4x^3z}{3y^2}$

5. $\dfrac{2xy}{5}$

6. $\dfrac{3x^3z}{5y^3}$

Additional Exercises 7.3

1. $\dfrac{6x-4}{3}$

2. $\dfrac{x+5}{7}$

3. $\dfrac{x+17}{x}$

4. $\dfrac{1}{3}$

5. $\dfrac{1}{4}$

6. $\dfrac{x+6}{x+5}$

7. $\dfrac{x+10}{x+4}$

8. $\dfrac{4}{x+8}$

9. $\dfrac{6}{x-9}$

10. x

11. $\dfrac{3x+13}{x+5}$

12. $\dfrac{3}{x-3}$

13. $\dfrac{1}{x-8}$

14. $\dfrac{1}{x+6}$

15. $28x^3y^6$

16. $40x^5y^7$

17. $(x+4)(x-4)$

18. $x(x-1)$

19. $(x-2)(x+3)(x+4)(x+2)$

20. $(x+1)(x+4)(x-5)$

9. $\dfrac{35p-12}{15p^2}$

10. $\dfrac{1}{x(x+2)}$

11. $\dfrac{2x^2-4}{x(x-2)}$

12. $\dfrac{1}{x+6}$

13. $\dfrac{1}{x+4}$

14. $\dfrac{1}{x+3}$

15. $\dfrac{2x^2-4x+5}{(x+2)(x-2)(x+3)}$

16. $\dfrac{13-7x}{(x+1)(x-3)(x+3)}$

17. $\dfrac{x^2-5x-2}{(x+2)(x+3)(x-1)}$

18. $\dfrac{-2x^2+9x-8}{(x-4)(x-2)(x+2)}$

19. $\dfrac{3x-4}{(x-1)(x-2)}$

20. $\dfrac{-x+9}{(x+3)(x-3)(x+3)}$

Additional Exercises 7.4

1. $\dfrac{14}{3x}$

2. $\dfrac{30x+7}{5x^2}$

3. $\dfrac{21x+6}{7x^2}$

4. $\dfrac{24x+7}{6x^2}$

5. $\dfrac{12y^2-5x}{10x^3y^3}$

6. $\dfrac{10x+7y}{2x^2y}$

7. $\dfrac{2xy+x}{y}$

8. $\dfrac{15a-11}{12a}$

Additional Exercises 7.5

1. $\dfrac{8}{7}$

2. 2

3. $\dfrac{27}{35}$

4. $\dfrac{56}{95}$

5. $-\dfrac{3xy^4}{8}$

6. $\dfrac{x-5}{2}$

7. $\dfrac{3x^4y^3}{16}$

Chapter 7 Answers

8. $\dfrac{3x^2(y+2)}{7y}$

9. $\dfrac{2}{5x}$

10. $\dfrac{x(x^2-2x+4)}{(x-2)(x^2+4)}$

11. $\dfrac{-x}{x^2+6x+9}$

12. $\dfrac{4x}{4x^2+9x+2}$

13. $\dfrac{x(x^3+8)}{x^4+16}$

14. $\dfrac{7x^2(y+6)}{3y}$

15. $\dfrac{x(x^3-5)}{x^4-10}$

16. $\dfrac{x}{2x^2-x-1}$

17. $\dfrac{11}{18}$

18. $\dfrac{6}{5}$

19. $\dfrac{2x+y}{3}$

20. $\dfrac{x-1}{x+1}$

Additional Exercises 7.6

1. $x = 10$
2. $x = -27$
3. $x = -\dfrac{5}{14}$
4. $n = 10$
5. $w = 7$
6. $x = 5$
7. $x = 18$
8. $x = 63$
9. $x = 6$
10. $y = \dfrac{9}{4}$

11. no solution
12. no solution
13. $x = 2$
14. $x = 4,\ x = -4$
15. $x = 4$
16. $x = -3$
17. $x = -5$
18. $x = 24$
19. $x = 80$
20. $x = 7$

Additional Exercises 7.7

1. 6 and 8
2. 5
3. 2.8 mph
4. 44 mph and 57 mph
5. 4 hr
6. 58 mph
7. $7\dfrac{23}{31}$ hr
8. 7.5 hr and 15 hr
9. $2\dfrac{2}{5}$ hr
10. 2 hr
11. $\dfrac{36}{5}$
12. 225 mph
13. 48 mph and 52 mph
14. 52 mph
15. $3\dfrac{15}{16}$ hr
16. 6 hours
17. 4 and 10
18. $\dfrac{12}{5}$
19. 262.5 mph
20. 46 mph and 47 mph

Chapter 7 Answers

Additional Exercises 7.8

1. 76

2. 16 watts

3. 4

4. $\dfrac{5}{81}$

5. $p = kmn^2$

6. 4.8 in.

7. 21.01 foot-candles

8. 36 ohms

9. 2826 square inches

10. $h = kfg^3$

11. $p = \dfrac{k}{q}$

12. $\dfrac{3}{4}$

13. 3456

14. 100

15. 125

16. 400 cubic centimeters

17. $\dfrac{1}{130}$

18. $\dfrac{1}{54}$

19. 120

20. 1050 large pizzas

Chapter 7 Test Form A

1. -1

2. $x - 1$

3. $\dfrac{a^2 + 2a - 63}{a - 3}$

4. $\dfrac{x^2 + 10x + 21}{x^2 - 7x + 10}$

5. -3

6. $\dfrac{x - 4y}{4}$

7. $\dfrac{x^2 + 6x + 9}{x^2 - 4}$

8. $3y$

9. $\dfrac{5x + 12}{x + 3}$

10. $\dfrac{-x - 8}{x^2 + 2x}$

11. $\dfrac{x + 6}{3x}$

12. $\dfrac{2}{x^2 - 4}$

13. $\dfrac{9}{5}$

14. $\dfrac{(5 - y)x^2}{(y + x)y}$

15. $x = -8$

16. $x = 28$

17. no solution

18. $y = 12$

19. $3\dfrac{1}{3}$ hours, or 3 hours and 20 minutes

20. 2 hours at each speed

Chapter 7 Test Form B

1. $\dfrac{1}{x + 2}$

2. $-(x + 2)$, or $-x - 2$

3. $\dfrac{6x^3 z}{y}$

4. $\dfrac{m}{3}$

5. $\dfrac{4(x - 1)}{x - 3}$

6. $\dfrac{z + 3}{z - 5}$

Chapter 7 Answers

7. $\dfrac{1}{x-5}$

8. $\dfrac{2z^2+3z-15}{z(z-5)(z-4)}$

9. $\dfrac{3y^2-5}{9x^2y^6}$

10. $\dfrac{4}{x+1}$

11. $\dfrac{3m^2+13m+8}{3(3m+4)(m-2)}$

12. $\dfrac{35}{11}$

13. $\dfrac{z^2(1-y)}{y(y+z)}$

14. $x=-\dfrac{73}{5}$

15. $x=-4,\ x=16$

16. no solution

17. $x=\dfrac{2}{5}$

18. 120 pounds per square foot

19. $8\dfrac{4}{7}$ hours

20. 60 miles per hour

7. $\dfrac{x^2+3x+2}{x-5}$

8. $5y$

9. $\dfrac{10x+16}{x+2}$

10. $\dfrac{-14x+5}{3x^2+6x}$

11. $\dfrac{27+5x}{15x}$

12. $\dfrac{2}{x^2+3x+2}$

13. $\dfrac{18}{5}$

14. $\dfrac{x^2(5+y)}{y(y-4x)}$

15. $x=2$

16. no solution

17. $x=-\dfrac{8}{17}$

18. $y=8$

19. 15 days

20. Porsche $=\dfrac{17}{25}$ hours, or 40.8 minutes

    Model T $=2\dfrac{4}{15}$ hours, or 2 hours and 16 minutes

Chapter 7 Test Form C

1. $-\dfrac{1}{5}$

2. $\dfrac{4}{x+2}$

3. $\dfrac{a+2}{a^2+7a+12}$

4. $\dfrac{x^2-12x+20}{x^2+11x+10}$

5. -4

6. $\dfrac{3x-2y}{8}$

Chapter 7 Test Form D

1. $\dfrac{2x}{x+1}$

2. $2(x-1)$

3. $a+1$

4. $\dfrac{x^2-8x+16}{x^2-3x-28}$

5. 3

6. $\dfrac{x+3y}{3}$

7. $\dfrac{-x^2+2x-1}{x^2-4}$

8. $-4y$

Chapter 7 Answers

9. $\dfrac{7x+10}{x+2}$

10. $\dfrac{-16x-7}{2x(x+1)}$

11. $\dfrac{84+5x}{60x}$

12. $\dfrac{1}{x^2+4x+3}$

13. $\dfrac{19}{67}$

14. $\dfrac{x^2(7+y)}{y(y-2x)}$

15. $x=-35$

16. no solution

17. $x=-8\dfrac{1}{3}$

18. $y=7.5$

19. Approximately $13\dfrac{1}{3}$ hours, or 13 hours and 20 minutes

20. 12 miles per hour

Chapter 7 Test Form E

1. -3

2. $\dfrac{x+2}{x+3}$

3. $\dfrac{21a^2}{2b^2c^7}$

4. $\dfrac{x^2+10x+25}{x^2-x-20}$

5. $\dfrac{3a-10}{8}$

6. $\dfrac{x-2y}{5}$

7. $\dfrac{1}{x-2}$

8. $3y$

9. $\dfrac{3x-23}{x-7}$

10. $\dfrac{-3}{2x^2-10x}$

11. $\dfrac{8}{x^2-2x-3}$

12. $\dfrac{6y^2+5}{xy^3}$

13. $7x$

14. $\dfrac{x^2(4-y)}{y(y+2x)}$

15. $x=-15$

16. no solution

17. $x=5$

18. $y=40$

19. $13\dfrac{1}{19}$ miles per hour

20. 2.4 hours, or 2 hours and 24 minutes

Chapter 7 Test Form F

1. $-\dfrac{3x}{x+2}$

2. $\dfrac{x-7y}{7x-y}$

3. $\dfrac{a^2+10a+21}{a-2}$

4. $\dfrac{3x}{x+3}$

5. $\dfrac{-5(x+3)}{2}$

6. $24(m-n)$

7. $\dfrac{2m-n}{3m-2n}$

8. $3y$

9. $\dfrac{9x-21}{x-2}$

10. $\dfrac{-2x-7}{x^2+7x}$

11. $\dfrac{2x+3}{4x}$

Chapter 7 Answers

12. $\dfrac{-3}{x^2 + 6x + 5}$

13. $\dfrac{188}{15}$ or $12\dfrac{8}{15}$

14. $\dfrac{x(5x - y)}{y(y - 3x)}$

15. $x = -\dfrac{8}{5}$

16. no solution

17. $x = 18$

18. $y = 12.5$

19. 3 miles per hour

20. 2 hours

Chapter 7 Test Form G

1. c
2. c
3. a
4. b
5. b
6. a
7. b
8. c
9. a
10. b
11. a
12. a
13. d
14. d
15. c
16. d
17. a
18. a
19. b
20. d

Chapter 7 Test Form H

1. a
2. c
3. b
4. d
5. a
6. c
7. c
8. d
9. b
10. c
11. a
12. c
13. b
14. a
15. d
16. c
17. c
18. a
19. b
20. d

Chapter 8 Answers

Chapter 8 Pretest Form A

1.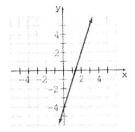

2.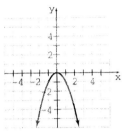

3.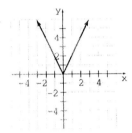

4. $\left\{-3,-2,-1,0,1,2,\ldots\right\}$

5. $\left(-2,7\right]$

6.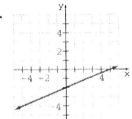

7. $\{-7, 2, 4, 8\}$

8. $\{-7, 2, 9\}$

9. yes

10. no

11. yes

12. 8

13. $-\frac{2}{3}; \left(0, 4\right)$

14. $24,750

15. -2

16. yes

17. $y = 3x - 7$

18. $3x^2 + x + 4$

19. 28

20. 1

Chapter 8 Pretest Form B

1.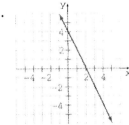

2.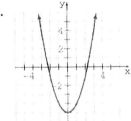

3.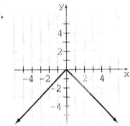

4. $\left\{x \mid -2 < x \le 3\right\}$

5. $\left(3, 12\right)$

6.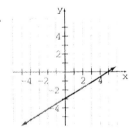

7. $\left\{-4, -3, 5, 7\right\}$

8. $\{-1, 2, 3\}$

9. yes

10. yes

11. no

12. 3

13. $30,000

Chapter 8 Answers

14. $\dfrac{7}{6}$; $(0, -2)$

15. $-\dfrac{11}{5}$

16. parallel

17. $y = -3x - 2$

18. $4x^2 + x - 8$

19. 6

20. 24

Mini-Lecture 8.1

1)

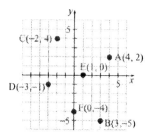

3a)

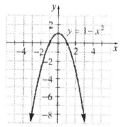

3b)

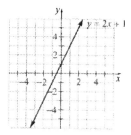

3c)

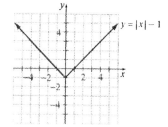

Mini-Lecture 8.5

3)

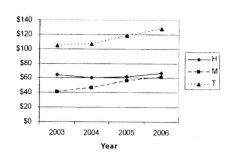

4)

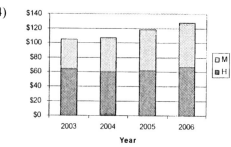

5)

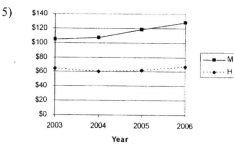

Additional Exercises 8.1

1. $(-2, -4)$

2.

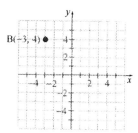

3. $A(-2, 1)$; $B(3, 3)$; $C(4, -1)$; $D(-4, -5)$

4.

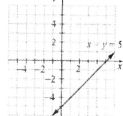

5.

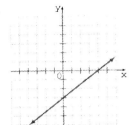

6.

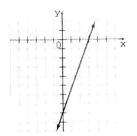

7.

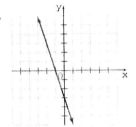

8.

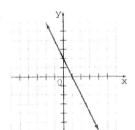

9.

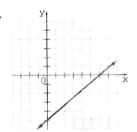

10.

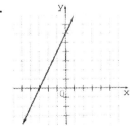

11.

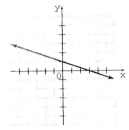

12.

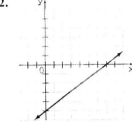

13.

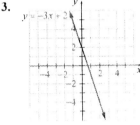

14.

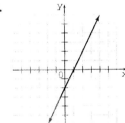

15.

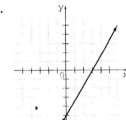

16.

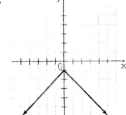

Chapter 8 Answers

17.

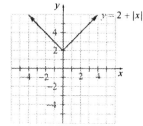

18.

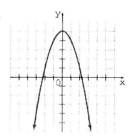

19.

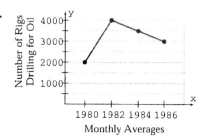

20. August 14th; August 15th; August 18th;
August 20th

Additional Exercises 8.2

1. D: $\mathbb{R}$ or $(-\infty, \infty)$; R: $\{y | y \geq 1\}$ or $[1, \infty)$

2. $\{-2, -1, 1\}$

3. $\{-5, 2, 7\}$

4. $\{y | -7 \leq y \leq 7\}$; $[-7, 7]$

5. no

6. yes

7. yes

8. no

9. 0

10. −21

11. −13

12. 16

13. $5.95

14. 144°

15. 8.8

16. no

17. yes

18. −1

19. 6

20. 12

Additional Exercises 8.3

1.

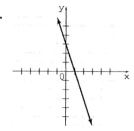

2.

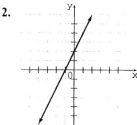

3.

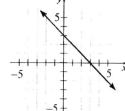

4.

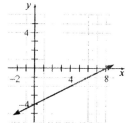

5.

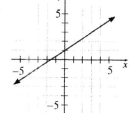

6.

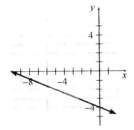

7. $y = 800 + 0.01x$; $7600

8.

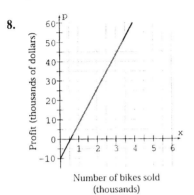

Number of bikes sold
(thousands)

9. approx. 600 bikes

10. approx. 2,222 bikes

11. $2x - 3y = -15$

12. $3x + 4y = 7$

13.

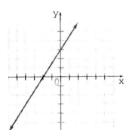

14.

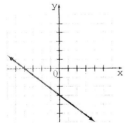

15.

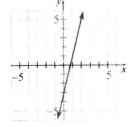

16.

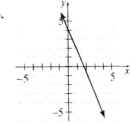

17. approx. $43,300

18.

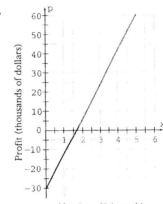

Number of bikes sold
(thousands)

19. approx. 1700 bikes

20. approx. 4200 bikes

Additional Exercises 8.4

1. $-\dfrac{3}{2}$

2. $\dfrac{3}{7}$

3. $m = -3$; $b = -18$

4.

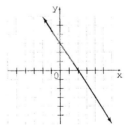

5.

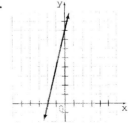

Chapter 8 Answers

6. $-\dfrac{4}{9}$

7. $m = \dfrac{7}{4}; \; b = -\dfrac{1}{2}$

8. $y = 4x + 5$

9.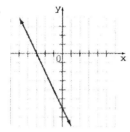

10. yes

11. no

12. perpendicular

13. $y = -5x - 37$

14. $2x + y = 0$

15. $y = \dfrac{6}{7}x + \dfrac{23}{7}$

16. $y = -\dfrac{2}{3}x - \dfrac{11}{3}$

17. $y = -\dfrac{5}{4}x - \dfrac{37}{4}$

18. no

19. yes

20. perpendicular

Additional Exercises 8.5

1. $4 + x$

2. $x^2 + 8x + 4$

3. $\dfrac{4x^2 - 9x + 3}{x^3}$

4. $x^2 + 10x + 12$

5. $3x^4 + 4x^3$

6. $x^3 - x^2 - x + 1$

7. $-x^2 + x + 6$

8. $\dfrac{5x^2 - 6x + 5}{x^4}$

9. $x^2 + 9x + 1$

10. $7x^4 - 4x^3$

11. $-11x^2 + 7x + 3$

12. $-x^3 - x^2 + x + 1$

13. $\dfrac{9x^2 - 4x + 7}{x^3}$

14. $x^2 - x - 4$

15. $-3x^5 + 5x^3$

16. $\left\{ x \mid x \neq -3 \right\}$

17. $\left\{ y \mid y \geq 8 \right\}$

18. $\mathbb{R}$

19. $\left\{ x \mid x \neq 0 \right\}$

20. $\left\{ y \mid y \geq -2 \right\}$

Chapter 8 Test Form A

1.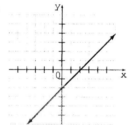

2.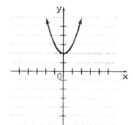

3.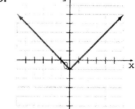

Chapter 8 Answers

4. $\{x \mid 3 < x \leq 17\}$

5. $(-3, \infty)$

6. no

7. D: $\{-1, 2, 3\}$;
   R: $\{-5, 4, 6, 7\}$

8. no

9. D: $\{x \mid x \geq -1\}$;
   R: $\mathbb{R}$

10.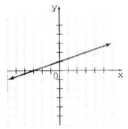

11.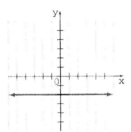

12. $\dfrac{5}{4}$

13. slope $= \dfrac{11}{2}$; y-intercept is $(0, -6)$

14. perpendicular since slopes are negative reciprocals

15. $y = \dfrac{3}{2}x + 11$

16. $x + 2y = -3$

17. 12,720 books

18. 12

19. $x^2 + 3x - 10$

20. $\dfrac{x^2 + x - 6}{2x - 4}$

Chapter 8 Test Form B

1. any set of ordered pairs

2. $\{x \mid -10 \leq x < -5\}$

3.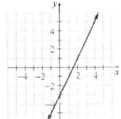

4.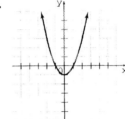

5.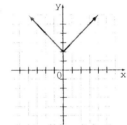

6. yes

7. D: $\{2, 3, 4, 5\}$;
   R: $\{6\}$

8. yes

9. D: {Paul, Dan, Doyle}; R: {31, 35}

10. slope is $-\dfrac{4}{5}$; y-intercept $\left(0, -\dfrac{4}{5}\right)$

11.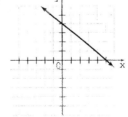

Chapter 8 Answers

12.

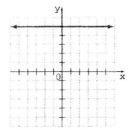

13. slope $= \dfrac{1}{5}$

14. neither; since slopes are not the same, nor negative reciprocals

15. $y = -\dfrac{3}{2}x + 3$

16. $3x + 2y = -5$

17. 13,637 books

18. 12

19. $x^2 + 5x - 9$

20. $\dfrac{x^2 + 2x}{3x - 9}$

Chapter 8 Test Form C

1. a relation in which each element of domain corresponds to exactly one element of the range.

2. independent variable

3.

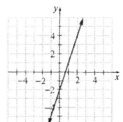

4.

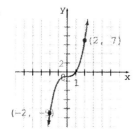

5.

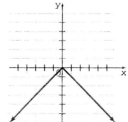

6. yes

7. D: {−2, 0, 1, 2};
R: {−1, 3, 5}

8. no

9. D: $\left\{x \mid -3 \le x \le 3\right\}$;
R: $\left\{y \mid -3 \le y \le 3\right\}$

10.

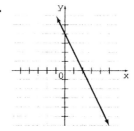

11.

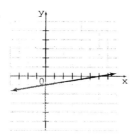

12.

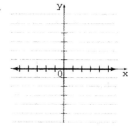

13. $\left\{\ldots, -6, -5, -4, -3\right\}$

14. slope = 4

15. slope $= -\dfrac{4}{3}$; y-intercept is $\left(0, -\dfrac{7}{3}\right)$

16. parallel; the slopes are equal and the y-intercepts are different

17. $y = -3x - 12$

18. $x + y = -8$

19. −10

20. 24

21. $-\dfrac{9}{4}$

22. $\dfrac{1}{2}x^3 + 2x^2 - \dfrac{1}{2}x - 2$

23. $\mathbb{R}$

24. $\left\{x \mid x \neq \pm 1\right\}$

25. May and June

Chapter 8 Test Form D

1. dependent variable

2. If a vertical line cannot be drawn to intersect the graph at more than one point, the graph represents a function.

3.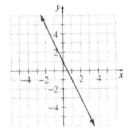

4.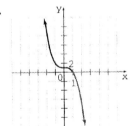

5.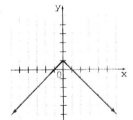

6. no

7. D: {−4, −3, −2};
R: {0, 2, 4, 6}

8. no

9. D: {Sam, Will, Rita};
R: {Bagel, Pasta, Grapes}

10.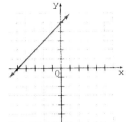

11.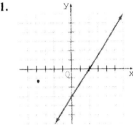

12.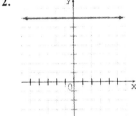

13. $\left\{1, 2, 3, 4, 5, 6\right\}$

14. slope $= -\dfrac{4}{3}$

15. slope $= -\dfrac{7}{9}$; y-intercept $\left(0, \dfrac{10}{9}\right)$

16. perpendicular since the slopes are negative reciprocals

17. $y = 2x - 6$

18. $3x + y = 4$

19. −78

20. 2

21. $x^2 + 2x - 4$

22. $\mathbb{R}$

23. $\left\{x \mid x \neq \pm 3\right\}$

24. $\left[5, \infty\right)$

25. 2 inches

Chapter 8 Answers

Chapter 8 Test Form E

1. false

2. $f(x) = 2x - 7$

3.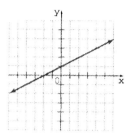

4.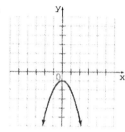

5.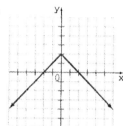

6. yes

7. D: $\{2, 4, 6, 8\}$
   R: $\{-5, -4, -3, -2\}$

8. yes

9. D: $\mathbb{R}$
   R: $\{y | y \geq -3\}$

10.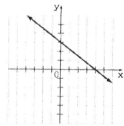

11.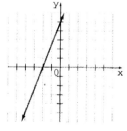

12.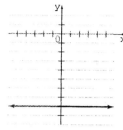

13. $\left(-\infty, -9\right)$

14. slope $= \frac{5}{3}$

15. slope $= -\frac{7}{2}$; y-intercept $(0, 4)$

16. perpendicular, since their slopes are opposite reciprocals of each other

17. $y = 6x - 33$

18. $x + 3y = 9$

19. -3

20. -72

21. 15

22. 250,000 bagels

23. $\mathbb{R}$

24. $\{x | x \neq 0\}$

25. Washington

Chapter 8 Test Form F

1. vertical line test

2. true

3. $y = 2x + 6$

4. $\{y | y \leq 2\}$

5. $\mathbb{R}$

6. $y = 3$

7. $\left(-0.5, 4.25\right]$

8. no

9. yes

10. $D : \mathbb{R}$; $R : \mathbb{R}$

11. $D : \{x | x \neq 0\}$; $R : \{y | y \neq 0\}$

12. $\left(\frac{3}{5}, 0\right)$; $(0, -3)$

13. $(-8, 0)$; $(0, 4)$

14. -2

15. $\frac{5}{4}; \left(0, -\frac{7}{4}\right)$

16. no

17. $y = \frac{2}{3}x + \frac{17}{3}$

18. $x + 2y = 4$

19. -5

20. -2

21. $\mathbb{R}$

22. $\left\{x \mid x \neq \pm 2\right\}$

23. slope = 3; y-intercept is $(0, -3)$

24. 250,000 bagels

25. May and June

Chapter 8 Test Form G

1. b
2. c
3. a
4. d
5. c
6. a
7. b
8. d
9. c
10. a
11. b
12. a
13. d
14. c
15. a
16. a
17. b
18. d
19. c
20. a
21. a
22. c
23. b

24. c
25. b

Chapter 8 Test Form H

1. a
2. a
3. c
4. b
5. d
6. a
7. c
8. d
9. a
10. c
11. a
12. c
13. d
14. d
15. c
16. d
17. a
18. b
19. c
20. d
21. a
22. b
23. b
24. c
25. d

Cumulative Review 1–8 Test Form A

1. 252

2. 98

3. $\dfrac{31}{4}$

4. $V = \dfrac{4R - SM}{2}$

5. $y = \dfrac{7}{2}x - \dfrac{15}{2}$

Chapter 8 Answers

6. $9x^2 - 5x + 4$

7. $\dfrac{81x^8}{y^8}$

8. $10x^3 - 19x^2 + 20x - 21$

9. $(w-2)(4w-5)$

10. $\left\{-1, \dfrac{5}{3}\right\}$

11. $\dfrac{x-5}{2x+1}$

12. $\dfrac{2x^2 + 3x + 10}{x^2 + 2x - 8}$

13. $\dfrac{10x + 51}{(x-5)(x+3)(x+6)}$

14. $x = -4$

15. $x = -\dfrac{28}{3}$

16. perpendicular; the slopes are opposite reciprocals

17. not a function; D: $\{x \mid x \geq 2\}$, R: $\mathbb{R}$

18. $x^2 - x + 5$

19. 70

20. 8.066×10^{12} cubic feet

14. b

15. a

16. d

17. c

18. c

19. b

20. d

Cumulative Review 1–8 Test Form B

1. d
2. d
3. b
4. a
5. **d**
6. c
7. a
8. c
9. b
10. a
11. b
12. d
13. c

Chapter 9 Answers

Chapter 9 Pretest Form A

1. (b)
2. inconsistent; no solution
3. one solution
4. no solution
5. infinite number of solutions
6. $(-2, -3)$

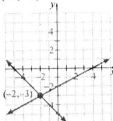

7. $(3, 2)$

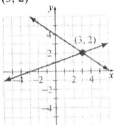

8. $(-1, 0)$
9. $(-4, 1)$
10. $(-4, -5)$
11. $(-8, 13)$
12. $(3, -2, 3)(-3, -3)$
13. $\begin{bmatrix} 1 & 1 & 3 & | & 5 \\ 0 & 1 & -4 & | & -10 \\ 1 & 2 & 1 & | & -1 \end{bmatrix}$
14. $\begin{bmatrix} -4 & 3 & | & 11 \\ 5 & 2 & | & -8 \end{bmatrix}$
15. $(-2, 1)$
16. -4
17. 43
18. $(-4, -7)$
19. 80 lbs of peanuts and 10 lbs of cashews
20. 7.5 liters of 60% chlorine solution and 2.5 liter of 20% chlorine solution

Chapter 9 Pretest Form B

1. (b)
2. consistent; one solution
3. no solution
4. infinite number of solutions
5. one solution
6. $(-1, 2)$;

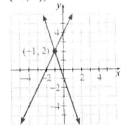

7. $(2, -3)$;

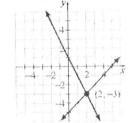

8. $(-2, 1)$
9. inconsistent, no solution
10. $(1, -3)$
11. $(-2, 0)$
12. $(7, -2, 1)$
13. $\begin{bmatrix} 1 & 2 & -1 & | & 0 \\ 0 & -7 & 2 & | & -3 \\ 2 & -1 & 3 & | & 5 \end{bmatrix}$
14. $\begin{bmatrix} 1 & 3 & | & 15 \\ 4 & -3 & | & 10 \end{bmatrix}$
15. $\left(5, \dfrac{10}{3}\right)$
16. 2
17. 25
18. $\left(\dfrac{1}{2}, -5\right)$
19. $3196 at 9% and $6809 at 10%
20. 300 miles

Chapter 9 Answers

Additional Exercises 9.1

1. b
2. c
3. b
4. No
5. No
6. Yes
7. Consistent; one solution
8. Inconsistent; no solution
9. Dependent; infinite number of solutions
10. One solution
11. No solution
12. One solution
13. Infinite number of solutions
14. No solution
15. One solution
16. Dependent

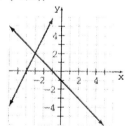

17. $(-3, 2)$;

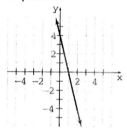

18. $(2, 2)$;

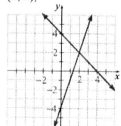

19. Inconsistent;

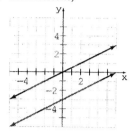

20. Dependent;

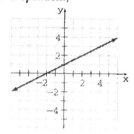

Additional Exercises 9.2

1. $(5, -4)$
2. $(2, -3)$
3. $(9, -8)$
4. No solution
5. $(7, -2)$
6. $(-1, -2)$
7. $\left(\dfrac{58}{101}, \dfrac{64}{101}\right)$
8. No solution
9. $(8, 8)$
10. $\left(\dfrac{3}{4}, -\dfrac{9}{8}\right)$
11. $(-2, 3)$
12. $(5, 4)$
13. $(4, 1)$
14. $\left(-\dfrac{1}{12}, -\dfrac{7}{12}\right)$
15. No solution
16. $(3, -9)$
17. Infinite number of solutions
18. $(-1, 3)$
19. No solution
20. $(3, -4)$

Chapter 9 Answers

Additional Exercises 9.3

1. $(7, -3)$
2. $(-4, 0)$
3. $(-1, 5)$
4. $\left(\dfrac{3}{10}, \dfrac{4}{5}\right)$
5. $(3, 2)$
6. $\left(-\dfrac{8}{7}, -\dfrac{13}{7}\right)$
7. $(-7, -3)$
8. $(-3, 5)$
9. $(0, 1)$
10. $(1, -8)$
11. $\left(\dfrac{13}{8}, \dfrac{15}{32}\right)$
12. $(-1, -5)$
13. Infinite number of solutions
14. No solution
15. $\left(-\dfrac{3}{46}, \dfrac{39}{46}\right)$
16. $\left(\dfrac{6}{7}, \dfrac{1}{7}\right)$
17. $(8, -3)$
18. Infinite number of solutions
19. $(7, -1)$
20. $(0, 0)$

Additional Exercises 9.4

1. $(2, 2, 5)$
2. $(5, -1, 5)$
3. $(4, -2, 0)$
4. $(3, -2, 1)$
5. $(-3, -5, 4)$
6. $(4, 1, 5)$
7. $\left(-\dfrac{1}{3}, -\dfrac{1}{3}, \dfrac{1}{3}\right)$
8. $(1, -1, 2)$

9. $\left(-\dfrac{3}{5}, \dfrac{1}{5}, \dfrac{1}{5}\right)$
10. $(2, 3, -5)$
11. $(2, -2, 3)$
12. $(-6, -2, -1)$
13. inconsistent
14. inconsistent
15. dependent
16. dependent
17. inconsistent
18. inconsistent
19. dependent
20. dependent

Additional Exercises 9.5

1. 60 tickets
2. 70 pounds of peanuts, 10 pounds of cashews
3. 30 dimes and 14 quarters
4. 20 pounds of peanuts, 40 pounds of cashews
5. 475 main floor tickets; 650 balcony tickets
6. $15,000 at 6% and $9,000 at 12%
7. $12,000 at 12% and $14,000 at 7%
8. $15,000 at 7% and $12,000 at 9%
9. 7 cookies
10. 29
11. $37°$
12. 10, -20 and 30
13. 7 cookies
14. 31, 35, and 93
15. $15,000 in mutual funds, $30,000 in bonds, $25,000 in the food franchise
16. $50°$
17. 4 cookies
18. 31, 38, 217
19. $52°$
20. $57°$

Chapter 9 Answers

Additional Exercises 9.6

1. $(-2, 3)$
2. $(2, 2)$
3. inconsistent system
4. inconsistent system
5. $(4, -4)$
6. $(5, -5)$
7. $(2, 2)$
8. $(3, 2)$
9. dependent system
10. $\left(\dfrac{6}{13}, -\dfrac{5}{13}\right)$
11. $(2, 5, -4)$
12. $(4, -2, 4)$
13. $(5, 0, -4)$
14. $(7, -3, -2)$
15. $(-2, -2, -5)$
16. $(1, -2, -1)$
17. $(-6, 3, 6)$
18. $(-2, 6, 4)$
19. dependent system
20. inconsistent system

Additional Exercises 9.7

1. 10
2. -26
3. -24
4. -31
5. 10
6. -5
7. 0
8. 0
9. 2
10. 30
11. $\left(-\dfrac{9}{5}, \dfrac{16}{25}\right)$
12. $\left(\dfrac{1}{5}, -2\right)$
13. $(5, -4)$

14. dependent system
15. $(-3, 3, -2)$
16. inconsistent system
17. inconsistent system
18. $(2, 3, 3)$
19. $(1, 0, -5)$
20. $(-3, 1, 3)$

Chapter 9 Test Form A

1. (c)
2. consistent; one solution
3. no solution
4. infinitely many solutions
5. one solution
6. $(3, -1)$;

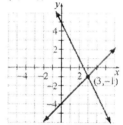

7. $(-2, 0)$;

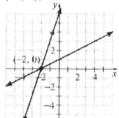

8. $(-2, 2)$
9. $\left(\dfrac{5}{2}, \dfrac{5}{4}\right)$
10. $(-6, 2)$
11. $\left(-\dfrac{1}{2}, 5\right)$
12. $(1, -1, 2)$
13. $\begin{bmatrix} 1 & -1 & 1 & | & 6 \\ 2 & 3 & 2 & | & 2 \\ 0 & 5 & 4 & | & -2 \end{bmatrix}$

14. $\begin{bmatrix} 1 & 1 & 1 & | & 4 \\ 0 & -3 & -2 & | & -3 \\ 2 & -1 & -2 & | & -1 \end{bmatrix}$

15. $(2, 7)$

16. 33

17. -10

18. $(1, -1)$

19. 42 L of 2%, 18 L of 6%

20. $(4, 2, -1)$

Chapter 9 Test Form B

1. (c)

2. inconsistent; no solution

3. infinite number of solutions

4. no solution

5. one solution

6. $(-3, 5)$;

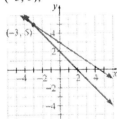

7. $(2, 2)$

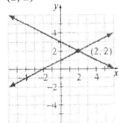

8. $(1, 3)$

9. $(2, -2)$

10. dependent; infinite number of solutions

11. $(-2, 0)$

12. $(5, -2, 1)$

13. $\begin{bmatrix} 2 & -3 & 1 & | & 5 \\ 1 & 3 & 8 & | & 22 \\ 3 & -1 & 2 & | & 12 \end{bmatrix}$

14. $\begin{bmatrix} 1 & 1 & 1 & | & 2 \\ 0 & -3 & -2 & | & 0 \\ 3 & 2 & 1 & | & 2 \end{bmatrix}$

15. $(2, 3)$

16. 44

17. 4

18. $\left(\dfrac{22}{7}, \dfrac{2}{7} \right)$

19. 12, 27, 18

20. $1\frac{1}{2}$ L of 5%, $4\frac{1}{2}$ L of 25%

Chapter 9 Test Form C

1. (a)

2. consistent; one solution

3. no solution

4. one solution

5. infinite number of solutions

6. $(-2, -3)$;

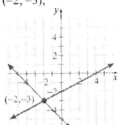

7. $(4, -2)$;

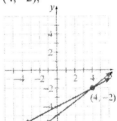

8. dependent; infinite number of solutions

9. $(2, 3)$

10. inconsistent; no solution

11. $(-1, -1)$

12. $(-3, -2, 4)$

Chapter 9 Answers

13. $\begin{bmatrix} 6 & -4 & 5 & | & 31 \\ 5 & 2 & 2 & | & 13 \\ 1 & 1 & 1 & | & 2 \end{bmatrix}$

14. $\begin{bmatrix} 1 & 1 & 1 & | & 6 \\ 0 & 1 & 1 & | & 5 \\ 0 & 0 & 4 & | & 12 \end{bmatrix}$

15. $(2, -1)$

16. -33

17. 29

18. $\left(\dfrac{8}{3}, \dfrac{1}{2}\right)$

19. 37 ft, 113 ft

20. $6000 at 10%, $3000 at 8%

Chapter 9 Test Form D

1. (d)

2. inconsistent; no solution

3. one solution

4. infinite number of solutions

5. no solution

6. $(2, 0)$;

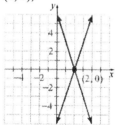

7. $(-2, -3)$;

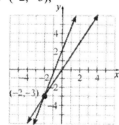

8. $(2, 3)$

9. $(4, 5)$

10. $(-1, -2)$

11. dependent; infinite number of solutions

12. $(2, -4, 16)$

13. $\begin{bmatrix} 3 & -4 & 1 & | & 7 \\ 5 & 1 & -3 & | & 11 \\ 0 & 7 & 2 & | & -5 \end{bmatrix}$

14. $\begin{bmatrix} 1 & -2 & -1 & | & -2 \\ 0 & 11 & 5 & | & 16 \\ 0 & 3 & 1 & | & 4 \end{bmatrix}$

15. $(4, 8)$

16. 4

17. 7

18. $(2, 4)$

19. 30 nickels, 22 dimes

20. 30 L of 5%, 70 L of 15%

Chapter 9 Test Form E

1. (b)

2. inconsistent; no solution

3. infinite number of solutions

4. no solution

5. one solution

6. $(-2, -2)$;

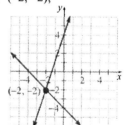

7. $(3, 1)$;

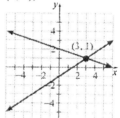

8. dependent; infinite number of solutions

9. $(5, 4)$

10. no solution

11. $(2, -3)$

12. $\left(\dfrac{1}{3}, -\dfrac{1}{3}, 1\right)$

686

13. $\begin{bmatrix} -2 & -1 & -1 & | & -3 \\ 3 & -2 & -2 & | & -5 \\ -1 & 1 & 0 & | & 0 \end{bmatrix}$

14. $\begin{bmatrix} 1 & 3 & -2 & | & 2 \\ 0 & 1 & 4 & | & 5 \\ 0 & 0 & -25 & | & -25 \end{bmatrix}$

15. $(3, 0)$

16. 32

17. -10

18. $(0, -4)$

19. boat 16 mph, current at 4 mph.

20. $8, 21, -3$

Chapter 9 Test Form F

1. (b)

2. inconsistent; no solution

3. no solution

4. one solution

5. infinite number of solutions

6. $(1, 3)$

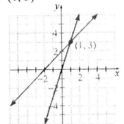

7. $(-4, -3)$

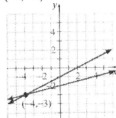

8. $(-5, 7)$

9. $\left(-2, -\dfrac{7}{2}\right)$

10. $(2, 4)$

11. dependent; infinitely many solutions

12. $(5, -3, 2)$

13. $\begin{bmatrix} 1 & 1 & 0 & | & 6 \\ 1 & -1 & 1 & | & 3 \\ 1 & 0 & -1 & | & 5 \end{bmatrix}$

14. $\begin{bmatrix} 1 & 5 & 0 & | & 6 \\ 0 & 2 & 1 & | & 3 \\ 0 & -20 & 3 & | & -17 \end{bmatrix}$

15. $(-3, -4)$

16. 22

17. -32

18. $\left(\dfrac{19}{13}, \dfrac{5}{13}\right)$

19. 20 quarts

20. 200

Chapter 9 Test Form G

1. c

2. a

3. c

4. b

5. a

6. d

7. c

8. b

9. d

10. d

11. b

12. c

13. a

14. d

15. d

16. c

17. b

18. a

19. a

20. c

Chapter 9 Answers

Chapter 9 Test Form H

1. d
2. a
3. a
4. c
5. b
6. b
7. d
8. a
9. c
10. b
11. a
12. d
13. c
14. b
15. c
16. b
17. c
18. a
19. a
20. d

Chapter 10 Answers

Chapter 10 Pretest Form A

1. $\{1, 2, 3, 4, 5, 6\}$

2. $\varnothing$

3. $x \geq -2$;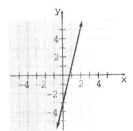

4. $x > 3$;

5. $[-2, 7)$

6. $\{x \mid x > 0\}$

7. $(-5, 4]$

8. $\{-9, 4\}$

9. $\{x \mid x < -5 \text{ or } x > 3\}$

10. $\left\{ x \mid -\dfrac{17}{2} < x < \dfrac{5}{2} \right\}$

11. $\left\{ \dfrac{5}{6}, \dfrac{7}{2} \right\}$

12. $1.25 + 0.75x \leq 4.25$; at most 4 hours

13.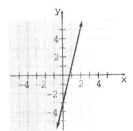

14.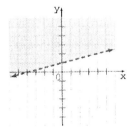

15.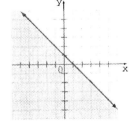

16.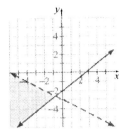

17.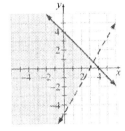

18.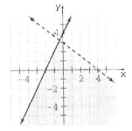

19.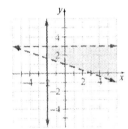

20.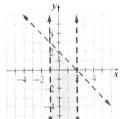

Chapter 10 Pretest Form B

1. $\{1, 3, 4, 5, 6, 7, 8, 9\}$

2. $\{5, 7\}$

3. $x \leq 9$;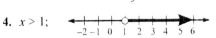

4. $x > 1$;

5. $[-4, 5)$

6. $\{x \mid x \leq 3 \text{ or } x > 4\}$

Chapter 10 Answers

7. [−3, 17)

8. {−2, 5}

9. $\{x \mid -10 \le x \le 2\}$

10. $\{x \mid x < -7 \ \text{or} \ x > 3\}$

11. $\left\{-\dfrac{3}{2}, \dfrac{1}{2}\right\}$

12. $\dfrac{90 + 87 + 96 + 79 + x}{5} \ge 90$; 98 or higher

13.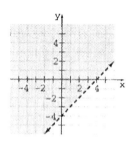

14.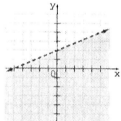

15.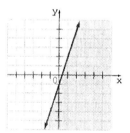

16.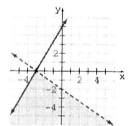

17.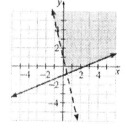

18.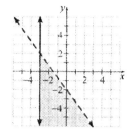

19.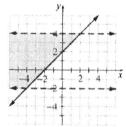

20.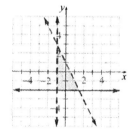

Mini-Lecture 10.3

1a)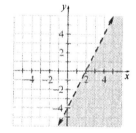

1b)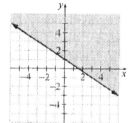

1c)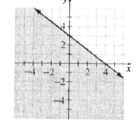

2a)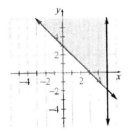

2b)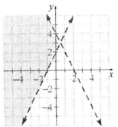

2c)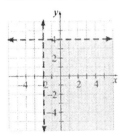

3a)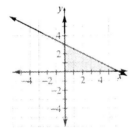

3b)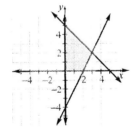

4a)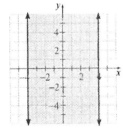

4b)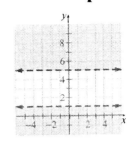

5a)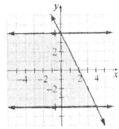

5b)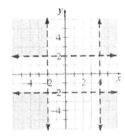

Additional Exercises 10.1

1. $A \cup B = \{3, 7, 8, 9, 10, 12, 17\}$;
   $A \cap B = \{7, 10\}$

2. $A \cup B = \{7, 9, 11, 13, ...\}$;
   $A \cap B = \{9, 11, 13, 15\}$

3. (a)

   (b) $[2, \infty)$

   (c) $\{x \mid x \geq 2\}$

4. (a)

   (b) $\left(-\infty, -\dfrac{4}{3}\right)$

   (c) $\left\{x \mid x < -\dfrac{4}{3}\right\}$

Chapter 10 Answers

5. (a)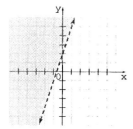

   (b) $\left[-5, \dfrac{2}{3}\right)$

   (c) $\left\{x\,\middle|\,-5 \le x < \dfrac{2}{3}\right\}$

6.

7.

8.

9. $\left(-\infty, 5\right]$

10. $(-27, -24)$

11. $\left[-\dfrac{41}{4}, -\dfrac{25}{4}\right)$

12. $\left\{x\,\middle|\,-3 < x \le -1\right\}$

13. $\left\{x\,\middle|\,-7 < x < \dfrac{3}{2}\right\}$

14. $\left\{x\,\middle|\,-1 \le x \le 3\right\}$

15. $\left\{x\,\middle|\,x > -12\right\}$

16. $\left\{x\,\middle|\,2 < x < 10\right\}$

17. $\left\{x\,\middle|\,x < 8\right\}$

18. $x \ge 353$ cm

19. $x \ge 190$ cm

20. $55 \le x \le 100$

Additional Exercises 10.2

1. $\{1, 5\}$

2. $\{1, -3\}$

3. $\left\{x\,\middle|\,6 < x < 8\right\}$

4. $\left\{x\,\middle|\,\dfrac{1}{8} \le x \le \dfrac{5}{8}\right\}$

5. $\left\{x\,\middle|\,-2 \le x \le 1\right\}$

6. $\left\{x\,\middle|\,-\dfrac{1}{3} < x < 3\right\}$

7. $\left\{x\,\middle|\,x < -6 \text{ or } x > 10\right\}$

8. $\left\{x\,\middle|\,x > 5 \text{ or } x < 4\right\}$

9. $\left\{x\,\middle|\,x < -4 \text{ or } x > \dfrac{2}{3}\right\}$

10. $\left\{x\,\middle|\,x < -\dfrac{4}{5} \text{ or } x > \dfrac{8}{5}\right\}$

11. $\varnothing$

12. $\mathbb{R}$

13. $\mathbb{R}$

14. $\varnothing$

15. $\left\{-\dfrac{7}{3}, -\dfrac{77}{3}\right\}$

16. $\{-6, 12\}$

17. $\{3\}$

18. $\{2, 16\}$

19. $\left\{-\dfrac{40}{3}, -\dfrac{120}{13}\right\}$

20. $\left\{\dfrac{15}{4}, -\dfrac{75}{14}\right\}$

Additional Exercises 10.3

1.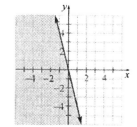

2.

3.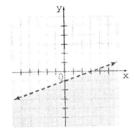

4.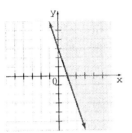

5.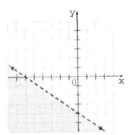

6.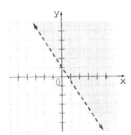

7.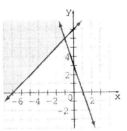

8.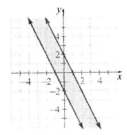

9.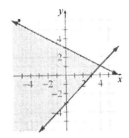

10.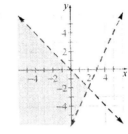

11.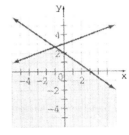

12.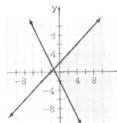

13.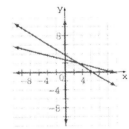

14.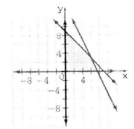

Chapter 10 Answers

15.

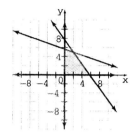

16.

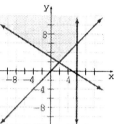

17.

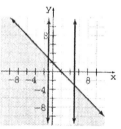

18.

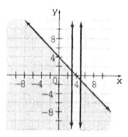

19.

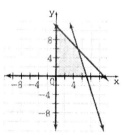

20.

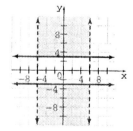

Chapter 10 Test Form A

1. $A \cup B = \{2, 4, 6, 8, 11\}$; $A \cap B = \{2, 4, 6\}$

2. $A \cup B = \{0, 2, 3, 4, 6, 8, 9\}$; $A \cap B = \{0, 6\}$

3. $x \le 8$

   [number line with closed dot at 8, shaded to the left; marks 0 1 2 3 4 5 6 7 8]

4. $-5 < x < 3$

   [number line with open circles at -5 and 3, shaded between; marks -5 -4 -3 -2 -1 0 1 2 3]

5. $\left[-6, -\dfrac{3}{2} \right]$

6. $[-4, \infty)$

7. $\left(\dfrac{8}{3}, \infty \right)$

8. $\{-10, 2\}$

9. $\left\{ x \,\middle|\, -\dfrac{2}{3} \le x \le \dfrac{26}{21} \right\}$

10. $\left\{ x \,\middle|\, x < -5 \text{ or } x > 1 \right\}$

11. $\left\{ \dfrac{2}{7}, 12 \right\}$

12. $2(x + 12) \ge 416$; The length is at least 196 cm.

13.

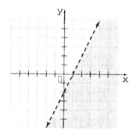

14.

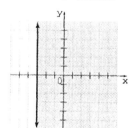

15.

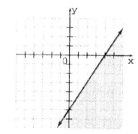

694

Chapter 10 Answers

16.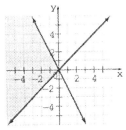

17.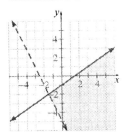

18.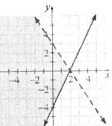

19.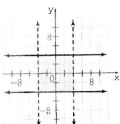

20.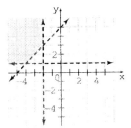

5. $[-14, 21]$

6. $(-\infty, 0) \cup (6, \infty)$

7. $[-1, 2]$

8. $\{-13, 5\}$

9. $\{x \mid -1 \le x \le 2\}$

10. $\{x \mid x > 3 \text{ or } x < -2\}$

11. $\left\{-\dfrac{6}{7}, 54\right\}$

12. $70x + 150 \le 1200$; Harold can take at most 15 boxes on the elevator.

13.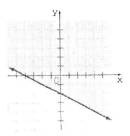

14.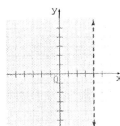

15.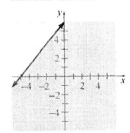

16.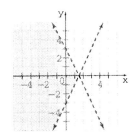

Chapter 10 Test Form B

1. $A \cup B = \{-1, 0, 1, e, i, \pi\}$; $A \cap B = \{-1, 0, 1\}$

2. $A \cup B = \{1, 2, 4, 6, 8, 10, 16\}$; $A \cap B = \{2, 4, 8\}$

3. $x < 8$

4. $-4 < x \le 3$

Chapter 10 Answers

17.

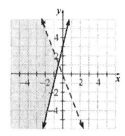

18.

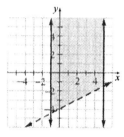

19.

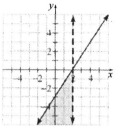

20.

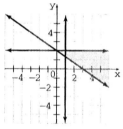

Chapter 10 Test Form C

1. $A \cup B = \{-3,-1,1,3,5,7,9\}$; $A \cap B = \{1,3,5\}$

2. $A \cup B = \{...,-3,-2,-1,0,1,2,3,...\}$;
    $A \cap B = \{2,4,6,8,...\}$

3. $x < -2$

    ```
    ←——————○+++++++++→
    -4 -3 -2 -1  0  1  2  3  4
    ```

4. $-3 < x \le 2$

    ```
    ←++○——————●++→
    -4 -3 -2 -1  0  1  2  3  4
    ```

5. $(0, 1]$

6. $(-\infty, 6]$

7. $(2, \infty)$

8. $\{-4, 10\}$

9. $\varnothing$

10. $\left\{ x \,\middle|\, x < \dfrac{2}{3} \text{ or } x > 4 \right\}$

11. $\left\{ -6, -\dfrac{9}{4} \right\}$

12. $80 \le \dfrac{83+71+96+x}{4} < 90$; Tracy must get a
    70 or higher.

13.

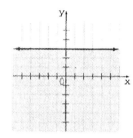

14.

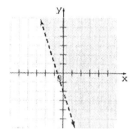

15.

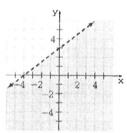

16.

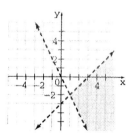

17.

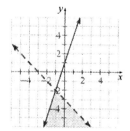

696

18.

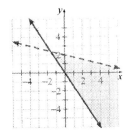

19.

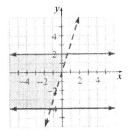

20.

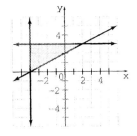

Chapter 10 Test Form D

1. $A \cup B = \{1, 3, 4, 5, 6, 8, 9\}$; $A \cap B = \{4\}$

2. $A \cup B = \{0, 1, 2, 3, 4, 5, 6, 7\}$; $A \cap B = \{3, 4\}$

3. $x \geq -7$

4. $-4 < x \leq 3$

5. $[-4, -2)$

6. $[-2, \infty)$

7. $[-4, 3)$

8. $\{3, 7\}$

9. $\{x \mid -4 \leq x \leq 1\}$

10. all real numbers

11. $\{-44, 4\}$

12. $5(40) + 0.10x \leq 300$; She can drive up to 1000 miles

13.

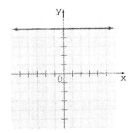

14.

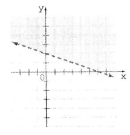

15.

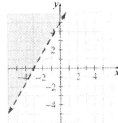

16.

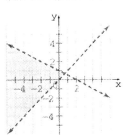

17.

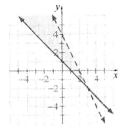

18.

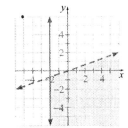

Chapter 10 Answers

19.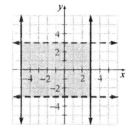

20.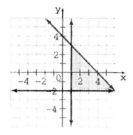

Chapter 10 Test Form E

1. $A \cup B = \{0, 1, 2, 3, 4, \ldots\}$; $A \cap B = \{\ \}$

2. $A \cup B = \{-7, -5, -4, -3, -1, 1, 2, 3, 5, 7\}$;
   $A \cap B = \{-7, -1, 5\}$

3. $x < -2$

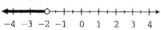

4. $-4 < x \leq 3$

5. $[-17, 7]$

6. $(-\infty, 2]$

7. $[-4, 4)$

8. $\{1, 3\}$

9. $\{x | -5 \leq x \leq 2\}$

10. all real numbers

11. $\left\{ \dfrac{12}{7}, \dfrac{16}{5} \right\}$

12. $200 + 0.05x \geq 400$; She must have weekly sales of at least $4000.

13.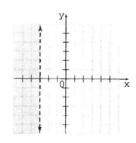

14.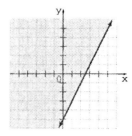

15.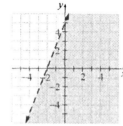

16.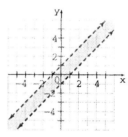

17.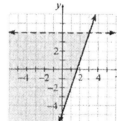

18.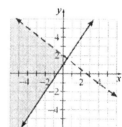

19.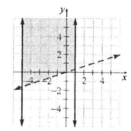

698

20.

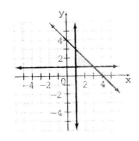

Chapter 10 Test Form F

1. $A \cup B = \{0, 1, 2, 3, 4, 5, 6, 7, 8, 9\}$; $A \cap B = \{\ \}$

2. $A \cup B = \{-3, -2, -1, 0, 1, 2, 3\}$; $A \cap B = \{0, 1\}$

3. $x < 25$

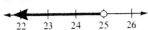

4. $-2 \le x \le 3$

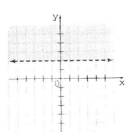

5. $[-55, 1]$

6. $(-\infty, 3)$

7. $(-\infty, 2)$

8. $\left\{\dfrac{1}{2}, \dfrac{7}{6}\right\}$

9. $\varnothing$

100. $\left\{x \middle| x \le -2 \ or \ x \ge \dfrac{8}{5}\right\}$

11. $\{-11, 3\}$

12. $2(x + 38) \ge 300$; The length is at least 112 cm.

13.

14.

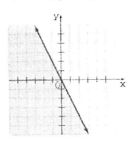

15.

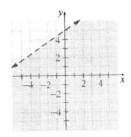

16.

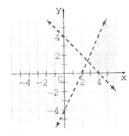

17.

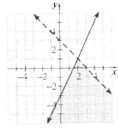

18.

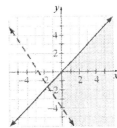

19.

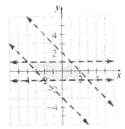

20.

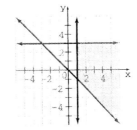

Chapter 10 Answers

Chapter 10 Test Form G

1. d
2. a
3. b
4. a
5. c
6. a
7. d
8. b
9. a
10. d
11. c
12. d
13. b
14. c
15. c
16. a
17. d
18. d
19. c
20. b

Chapter 10 Test Form H

1. b
2. d
3. d
4. b
5. b
6. d
7. a
8. c
9. c
10. b
11. a
12. d
13. a
14. c
15. b

16. d
17. c
18. a
19. a
20. c

Cumulative Review 1–10 Test Form A

1.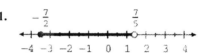

2. 41

3. $\dfrac{4a^2 b^2 c^6}{9}$

4. 2.94×10^{-4}

5. Aug. 12 and Aug. 19

6. $178°$

7. $2°$

8. 1

9. $\dfrac{5}{3}$

10.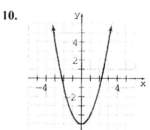

11.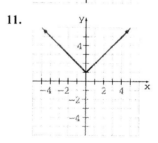

12. function

13. $y = \dfrac{3}{2}x - 4$

14. -25.5

15. $(3, -4, -4)$

16. 3.125 liters

17. $x^2 + 10x + 12$

18. $\left(4x^2+9y^2\right)\left(2x+3y\right)\left(2x-3y\right)$

19. $x=3$

20. no solution

Cumulative Review 1–10 Test Form B

1. c
2. b
3. d
4. c
5. b
6. a
7. d
8. a
9. b
10. a
11. d
12. c
13. a
14. d
15. b
16. c
17. d
18. b
19. a
20. d

Chapter 11 Answers

Chapter 11 Pretest Form A

1. 11
2. –2
3. $|m+4|$
4. $5|b|$
5. $53^{1/3}$
6. $(3a)^{1/4}$
7. $\sqrt[4]{3x}$
8. $\sqrt[4]{\dfrac{1}{2}x^3 y}$
9. $\dfrac{\sqrt[5]{3}}{2}$
10. $x^4 y \sqrt[3]{18}$
11. $4\sqrt{5c}$
12. $6b$
13. $9\sqrt{3}$
14. –2
15. $-3-2\sqrt{3}$
16. $-12+34i$
17. $-\dfrac{21}{2}-\dfrac{3}{2}i$
18. 16
19. 1, 9
20. no solutions

Chapter 11 Pretest Form B

1. 9
2. –10
3. 17
4. $|x+9|$
5. $x^{9/2}$
6. $m^{4/5}$
7. $\sqrt[6]{a}$

8. $\sqrt[5]{(5m-n)^8}$ or $\left(\sqrt[5]{5m-n}\right)^8$
9. 36
10. $\sqrt[6]{x}$
11. $3x^5 y^{15}\sqrt{5y}$
12. $\dfrac{x^2 y^3 \sqrt{3xz}}{6z}$
13. $6\sqrt{6}$
14. $3\sqrt{2}-2\sqrt{3}+\sqrt{6}-2$
15. $\sqrt{6}-\sqrt{5}$
16. $1+10i$
17. $4i$
18. 4
19. 9
20. 75 feet

Additional Exercises 11.1

1. 12
2. $\dfrac{5}{9}$
3. 0.6
4. 1.5
5. not a real number
6. –4
7. 29
8. 0.14
9. $\left|5x^2 - y\right|$
10. $|4x+2|$
11. $\left|y^8\right|$
12. $|x-5|$
13. $|2a-7b|$
14. 4
15. 3.162
16. 2.924

17. 2

18. $x > -1$

19. about 43.95 feet per second

20.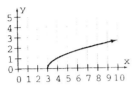

Additional Exercises 11.2

1. $x^{4/9}$

2. $y^{3/5}$

3. $\sqrt[3]{\left(12b^5\right)^2}$ or $\left(\sqrt[3]{12b^5}\right)^2$

4. $\sqrt{3}$

5. x^2

6. x^7

7. $a^4 b^2$

8. $\sqrt[10]{x^7}$

9. $\dfrac{8}{15}$

10. $\dfrac{2}{11}$

11. $x^{15/4}$

12. $\dfrac{1}{x}$

13. $64x^{8/15}$

14. $-\dfrac{20}{y^{1/4}} + 4y^{13/8}$

15. 2.91

16. $x^{1/2}\left(1+x\right)$

17. $\dfrac{x+1}{x^6}$

18. about 7.31 milligrams

19. about 10.51 milligrams

20. $x > 5$

Additional Exercises 11.3

1. $4\sqrt{11}$

2. $3\sqrt[4]{5}$

3. $4a^2 b\sqrt[3]{3a}$

4. $xy^2 \sqrt[3]{x^2 y^2}$

5. $6xy^2 \sqrt{6y}$

6. $3ab\sqrt[5]{5ab^3}$

7. $10a^{50}$

8. $x^2 y^4 \sqrt{y}$

9. $75\sqrt{7}$

10. $6x^2 y^2 \sqrt{3y}$

11. $5x^4 y^4 \sqrt[3]{3x^2 y}$

12. $x^2 y \sqrt[3]{49x^2 y}$

13. $2a^2 b^4 \sqrt[5]{4a^3 b^3}$

14. $2x^3 y^2 \sqrt{2y}$

15. 4

16. $\dfrac{4}{9}$

17. $5x$

18. $\dfrac{2}{y}$

19. $\dfrac{\sqrt[3]{5y}}{3x^3}$

20. $\dfrac{xy^2}{2}$

Additional Exercises 11.4

1. $-3\sqrt[4]{9}$

2. $7\sqrt[3]{y}$

3. $7 - 4\sqrt{x}$

4. $11\sqrt{5}$

Chapter 11 Answers

5. $5x\sqrt{7y}$

6. $2\sqrt[3]{10}$

7. $7\sqrt[3]{x} - x\sqrt[3]{x}$

8. $-2\sqrt[3]{x} - 3x\sqrt[3]{x}$

9. $4x^2 y^2 \sqrt{6y}$

10. $2a^2 b^4 \sqrt[5]{4a^3 b^3}$

11. $x^2 y \sqrt[3]{49x^2 y}$

12. 24

13. $6y\sqrt{2} - y^2 \sqrt{6}$

14. $\sqrt{2} - 18$

15. $-6x - 47\sqrt{xy} + 8y$

16. $32 - 10\sqrt{7}$

17. $7 - \sqrt[3]{10} - \sqrt[3]{100}$

18. $x - 7\sqrt[3]{x} - 2\sqrt[3]{x^2} + 14$

19. $P = 28\sqrt{3}$, $A = 120$

20. $P = 23\sqrt{2}$, $A = 22$

9. $\dfrac{2y^3 \sqrt[3]{x^2}}{x}$

10. $\dfrac{2\sqrt{7}}{7}$

11. $\dfrac{25\sqrt{2}}{4}$

12. $\dfrac{41\sqrt{5}}{10}$

13. $\dfrac{15 + 3\sqrt{6}}{19}$

14. $\dfrac{x - \sqrt{2x}}{x - 2}$

15. $\dfrac{14 - 5\sqrt{3}}{11}$

16. $\dfrac{13 - 4\sqrt{3}}{11}$

17. $\sqrt[12]{x}$

18. $x^2 \sqrt[6]{xy^5}$

19. 7 feet

20. 30 inches

Additional Exercises 11.5

1. $\dfrac{h\sqrt{6}}{6}$

2. $5\sqrt{2}$

3. $\dfrac{\sqrt{6}}{8}$

4. $\dfrac{\sqrt{3x}}{x}$

5. $\sqrt{10}$

6. $\dfrac{\sqrt{6x}}{x}$

7. $\dfrac{x^2 y^2 \sqrt{yz}}{2z}$

8. $\dfrac{\sqrt[3]{9r^2 s^2}}{3s^4}$

Additional Exercises 11.6

1. 21

2. 60

3. 225

4. 2

5. -5

6. 3, 4

7. 7

8. no real solution

9. $0, \dfrac{1}{4}$

10. 36

11. no real solution

12. $0, \dfrac{100}{81}$

13. $x = 8$

Chapter 11 Answers

14. $x = 2$

15. $x = 1$

16. $H = \dfrac{u^2 a}{R}$

17. $y = \dfrac{qx}{r^2}$

18. about 9.42 seconds

19. $\ell = \dfrac{4T^2}{3\pi^2}$

20. about 3.38 feet

Additional Exercises 11.7

1. $-7 + 10i$

2. $-5i$

3. $2 - 4i$

4. $-9 + 18i$

5. $11i\sqrt{3}$

6. $11i\sqrt{2}$

7. $7 + 5i\sqrt{7}$

8. $-12 + 20i$

9. $12 + 33i$

10. $-19 - 35i$

11. $-41 + 68i$

12. $8 + \dfrac{29}{4}i\sqrt{2}$

13. $-\dfrac{7i}{4}$

14. $4 - \dfrac{3i}{2}$

15. $\dfrac{5}{82} - \dfrac{37i}{82}$

16. $\dfrac{\sqrt{2} + i\sqrt{14}}{56}$

17. $-i$

18. 1

19. $3 - 4i$

20. $-7 + i$

Chapter 11 Test Form A

1. 5

2. $\dfrac{5}{12}$

3. $|2x - 3|$

4. $\left(\sqrt[5]{7b^2 c}\right)^3$

5. $x^{3/5} y^{2/5}$

6. x^3

7. $5\sqrt{3}$

8. $-2x^3 y^3 z^6 \sqrt{5y}$

9. $\dfrac{x^3 \sqrt[4]{20}}{3}$

10. $2\left(\sqrt{2} - \sqrt{3}\right)$

11. $x - y^2$

12. $3x^3 y^4 \sqrt[4]{2xy^3}$

13. $\dfrac{x\sqrt{13}}{13}$

14. $5\sqrt{2} - 5$

15. $\dfrac{c + \sqrt{cd} - \sqrt{2cd} - d\sqrt{2}}{c - d}$

16. $-\dfrac{301\sqrt{2}}{20}$

17. $\dfrac{\left(6\sqrt{3} - \sqrt{15}\right) - \left(6\sqrt{5} + 3\right)i}{4}$

18. 1

19. $\dfrac{9}{16}$

20. $21 - 6i$

Chapter 11 Answers

Chapter 11 Test Form B

1. $\dfrac{1}{3}$

2. $|5x - 8|$

3. $\dfrac{1}{\sqrt[6]{7x^2 + 2y^3}}$

4. y^3

5. $3y^{3/2}$

6. $-9z^3 + 9$

7. $2\sqrt[4]{5}$

8. $2x^2 y^5 \sqrt[4]{3x^3 y}$

9. $5a^4 b^4 \sqrt{3ab}$

10. $9\sqrt{5}$

11. $9a - 49b$

12. $15 + 3\sqrt{5}$

13. $\dfrac{2y^3 z \sqrt{15xz}}{3x}$

14. $5\sqrt{6} - 5\sqrt{5}$

15. $\dfrac{2\sqrt{x+2} + 6}{x - 7}$

16. $\dfrac{-16\sqrt{6}}{3}$

17. $\dfrac{1}{4}$

18. -1

19. $4\sqrt{5} + 3i\sqrt{3}$

20. $\dfrac{\left(5\sqrt{10} - 2\sqrt{15}\right) + \left(10\sqrt{2} + 5\sqrt{3}\right)i}{45}$

Chapter 11 Test Form C

1. 6

2. $|x - 5|$

3. $-\dfrac{3x}{y^3}$

4.

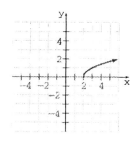

5. $x^{2/3} y^{1/3}$

6. $2a^4 b^3 \sqrt{6a}$

7. $7xy\sqrt{2}$

8. $x\sqrt{5x}$

9. $\dfrac{\sqrt{21x}}{3x}$

10. $2 + 2\sqrt{2}$

11. $10\sqrt{2}$

12. $22\sqrt{x}$

13. $10 - 4\sqrt{2}$

14. $\sqrt[6]{xy^2}$

15. $\sqrt[6]{x^5}$

16. 66

17. 4

18. $18 + 16i$

19. $\dfrac{-4 + 7i}{5}$

20. $3 - 10i$

Chapter 11 Test Form D

1. 9

2. $|2x - 1|$

3. $x^{3/4}$

4.

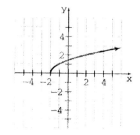

5. $x^{1/6} y^{5/6}$

6. $3xyz^2 \sqrt{5yz}$

7. $6y\sqrt{x}$

8. $b^2 \sqrt{3a}$

9. $\dfrac{\sqrt{30}}{5}$

10. $6 + 3\sqrt{3}$

11. $7\sqrt[3]{3}$

12. $9ab^2 \sqrt{3ab}$

13. $23 - 9\sqrt{5}$

14. $6x^3 y^2 \sqrt{2yz}$

15. $\sqrt[12]{x^{11}}$

16. 1

17. 5

18. $21 + 22i$

19. $\dfrac{-7 + 4i}{6}$

20. $3 - 27i$

Chapter 11 Test Form E

1. $\dfrac{1}{8}$

2. $|x + 1|$

3. $\dfrac{1}{xy^6}$

4.

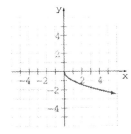

5. $x^{9/10} y^{7/10} z^{3/10}$

6. $-2xy^3 \sqrt[3]{x^2}$

7. $9x^2 y$

8. $5ab^2 \sqrt{a}$

9. $\dfrac{\sqrt[3]{21}}{3}$

10. $\dfrac{12 - 4\sqrt{2}}{7}$

11. $2\sqrt{2}$

12. $8x\sqrt{2}$

13. $23 - 10\sqrt{7}$

14. $\sqrt[6]{x^2 y^4}$

15. $\sqrt[4]{x}$

16. 32

17. 2

18. $-7 + 22i$

19. $-\dfrac{5 + 4i}{2}$

20. $-24 - 20i$

Chapter 11 Test Form F

1. $\dfrac{4}{9}$

2. $|x - 10|$

3. $x^{12/5}$

4.

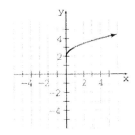

5. $\sqrt[5]{x^3 y^2}$

6. $5xy^3 \sqrt{2}$

7. 6

8. $2y\sqrt[3]{x}$

9. $\dfrac{5\sqrt[3]{a^2}}{a}$

Chapter 11 Answers

10. $2\sqrt{5} - 4$

11. $9\sqrt{3}$

12. $33\sqrt{x}$

13. -10

14. $\sqrt[6]{x^4 y^3}$

15. $\sqrt[6]{x}$

16. 2

17. 1

18. $7 + 22i$

19. $\dfrac{2 - i}{2}$

20. $6 - 5i$

Chapter 11 Test Form G

1. b
2. b
3. c
4. d
5. c
6. b
7. d
8. a
9. b
10. b
11. a
12. c
13. b
14. d
15. c
16. c
17. c
18. d
19. d
20. c

Chapter 11 Test Form H

1. d
2. a
3. a
4. b
5. a
6. d
7. b
8. c
9. c
10. a
11. d
12. b
13. a
14. c
15. b
16. a
17. b
18. a
19. c
20. b

708

Chapter 12 Answers

Chapter 12 Pretest Form A

1. $2, -4$

2. $\dfrac{-5 \pm \sqrt{5}}{10}$

3. $-3 \pm \sqrt{5}$

4. a single real solution

5. distinct real solutions

6. no real solutions

7. $h = \sqrt{d^2 - l^2 - w^2}$

8. $x^2 - 8x + 15 = 0$

9.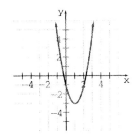

10. upward

11. $x = \dfrac{1}{2}$

12. $\left(\dfrac{1}{2}, \dfrac{-25}{4} \right)$

13. $(-2, 0)$ and $(3, 0)$

14.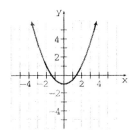

15.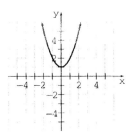

16. $(-\infty, -5) \cup (5, \infty)$

17. 256 feet

18. 8 seconds

19. $[0, 3)$

20. $(-\infty, -2) \cup (2, 18]$

Chapter 12 Pretest Form B

1. $-6 \pm 2\sqrt{10}$

2. $2, -10$

3. $2 \pm i\sqrt{3}$

4. a single real solution

5. two distinct real solutions

6. no real solutions

7. $a = \sqrt{P^2 - b}$

8. $f(x) = 5x^2 + 7x - 6$

9.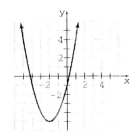

10. downward

11. $x = 2$

12. $(2, 4)$

13. $(0, 0)$ and $(4, 0)$

14.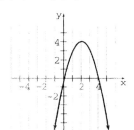

15. $-5 < x < 6$

16. $\left[-\dfrac{1}{2}, 2 \right]$

17. 192

18. \$29,864

19. $(-\infty, -4] \cup (2, 3]$

20. $(-\infty, -6) \cup (5, 27]$

Chapter 12 Answers

Mini–Lecture 12.5

1. a. $f(x) = -x^2 + 2x + 8$

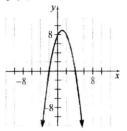

b. $f(x) = 4x^2 - 12x + 9$

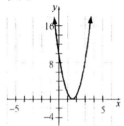

c. $f(x) = x^2 + 2x + 2$

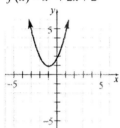

d. $f(x) = \frac{1}{6}x^2 + x$

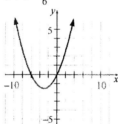

Additional Exercises 12.1

1. $\dfrac{25}{4}$

2. $\dfrac{1}{4}$

3. 81

4. $14x$

5. $\dfrac{-3 \pm \sqrt{15}}{3}$

6. $-3 \pm i\sqrt{5}$

7. $-3, 0$

8. $\dfrac{2 \pm \sqrt{5}}{2}$

9. $\dfrac{9}{4}$

10. $\dfrac{49}{4}$

11. $6x, 9$

12. 64

13. $\dfrac{-7 \pm \sqrt{61}}{6}$

14. $-2 \pm i\sqrt{3}$

15. $\dfrac{1 \pm \sqrt{13}}{2}$

16. $\dfrac{-2 \pm \sqrt{5}}{2}$

17. $\dfrac{1 \pm i\sqrt{39}}{4}$

18. $1, 2$

19. $\dfrac{-1 \pm i\sqrt{3}}{2}$

20. $7.00\,\%$

Additional Exercises 12.2

1. $\dfrac{-q \pm \sqrt{q^2 - 4pr}}{2p}$

2. $\dfrac{-b \pm \sqrt{b^2 - 4ac}}{2a}$

3. $\dfrac{1 \pm \sqrt{5}}{2}$

4. $-7 \pm \sqrt{3}$

5. $\dfrac{5 \pm \sqrt{37}}{6}$

6. $\dfrac{-5 \pm \sqrt{165}}{14}$

7. $4x^2 + 11x - 20 = 0$

Chapter 12 Answers

8. $6x^2 + 5x - 6 = 0$

9. $7x^2 + 30x + 8 = 0$

10. $3x^2 + 4x - 15 = 0$

11. no real solution

12. single unique

13. two unequal real roots

14. two unequal imaginary roots

15. $\dfrac{-h \pm \sqrt{h^2 - 4gk}}{2g}$

16. $\dfrac{1}{2}, -\dfrac{1}{5}$

17. $\dfrac{5 \pm \sqrt{13}}{2}$

18. $-9 \pm \sqrt{2}$

19. $\dfrac{3 \pm \sqrt{6}}{3}$

20. 120

Additional Exercises 12.3

1. $\dfrac{1}{4}, \dfrac{9}{4}$

2. $\dfrac{3 \pm \sqrt{30}}{9}$

3. $7, -7$

4. $-\dfrac{1}{2}, \dfrac{5}{2}$

5. $b = \pm\sqrt{\dfrac{3Z}{s}}$

6. $f = \pm\sqrt{\dfrac{c - 3d}{7}}$

7. $-6, -2$

8. $\dfrac{3 \pm \sqrt{15}}{7}$

9. $0, 6$

10. $1 \pm \dfrac{\sqrt{3}}{3}$

11. $f = \pm\sqrt{\dfrac{6A}{g}}$

12. $m = \pm\sqrt{\dfrac{j - 8k}{5}}$

13. 4 ft

14. length = 13 ft; width = 3 ft

15. 3 seconds

16. $10 \pm 2\sqrt{5}$

17. $400

18. 7.2 mi/hr

19. 1500

20. (a) 4 seconds
    (b) 338 ft

Additional Exercises 12.4

1. $-1, -\dfrac{1}{8}$

2. $-\dfrac{1}{5}, -\dfrac{1}{8}$

3. $\pm 3, \pm i$

4. $\pm 1, \pm 3$

5. $1, 625$

6. $-1, 27$

7. $36, 49$

8. $49, 100$

9. $-\dfrac{1}{3}, -1$

10. $-\dfrac{1}{2}, \dfrac{1}{5}$

11. $\pm 1, \pm\sqrt{10}$

12. $\pm 1, \pm\sqrt{15}$

13. $16, 256$

14. $\pm\sqrt{\dfrac{1}{5}}, \pm\sqrt{\dfrac{1}{2}}$

15. $9, 100$

16. $9, 121$

17. $\pm 1, \pm i\sqrt{3}$

18. no solution

19. $4, 16$

20. $\sqrt[3]{3}, -1$

Chapter 12 Answers

Additional Exercises 12.5

1. $x = -2; (-2,-3); x = -2 \pm \sqrt{3}$

2. $x = 2; (2,-5); x = 2 \pm \sqrt{5}$

3. $x = 1; (1,-2); x = 0, 2$

4. 1, 3

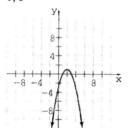

5.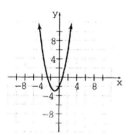

6.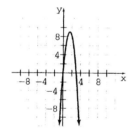

7.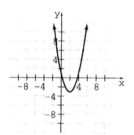

8. 1, 3

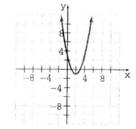

9.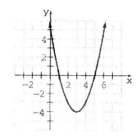

10.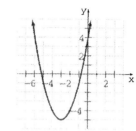

11.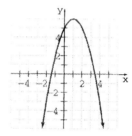

12.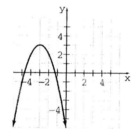

13.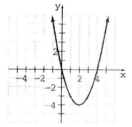

14.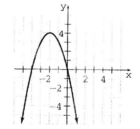

Chapter 12 Answers

15.

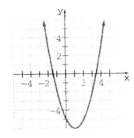

16.

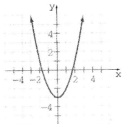

17.

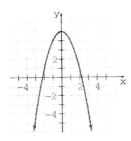

18. $y = -2(x+1)^2 + 1$

19. $y = 2(x - 3/2)^2 + \frac{5}{2}$

20. a square 36 inches on a side

Additional Exercises 12.6

1. $x \le -\frac{4}{3}$ or $x \ge 4$

2. ![number line] -4 -3 -2 -1 0 1 2 3 4

3. $-9 < x < -6$

4. $\left\{ x \middle| x \le -1 \text{ or } x \ge \frac{5}{2} \right\}$

5. $\left\{ x \middle| -8 < x < -3 \text{ or } x > 1 \right\}$

6. $\left\{ x \middle| -5 < x < -3 \text{ or } x > 2 \right\}$

7. $\left\{ x \middle| -7 \le x < 3 \right\}$

8. ![number line] -5 1 6; -8 -6 -4 -2 0 2 4 6 8

9. (a) $(-\infty, -3) \cup (-2, 1)$
   (b) $(-3, -2) \cup (1, \infty)$

10. $\left\{ x \middle| x < -2 \text{ or } 0 < x < 3 \right\}$

11. $-\frac{3}{5} \le x \le 8$

12. ![number line] -6 7; -8 -6 -4 -2 0 2 4 6 8

13. $-5 < x < -2$

14. $\left\{ x \middle| -2 < x < -1 \text{ or } 1 < x < 2 \right\}$

15. $\left\{ x \middle| -9 < x < -2 \text{ or } x > 4 \right\}$

16. $\left\{ x \middle| -8 < x < -5 \text{ or } x > 3 \right\}$

17. $\left\{ x \middle| -2 \le x < 7 \right\}$

18. ![number line] -3 3 6; -8 -6 -4 -2 0 2 4 6 8

19. (a) $(-\infty, -3) \cup (1, 2)$
    (b) $(-3, 1) \cup (2, \infty)$

20. $\left\{ x \middle| x < -5 \text{ or } -3 < x < -1 \right\}$

Chapter 12 Test Form A

1. 1, 5

2. 3, 6

3. $-4, -2$

4. $-3, \frac{1}{2}$

5. one real solution

6. $f(x) = x^2 + 8x + 15$

7. $r = \sqrt{\dfrac{Gm_1 m_2}{F}}$

8. $\pm \frac{1}{3}, \pm 1$

9. 4

10. $0, \pm i$

11. $(16, 0), (81, 0)$

12. $x^4 - x^2 - 6 = 0$

13. downward

14. $x = -1$

15. $(-1, 25)$

16. $(-6, 0), (4, 0)$

713

Chapter 12 Answers

17.

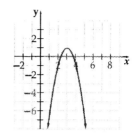

18. $y = \dfrac{5}{2}x^2 - 10x + 7$

19. $\{x \mid x < -6 \text{ or } x > 4\}$

20. $(-\infty, -3) \cup [-1, 5]$

Chapter 12 Test Form B

1. $-1, 4$

2. $-5, 9$

3. $0, 3$

4. $2 \pm 2i$

5. no real solutions

6. $f(x) = x^2 - 5$

7. $b = \sqrt{c^2 - a^2}$

8. $\pm\sqrt{6}, \pm i\sqrt{5}$

9. 4

10. $\dfrac{1}{4}$

11. $\left(-\dfrac{2}{3}, 0\right)$

12. $1 \pm i$

13. upward

14. $(0, 3)$

15. $\left(-\dfrac{2}{3}, \dfrac{5}{3}\right)$

16. no x–intercepts

17. $y = -x^2 + 4x + 5$

18.

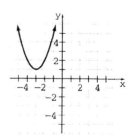

19. $\{y \mid -6 < y \le -2\}$

20. $[-1, 3) \cup [6, \infty)$

Chapter 12 Test Form C

1. $-10, 8$

2. $-\dfrac{3}{4}, \dfrac{5}{4}$

3. $-\dfrac{1}{3}, \dfrac{2}{5}$

4. $\dfrac{2 \pm i\sqrt{10}}{2}$

5. two distinct real solutions

6. $f(x) = 6x^2 - 5x + 1$

7. $l = \pm\sqrt{d^2 - w^2 - h^2}$

8. $\pm 2, \pm i\sqrt{3}$

9. $1, 9$

10. $0, \pm\sqrt{5}$

11. ± 64

12. $x^4 - 2x^2 - 3 = 0$

13. upward

14. $x = 2$

15. $(2, -1)$

16. $(1, 0)$ and $(3, 0)$

17.

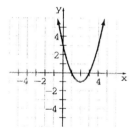

18.

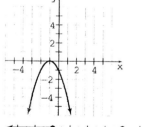

19.

20.

Chapter 12 Answers

21. $(-\infty, 1] \cup \left[\dfrac{5}{2}, \infty\right)$

22. $\left\{x \middle| x \le 1 \text{ or } x \ge \dfrac{5}{2}\right\}$

23. 11, 17 and $-17, -11$

24. 4 ft by 6 ft

25. $[-3, 2) \cup [5, \infty)$

20.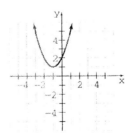

21. $(-\infty, -1) \cup [2, \infty)$

22. $[-2, -1) \cup [3, \infty)$

23. 5, 7 and $-7, -5$

24. 3%

25. 5 ft

Chapter 12 Test Form D

1. 12, -8

2. $\dfrac{7}{3}, \dfrac{11}{3}$

3. 6, -5

4. $\dfrac{5}{2}, 1$

5. 2 distinct real solutions

6. $f(x) = x^2 - 3x - 28$

7. $r = \pm\sqrt{\dfrac{A}{\pi}}$

8. $\pm 2, \pm i$

9. 49

10. $\pm i, \pm 1$

11. $-1, 125$

12. $x^4 + x^2 - 2 = 0$

13. downward

14. $x = 2$

15. $(2, -1)$

16. none

17.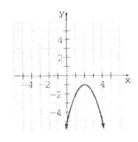

18. $y = x^2 + 4x + 2$

19.

Chapter 12 Test Form E

1. 7, -15

2. $-\dfrac{36}{5}, -\dfrac{24}{5}$

3. $\dfrac{2}{3}, 1$

4. $1 \pm \dfrac{i}{2}$

5. single unique solution

6. $f(x) = x^2 - 5x - 24$

7. $f_y = \pm\sqrt{f^2 - f_x^2}$

8. $\pm 2, \pm 3i$

9. 16

10. $0, \pm\sqrt{2}\, i$

11. 8, -216

12. $x^4 - 3x^2 - 4 = 0$

13. upward

14. $x = -1$

15. $(-1, 1)$

16. none

17.

18. $y = 3x^2 + 1$

Chapter 12 Answers

19.

$$\text{—} \quad \underset{-8\ -6\ -4\ -2\ \ 0\ \ 2\ \ 4\ \ 6\ \ 8}{\overset{-5 \qquad\quad 4}{\circ\!-\!\!-\!\!-\!\!-\!\!-\!\!-\!\!-\!\!\circ}}$$

20.

$$\underset{-8\ -6\ -4\ -2\ \ 0\ \ 2\ \ 4\ \ 6\ \ 8}{\overset{-5 \qquad\qquad 5}{\bullet\!-\!\!-\!\!-\!\!-\!\!-\!\!-\!\!\bullet}}$$

21. $\left(-4, \dfrac{1}{2}\right)$

22. $\left\{ x \middle| -4 < x < \dfrac{1}{2} \right\}$

23. 12, 14 and $-14, -12$

24. $(-3, 1] \cup [5, \infty)$

25. 5 seconds

Chapter 12 Test Form F

1. $x = -9, -5$

2. $x = \dfrac{8}{3}, -\dfrac{2}{3}$

3. $x = \dfrac{2}{3}, -\dfrac{1}{4}$

4. $x = \dfrac{-1 \pm 3i}{2}$

5. 1

6. $f(x) = x^2 - 6x - 16$

7. $r = \sqrt{\dfrac{2S + \pi h^2}{4\pi}} - \dfrac{h}{2}$

8. $x = \pm\sqrt{2}, \pm i$

9. $x = 49$

10. $x = \pm\sqrt{3}, \pm 1$

11. $x = -8, \dfrac{1}{8}$

12. $x^4 - 18x^2 - 175 = 0$

13. $y = -x^2 - 2x$

14. $(0, 0)$

15. $(-1, 1)$

16. $x = 0, -2$

17. $x = -1$

18.

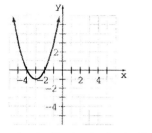

19.

$$\underset{-4\ -3\ -2\ -1\ \ 0\ \ 1\ \ 2\ \ 3\ \ 4}{\overset{}{\circ\!-\!\!-\!\!-\!\!-\!\!-\!\!-\!\!\circ}}$$

20.

$$\underset{-8\ -6\ -4\ -2\ \ 0\ \ 2\ \ 4\ \ 6\ \ 8}{\overset{-4 \qquad\quad 3}{\bullet\!-\!\!-\!\!-\!\!-\!\!-\!\!-\!\!\bullet}}$$

21. $\left[-3, \dfrac{1}{2}\right]$

22. $\left\{ x \middle| -3 \le x \le \dfrac{1}{2} \right\}$

23. $x = \dfrac{9}{2}, 8$

24. $[-3, 1.5) \cup [2, \infty)$

25. *2 seconds*

Chapter 12 Test Form G

1. c

2. c

3. c

4. b

5. a

6. d

7. c

8. a

9. d

10. b

11. c

12. d

13. a

14. b

15. c

16. d

17. b

18. a

Chapter 12 Answers

Chapter 12 Test Form H

19. d
20. c
21. c
22. d
23. c
24. b
25. b

Chapter 12 Test Form H

1. c
2. d
3. a
4. c
5. c
6. b
7. d
8. c
9. a
10. a
11. d
12. b
13. c
14. d
15. c
16. d
17. b
18. a
19. a
20. b
21. a
22. d
23. d
24. a
25. c

Cumulative Review 1–12 Test Form A

1. $-\dfrac{19}{7}$

2. about 2.024×10^6 bushels

3. $x = -44$

4. $\{x \mid x > 9 \text{ or } x < -1\}$

5. $x = 6$

6. yes

7. $D = \{x \mid x \text{ is a real number}\}$
   $R = \{y \mid y \geq 2\}$

8.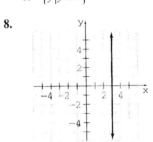

9.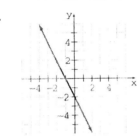

10. $y + 1 = \dfrac{2}{7}(x - 3)$

11. $(1, 2)$

12. 22

13. $xy(x - y)(x + y)\left(x^2 + xy + y^2\right)\left(x^2 - xy + y^2\right)$

14. $-x^2 - x + 6$

15. $x^3 - 2x^2 - 4x + 8$

16. $\dfrac{5}{2}$ sec

17. $-2 < x < \dfrac{3}{2}$

18. 20.35 foot–candles

19. $\dfrac{44}{89} + \dfrac{17}{89} i$

20. $\dfrac{7}{2} \pm \dfrac{\sqrt{11}}{2} i$

Chapter 12 Answers

Cumulative Review 1–12 Test Form B

1. b
2. d
3. c
4. b
5. a
6. a
7. c
8. b
9. d
10. a
11. c
12. a
13. a
14. b
15. d
16. c
17. b
18. c
19. a
20. d

Chapter 13 Answers

Chapter 13 Pretest Form A

1. yes

2. $D:\{1,2,3,4\}; R:\{1,4,9,16\}$

3. $\{(1, 1)(4, 2)(9, 3)(16, 4)\}$

4. $D:\{1,4,9,16\}; R:\{1,2,3,4\}$

5. 7

6. 24

7. $2x^2 - 1$

8. $f^{-1}(x) = \dfrac{5x + 4}{1 - x}$

9.

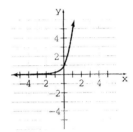

10.

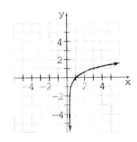

11. $\log_3 17 = x$

12. 49

13. -3

14. 8

15. $\log_b \dfrac{x+1}{x^3}$

16. $\frac{2}{3}$

17. 5

18. -1, -3

19. $\dfrac{3 + \sqrt{5}}{2}$

20. 1.7712

Chapter 13 Pretest Form B

1. yes

2. D: $\{-6, -5, 2, 1\}$
   R: $\{5, 4, -3, -1\}$

3. $\{(5, -6), (4, -5), (-3, 2), (-1, 1)\}$

4. D: $\{5, 4, -3, -1\}$
   R: $\{-6, -5, 2, 1\}$

5. $(f \circ g)(x) = x - 3$

6. $(g \circ f)(x) = \sqrt{x^2 - 3}$

7. 1

8. $f^{-1}(x) = \dfrac{9}{x + 4}$

9.

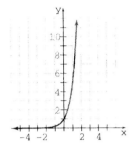

10.

11. $\log_5 x = 2.3$

12. $10^4 = x; 10,000$

13. $49^y = 7; \dfrac{1}{2}$

14. $3\log_6 x + \log_6(x+1) - \log_6 y$

15. $\log_2 \dfrac{(x+7)^4}{\sqrt[5]{x}}$

16. $\frac{4}{9}$

17. 10

18. 6

19. 1.7712

20. 1.9129

Chapter 13 Answers

Mini–Lecture 13.2

1. a. $y = 4^x$

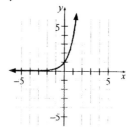

b. $y = \left(\dfrac{3}{7}\right)^x$

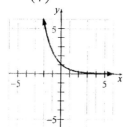

c. $y = 3 \cdot (2^x)$

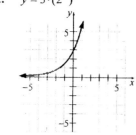

d. $y = 5^x + 2$

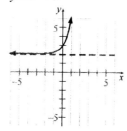

Mini–Lecture 13.3

1. a. $y = \log_4 x$

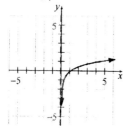

b. $y = \log_{1/7} x$

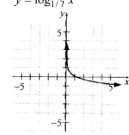

c. $y = 3 \cdot \log_2 x$

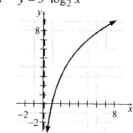

d. $y = \log_5 x + 2$

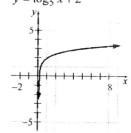

Additional Exercises 13.1

1. $4 + 3x^3$

2. $(f \circ g)(x) = x - 6$

3. 24

4. -27

5. no

6. yes

7. yes

8. (a) no
   (b) NA

9. (a) yes
   (b) $f^{-1}(x) = \sqrt[3]{\dfrac{x+3}{5}}$

10. (a) yes
    (b) $f^{-1}(x) = x^2 + 4, \ x \geq 0$

720

11. $f^{-1}(x) = \dfrac{x+3}{2}$;

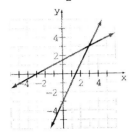

Additional Exercises 13.2

1.

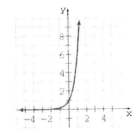

12. $f^{-1}(x) = \dfrac{x-1}{3}$;

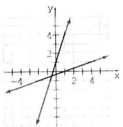

2.

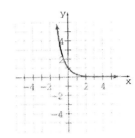

3.

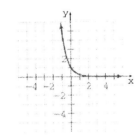

13. $f^{-1}(x) = x^3 + 1$;

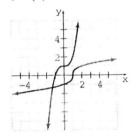

14. $f^{-1}(x) = x^2 - 4,\ x \geq 0$;

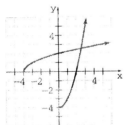

4.

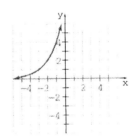

5.

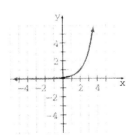

15. $f^{-1}(x) = \dfrac{x-3}{4}$

16. $\left(f^{-1} \circ f\right)(x) = \dfrac{(4x+3)-3}{4} = \dfrac{4x}{4} = x$

17. $f^{-1}(x) = x^3 - 5$

18. $\left(f^{-1} \circ f\right)(x) = \left(\sqrt[3]{x+5}\right)^3 - 5 = (x+5-5) = x$

19. no

20. yes

6.

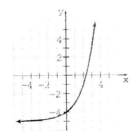

Chapter 13 Answers

7.

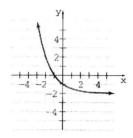

8.

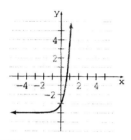

9. $920

10. $554

11. $1950.

12. $816.31

13. $3870.11

14. $3250.

15. 27.43 g

16. 21.41 g

17. 74.27 g

18. 11.14 g

19. 7200

20. $4113.21

Additional Exercises 13.3

1.

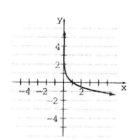

2.

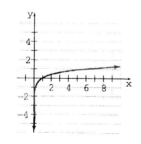

3.

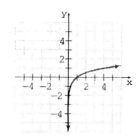

4.

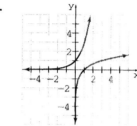

5.

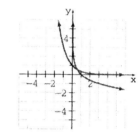

6.

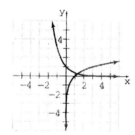

7. $\log_3 81 = 4$

8. $\log_5 f = d$

9. $y = \log_{9/8} x$

10. $y = \log_x \dfrac{2}{7}$

11. $k = \log_4 m$

12. $y = \log_{7/8} x$

13. $y = \log_x z$

14. $\left(\dfrac{1}{3}\right)^2 = \dfrac{1}{9}$

15. $10^4 = 10{,}000$

16. 0

17. −3

Chapter 13 Answers

18. -5

19. $\dfrac{1}{2}$

20.

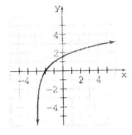

Additional Exercises 13.4

1. $\log_p Q - \log_p 5$

2. $\log_a 7 + \log_a x + 3\log_a y - 5\log_a z$

3. $\log_a 5 + \log_a x + 3\log_a y - 4\log_a z$

4. $\log_5 x + \log_5(x+6) - 2\log_5 x$

5. $\log_3 x + \log_3(x+5) - 6\log_3 x$

6. $\log_b x + \dfrac{2}{7}\log_b y - \dfrac{6}{7}\log_b z$

7. $\dfrac{1}{6}\log_b x + \dfrac{1}{8}\log_b y$

8. $5\log_8 x + 4\log_8(x-3)$

9. $\log_2 \dfrac{(x+2)^3 (x+4)^7}{\sqrt{x}}$

10. $\log_b\left(\dfrac{x^2}{y^7}\right)$

11. $\dfrac{7}{6}\log_5 x$

12. $\log_9 x\sqrt{x+2}$

13. $\log_a \dfrac{4xy^2}{z^5}$

14. $\log_5 \dfrac{(x+6)^4 (x+4)^3}{\sqrt{x}}$

15. $\log_2\left(x\sqrt{y}\right)$ or $\log_4\left(x^2 y\right)$

16. 1.342

17. 0.6020

18. 5

19. 12

20. $\dfrac{1}{6}$

Additional Exercises 13.5

1. -4.0269

2. 3.5940

3. 792

4. 9920

5. 29.0

6. 0.0823

7. 1.00

8. 0.994

9. 108

10. 0.000781

11. 2.8573

12. 1.4624

13. -0.1871

14. $67,800$

15. 0.00315

16. -4

17. 5

18. 8.7

19. 3.8

20. 26.5

Additional Exercises 13.6

1. 6

2. $\dfrac{3}{2}$

3. 5 ·

4. $\dfrac{3}{4}$

5. 2.21

6. 2.11

7. 4.45

8. 20

9. 1.79

10. 12

Chapter 13 Answers

11. 1.40

12. 6

13. 5.02 hr

14. 4.92 yr

15. 12.69 yr

16. 5.49 hr

17. 8.53 yr

18. 6.03 yr

19. 5.1823

20. $429.95

Additional Exercises 13.7

1. 6.4394

2. −0.4620

3. 0.199

4. 7.51

5. 1.681

6. 2.164

7. − 0.3604

8. 1.2468

9. 5

10. 2

11. $P \approx 18{,}610.2$

12. 0.1205

13. 0.1871

14. $x = e^t (y + 3)$

15. $x = \dfrac{e^t}{y - 5}$

16. $x = \sqrt[3]{e^t (y - 1)^2}$

17. $11,843.77

18. (a) 1138.2 grams
    (b) 17.8 yr

19. 15,021

20. $\approx 7\dfrac{1}{3}$ years

Chapter 13 Test Form A

1. yes

2. $x^2 + 4x$

3. $x^2 + 8x + 10$

4. $f^{-1}(x) = -\dfrac{1}{3}x + 2$

5.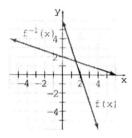

6.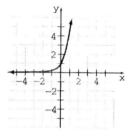

7.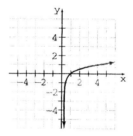

8. $\log_{81} 9 = \dfrac{1}{2}$

9. −2

10. $\dfrac{1}{2}\log_8 x - \log_8 13$

11. $\log_7 \dfrac{(a + 3)^5 (a - 1)^2}{\sqrt{a}}$

12. $\log_6 \dfrac{(x + 3)^5}{8(x - 4)^2}$

13. $\dfrac{1}{2}$

Chapter 13 Answers

14. $\dfrac{1}{6}$

15. 0.0713

16. $\dfrac{1}{2}$

17. 1

18. 4

19. 1.72

20. 1.4371

Chapter 13 Test Form B

1. no

2. $(f \circ g)(x) = x^2 + 10x + 26$

3. 22

4. $f^{-1}(x) = x^2 - 3; x \geq 0$

5.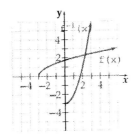

6.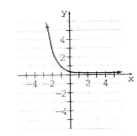

7.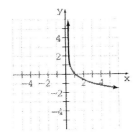

8. $\log_a b = n$

9. $\dfrac{1}{81}$

10. $6\log_3 d - 4\log_3(a-5)$

11. $\log_5 \sqrt{\dfrac{x-4}{x}}$

12. $\log_6 \dfrac{81}{(x+3)^2 x^4}$

13. 10

14. 125

15. 2.2046

16. -6

17. $\sqrt{13}$

18. 243

19. 60.3

20. 4.0871

Chapter 13 Test Form C

1. yes

2. $(f \circ g)(x) = x + 1$

3. $(g \circ f)(x) = \sqrt{x^2 + 1}$

4. $f^{-1}(x) = \dfrac{1}{4}x - \dfrac{1}{2}$

5.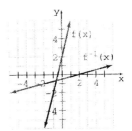

6.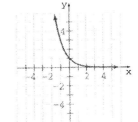

Chapter 13 Answers

7.

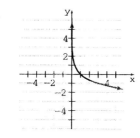

8. $\log_5 74 = x$

9. $x = 2^4; \; x = 16$

10. $\dfrac{5}{6}\log_2 x - \log_2(x-1)$

11. $\log_6 \dfrac{(x+1)(x-1)^5}{(x-2)^7}$

12. $\log_{12} \dfrac{x^3}{(x-3)^4(x-8)^5}$

13. 2

14. 4.7

15. 140

16. 0.661

17. $x = -\dfrac{5}{2}$

18. $x = 11, \; x = -5$

19. $x = -1, x = \dfrac{3}{2}$

20. -0.0588

21. 3.7549

22. $k = \dfrac{1}{t}\ln\left(\dfrac{100}{P_0}\right)$

23. $y = \dfrac{x-4}{3}$

24. \$9933.47

25. 19.8 years

Chapter 13 Test Form D

1. no

2. $(f \circ g)(x) = \sqrt{x^2 - 2}$

3. $(g \circ f)(x) = x - 2$

4. $f^{-1}(x) = \dfrac{1}{2}x + 2$

5.

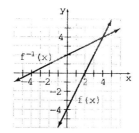

6.

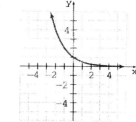

7.

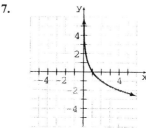

8. $\log_b N = L$

9. $64^y = \dfrac{1}{8}; \; y = -\dfrac{1}{2}$

10. $\dfrac{1}{2}\log_9 x + 3\log_9(x-2) - 5\log_9(x+1)$

11. $\log_{12} \dfrac{25(x-7)^3}{x^7}$

12. $\log_2 \dfrac{(x+11)}{(x^2-64)^5}$

13. 1

14. 13.2

15. 156.

16. 0.793

17. $x = \dfrac{1}{4}$

18. $x = 8, \; x = -6$

19. $x = -3$

20. 2.0123

21. 0.2248

22. $t = -50\ln\left(\dfrac{3}{4}\right) \approx 14.3841$

23. $y = x^2 + x - 6$

24. ≈ 17.4 years

25. 17.22 million

Chapter 13 Test Form E

1. yes

2. $(f \circ g)(x) = x^2 - 16x + 73$

3. $(g \circ f)(x) = x^2 + 1$

4. $f^{-1}(x) = \dfrac{\sqrt[3]{4x+16}}{2}$

5.

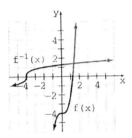

6.

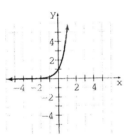

7.

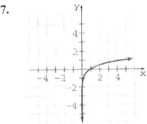

8. $\log_{10} 10{,}000 = 4$

9. $9^x = \dfrac{1}{3}$; $x = -\dfrac{1}{2}$

10. $3\log_7 x - 3\log_7 5$

11. $\log_8 \dfrac{x^8(x-4)^8}{(x-1)^3}$

12. $\log_{1/2} \dfrac{(x+1)^4}{125(x-2)^5}$

13. 18.3

14. 2

15. 2.46

16. 0.716

17. $x = -\dfrac{3}{2}$

18. $x = -8$

19. $x = 6$

20. 1.7355

21. -0.6700

22. $t \approx 69.976$

23. $y = e^3(x+5)$

24. \$9160.

25. 197.6 grams

Chapter 13 Test Form F

1. yes

2. $x - 4$

3. $\sqrt{x^2 - 4}$

4. $x^3 + 4$

5. D:$\mathbb{R}$, R:$\mathbb{R}$

6. $y = \left(\dfrac{1}{5}\right)^x$

7. $y = \log_{1/5} x$

8. $\log_x(y+1) = z$

9. $x^{\frac{1}{3}} = 5$; $x = 125$

10. $\log_{1/3}(x+4) + 3\log_{1/3}(x-2)$
    $- \left(8\log_{1/3} x + 4\log_{1/3} 2\right)$

11. $\log_2 \dfrac{x^5(x+1)^3}{(x-1)^8}$

12. $\log_{1/2} \dfrac{(x+1)^2}{x(x-1)^3}$

13. 8.2

14. 1

15. 3.49

16. 4.20

Chapter 13 Answers

17. $x = -\dfrac{1}{2}$

18. $x = 13$

19. $x = 3$

20. -0.0433

21. 3.0775

22. 10.217

23. $y = 8.1(x + 4)$

24. 11.58 years

25. $35.6\,\%$

Chapter 13 Test Form G

1. c
2. b
3. d
4. a
5. a
6. a
7. c
8. b
9. d
10. a
11. c
12. b
13. b
14. d
15. a
16. b
17. a
18. b
19. c
20. d
21. d
22. a
23. b
24. c
25. d

Chapter 13 Test Form H

1. d
2. d
3. d
4. c
5. a
6. b
7. d
8. b
9. c
10. d
11. b
12. a
13. c
14. c
15. c
16. c
17. c
18. d
19. a
20. d
21. b
22. c
23. a
24. b
25. d

Chapter 14 Answers

Chapter 14 Pretest Form A

1. 10

2. $\left(\dfrac{1}{2}, 4\right)$

3. $\left(-\dfrac{3}{2}, \dfrac{11}{2}\right), (1,3), \left(\dfrac{7}{2}, \dfrac{1}{2}\right)$

4. $4x = 3(y-3)^2 + 8$
   $9y = -2(x-2)^2 + 27$

5. $x = -2(y-1)^2 + 4$

6.
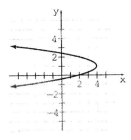

7. $(x-5)^2 + (y+4)^2 = 36$

8.
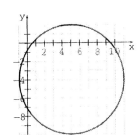

9.
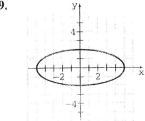

10. $(-1, 0)$

11. $y + 2 = \pm(x+1)$

12.
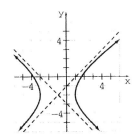

13. ellipse

14. parabola

15. hyperbola

16. circle

17. $\left(-\dfrac{1}{2}, \dfrac{1}{2}\right)$ and $(2, 8)$

18. $(2, 1), (-2, -1)$

19. $\dfrac{1}{2}\left(3 - \sqrt{5}\right), \dfrac{1}{2}\left(3 + \sqrt{5}\right)$

20. $(1, \sqrt{3}), (1, -\sqrt{3}), (-1, \sqrt{3}), (-1, -\sqrt{3})$

Chapter 14 Pretest Form B

1. $\sqrt{53} \approx 7.28$

2. $\left(-6, -\dfrac{1}{2}\right)$

3. $(8, 16)$

4. $y = 4(x-1)^2 - 1$
   $16x = -(y+1)^2 + 16$

5. $x = 2(y+1)^2 - 5$

6.
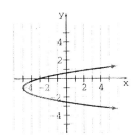

7. $(x-1)^2 + (y+1)^2 = 25$

8.
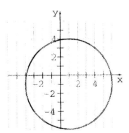

9.
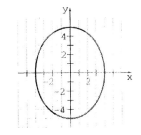

Chapter 14 Answers

10. $(8, -7)$

11. $y = \dfrac{2}{5}x$ and $y = -\dfrac{2}{5}x$

12.

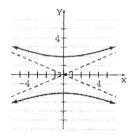

13. circle

14. hyperbola

15. parabola

16. $(6, 8)$ and $(0, -10)$

17. $(2, 2\sqrt{2}), (-2, 2\sqrt{2}), (2, -2\sqrt{2}), (-2, -2\sqrt{2})$

18. no solution

19. $(-1, -1)$

20. $l = 17$ ft., $w = 13$ ft.

Mini–Lecture 14.1

2. e.

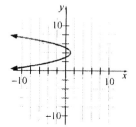

4. a.

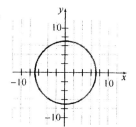

4. b.

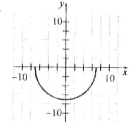

Mini–Lecture 14.2

1. e.

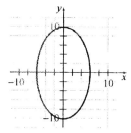

2. e.

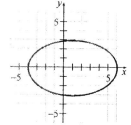

3. e.

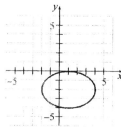

4. e.

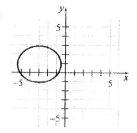

Mini–Lecture 14.3

1. e.

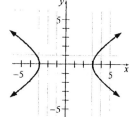

2. e.

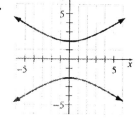

3. e.

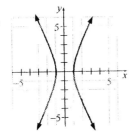

4. e.

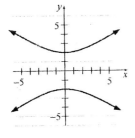

Additional Exercises 14.1

1.

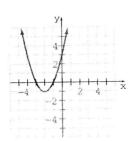

2.

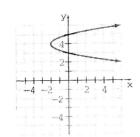

3.

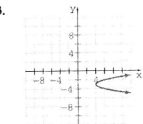

4. $x = (y+4)^2 - 1$

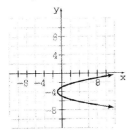

5. $y = (x-2)^2 - 6$

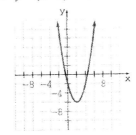

6. $y = 3(x-1)^2 + 1$

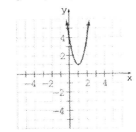

7. 5

8. $\sqrt{205} \approx 14.32$

9. $\sqrt{13} + 2\sqrt{5} + \sqrt{29} \approx 13.46$

10. $(-3, 4)$

11. $(-3.5, 2.5)$

12. $(5, 9), (8, 13), (11, 17)$

13. $x^2 + y^2 = 36$

14. $(x+3)^2 + (y+1)^2 = 4$

15. $(y+2)^2 + 3x = 15$

16. $(x-3)^2 + (y+2)^2 = 16$

17.

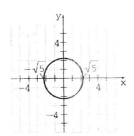

18.

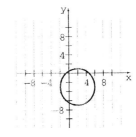

Chapter 14 Answers

19.

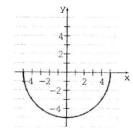

20. $(x+3)^2 + (y+5)^2 = 9$

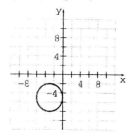

Additional Exercises 14.2

1.

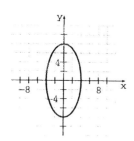

2.

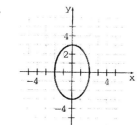

3. $\dfrac{x^2}{1} + \dfrac{y^2}{16} = 1$

4. $(0, 0)$

5.

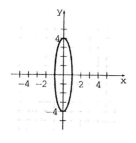

6. $\dfrac{(x+1)^2}{5} + \dfrac{(y-2)^2}{20} = 1$

7. $(-1, 2)$

8. $(1, 0)$ and $(-3, 0)$

9. $(0, 6)$ and $(0, -2)$

10.

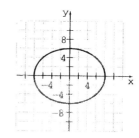

11. $(-2, 3)$

12.

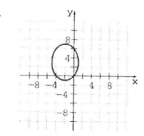

13. $\dfrac{(x+4)^2}{16} + \dfrac{(y+3)^2}{4} = 1$

14. $(-4, -3)$

15.

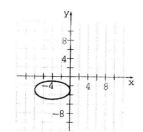

16. $\dfrac{(x-1)^2}{16} + \dfrac{(y+2)^2}{9} = 1$

17. $(1, -2)$

18. $y = \dfrac{3\sqrt{3} - 4}{2} \approx 0.598$

19. 8

20. 12π

Chapter 14 Answers

Additional Exercises 14.3

1. $y = \pm\frac{3}{2}x$

2.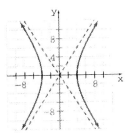

3. $y = \pm x$

4.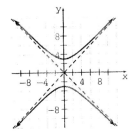

5. $y = \pm\frac{3}{2}x$

6.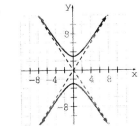

7. $\frac{x^2}{16} - \frac{y^2}{9} = 1$

8. $y = \pm\frac{3}{4}x$

9. $(4, 0)$ and $(-4, 0)$

10.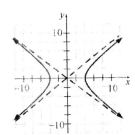

11. $\frac{y^2}{5} - \frac{x^2}{20} = 1$

12. $y = \pm\frac{1}{2}x$

13. $(0, \sqrt{5})$ and $(0, -\sqrt{5})$

14. $y = \frac{5}{2}$

15. hyperbola

16. ellipse

17. hyperbola

18. circle

19. hyperbola

20. parabola

Additional Exercises 14.4

1. $(0, 7); (7, 0)$

2. $(4, 5); \left(-\frac{31}{5}, \frac{8}{5}\right)$

3. $(0, -4), (5, 0)$

4. $\left(\sqrt{13}, 6\right); \left(-\sqrt{13}, 6\right); (0, -7)$

5. $(0, 5); (5, 0)$

6. $(4, 1); \left(-\frac{8}{5}, \frac{19}{5}\right)$

7. $(0, 2); (3, 0)$

8. $(1, 3); (4, 2)$

9. $\left(8\sqrt{2}, 4\right); \left(8\sqrt{2}, -4\right); \left(-8\sqrt{2}, 4\right);$ $\left(-8\sqrt{2}, -4\right)$

10. $(1, 1); (1, -1); (-1, 1); (-1, -1)$

11. $(0, -8); (0, 8)$

12. $(4, 2); (4, -2); (-4, 2); (-4, -2)$

13. $(4, 3); (4, -3); (-4, 3); (-4, -3)$

14. $\left(3\sqrt{6}, 3\right); \left(3\sqrt{6}, -3\right); \left(-3\sqrt{6}, 3\right); \left(-3\sqrt{6}, -3\right)$

15. $(1, \sqrt{2}); (1, -\sqrt{2})$

16. 8.5 ft and 10 ft

17. 27 ft

18. 6 in. and 8 in.

19. 24 units

20. $2\sqrt{2}$ and $3\sqrt{2}$

Chapter 14 Answers

Chapter 14 Test Form A

1.

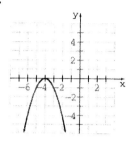

2. $y = (x+1)^2 - 8$

3. $x = 3(y-2)^2 - 48$

4. 10

5. $(7, 12)$

6. $(x+5)^2 + (y-2)^2 = 1$

7. $(x+1)^2 + (y-2)^2 = 3^2$

8.

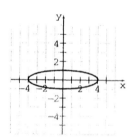

9. $\dfrac{(x+2)^2}{25} + \dfrac{(y-4)^2}{4} = 1$

10. 10π

11. $\dfrac{y^2}{1} - \dfrac{x^2}{25} = 1; \ y = \pm \dfrac{1}{5}x$

12. parabola

13. ellipse

14. circle

15. hyperbola

16. $(0, -2), \left(\dfrac{8}{5}, \dfrac{-6}{5}\right)$

17. $5 + 2\sqrt{10} + 3\sqrt{5} \approx 18.03$ units

18. $\left(\sqrt{15}, 1\right), \left(-\sqrt{15}, 1\right), \left(\sqrt{15}, -1\right), \left(-\sqrt{15}, -1\right)$

19. 13 and 21 units

20. 3 and 10, – 3 and – 10

Chapter 14 Test Form B

1.

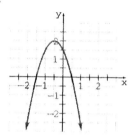

2. $x = -\left(y + \dfrac{5}{2}\right)^2 + \dfrac{9}{4}$

3. $y = \left(x + \dfrac{7}{2}\right)^2 - \dfrac{9}{4}$

4. $\dfrac{\sqrt{265}}{4}$

5. $\left(\dfrac{9}{4}, \dfrac{15}{4}\right)$

6. $(x+6)^2 + (y+1)^2 = 5$

7. $\left(x - \dfrac{1}{2}\right)^2 + \left(y + \dfrac{3}{2}\right)^2 = 4$

8.

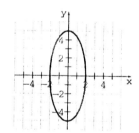

9. $\dfrac{(x-4)^2}{16} + \dfrac{(y+1)^2}{25} = 1$

10. 20π

11. $\dfrac{y^2}{25} - \dfrac{x^2}{64} = 1; \ y = \pm \dfrac{5}{8}x$

12. circle

13. ellipse

14. hyperbola

15. hyperbola

16. $(-1, 3), (1, 3)$

17. 43.5 square units

18. $(3, 2), (3, -2), (-3, 2), (-3, -2)$

19. no real solution

20. 6 cm and 8 cm

Chapter 14 Test Form C

1. $\sqrt{5} \approx 2.24$

2. $(-2, 7)$

3. $(3, -4)$

4.

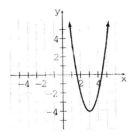

5. $x = 2(y + 2)^2 - 1$

6.

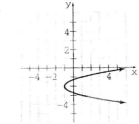

7. $\dfrac{(x-3)^2}{3} + \dfrac{(y+2)^2}{2} = 1$

8. $(3, -2)$

9. $\sqrt{6}\,\pi$

10. $(x+1)^2 + (y-5)^2 = 20$

11. $\dfrac{x^2}{64} + \dfrac{y^2}{25} = 1$

12. $(8, 0)$ and $(-8, 0)$

13. $(0, 5)$ and $(0, -5)$

14.

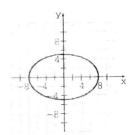

15. $\dfrac{y^2}{9} - \dfrac{x^2}{36} = 1$

16. $y = \pm\dfrac{1}{2}x$

17.

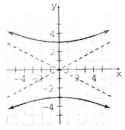

18. $\left(\sqrt{2}, -5\right); \left(-\sqrt{2}, -5\right); \left(\sqrt{11}, 4\right); \left(-\sqrt{11}, 4\right)$

19. 7 and 24

20. 22

Chapter 14 Test Form D

1. $\sqrt{170} \approx 13.04$

2. $(2.5, 0)$

3. $y = x - 1$

4. $20\sqrt{2}$

5. $x = (y - 2)^2 - 2$

6.

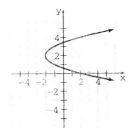

7. $5(x-1)^2 + (y-2)^2 = 9$

8. $\dfrac{9\sqrt{5}}{5}\pi$

9. $(x-3)^2 + (y+7)^2 = 25$

Chapter 14 Answers

10. $(3, -7)$; $r = 5$

11. $\dfrac{x^2}{16} + \dfrac{y^2}{25} = 1$

12. $(4, 0)$ and $(-4, 0)$

13. $(0, 5)$ and $(0, -5)$

14.

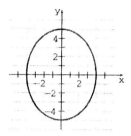

15. $\dfrac{y^2}{49} - \dfrac{x^2}{4} = 1$

16. $y = \pm\dfrac{7}{2}x$

17.

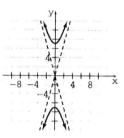

18. $(0, \sqrt{2}), (0, -\sqrt{2})$

19. $(0, -8)$; $\left(\sqrt{15}, 7\right)$; $\left(-\sqrt{15}, 7\right)$

20. 6 ft, 8 ft

Chapter 14 Test Form E

1. 5

2. $(4, 5), (5, 2), (6, -1)$

3. $(-5, -1)$

4.

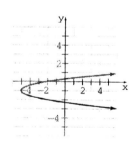

5. $x = 2(y - 2)^2 - 3$

6.

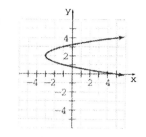

7. $(x + 1)^2 + (y - 3)^2 = 64$

8.

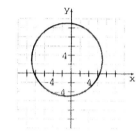

9. $(2, -3)$

10. 8π

11. $\dfrac{x^2}{25} + \dfrac{y^2}{9} = 1$

12. $(5, 0)$ and $(-5, 0)$

13. $(0, 3)$ and $(0, -3)$

14.

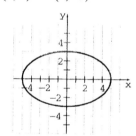

15. $\dfrac{x^2}{25} - \dfrac{y^2}{4} = 1$

16. $y = \pm\dfrac{2}{5}x$

17.

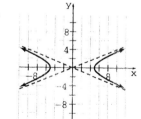

18. $(8, 6)$; $(-6, -8)$

19. $(\sqrt{3}, 3)$; $(-\sqrt{3}, 3)$

20. 7 meters by 9 meters

Chapter 14 Answers

Chapter 14 Test Form F

1. $\sqrt{113}$ units

2. $\left(1, -\dfrac{3}{2}\right)$

3. $(8, 7)$

4. $x = 2(y-2)^2 - 5$

5. $y = (x+3)^2 - 4$

6. hyperbola

7. $(x-9)^2 + (y+7)^2 = 144$

8. $(x+2)^2 + (y-3)^2 = 16$

9. $(x-5)^2 + (y-2)^2 = 4$

10. center $(5, 2)$; $r = 2$

11. $(-5, -2)$

12. 4π

13. $(0, 2)$ and $(0, -2)$

14.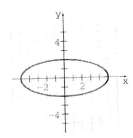

15. $\dfrac{x^2}{4} - \dfrac{y^2}{49} = 1$

16. $y = \pm\dfrac{7}{2}x$

17.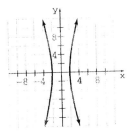

18. $(0, -5); (3, 4); (-3, 4)$

19. $(3\sqrt{6}, 3); (3\sqrt{6}, -3); (-3\sqrt{6}, 3); (-3\sqrt{6}, -3)$

20. 21

Chapter 14 Test Form G

1. a
2. c
3. d
4. b
5. d
6. a
7. b
8. c
9. b
10. b
11. a
12. b
13. d
14. d
15. c
16. a
17. c
18. b
19. a
20. c

Chapter 14 Test Form H

1. d
2. a
3. c
4. b
5. a
6. c
7. d
8. b
9. d
10. c
11. a
12. b
13. d
14. a
15. c

Chapter 14 Answers

16. a

17. c

18. c

19. d

20. b

Cumulative Review 1–14 Test Form A

1. $5\sqrt{2}$

2. $x = \dfrac{6}{5}$

3. $\dfrac{3y^3}{x^3 z}$

4. $(x+2)(x-2)(x-3)$

5. $x = \dfrac{11}{3}$ or $x = -5$

6. i

7. Domain: $x \geq -1$
Range: $y \geq -2$

8.

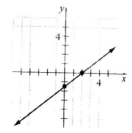

9. $\dfrac{5}{(x-3)(x-2)}$

10. $y = \dfrac{3}{2}x + 6$

11. $(x,y) = \left(\dfrac{23}{9}, \dfrac{4}{9}\right)$

12. $x = 46$

13. -8

14. $(f \circ g)(x) = 3x^2 + 3x - 19$

15. $x = 4$

16. $\approx \$2.19$ per wicket

17. $x = 2 - \sqrt{6}$, $x = 2 + \sqrt{6}$

18. $\begin{bmatrix} 2 & 1 & 0 & | & 4 \\ 1 & -3 & 1 & | & 3 \\ 5 & 2 & -1 & | & 7 \end{bmatrix}$

19. $y = 3x + 1$

20. $(3, -2)$; $\sqrt{6}\,\pi$

Cumulative Review 1–14 Test Form B

1. d

2. c

3. b

4. d

5. d

6. c

7. b

8. b

9. b

10. a

11. a

12. d

13. b

14. a

15. b

16. c

17. a

18. a

19. a

20. c

Chapter 15 Answers

Chapter 15 Pretest Form A

1. arithmetic
2. geometric
3. neither
4. 3, 6, 12, 24
5. 3, 5, 7, 9
6. 1.5, 3, 4.5, 6
7. $a_n = 4n + 1$
8. $2\left(-\dfrac{1}{2}\right)^{n-1}$
9. 32
10. 90
11. $-\dfrac{1}{9}$
12. 312.48
13. $\dfrac{25}{99}$
14. $\dfrac{817}{990}$
15. 3, 6, 11, 18
16. 38
17. 9
18. $\dfrac{9}{5}$
19. $x^3 + 6ax^2 + 12a^2x + 8a^3$
20. $16x^4 + 96x^3y + 216x^2y^2 + 216xy^3 + 81y^4$

Chapter 15 Pretest Form B

1. geometric
2. neither
3. arithmetic
4. 15, 11, 7, 3
5. $7, 4, 3, \dfrac{5}{2}$
6. 48, 24, 12, 6
7. $a_n = 1 + (n-1)\dfrac{1}{2} = \dfrac{1}{2}n + \dfrac{1}{2}$
8. $a_n = 4(2)^{n-1}$

9. 72
10. -300
11. $\dfrac{4}{729}$
12. $\dfrac{46,655}{6}$
13. $\dfrac{8}{11}$
14. $\dfrac{41}{110}$
15. $-1 + 14 + 39 + 74$
16. 126
17. 50
18. $\dfrac{150}{7}$
19. $x^4 - 12x^3 + 54x^2 - 108x + 81$
20. $81x^4 + 216x^3y + 216x^2y^2 + 96xy^3 + 16y^4$

Additional Exercises 15.1

1. 0, 3, 8, 15, 24
2. $-\dfrac{2}{3}, -\dfrac{1}{4}, 0, \dfrac{1}{6}, \dfrac{2}{7}$
3. 118
4. 18
5. 198
6. $-\dfrac{49}{17}$
7. $-3; -9$
8. $7; \dfrac{49}{3}$
9. 21, 34, 55 (add the two previous terms)
10. $\dfrac{1}{25}, \dfrac{1}{36}, \dfrac{1}{49}$
11. $-5 - 11 - 21 = -37$
12. $-1 + 5 + 15 = 19$
13. $\dfrac{1}{8} + \dfrac{1}{16} + \dfrac{1}{32} + \dfrac{1}{64} = \dfrac{15}{64}$
14. $\dfrac{1}{3} + \dfrac{1}{9} + \dfrac{1}{27} + \dfrac{1}{81} = \dfrac{40}{81}$

Chapter 15 Answers

15. $\displaystyle\sum_{n=1}^{5}\left(n^2-5\right)$

16. $\displaystyle\sum_{n=1}^{4}\dfrac{n-1}{n^2+2}$

17. 7

18. 55

19. 12

20. 87

Additional Exercises 15.2

1. $-6, -4, -2, 0, 2;\ a_n=-6+2(n-1)$

2. $\dfrac{7}{2}, 3, \dfrac{5}{2}, 2, \dfrac{3}{2};\ a_n=\dfrac{7}{2}-\dfrac{1}{2}(n-1)$

3. -50

4. -2

5. 3

6. $\dfrac{22}{9}$

7. 17

8. -7

9. 76

10. 826

11. 275

12. -473

13. 75

14. -60

15. $\dfrac{85}{9}$

16. 98

17. 58,310

18. 29

19. 4292

20. 3225 seats

Additional Exercises 15.3

1. $5, -10, 20, -40, 80$

2. $64, 32, 16, 8, 4$

3. 5120

4. $\dfrac{125}{2}$

5. $-\dfrac{81}{512}$

6. $\dfrac{729}{2}$

7. $-26{,}240$

8. $\dfrac{4118}{729}$

9. -171

10. $r=\dfrac{3}{4}$

11. $a_n=7\left(\dfrac{3}{4}\right)^{n-1}$

12. $r=\dfrac{5}{2}$

13. $a_n=-\dfrac{2}{3}\left(\dfrac{5}{2}\right)^{n-1}$

14. 9

15. 1

16. $\dfrac{2}{9}$

17. $-\dfrac{1}{3}$

18. $\dfrac{63}{110}$

19. $\dfrac{23}{33}$

20. \$457,435.52

Additional Exercises 15.4

1. 56

2. 35

3. 210

4. 1

5. 165

6. 165

7. $a^4-12a^3b+54a^2b^2-108ab^3+81b^4$

8. $x^4+16x^3y+96x^2y^2+256xy^3+256y^4$

9. $s^5 - 5s^4w + 10s^3w^2 - 10s^2w^3 + 5sw^4 - w^5$

10. $625x^4 - 500x^3y + 150x^2y^2 - 20xy^3 + y^4$

11. $81a^4 - 108a^3b + 54a^2b^2 - 12ab^3 + b^4$

12. $x^4 - 8x^3y + 24x^2y^2 - 32xy^3 + 16y^4$

13. $16x^4 + 96x^3y + 216x^2y^2 + 216xy^3 + 81y^4$

14. $1296x^4 - 864x^3y + 216x^2y^2 - 24xy^3 + y^4$

15. $f^5 + 5f^4g + 10f^3g^2 + 10f^2g^3$

16. $b^6 + 6b^5c + 15b^4c^2 + 20b^3c^3$

17. $2^{10}x^{10} + (10)2^9x^9y + (45)2^8x^8y^2 + 120(2^7)x^7y^3$

18. $p^{11} + 22p^{10}q + 220p^9q^2 + 1320p^8q^3$

19. $x^8 - 24x^7y + 252x^6y^2 - 1512x^5y^3$

20. $\dfrac{x^7}{2187} + \dfrac{7x^6y}{1458} + \dfrac{21x^5y^2}{972} + \dfrac{35x^4y^3}{648}$

Chapter 15 Test Form A

1. $4, -8, 16, -32, 64$

2. 99

3. $\dfrac{1}{5}, \dfrac{226}{105}$

4. $\dfrac{1}{2} + 2 + \dfrac{9}{2} + 8 = 15$

5. $2 + \dfrac{5}{3} + \dfrac{3}{2} + \dfrac{7}{5} + \dfrac{4}{3} = 7.9$

6. 18

7. $a_{13} = 5$

8. $s_8 = \dfrac{44}{3}, d = \dfrac{1}{21}$

9. $\dfrac{50}{19,683}$

10. -1365

11. $\dfrac{4}{3}$

12. 4

13. $16x^4 + 8x^3 + 6x^2 + \dfrac{3}{2}x + \dfrac{1}{8}$

14. $x^5 - 10x^4 + 40x^3 - 80x^2 + 80x - 32$

15. $2187x^7 + 10206x^6 + 20412x^5 + 22680x^4$

16. $-25, -33, -41$

17. $\dfrac{9}{5}, \dfrac{12}{5}, 3, \dfrac{18}{5}; a_{10} = \dfrac{36}{5}, s_{10} = 45$

18. $n = 29, s_{29} = 1479$

19. $\dfrac{125}{333}$

20. $\dfrac{181}{330}$

Chapter 15 Test Form B

1. $1, 3, 9, 27, 81$

2. $\dfrac{10}{9}$

3. $\dfrac{1}{2}, 7$

4. $3 + 6 + 11 + 20 + 37 = 77$

5. $2 + 3 + 4 = 9$

6. 9

7. $a_{13} = 2$

8. $s_{11} = 407, d = 6$

9. 5120

10. $\dfrac{3279}{5}$

11. $\dfrac{12}{5}$

12. $\dfrac{25}{2}$

13. $8x^3 - 36x^2 + 54x - 27$

14. $-x^5 + 10x^4 - 40x^3 + 80x^2 - 80x + 32$

15. $512x^9 + \dfrac{2304}{5}x^8y + \dfrac{4608}{25}x^7y^2 + \dfrac{5376}{125}x^6y^3$

16. $-\dfrac{1}{24}, \dfrac{1}{48}, -\dfrac{1}{96}$

17. $35, 38, 41, 44; a_{10} = 62, s_{10} = 485$

18. $n = 11, s_{11} = -352$

19. $\dfrac{742}{999}$

20. $\dfrac{1741}{9990}$

Chapter 15 Answers

Chapter 15 Test Form C

1. neither

2. $-3, -\dfrac{1}{2}, -\dfrac{1}{9}, 0$

3. $\dfrac{1}{4}, -\dfrac{1}{12}, \dfrac{1}{36}, -\dfrac{1}{108}$

4. $0; 8$

5. $\dfrac{1}{2} + 1 + \dfrac{3}{2} + 2; 5$

6. $5 + 12 + 27 + 58 + 121 = 223$

7. $d = 2$

8. $s_4 = 6; d = -1$

9. 8

10. $a_5 = \dfrac{81}{4}$

11. $r = \dfrac{1}{2}, a_1 = 2304$

12. $s_3 = \dfrac{19}{3}$

13. $-\dfrac{32}{3}$

14. $s_\infty = -\dfrac{32}{5}$

15. $\dfrac{26}{99}$

16. $\dfrac{211}{495}$

17. $\dfrac{3}{2}$

18. $x^5 - 5x^4 y + 10x^3 y^2 - 10x^2 y^3 + 5xy^4 - y^5$

19. $x^5 + 10x^4 + 40x^3 + 80x^2 + 80x + 32$

20. $16x^4 - 96x^3 y + 216x^2 y^2 - 216xy^3 + 81y^4$

Chapter 15 Test Form D

1. neither

2. $5, 11, 17, 23$

3. $7, -\dfrac{7}{5}, \dfrac{7}{25}, -\dfrac{7}{125}$

4. $26.5, 25, 23.5, 22$

5. $2; \dfrac{49}{18}$

6. $0 + \dfrac{3}{5} + \dfrac{8}{10}; \dfrac{7}{5}$

7. $19 + 65 + 211 + 665 = 960$

8. $d = -2$

9. $s_4 = 5; d = -\dfrac{1}{2}$

10. 7

11. $a_4 = \dfrac{343}{8}$

12. $s_3 = \dfrac{15}{4}$

13. $-\dfrac{9}{5}$

14. 9

15. $\dfrac{35}{99}$

16. $\dfrac{41}{396}$

17. $-\dfrac{5}{2}$

18. $x^4 - 12x^3 + 54x^2 - 108x + 81$

19. $16x^4 + 16x^3 + 6x^2 + x + \dfrac{1}{16}$

20. $16x^4 + 96x^3 + 216x^2 + 216x + 81$

Chapter 15 Test Form E

1. geometric

2. $-1, 1, \dfrac{7}{3}, \dfrac{7}{2}$

3. $145, 141, 137, 133$

4. $-\dfrac{3}{4}, -\dfrac{9}{4}, -\dfrac{27}{4}, -\dfrac{81}{4}$

5. $\dfrac{1}{2}; \dfrac{23}{10}$

6. $2 + \dfrac{5}{2} + \dfrac{10}{3}; \dfrac{47}{6}$

Chapter 15 Answers

7. $\dfrac{9}{8} + 1 + \dfrac{25}{32} + \dfrac{36}{64} = \dfrac{111}{32}$

8. $d = 1.5$

9. $s_3 = 6; d = -2$

10. 11

11. $a_4 = -\dfrac{2}{9}$

12. $a_7 = -54$

13. $s_4 = \dfrac{40}{9}$

14. $\dfrac{42}{5}$

15. $s_\infty = 1$

16. $\dfrac{542}{990} = \dfrac{271}{495}$

17. $\dfrac{7}{330}$

18. -2

19. $x^4 - 8x^3 + 24x^2 - 32x + 16$

20. $27a^3 + 54a^2b + 36ab^2 + 8b^3$

Chapter 15 Test Form F

1. arithmetic

2. neither

3. $\dfrac{3}{2}, \dfrac{6}{5}, \dfrac{9}{10}, \dfrac{12}{17}$

4. $-2, 3, -\dfrac{9}{2}, \dfrac{27}{4}$

5. 11.5, 10, 8.5, 7

6. 3; 21

7. $0 + (-3) + (-8) + (-15); -26$

8. $\dfrac{10}{6} + \dfrac{12}{7} + \dfrac{14}{8} + \dfrac{16}{9} = \dfrac{1741}{252}$

9. $d = 2$

10. $-\dfrac{3}{2}$

11. $s_4 = 14; d = 3$

12. $a_1 = 5.5; d = -1.5$

13. $\dfrac{3}{8}$

14. $\dfrac{37}{24}$

15. 7

16. $\dfrac{15}{4}$

17. $\dfrac{7}{37}$

18. $\dfrac{57}{110}$

19. $8x^3 + 12x^2y^2 + 6xy^4 + y^6$

20. $x^4 - 16x^3 + 96x^2 - 256x + 256$

Chapter 15 Test Form G

1. d

2. b

3. d

4. b

5. a

6. c

7. a

8. c

9. b

10. d

11. b

12. d

13. c

14. c

15. a

16. c

17. d

18. d

19. a

20. c

Chapter 15 Answers

Chapter 15 Test Form H

1. d
2. a
3. b
4. b
5. b
6. b
7. a
8. d
9. a
10. c
11. a
12. d
13. c
14. c
15. d
16. b
17. d
18. b
19. c
20. c

Final Exam Answers

Final Exam Form A

1. $1, 2, 6, 9$
2. $0, 1, 2, 6, 9$
3. $A \cup B = \{2, 4, 6, 8, 11\}$
4. -8
5. 52
6. $\pm\sqrt{2}$
7. -6
8. $2, 10$
9. $(1, 2, -3)$
10. $486,000,000$
11. $-2, 0, 2$
12. -3
13. $\dfrac{1}{a-2}$
14. $\dfrac{9}{16b^2}$
15. 11
16. 68
17. $4\sqrt{2}$
18. $(6,5), (5,6), (4,7)$
19. $(3x-4)^2$
20. $-a(a-5)(a+5)$
21. $x^2 + 9x + 24 + \dfrac{67}{x-3}$
22. $9w^2 - 12wx + 4x^2$
23. $\dfrac{3}{5} + \dfrac{4}{5}i$
24. $1, 3$
25. $\dfrac{1}{2}$
26. $(f \circ g)(x) = 4x^2 + 4x - 5$
27. $\left[-\dfrac{1}{3}, \dfrac{11}{3}\right]$
28. $\left(-\infty, -\dfrac{3}{2}\right] \cup [5, \infty)$
29. $x - 4y = 0$

30.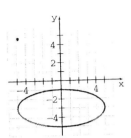

31.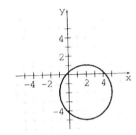

32. $(1, -2)$
33. 4π
34. 14
35. 7
36. $\log_b \dfrac{x(x+2)}{8}$
37. $\dfrac{625}{4}$
38. $\dfrac{625}{6}$
39. $a_1 = -7, \ d = 2.5$
40. $128a^7 + 2240a^6b + 16,800a^5b^2 + 70,000a^4b^3$

Final Exam Form B

1. $-3, -\dfrac{1}{2}, -1, 0, 1, 2, \dfrac{5}{3}, \sqrt{7}, 3.25, 6, 9$
2. $\sqrt{7}$
3. $A \cap B = \{4, 6\}$
4. $\{3, 7\}$
5. $(-\infty, -2) \cup (-1, \infty)$
6. $\dfrac{3}{2}$
7. 37
8. $(-2, 5)$
9. -5

Final Exam Answers

10. $\dfrac{1+\sqrt{6}+\sqrt{2}+\sqrt{3}}{2}$

11. $-\dfrac{6}{25}-\dfrac{17}{25}i$

12. $\dfrac{a^2b+3a}{a+b^2}$

13. 5

14. $\dfrac{x+3}{2}$

15. $\dfrac{4}{9}$

16. $\dfrac{1}{5r^8}$

17. 1.5×10^{-3}

18. 0.06

19. 10

20. $-8, 2$

21. $\left\{-\dfrac{3}{2},\dfrac{3}{2}\right\}$

22. $3, -3, 1, -1$

23. $0, -8$

24. $1\pm\dfrac{\sqrt{5}}{5}$

25. $(-4, 1)$

26. -8

27. $[2,\infty)$

28. $2x+y=3$

29. $2500\ \text{ft}^2$

30.
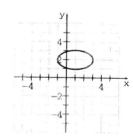

31.
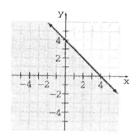

32. $2\sqrt{73}$

33. $(5, 9)$

34. $a_1=8,\ d=-1.5$

35. 95

36. $\dfrac{18}{55}$

37. $\log_b\dfrac{z}{x^2y^3}$

38. 3

39. $\dfrac{2}{3}$

40. $32x^5+240x^4+720x^3+1080x^2+810x+243$

Final Exam Form C

1. $\sqrt{72}$

2. $\sqrt{-8}, 3i$

3. $A\cap B=\left\{\dfrac{3}{2}\right\}$

4. $x=4$

5. $x=\dfrac{3}{4}$

6. $x=\pm\dfrac{\sqrt{10}}{5},\ x=\pm i\sqrt{3}$

7. $0, -7$

8. $x=6\pm\sqrt{31}$

9. $(-\infty, -12]$

10. $\left\{x\middle|-\dfrac{6}{5}<x<2\right\}$

11. $(-3, 2)$

12. $y=-3x+2$

13. $(2, 2)$

14. $(1, 1, -1)$

15. 0

16. 87

17. $6x^2 - 7x - 20$

18. $27 - 15i$

19. $(x+3)(x-3)(x^2+9)$

20. $\dfrac{3x+1}{x^2-5x+25}$

21. $(a-4b)(2a-b)$ or $2a^2 - 9ab + 4b^2$

22. $\dfrac{5}{2} + \dfrac{1}{2}i$

23. $4x - 5 + \dfrac{12}{2x+1}$

24. $\dfrac{2a\sqrt{2a^2c}}{bc^2}$

25. $7\sqrt{5}$

26. $x = -1/7$

27. $4^4 = x;\; x = 256$

28. $\log_3 \dfrac{(x+1)^5(x-4)^6}{x^3}$

29. $x = 14$

30. $x = 4$

31. $\dfrac{1.1045884}{0.62747356} = 1.7603744$

32. 38

33. 54

34. $\dfrac{24}{5}$

35.

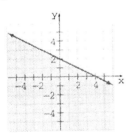

36.

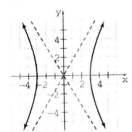

37. $15°, 45°, 120°$

38. \$3380.48

39. $512x^9 + 2304x^8y + 4608x^7y^2 + 5376x^6y^3$

40. $\dfrac{9}{37}$

Final Exam Form D

1. $-4, -2.3\overline{7}, -\sqrt{36}, \dfrac{2}{7}$

2. $\sqrt{5}$

3. $A \cap B = \{\ \}$

4. $x = -1$

5. $x = 1,\; x = \dfrac{3}{2}$

6. $x = \pm 2, \pm i$

7. $x = \pm \dfrac{a\sqrt{b^2 - y^2}}{b}$

8. $x = 4 + \sqrt{3},\; x = 4 - \sqrt{3}$

9. $x = 0,\; x = 10$

10. $\{x \mid -5 < x < -2\}$

11. $\left(-\infty, -\dfrac{3}{2}\right] \cup [5, \infty)$

12. $y = 3x + 14$

13. $(3, -3)$

14. $(2, -1, 2)$

15. 1

16. -24

17. $x^3 + 8y^3$

18. $57 - 51i$

19. $(3a + 10b)(9a^2 - 30ab + 100b^2)$

20. $\dfrac{x+7}{25x^2 - 5x + 1}$

21. $\dfrac{a - 4b}{2a - 3b}$

22. $\dfrac{-3x^2 + 7x + 4}{x^2 + x - 6}$

23. $5x - 13 + \dfrac{16}{x+2}$

Final Exam Answers

24. $-\dfrac{2x^2 \sqrt[3]{625y^3}}{5y^3z^4}$

25. $9\sqrt{6}$

26. $1 - i$

27. $\log_7 \dfrac{(x+3)^3}{x(x-1)^5}$

28. $x = 1$

29. $x = 5$

30. $\dfrac{1.153144}{0.773976} = 1.489896$

31. 25

32. 24

33. $(f \circ g)(x) = \sqrt{4x^2 + 20x + 26}$

34.

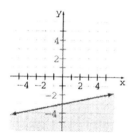

35.

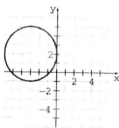

36. $30°, 50°, 100°$

37. 11.6 years

38. $a_1 = -5, \ d = 3$

39. $x^{10} - 40x^9y + 720x^8y^2 - 7680x^7y^3$

40. $\dfrac{141}{1100}$

Final Exam Form E

1. $\sqrt{-3}, i, \sqrt{-4}$

2. $-\sqrt{3}$

3. $A \cap B = \{\ \}$

4. $x = 3$

5. $x = \dfrac{7}{3}$

6. $x = \pm 3, \ x = \pm i\sqrt{3}$

7. $-3, \ -\dfrac{5}{3}$

8. $x = 4 \pm \sqrt{13}$

9. $(-\infty, 29]$

10. $\{x \mid -4 < x < -1\}$

11. $(-\infty, -3) \cup (8, \infty)$

12. $y = 2x + 6$

13. $(-2, 1)$

14. $(1, 1, 2)$

15. -1

16. 85

17. $8x^3 - 27y^3$

18. $32 - 60i$

19. $(5x + 4y)(25x^2 - 20xy + 16y^2)$

20. $\dfrac{5x+1}{x^2 - 7x + 49}$

21. $(3a - b)(2a - 5b)$ or $6a^2 - 17ab + 5b^2$

22. $6\sqrt{5}; \ (-1,1)$

23. $3x + 2$

24. $\dfrac{3x\sqrt[4]{8x}}{2y^7}$

25. $6\sqrt{5}$

26. $3 - 2i$

27. $\log_8 \dfrac{(2x-1)^5}{x^2(x+2)^3}$

28. $\dfrac{1.1045884}{0.62747356} = 1.7603744$

29. $x = 12$

30. $a_1 = 11, \ d = -2$

31. 45

32. 25

33. $-\dfrac{15}{4}$

34. $\dfrac{203}{330}$

35.

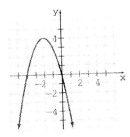

36.

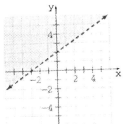

37.

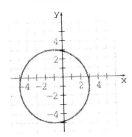

38. $40°, 50°, 90°$

39. 14.6 years

40. $1024x^{10} - 5120x^9y + 11,520x^8y^2 - 15,360x^7y^3$

Final Exam Form F

1. $\left\{ -2.3\overline{7}, -\sqrt{36} \right\}$

2. $\left\{ \sqrt{8}, \dfrac{\pi}{7} \right\}$

3. $\{ \ldots, -3, -2, -1 \}$

4. $x = -2$

5. $x = \dfrac{1}{2}, 2$

6. $x = \pm\dfrac{\sqrt{2}}{2}, \pm\sqrt{3}\, i$

7. $x = \dfrac{11}{3}, -2$

8. $x = 6 \pm 4\sqrt{2}$

9. $(-\infty, 5]$

10. $\left\{ x \mid -4 < x < -\dfrac{8}{3} \right\}$

11. $[-2, 7]$

12. $y = -3x + 3$

13. $(-2, 1)$

14. $(-1, 2, 2)$

15. $(-2, 7), \left(\dfrac{8}{9}, \dfrac{59}{27} \right)$

16. -10

17. $3x^2 + 13x - 10$

18. $-12\, i$

19. $(x + 4)(x - 4)(x^2 + 16)$

20. $\dfrac{5x + 1}{x^2 - 3x + 9}$

21. $4(a - 7b)(a^2 + 2ab + 4b^2)$

22. $5 - 3i$

23. $5x - 8 + \dfrac{16}{x + 1}$

24. $\dfrac{-5x\sqrt[3]{x^2}}{2y^3z^5}$

25. $5\sqrt{2}$

26. $\log_3 81 = 4$

27. $\log_8 \dfrac{x^3}{(x + 4)^2(2x - 1)^7}$

28. $x = \dfrac{9}{2}$

29. $\dfrac{1.153144}{0.773976} = 1.489896$

30. $10; (2, -6)$

31. 53

32. $\dfrac{25}{3}$

33. $\dfrac{10}{7}$

Final Exam Answers

34.

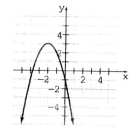

35.

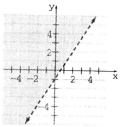

36.

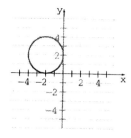

37. $30°, 60°, 90°$

38. 27.7 years

39. $-108,864\,a^5 b^3$

40. $a_1 = 17, \ d = -5$

Final Exam Form G

1. c
2. a
3. b
4. a
5. a
6. d
7. c
8. b
9. b
10. d
11. b
12. a
13. c

14. d
15. a
16. b
17. b
18. d
19. c
20. a
21. a
22. d
23. a
24. c
25. b
26. d
27. c
28. b
29. a
30. c
31. d
32. b
33. a
34. b
35. c
36. c
37. a
38. a
39. b
40. c

Final Exam Form H

1. b
2. c
3. b
4. a
5. a
6. c
7. d

8. b
9. b
10. a
11. d
12. a
13. b
14. b
15. a
16. a
17. c
18. a
19. b
20. d
21. a
22. c
23. d
24. b
25. c
26. a
27. c
28. d
29. b
30. c
31. c
32. c
33. d
34. b
35. a
36. d
37. b
38. a
39. c
40. d